Genomics and Clinical Diagnostics

Genomics and Clinical Diagnostics

Edited by

David Whitehouse
University of Hertfordshire, Hatfield, UK
Email: d.whitehouse@herts.ac.uk

and

Ralph Rapley
University of Hertfordshire, Hatfield, UK
Email: r.rapley@herts.ac.uk

ROYAL SOCIETY
OF CHEMISTRY

Print ISBN: 978-1-78262-821-7
EPUB ISBN: 978-1-78801-680-3

A catalogue record for this book is available from the British Library

The Royal Society of Chemistry is a charity, registered in England and Wales, Number 207890, and a company incorporated in England by Royal Charter (Registered No. RC000524), registered office: Burlington House, Piccadilly, London W1J 0BA, UK, Telephone: +44 (0) 20 7437 8656.

Visit our website at www.rsc.org/books

Printed in the United Kingdom by CPI Group (UK) Ltd, Croydon, CR0 4YY, UK

Preface

Since the completion of the human genome project some 15 years ago much promise has been made of the potential to harness the information to enhance the field of clinical diagnostics. Indeed, numerous genomics projects have now been initiated, many of which are already yielding important information about pathologies at the molecular level. This new and exciting area of biomedical science has also given rise to the possibility of intervention with smart drugs and biological and chemical molecules designed to specific targets. We are now in the post-genomics era and the ability to rapidly sequence genomes, identify targets and design and implement many advanced therapeutic strategies is a clear reality. This has been made possible by the enormous advances in technologies such as next generation sequencing to identify clinically relevant nucleic acids and the development of bioinformatics to analyse the resulting data. **Genomics and Clinical Diagnostics** brings together, in a number of chapters, coverage of some of the key technologies that have been recently developed and an overview of some of the diseases to which these techniques have been applied. The book begins with chapters that set the scene, from an introduction to nucleic acids and their analysis through to key elements of genes and genomes to the coverage of molecular cloning and sequencing technologies. This theme is continued and extended in the next series of chapters that cover proteomics, bioinformatics and further advanced molecular diagnostic techniques. The newly expanding

Genomics and Clinical Diagnostics
Edited by David Whitehouse and Ralph Rapley

Published by the Royal Society of Chemistry, www.rsc.org

area of nanoparticles in diagnostics is covered and molecular microbial diagnostics completes the initial coverage of technologies. The theme of the book then extends into chapters that concentrate on the application of genomics to a number of diseases such as connective tissue disorders, blood disorders, microbial and parasite disease. This is completed with coverage or rare diseases and biomarker analysis in oncology. Rather than a comprehensive collection of all diseases in which molecular analysis is undertaken we have identified key areas that we believe show the power of the techniques and they way in which they can be implemented. We believe that the coverage of many of the key areas will serve as a solid foundation for those embarking on or engaged in studies of this exciting and fast-moving area. This book should be of particular interest to students of biomedical and clinical science and scientists who require a sound introduction to this rapidly advancing and expanding area.

David Whitehouse
Ralph Rapley

Contents

Genomics and Clinical Diagnostics
Edited by David Whitehouse and Ralph Rapley

Published by the Royal Society of Chemistry, www.rsc.org

1 Nucleic Acid Structure and Basic Analysis

Ralph Rapley

University of Hertfordshire, School of Life and Medical Sciences, College Lane, Hatfield, AL10 9AB, UK
*E-mail: r.rapley@herts.ac.uk

1.1. Introduction

Major advances have been made in gene analysis and genomics in recent years and this has been accelerated by the continued development and refinement of methods and techniques for studying nucleic acids. One major area of current research is the identification and diagnosis of diseases that are multifactorial in nature. There are numerous diseases where analysis of genomes has provided insights into the disease and one particularly notable example is oncology. Molecular genetic analysis of this area has allowed a discrete set of cellular genes, termed oncogenes and tumour suppressor genes, to be identified and characterised. These genes and the proteins and enzymes they encode are major components of cell signalling and the cell cycle and are intimately involved in many aspects cell regulation. The disruption of oncogenes and tumour suppressor genes contributes to the early changes required for cancers to develop. Identification and analysis of these genes and genomes has already provided information for use in diagnostics and prognostics and a number

Genomics and Clinical Diagnostics
Edited by David Whitehouse and Ralph Rapley

Published by the Royal Society of Chemistry, www.rsc.org

have been shown to be biomarkers of a particular cancer type. In a number of cancers well-defined molecular events have been correlated with mutations in oncogenes and therefore in the corresponding protein. It is already possible to screen and predict the outcome of some disease processes at an early stage, a point which itself raises significant ethical dilemmas. The application of molecular biology has allowed understanding of cellular processes both in normal and disease states. The advent of this type of analysis has given rise to the development of personalised or precision medicine and there is now great promise in further developments in drug discovery and molecular gene therapy. A number of genetically engineered therapeutic proteins and enzymes have been developed and are already having an effect on disease management. In addition the correction of disorders at the gene level using gene therapy is also under way. Perhaps one of the most startling applications of molecular biology to date is indeed gene editing and the development of gene modifications methods such as the clustered regularly interspaced short palindromic repeats (CRISPR)–CRISPR-associated system 9 (cas9) system, which may have a profound effect on treating genetic-based diseases. In considering the potential utility of molecular biology techniques it is important to understand the basic structure of nucleic acids and gain an appreciation of how this dictates the function *in vivo* and *in vitro*. Indeed, many techniques used in molecular biology mimic in some way the natural functions of nucleic acids, such as replication and transcription. This chapter is intended to provide an overview of the general features of nucleic acid structure and function and describe some of the basic methods used in their isolation and analysis.

1.2. Structure of Nucleic Acids

1.2.1. Primary Structure of Nucleic Acids

DNA and RNA are macromolecular structures composed of regular repeating polymers formed from nucleotides.[1] These are the basic building blocks of nucleic acids and are derived from nucleosides, which are composed of two elements: a five-membered pentose carbon sugar (2-deoxyribose in DNA and ribose in RNA), and a nitrogenous base (Figure 1.1). The carbon atoms of the sugar are designated 'prime' (1′, 2′, 3′, *etc.*). To distinguish them from the carbons of the nitrogenous bases, of which there are two types, either a purine or a pyrimidine. A nucleotide, or nucleoside phosphate, is formed by the attachment of a phosphate to the 5′ position of a nucleoside by an ester linkage. Such nucleotides can be joined together by the

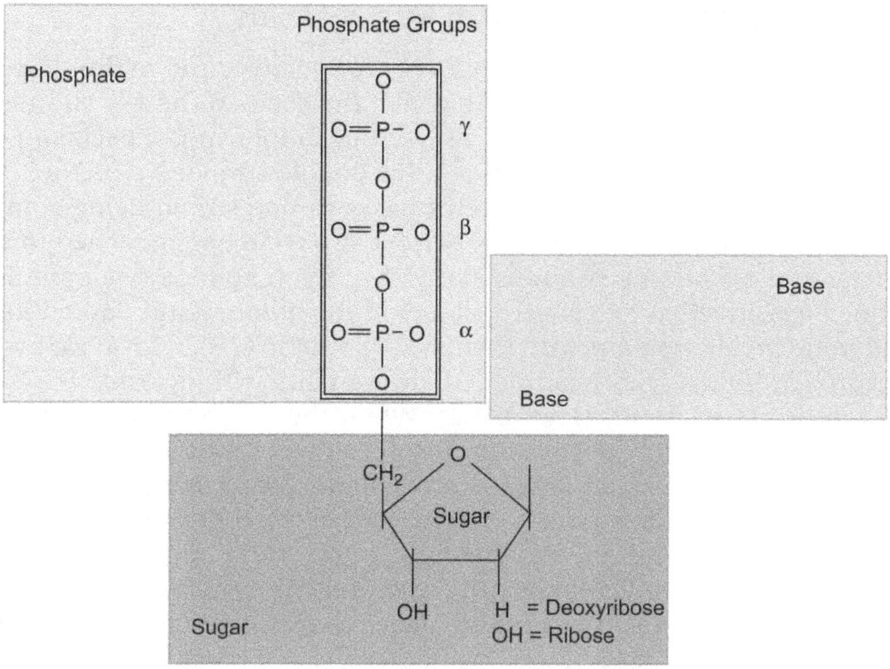

Figure 1.1 Representation of a deoxynucleoside triphosphate indicating the three components of a sugar, triphosphate and a base. The base can be either A C G or T. In RNA the 2′ carbon has an OH whereas it is deoxy in DNA. Reproduced from ref. 22 with permission from the Royal Society of Chemistry.

formation of a second ester bond by reaction between the phosphate of one nucleotide and the 3′ hydroxyl of another, thus generating a 5′ to 3′ phosphodiester bond between adjacent sugars; this process can be repeated indefinitely to give long polynucleotide molecules. DNA has two such polynucleotide strands. However, since each strand has both a free 5′ hydroxyl group at one end, and a free 3′ hydroxyl at the other end, each strand has a polarity or directionality. The polarities of the two strands of the molecule are in opposite directions, and thus DNA is described as an 'anti-parallel' structure.

The purine bases (composed of fused five- and six-membered rings), adenine (A) and guanine (G), are found in both RNA and DNA, as is the pyrimidine (a single six-membered ring) cytosine (C). The other pyrimidines are each restricted to one type of nucleic acid: uracil (U) occurs exclusively in RNA, whilst thymine (T) is limited to DNA. Thus it is possible to distinguish between RNA and DNA on the basis of the presence of ribose and uracil in RNA, and deoxyribose and thymine in DNA. However, it is the sequence of bases along the structure, which distinguishes one DNA (or RNA) from another.

1.2.2. Secondary Structure of Nucleic Acids

The two polynucleotide chains in DNA are usually found in the shape of a right-handed double helix, in which the bases of the two strands lie in the centre of the molecule, with the sugar–phosphate backbones on the outside.[2] A crucial feature of this double-stranded structure is that it depends on the sequence of bases in one strand being complementary to those in the other strand. A purine base attached to a sugar residue on one strand is always hydrogen bonded to a pyrimidine base attached to a sugar residue on the other strand. Moreover, adenine (A) always pairs with thymine (T) or uracil (U) in RNA, *via* two hydrogen bonds, and guanine (G) always pairs with cytosine (C) by three hydrogen bonds (Figure 1.2). When these conditions are met a stable double-helical structure results in which the backbones of the two strands are, on average, a constant distance apart. Thus, if the sequence of one strand is known, that of the other strand can be deduced. The strands are designated as plus (+) and minus (−) and an RNA molecule complementary to the minus (−) strand is synthesised during transcription. The base sequence may cause significant local variations in the shape of the DNA molecule, and these variations are vital for specific interactions between the DNA and various proteins to take place. Although the three-dimensional structure of DNA may vary it generally adopts a double helical structure termed the b form or b-DNA *in vivo*.

It has been well recognized for some time that DNA as a structure may be chemically modified without the underlying DNA sequence

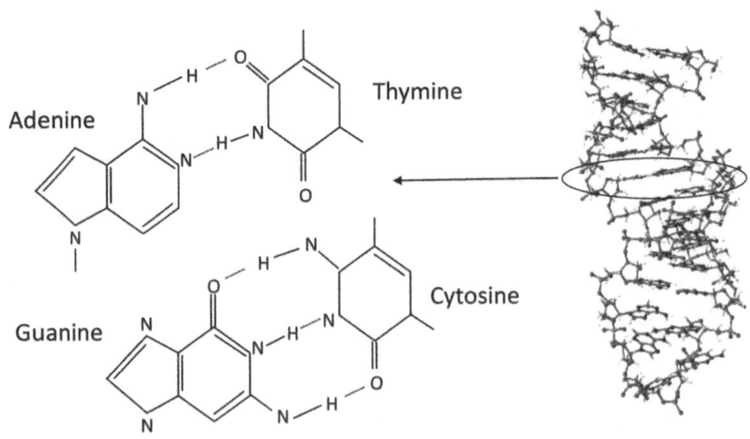

Figure 1.2 Representation of the four bases in DNA and their complementary base pairing, A–T and C–G through hydrogen bonds. The right hand side depicts a DNA double helix. Reproduced from ref. 22 with permission from the Royal Society of Chemistry.

being altered. One of the most important modifications is the addition of a methyl (CH_3) group to cytosine, termed DNA methylation and catalysed by DNA methyltransferases. This results in what some describe as the fifth base in DNA. Approximately 1.5% of human DNA is methylated (termed the epigenome) and the methylation status appears to have a profound effect on gene expression, where hypomethylation appears to promote gene expression. This feature of gene expression control is termed epigenetics and is a complex process which is also extended to the modification of histone proteins and some small RNA molecules involved in gene expression control. Importantly, epigenetics appears to play a role in disease states, such as cancers and certain neurological diseases, which may lead to a new means of future treatment.

1.2.3. Denaturation of Double-stranded DNA

The two anti-parallel strands of DNA are held together only by the weak forces of hydrogen bonding between complementary bases, and partly by hydrophobic interactions between adjacent, stacked base pairs, termed base-stacking. Little energy is needed to separate a few base pairs, and so, at any instant, a few short stretches of DNA will be opened up to the single-stranded conformation. However, such stretches immediately pair up again at room temperature, so the molecule as a whole remains predominantly double-stranded.

If, however, a DNA solution is heated to approximately 90 °C or above there will be enough kinetic energy to denature the DNA completely, causing it to separate into single strands. The temperature at which 50% of the DNA is melted is termed the melting temperature or T_m, and this depends on the nature of the DNA. If several different samples of DNA are melted, it is found that the T_m is highest for those DNA molecules that contain the highest proportion of cytosine and guanine, and T_m can actually be used to estimate the percentage (C + G) in a DNA sample. This relationship between T_m and (C + G) content arises because cytosine and guanine form three hydrogen bonds when base-paired, whereas thymine and adenine form only two. Because of the differential numbers of hydrogen bonds between A–T and C–G pairs those sequences with a predominance of C–G pairs will require greater energy to separate or denature them. The conditions required to separate a particular nucleotide sequence are also dependent on environmental conditions, such as salt concentration. If denatured DNA is cooled it is possible for the separated strands to re-associate, a process known as renaturation.

Strands of RNA and DNA will associate with each other, if their sequences are complementary, to give double-stranded, hybrid molecules. Similarly, strands of labelled RNA or DNA, when added to a denatured DNA preparation, will act as probes for DNA molecules to which they are complementary. This hybridisation of complementary strands of nucleic acids is a cornerstone for many molecular biology techniques and is very useful for isolating a specific fragment of DNA from a complex mixture. It is also possible for small single-stranded fragments of DNA (up to 40 bases in length), termed oligonucleotides, to hybridise to a denatured sample of DNA. This type of hybridisation is termed annealing and again is dependent on the base sequence of the oligonucleotide and the salt concentration of the sample.

1.3. Isolation and Separation of Nucleic Acids

1.3.1. Isolation of DNA

The use of DNA for analysis or manipulation usually requires that it is isolated and purified to a certain extent.[3] DNA is recovered from cells by the gentlest possible method of cell rupture to prevent the DNA from fragmenting by mechanical shearing. This is usually in the presence of ethylenediaminetetraacetic acid (EDTA) which chelates the Mg^{2+} ions needed for enzymes that degrade DNA, termed DNAse. Ideally, cell walls, if present, should be digested enzymatically (*e.g.* by lysozyme treatment of bacteria), and the cell membrane should be solubilised using detergent. If physical disruption is necessary it should be kept to a minimum and should involve cutting or squashing of cells, rather than the use of shear forces. Cell disruption (and most subsequent steps) should be performed at 4 °C, using glassware and solutions which have been autoclaved to destroy DNAse activity (Figure 1.3).

After release of nucleic acids from the cells, RNA can be removed by treatment with ribonuclease (RNAse) that has been heat-treated to inactivate any DNAse contaminants; RNAse is relatively stable to heat as a result of its disulphide bonds, which ensure rapid renaturation of the molecule on cooling. The other major contaminant, protein, is removed by shaking the solution gently with water-saturated phenol, or with a phenol–chloroform mixture, either of which will denature proteins but not nucleic acids. Centrifugation of the emulsion formed by this mixing produces a lower, organic phase, separated from the upper, aqueous phase by an interface of denatured protein. The

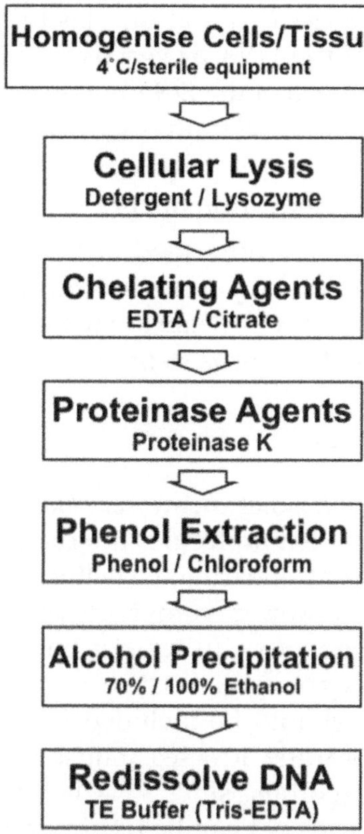

Figure 1.3 General steps involved in extracting DNA from cells or tissues. Reproduced from ref. 22 with permission from the Royal Society of Chemistry.

aqueous solution is recovered and deproteinised repeatedly, until no more material is seen at the interface. Finally, the deproteinised DNA preparation is mixed with two volumes of absolute ethanol and the DNA is allowed to precipitate out of solution in a freezer. After centrifugation, the DNA pellet is redissolved in a buffer containing EDTA to inactivate any DNAses present. This solution can be stored at 4 °C for at least a month. DNA solutions can be stored frozen, although repeated freezing and thawing tends to damage long DNA molecules by shearing.

The procedure described is suitable for total cellular DNA. If the DNA from a specific organelle or viral particle is needed, it is best to isolate the organelle or virus before extracting its DNA, since the recovery of a particular type of DNA from a mixture is usually rather difficult. Where a high degree of purity is required DNA may be subjected to density gradient ultracentrifugation through caesium chloride which is particularly useful for the preparation plasmid DNA. It is possible to check the integrity of the DNA by agarose gel electrophoresis and determine the concentration of the DNA by using the fact that 1 absorbance unit at a wavelength of 260 nm (A_{260}) equates to 50 µg ml^{-1} of DNA and thus:

$$50 \times A_{260} = \text{concentration of DNA sample } (\mu\text{g ml}^{-1})$$

Contaminants may also be identified in the sample by employing scanning UV-spectrophotometry from 200 nm to 300 nm. A ratio of A_{260}:A_{280} of approximately 1.8 indicates that the sample is free of protein contamination, which absorbs strongly at 280 nm.

1.3.2. Isolation of RNA

The methods used for RNA isolation are very similar to those described above for DNA; however, RNA molecules are relatively short and therefore less easily damaged by shearing, so cell disruption can be rather more vigorous. RNA is, however, vulnerable to digestion by RNAses, which are present endogenously in various concentrations in certain cell types and exogenously on fingers. Gloves should therefore be worn, and a strong detergent should be included in the isolation medium to immediately denature any RNAses. Subsequent deproteinization should be particularly rigorous, since RNA is often tightly associated with proteins. DNAse treatment can be used to remove DNA, and RNA can be precipitated by ethanol. One reagent which is commonly used in RNA extraction is guanidinium thiocyanate (GTC) which is both a strong inhibitor of RNAse and a protein denaturant. It is possible to check the integrity of an RNA extract by analysing it by agarose gel electrophoresis. The most abundant RNA species are the rRNA molecules. For prokaryotes these are 16S and 23S and for eukaryotes the molecules are slightly heavier at 18S and 28S. These appear as discrete bands following agarose gel electrophoresis and, importantly, if intact indicate that the other RNA components, such as mRNA, are likely to be intact also. Electrophoresis is usually carried out under denaturing conditions to prevent secondary structure formation in the RNA. The concentration of the RNA may be estimated by using UV-spectrophotometry. At 260 nm 1 absorbance unit equates to 40 µg ml^{-1} of RNA and therefore:

$$40 \times A_{260} = \text{concentration of RNA sample } (\mu\text{g ml}^{-1})$$

Contaminants may also be identified in the same way as for DNA by scanning UV-spectrophotometry, however in the case of RNA an A_{260}:A_{280} ratio of approximately 2 would be expected for a sample containing no contaminants.

In many cases it is desirable to isolate eukaryotic mRNA which constitutes only 2–5% of cellular RNA from a mixture of total RNA molecules. This may be carried out by affinity chromatography on oligo(dT)-cellulose columns. At high salt concentrations, the mRNA containing poly(A) tails binds to the complementary oligo(dT) molecules of the affinity column, and so mRNA will be retained; all other RNA molecules can be washed through the column by further high-salt solution. Finally, the bound mRNA can be eluted using a low concentration of salt. Nucleic acid species may also be subfractionated by more physical means such as electrophoretic or chromatographic separations based on differences in nucleic acid fragment sizes or physicochemical characteristics.

1.3.3. Automated Nucleic Acid Extraction

Many current molecular biology methods and their reagents can now be found in the form of a kit from a manufacturer, such as Sigma, or can be automated. The extraction of nucleic acids by these means is no exception. The advantage of their use lies in the fact that the reagents are standardised and quality control tested, providing a high degree of reliability. For example, glass bead preparations for DNA purification have been used increasingly and with reliable results. Small compact column-type preparations, such as Qiagen spin columns, are also used extensively in research and in routine DNA extraction. Essentially the same reagents for nucleic acid extraction may be used in a format that allows reliable and automated extraction. The process can also be automated with a low-throughput Qiacube system. Further methods are also available using kit-based extraction methods for RNA; these in particular have overcome some of the problems of RNA extraction, such as RNAse contamination. A number of fully automated nucleic acid extraction machines, such as the Qiasymphony system are now employed in areas where high throughput is required, *e.g.* clinical diagnostic laboratories. Here the raw samples, such as blood specimens, are placed in 96- or 384-well microtitre plates and these undergo a set computer-controlled processing pattern carried out robotically. Thus the samples are rapidly manipulated and extracted in approximately 45 min without any manual operations being undertaken. Nucleic acids can be extracted from a variety of samples, including blood, serum, frozen tissue sections, formalin-fixed paraffin-embedded (FFPE) tissue sections and biopsies among others, and all have their own unique challenges in the extraction process. Reliable and effective methods are a crucial element for further analysis, such as DNA sequencing and long-term storage in DNA banks.

1.3.4. Single-cell and Cell-free DNA Analysis

A number of methods have been developed for the extraction of nucleic acids in single cells or completely cell-free.[4] In addition to DNA, mRNA and microRNA (miRNA) and the various individual components have also been successfully isolated and assayed. Single-cell analysis relies on the preparation of single cells from a mixed population of cells and tissues. This can be achieved using a number of methods, although one of the most common is laser capture microdissection (LCM). Here, a few cells or single cells are typically prepared by a UV pulsed laser under automated control, or by manual microscopy of the tissue. The cells are removed from the surrounding tissue and captured for further analysis, such as nucleic acid amplification and DNA sequencing.

In addition to single-cell analysis cell-free DNA has been found in the circulation and has potential utility, especially in cases of fetal DNA analysis in the maternal bloodstream. This is proving useful as cell-free fetal DNA (cffDNA) is a very promising non-invasive method that can be used with various DNA analysis techniques to determine a number of genetic disorders. In a similar way circulating tumour cells (CTC) and circulating tumour DNA (ctDNA) may also be assayed for a number of diseases, including cancers. Of particular interest is the concept of the liquid biopsy using ctDNA. These are small fragments of DNA of approximately 170 base pairs (bp) originating from the tumour cells found in the bloodstream and which may be analysed using amplification methods, such as the polymerase chain reaction (PCR). This allows for the identification of biomarkers of the disease (*e.g.* KRAS) and mutations within them. However the collection of samples is not without problems, especially when standard methods of blood collection are used, *e.g.* heparin and when employing EDTA tubes. Furthermore white cell genomic DNA will be released from the samples if the methods are not optimised and this may interfere with the ctDNA analysis. Various methods of stabilisation have therefore been used to overcome these problems. The use of liquid biopsy is gaining interest as not only is it a non-invasive process but a number of studies have reported its ability to not only aid in diagnostics and screening but also to be of benefit in prognostics. It may further be an effective means of monitoring disease progression and response to treatment and a sensitive way of identifying minimal residual disease (MRD) found in reoccurrence of certain tumours.

1.3.5. Enzymes Used in Molecular Biology

The discovery and characterisation of a number of key enzymes has enabled the development of various techniques for the analysis and manipulation of DNA. In particular, the enzymes termed type II restriction endonucleases have come to play a key role in all aspects of molecular biology.[5] These enzymes recognise certain DNA sequences, usually 4–6 bp in length, and cleave them in a defined manner. The sequences recognised are palindromic or of an inverted repeat nature (Figure 1.4). That is they read the same in both directions on each strand. When cleaved they leave a flush-ended or staggered (also termed a cohesive-ended) fragment depending on the particular enzyme used (Table 1.1). An important property of staggered ends is that those produced from different molecules by the same enzyme are complementary (or 'sticky') and so will anneal to each other. The annealed strands are held together only by hydrogen bonding between complementary bases on opposite strands. Covalent joining of the ends of each of the two strands may be carried out using the enzyme DNA ligase. This is widely exploited in molecular biology to enable the construction of recombinant DNA *i.e.* the joining of DNA fragments from different sources. Approximately 500 restriction enzymes have been characterised that recognise over 100 different target sequences. A number of these, termed isoschizomers, recognise different target sequences but produce the same staggered ends or overhangs. In addition restriction enzymes may be applied to various situations in molecular biology as indicated in Figure 1.5.

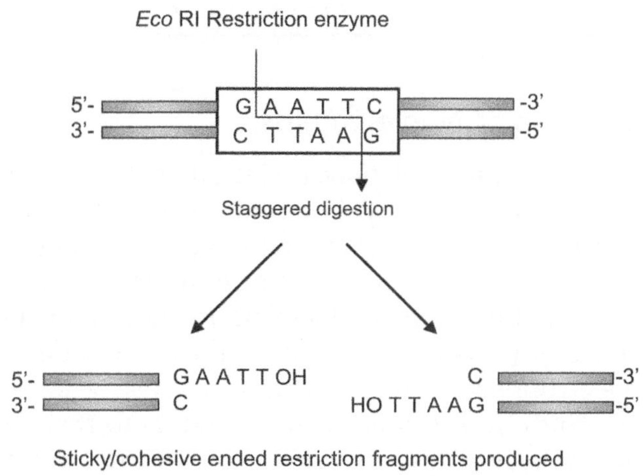

Figure 1.4 The cleavage of a DNA strand with a target site for the restriction enzyme *Eco*R1 indicating the ends of the DNA formed following digestion. Reproduced from ref. 22 with permission from the Royal Society of Chemistry.

Table 1.1 Examples of restrictions enzymes with four-, six- or eight-base recognition sequences. Reproduced from ref. 22 with permission from the Royal Society of Chemistry.

Name	Recognition Sequence	Digestion Products		
Four-nucleotide recognition sequence				
*Hae*III	5′-GGCC-3′ 3′-CCGG-5′	5′-GG 3′-CC	CC-3′ GG-5′	Blunt end digestion
*Hpa*II	5′-CCGG-3′ 3′-GGCC-5′	5′-C 3′-GGC	CGG-3′ C-5′	Cohesive end digestion
Six-nucleotide recognition sequence				
*Bam*HI	5′-GGATTC-3′ 3′-GGCC-5′	5′-G 3′-CCTAG	GATCC-3′ G-5′	
*Eco*RI	5′-GAATTC-3′ 3′-CTTAAG-5′	5′-G 3′-CTTAA	AATCC-3′ G-5′	
*Hind*III	5′-AAGCTT-3′ 3′-TTCGAA-5′	5′-A 3′-TTCGA	AGCTT-3′ A-5′	
Eight-nucleotide recognition sequence				
*Not*I	5′-GCGGCCGC-3′ 3′-CGCCGGCG-5′	5′-GC 3′-CGCCGG	GGCCGC-3′ CG-5′	

1.3.6. Separation of Nucleic Acids

Electrophoresis in agarose or polyacrylamide gels is the most usual way to separate DNA molecules according to size. The technique can be used analytically or preparatively and can be qualitative or quantitative. Large fragments of DNA, such as chromosomes, may also be separated by a modification of electrophoresis termed pulsed field gel electrophoresis (PFGE). The easiest and most widely applicable method is electrophoresis in horizontal agarose gels, followed by staining with ethidium bromide. This dye binds to DNA by insertion between stacked base pairs (intercalation), and it exhibits a strong orange or red fluorescence when illuminated with ultraviolet light. Very often electrophoresis is used to check the purity and intactness of a DNA preparation or to assess the extent of an enzymatic reaction

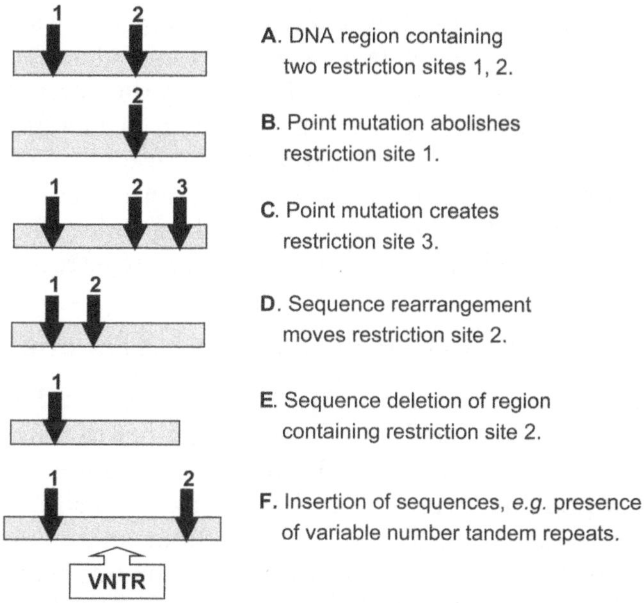

Figure 1.5 Indication of the application of restriction enzymes and the restriction fragment lengths generated in various situations. For example in A digestion with enzyme 1 and 2 results in a particular fragment size whereas in B a point mutation resulting in the abolition of enzyme 1 will result in a different fragment size. Reproduced from ref. 22 with permission from the Royal Society of Chemistry.

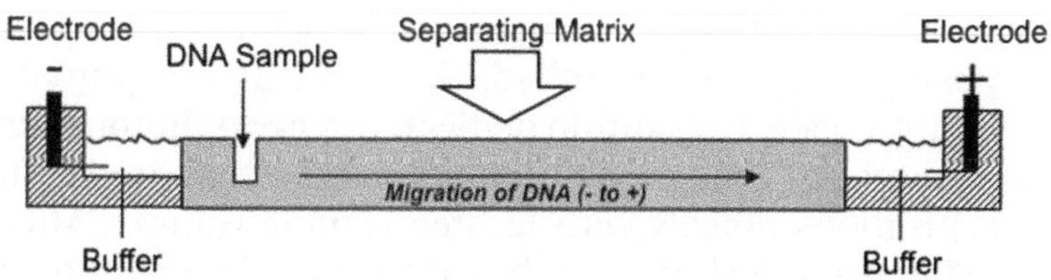

Figure 1.6 A schematic illustration of a typical horizontal agarose gel electrophoresis set up for the separation of nucleic acids. Reproduced from ref. 22 with permission from the Royal Society of Chemistry.

during, for example, the steps involved in the cloning of DNA (Figure 1.6). For such checks 'mini-gels' are particularly convenient, since they need little preparation, use small samples and provide results quickly. Agarose gels can be used to separate molecules larger than about 100 bp. For higher resolution or for the effective separation of shorter DNA molecules polyacrylamide gels are the preferred method.

When electrophoresis is used preparatively, the piece of gel containing the desired DNA fragment is physically removed with a scalpel. The

DNA may be recovered from the gel fragment in various ways. This may include crushing with a glass rod in a small volume of buffer, using agarase to digest the agarose leaving the DNA, or by the process of electro-elution. In this method the piece of gel is sealed in a length of dialysis tubing containing buffer and is then placed between two electrodes in a tank containing more buffer. Passage of an electrical current between the electrodes causes DNA to migrate out of the gel piece but it remains trapped within the dialysis tubing and can therefore be recovered easily.

1.3.7. Automated Analysis of Nucleic Acid Fragments

Gel electrophoresis remains the established method for the separation and analysis of nucleic acids. However, a number of automated systems using pre-cast gels and standardised reagents are available that are now very popular. These are especially useful in situations where a large number of samples or high-throughput analysis is required. In addition, technologies, such as the Agilent bioanalyser, have been developed that obviate the need to prepare electrophoretic gels. These employ microfluidic circuits constructed on small cassette units that contain interconnected micro-reservoirs. The sample is applied to one area and driven through microchannels under computer-controlled electrophoresis. The channels lead to reservoirs allowing, for example, incubation with other reagents, such as dyes, for a specified time. Electrophoretic separation is thus carried out in a microscale format. The small sample size minimises sample and reagent consumption and the units, being computer controlled, allow data to be captured within a very short timescale. In addition dedicated spectrophotometers such as the Thermofisher Scientific Nanodrop system can provide nucleic acid concentrations quickly with limited sample volume. Alternative methods of analysis, including high-performance liquid chromatography-based approaches, have gained in popularity, especially for DNA mutation analysis. Mass spectrometry methods traditionally used in protein analysis, such as matrix-assisted laser desorption and ionization-time of flight (MALDI-TOF) are also becoming increasingly used for nucleic acid analysis due to their rapidity and increasing reliability.

1.4. Restriction Mapping of DNA Fragments

1.4.1. Nucleic Acid Blotting Methods

Electrophoresis of DNA restriction fragments allows separation based on size to be carried out. However, it provides no indication as to the presence of a specific, desired fragment among the complex sample. This can be achieved by transferring the DNA from the intact gel onto a piece of nitrocellulose or nylon membrane placed in contact with it. This provides a more permanent record of the sample since DNA begins to diffuse out of a gel that is left for a few hours. First the gel is soaked in alkali to render the DNA single stranded. It is then transferred to the membrane so that the DNA becomes bound to it in exactly the same pattern as that originally on the gel. This transfer, named a Southern blot after its inventor Ed Southern, can be performed electrophoretically, or by drawing large volumes of buffer through both gel and membrane, thus transferring DNA from one to the other by capillary action.[6] The point of this operation is that the membrane can now be treated with a labelled DNA molecule, for example a gene probe. This single-stranded DNA probe will hybridise under the right conditions to complementary fragments immobilised onto the membrane (Figure 1.7). The conditions of hybridisation, including the temperature and salt concentration, are critical for this process to take place effectively. This is usually referred to as the stringency of the hybridisation and it is particular to each individual gene probe and for each sample of DNA. A series of washing steps with buffer is then carried out to remove any unbound

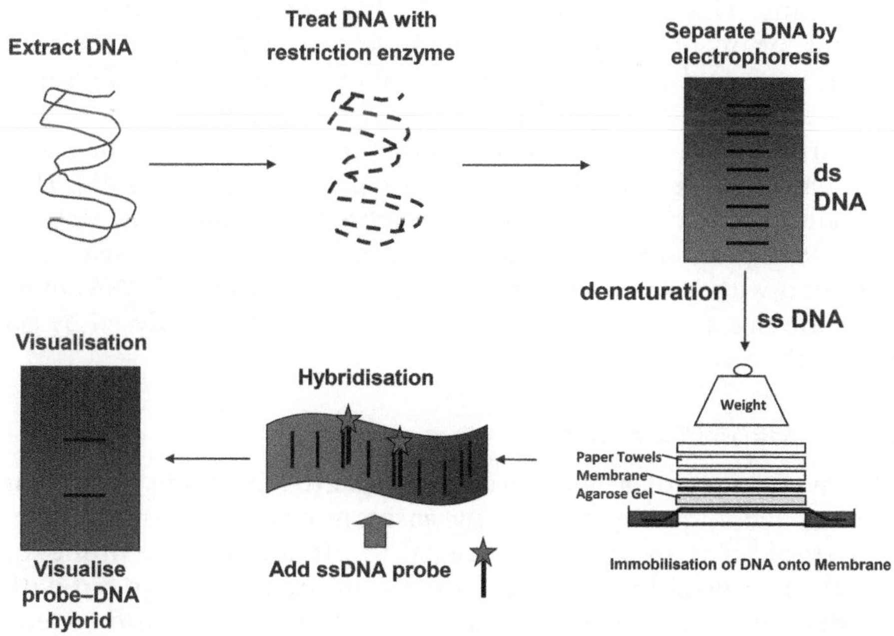

Figure 1.7 Steps involved in a Southern blot, note that the denaturation step can be achieved using sodium hydroxide before the blotting stage with a nylon membrane. Reproduced from ref. 22 with permission from the Royal Society of Chemistry.

probe and the membrane is developed, after which the precise location of the probe and its target may be visualised. It is also possible to analyse DNA from different species or organisms by blotting the DNA and then using a gene probe representing a protein or enzyme from one of the organisms. In this way it is possible to search for related genes in different species. This technique is generally termed zoo blotting.

1.4.2. RNA Blotting

The same basic process of nucleic acid blotting can be used to transfer RNA from gels onto similar membranes.[7] This allows the identification of specific mRNA sequences of a defined length by hybridisation to a labelled gene probe and is known as northern blotting. It is possible with this technique to not only detect specific mRNA molecules but also to quantify the relative amounts of the specific mRNA. It is usual to separate the mRNA transcripts by gel electrophoresis under denaturing conditions since this improves resolution and allows a more accurate estimation of the sizes of the transcripts. The format of the blotting may be altered from transfer from a gel to direct application to slots on a specific blotting apparatus containing the nylon membrane. This is termed slot or dot blotting and provides a convenient means of measuring the abundance of specific mRNA transcripts without the need for gel electrophoresis. It does not, however, provide information regarding the size of the fragments.

A further method of RNA analysis that overcomes the problems of RNA blotting is termed the ribonuclease protection assay (RPA). This is a solution-based method where a probe that is complementary to the mRNA of interest is bound to form a hybrid that is resistant to digestion with RNAse. Thus while other single-stranded RNA molecules are digested, the intact hybrids can be further analysed by gel electrophoresis.

1.4.3. Gene Probe Derivation

The availability of a gene probe is essential in many molecular biology techniques yet in many cases is one of the most difficult steps.[8] The information needed to produce a gene probe, or a primer as used in PCR, may come from many sources but with the development and sophistication of bioinformatics and genetic databases computer-based analysis is usually one of the first stages (see Chapter 5). In some cases, it is possible to use related proteins from the same gene family to gain information on the most useful

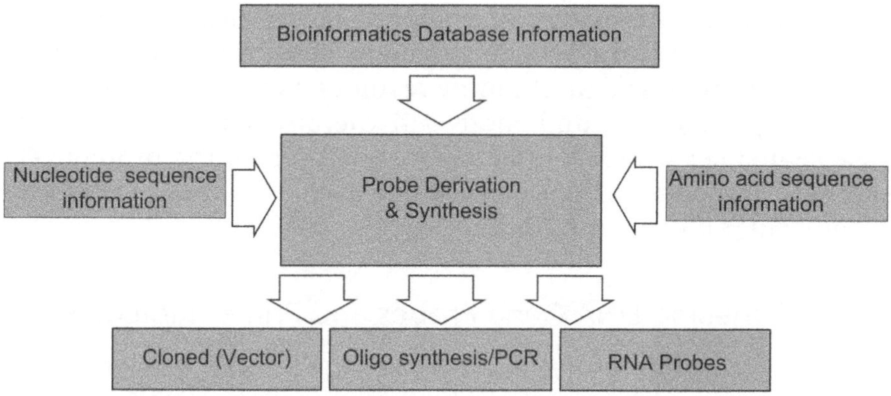

Figure 1.8 Various methods and applications in gene probe and primer production. Reproduced from ref. 22 with permission from the Royal Society of Chemistry.

DNA sequence. Similar proteins or DNA sequences but from different species may also provide a starting point with which to produce a so-called heterologous gene probe. Although in some cases probes are already produced and cloned it is possible, armed with a DNA sequence from a DNA database, to chemically synthesise a single-stranded oligonucleotide probe. This is usually undertaken by computer-controlled gene synthesisers, which link deoxynucleotide triphosphates (dNTPs) together based on a desired sequence. It is essential to carry out certain checks before probe production to determine that the probe is unique, is not able to self-anneal and that it is not self complementary, all of which may compromise its use (Figure 1.8).

Where little DNA information is available to prepare a gene probe it is possible, in some cases, to use the knowledge gained from analysis of the corresponding protein. Thus it is possible to isolate and purify proteins and sequence part of the N-terminal end of the protein. From our knowledge of the genetic code, it is possible to predict the various DNA sequences that could code for the protein, and then synthesise appropriate oligonucleotide sequences chemically. Due to the degeneracy of the genetic code most amino acids are coded for by more than one codon, therefore there will be more than one possible nucleotide sequence which could code for a given polypeptide. The longer the polypeptide, the greater the number of possible oligonucleotides which must be synthesised. Fortunately, there is no need to synthesise a sequence longer than about 20 bases, since this should hybridise efficiently with any complementary sequences and should be specific for one gene. Ideally, a section of the protein should be chosen which contains as many tryptophan and methionine residues as possible, since these have unique codons, and there will therefore be fewer possible base sequences which could code for that part of the protein. The synthetic oligonucleotides can then be used as probes in a number of molecular biology methods.

1.4.4. Labelling DNA Gene Probes and Primer Molecules

An essential feature of a gene probe (and possibly a primer used in PCR) is that it can be visualised by some means. In this way a gene probe that hybridises to a complementary sequence may be detected and used to identify that desired sequence from a complex mixture. There are two main ways of labelling gene probes. Fluorescent labelling is now a popular method for tagging nucleic acids and includes dyes such as fluorescein amidite (FAM) and digoxigenin-labelled nucleotides. Additionally, fluorescent dyes, such as 4′,6-diamidino-2-phenylindole (DAPI) Picogreen and Ribogreen, are commonly used. Radioactive labelling with 32 phosphorous (^{32}P), or for certain techniques 35 sulphur (^{35}S) and tritium (^{3}H), can also be used. These may be detected by the process of autoradiography where the labelled probe molecule, bound to sample DNA, located, for example, on a nylon membrane, is placed in contact with an X-ray-sensitive film. Following exposure, the film is developed and fixed just as a black and white negative and reveals the precise location of the labelled probe and therefore the DNA to which it has hybridised.

Non-radioactive fluorescent labels are increasingly being used to label DNA gene probes and now many have similar sensitivities which, when combined with their improved safety, have led to their widespread use.

The labelling systems are termed either direct or indirect. Direct labelling allows an enzyme reporter, such as alkaline phosphatase, to be coupled directly to the DNA. Although this may alter the characteristics of the DNA gene probe it offers the advantage of rapid analysis since no intermediate steps are needed. However indirect labelling is, at present, more popular. This relies on the incorporation of a nucleotide that has a label attached. At present three of the main labels in use are biotin, fluorescein and digoxigenin. These molecules are covalently linked to nucleotides using a carbon spacer arm of 7, 14 or 21 atoms. Specific binding proteins may then be used as a bridge between the nucleotide and a reporter protein, such as an enzyme. For example, biotin incorporated into a DNA fragment is recognised with a very high affinity by the protein streptavidin. This may either be coupled or conjugated to a reporter enzyme molecule, such as alkaline phosphatase. This is able to convert a colourless substrate *para* nitrophenol phosphate (PNPP) into the yellow compound *para* nitrophenol (PNP) and also offers a means of signal amplification. Alternatively, labels, such as digoxigenin, incorporated into DNA sequences may be detected by monoclonal antibodies, again conjugated to reporter molecules, including alkaline phosphatase. Thus, rather than the detection system relying on autoradiography which is necessary for radiolabels, a series of reactions resulting in either a colour, light or a chemiluminescence reaction takes place. This has important practical implications since autoradiography may take one to three days whereas colour and chemiluminescence reactions take minutes.

1.4.5. End-labelling of DNA Molecules

The simplest form of labelling DNA is by 5′ or 3′ end-labelling. 5′ end-labelling involves a phosphate transfer or exchange reaction where the 5′ phosphate of the DNA to be used as the probe is removed and in its place a labelled phosphate, traditionally ^{32}P, is added. This is usually carried out by using two enzymes, the first, alkaline phosphatase, is used to remove the existing phosphate group from the DNA. Following removal of the released phosphate from the DNA a second enzyme, polynucleotide kinase, is added which catalyses the transfer of a phosphate group (^{32}P labelled) to the 5′ end of the DNA. The newly labelled probe is then purified, usually by chromatography through a sephadex column, and may be used directly (Figure 1.9).

Using the other end of the DNA molecule, the 3′ end, is slightly less complex. Here a new dNTP which is labelled (*e.g.* $^{32}P\alpha dATP$ or biotin-labelled dNTP) is added to the 3′ end of the DNA by the enzyme terminal transferase. Although this is a simpler reaction a potential problem exists because a new nucleotide is added to the existing sequence and so the complete sequence of the DNA is altered, which may affect its hybridisation to its target sequence. End-labelling methods also suffer from the fact that only add one label is added to the DNA, so they are of a lower activity in comparison with methods that incorporate label along the length of the DNA. Fluorescent labels, such as FAM, are now used as an alternative to radiolabels and further developments in

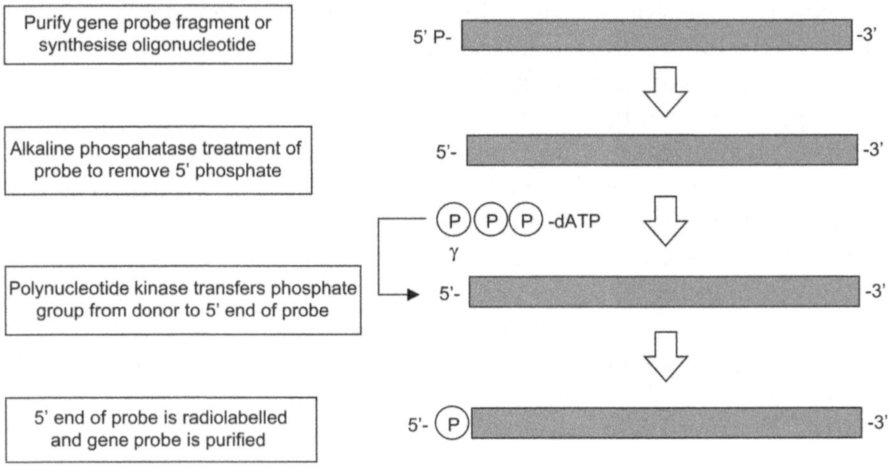

Figure 1.9 End-labelling of a gene probe at the 5′ end with alkaline phosphatase and polynucleotide kinase. Reproduced from ref. 22 with permission from the Royal Society of Chemistry.

probe technology, such as peptide nucleic acid (PNA), locked nucleic acid (LNA) or minor groove binding (MGB) probes, are also widely available.

1.4.6. Random Primer Labelling

The DNA to be labelled is first denatured and then placed under renaturing conditions in the presence of a mixture of many different random sequences of hexamers or hexanucleotides. These hexamers will, by chance, bind to the DNA sample wherever they encounter a complementary sequence and so the DNA will rapidly acquire an approximately random sprinkling of hexanucleotides annealed to it. Each of the hexamers can act as a primer for the synthesis of a fresh strand of DNA catalysed by DNA polymerase since it has an exposed 3′ hydroxyl group. The Klenow fragment of DNA polymerase is used for random primer labelling because it lacks a 5′–3′ exonuclease activity. This is prepared by cleavage of DNA polymerase with subtilisin, giving a large enzyme fragment which has no 5′–3′ exonuclease activity but which still acts as a 5′–3′ polymerase. Thus, when the Klenow enzyme is mixed with the annealed DNA sample in the presence of dNTPs, including at least one which is labelled, many short stretches of labelled DNA will be generated. In a similar way to random primer labelling the polymerase chain reaction may also be used to incorporate radioactive or non radioactive labels (Figure 1.10).

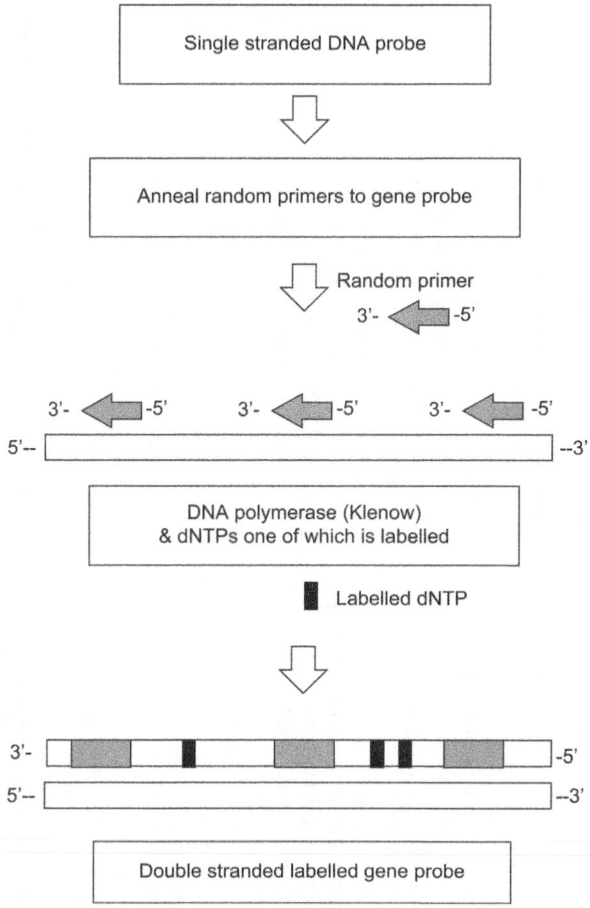

Figure 1.10 Random primer gene probe labelling. Random primers are incorporated and used as start points for DNA polymerase to synthesise a complementary strand of DNA whilst incorporating a labelled dNTP at complementary sites. Reproduced from ref. 22 with permission from the Royal Society of Chemistry.

1.5. The Polymerase Chain Reaction (PCR)

There have been a number of key developments in molecular biology, however, one that has had the most impact in recent years has been the polymerase chain reaction or PCR.[9] One of the reasons for the adoption of the PCR is the elegant simplicity of the reaction and relative ease of the practical manipulation steps. Frequently this is one of the first techniques used when manipulating DNA and has opened up the analysis of numerous cellular and molecular events in many disease processes (Table 1.2).

Table 1.2 Examples of applications of PCR in various fields of biosciences. Reproduced from ref. 22 with permission from the Royal Society of Chemistry.[a]

Field of study	Applications	Specific uses
DNA amplification	General molecular biology	Screening gene libraries
Bridge PCR	Next-generation sequencing	Template cluster preparation
Production/labelling	Gene probe production	Use with blots and hybridisations
Reverse transcriptase PCR (RT-PCR)	RNA analysis	Active latent viral infections
Scenes of crime	Forensic science	Analysis of DNA from blood
Microbial detection	Infection and disease monitoring	Strain typing and analysis (RAPDs)
Cycle sequencing	Sequence analysis	Rapid DNA sequencing
Referencing points in a genome	Genome mapping studies	Sequence tagged sites (STS)
mRNA analysis	Gene discovery	Expressed sequence tags(ESTs)
Detection of known mutations	Genetic mutation analysis	Screening for cystic fibrosis
Digital PCR (dPCR)	Quantification	Viral copy number analysis
Quantitative PCR (qPCR)	Quantification analysis	5′ nuclease (TaqMan assay)
Detection of unknown mutations	Genetic mutation analysis	Gel-based PCR methods (DGGE)
Production of novel proteins	Protein engineering	PCR mutagenesis
Retrospective studies	Molecular archaeology	Dinosaur DNA analysis
Sexing or cell mutation sites	Single-cell analysis	Sex determination of unborn
Studies on frozen sections	*In situ* analysis	Localisation of DNA and/or RNA

[a]Abbreviations: RT, reverse transcriptase; RAPDs, rapid amplification polymorphic DNA; DGGE, denaturing gradient gel electrophoresis.

PCR is used to amplify a precise fragment of DNA (the target) from a complex mixture of starting material, usually termed the template, which may be DNA from microbes, mouth swabs, blood, urine, tissue biopsy.[10] It does require the knowledge of some DNA sequence that flanks the fragment of DNA to be amplified (target DNA). From this information two oligonucleotide primers may be chemically synthesised each complementary to a stretch of DNA to the 3′ side of the target DNA, one oligonucleotide for each of the two DNA strands (Figure 1.11) that may be subsequently extended by a DNA polymerase.

One of the reasons for the global adoption of PCR is the elegant simplicity of the reaction and relative ease of the practical manipulation steps. Importantly, the process, which requires a cyclic process of heating and cooling of the individual reagents, is automated. The reagents are added to sample tubes with thin walls for efficient thermal transfer or to 96-well microtitre plates, although there are various formats available. In many cases most of the reagents are pre-mixed in what is termed a 'master mix', this minimises pipetting errors and is important since small volumes are usually used. The tubes are then placed in an automated thermal cycler. There are numerous makes and models of different configurations on the market that employ various methods for heating and cooling, the most common uses Peltier effect heating blocks and water or fan cooling. High-end thermal cyclers may also include a robotic sample preparation system and many can be programmed directly on the cycler or through a connected PC.

One problem with early PCR reactions was that the temperature needed to denature the DNA also denatured the DNA polymerase. However, the availability of a thermostable DNA polymerase enzyme isolated from the thermophilic bacterium *Thermus aquaticus* (found in hot springs) provided the means to automate the reaction. *Taq* DNA polymerase has a temperature optimum of 72 °C and survives prolonged exposure to temperatures as high as 96 °C and so is still active after each of the denaturation steps. For many applications PCR has replaced the traditional DNA cloning methods as it fulfils the same function, the production of large amounts of DNA from limited starting material; however, this is achieved in a fraction of the time needed to clone a DNA fragment.

1.5.1. Cycles and Steps in the PCR

PCR consists of three defined steps which are repeated a defined number of times or cycles. In the first cycle the double stranded 'high molecular weight' template DNA is denatured by heating the reaction mix to above 90 °C. Within the complex mass of DNA strands, the

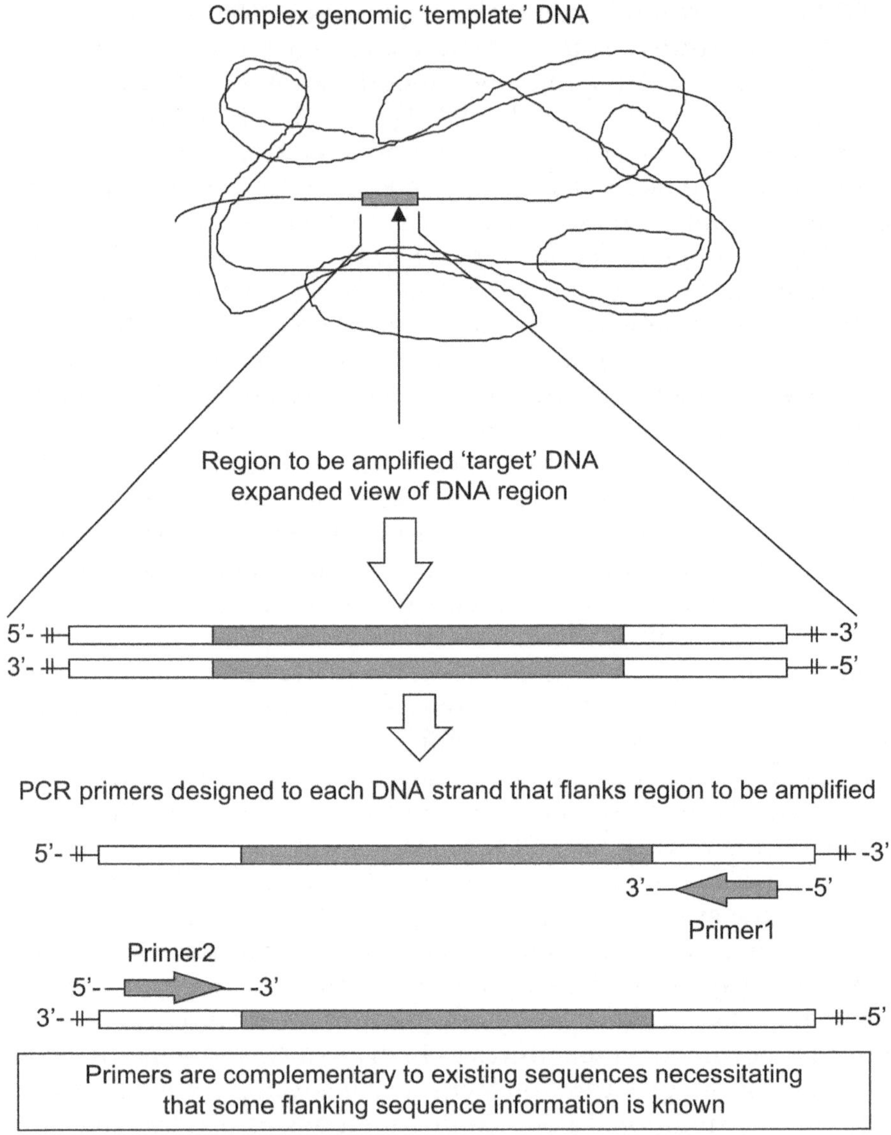

Figure 1.11 The location of PCR primers. PCR primers are designed based on sequences adjacent to the region to be amplified, allowing a region of DNA (*e.g.* a gene) to be amplified from a complex starting material of genomic template DNA. Reproduced from ref. 22 with permission from the Royal Society of Chemistry.

target region to be specifically amplified is thus made accessible and is single stranded. The temperature is then cooled to between 40 and 60 °C to allow the hybridisation of the two oligonucleotide primers, which are present in excess, to bind to complementary sites that flank

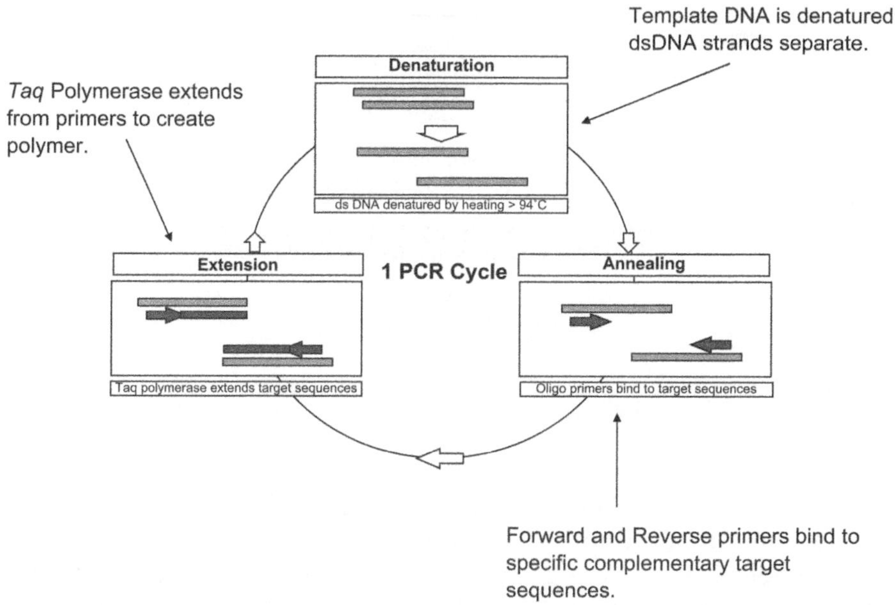

Figure 1.12 The steps involved in one cycle of the polymerase chain reaction. Denaturation of the template strands is followed by annealing of the primers to the template after which extension by *Taq* polymerase from the 3′ end of the primers takes place. Reproduced from ref. 22 with permission from the Royal Society of Chemistry.

the target DNA. This annealing step is unique for each PCR and thus has a defined temperature. The annealed oligonucleotides then act as primers or start points for DNA synthesis since they provide a free 3′ hydroxyl group for DNA polymerase. The DNA synthesis step is termed extension and carried out at 72 °C by a thermostable DNA polymerase, most commonly *Taq* DNA polymerase (Figure 1.12).

DNA synthesis proceeds from both of the primers until the new strands have been extended along and beyond the target DNA to be amplified. It is important to note that, since the new strands extend beyond the target DNA they will contain a region near their 3′ ends which is complementary to the other primer (Figure 1.13). Thus, if another round of DNA synthesis is allowed to take place not only the original strands will be used as templates but also the new strands. The products obtained from the new strands will have a precise length, delimited exactly by the two regions complementary to the primers. As the system is taken through successive cycles of denaturation, annealing and extension all the new strands will act as templates and so there will be an exponential increase in the amount of DNA produced. This is against a background of linear amplification of the long original template strands. The net effect is to selectively

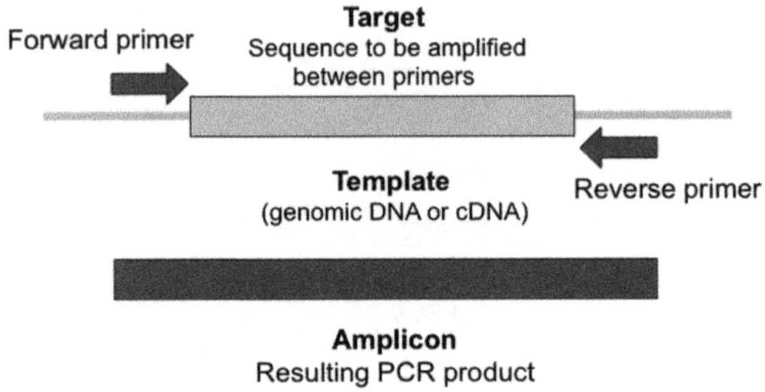

Figure 1.13 Terms associated with amplification components and resulting PCR products. Reproduced from ref. 22 with permission from the Royal Society of Chemistry.

amplify the target DNA and the primer regions flanking it, leading to the production of millions of identical copies.

For efficient annealing of the primers, the precise temperature at which the annealing occurs is critical and each PCR system has to be defined and optimised by the user.[11] One useful technique for optimisation of annealing temperature is called touchdown PCR where a specialised programmable thermocycler is used to incrementally decrease the temperature until the optimum annealing temperature is reached. Reactions that are not optimised may give rise to other misprimed DNA products in addition to the specific target or may not produce any amplified products at all. Another approach to reducing spurious non-specific amplification during the early stages of the reaction is termed 'hot start' PCR. In this method the reaction components are heated to the melting temperature before adding the polymerase. At one time hot start was achieved by introducing a physical wax barrier between the *Taq* polymerase and the remainder of the reaction components. The wax melted at the denaturation temperature allowing the *Taq* polymerase access to the reaction mix. More recently, modified polymerase enzyme systems have been developed that inhibit polymerisation at ambient temperature, either by the use of binding ligands, such as antibodies, or by the presence of bound inhibitors that dissociate only after a high-temperature activation step is performed. *Taq* polymerase has a number of limitations, such as the lack of proof reading resulting in misincorporation errors (some as high as 1×10^{-5} errors per base) and sensitivity to inhibitors, such as those found in blood *e.g.* haem, immunoglobulin M, or in extraction buffers (proteinase K, phenol *etc.*). It also leaves a residue at the end of the amplicon which can be useful for dA:dT cloning (see Chapter 3). Engineered versions of the enzyme, which overcome some of these problems, have been produced and the use of other enzymes, such as *Pfu* (from *Pyrococcus furiosus*) DNA polymerase, which have proof-reading ability, or combinations of enzymes can be especially useful when amplifying fragments of 5 kb or longer. Further cocktails of DNA polymerases can be used in the reaction to amplify PCR products of 20–30 kb. There have been a number of enhancements for the PCR, typically they try to improve the accessibility of the polymerase to the template, chemicals such as betaine, dimethyl sulphoxide (DMSO), T4 gene 32 protein and single-stranded binding protein (SSBP). One interesting approach is nanoparticle PCR or nanoPCR, where adding gold nanoparticles at nanomolar concentrations can increase the sensitivity of the reaction by as much as a 1000-fold. This may be due to a number of factors, including the enhanced thermal transfer properties of the particles and/or assisting with primer–template binding. Indeed the potential use of nanoparticles is changing the way PCR may be undertaken in the future and with increasing use of carbon nanotubes and quantum dots the specificity and sensitivity may be improved.[12]

1.5.2. PCR Primer Design

The specificity of the PCR lies in the design of the two oligonucleotide primers.[13] These are required to be complementary to sequences flanking the target DNA but must not be self complementary and form hairpins or inadvertently bind each other to form dimers, since both prevent authentic DNA amplification. They also have to be matched in their GC content, have similar annealing temperatures and be incapable of mispriming and therefore amplifying unwanted genomic sequences. Manual design of primers is time consuming and often hit or miss, although formulae such as the following are still used to derive the annealing temperature or T_a for each primer:

$$4(G + C) + 2(A + T) = T_m$$

where T_m is the melting temperature of the primer–target duplex and G, C, A and T are the numbers of the respective bases in the primer. In general the T_a is set 3–5 °C lower than the T_m. On occasions, secondary or primer dimer bands may be observed on the electrophoresis gel in addition to the authentic PCR products. In these situations touchdown or hot-start regimes may help. Alternatively, increasing the T_a closer to the T_m can improve the specificity of the reaction. The increasing use of bioinformatics resources, such as primer3, primer-blast at the National Center for Biotechnology Information (NCBI) and primer design assistant, in the design of primers makes the design and the selection of reaction conditions much more straightforward (see Chapter 5). These computer-based resources allow the sequences to be amplified, primer length, product size, GC content *etc.* to be input and, following analysis, provide a choice of matched primer sequences. Indeed the initial selection and design of primers without the aid of bioinformatics would now be unnecessarily time-consuming. Some of these methods are useful when designing a multiplex PCR where more than one set of primers is included in a PCR. Here multiple amplicons are produced but the cycling conditions and the primer sets require careful design and optimisation. Bioinformatic resources, such as primerplex, are also able to assist with the process of multiplex PCR design. Finally, before ordering or synthesising the primers it is wise to submit proposed sequences to a nucleotide sequence search program, such as Blast, which can be used to interrogate GenBank or other comprehensive public DNA sequence databases to increase confidence that the reaction will be specific for the intended target sequence.

1.5.3. Reverse Transcriptase PCR (RT-PCR)

RT-PCR is an extremely useful variation of standard PCR which enables the amplification of specific mRNA transcripts from biological samples without the need for the rigorous extraction procedures associated with mRNA purification which are required for conventional cloning purposes.[14] Conveniently, the dNTPs, buffer, *Taq* polymerase, oligonucleotide primers, reverse transcriptase (RT) and the RNA template are added together to the reaction tube. The reaction is heated to 37 °C, thus allowing the RT to work and enables the production of a cDNA copy of the RNA strands that anneal to one of the primers in the mix. Some thermostable DNA polymerases such as *Tth* (from *Thermus thermophilus*) have a reverse transcriptase activity when manganese is added to the reaction buffer. Thus a single step RT-PCR can be undertaken rather than two-step reaction as in the traditional method.

Following 'first strand synthesis' a standard PCR is carried out to amplify the cDNA product, resulting in 'second strand synthesis' and subsequently a double-stranded DNA (dsDNA) product. The choice of primer for the first strand synthesis depends on the experiment. If amplification of all mRNAs in the cell extract is required then an oligo dT primer that would anneal to all the poly-A tails can be used. If a specific cDNA is sought, then a coding-region-specific primer can be used. Alternatively, random hexamers may be used as described in Section 1.4.6 (Figure 1.14). RT-PCR has many applications, such as the assessment of transcript levels

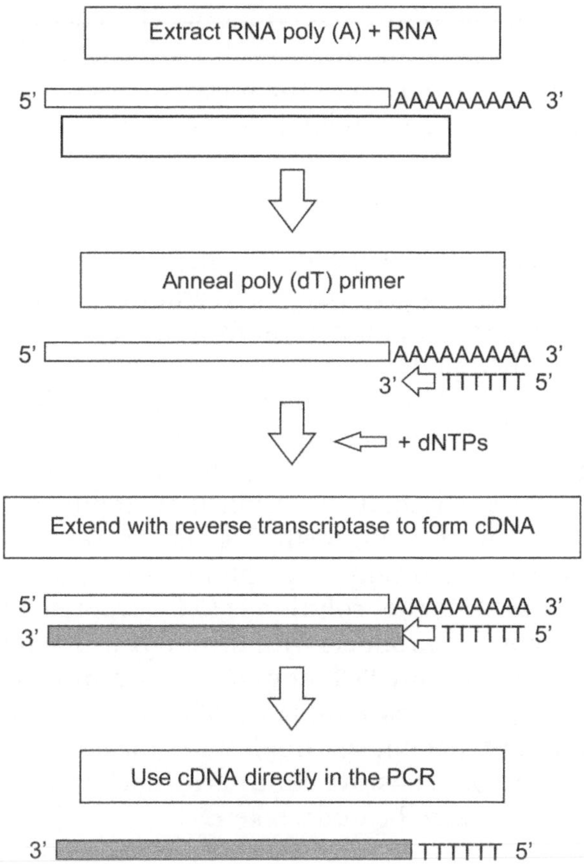

Figure 1.14 Reverse transcriptase PCR (RT-PCR). In RT-PCR mRNA is converted to complementary DNA (cDNA) using the enzyme reverse transcriptase. The cDNA is then used directly in the PCR. Reproduced from ref. 22 with permission from the Royal Society of Chemistry.

in different cells and tissues when analysed with quantitative PCR (qPCR), is widely used as a diagnostic tool and is especially useful in the fields of microbiology and virology as indicated in Table 1.2. A recent development in this area to try to quantitate rare alleles or viral nucleic acid copies is termed digital PCR (dPCR). Here the template is diluted or partitioned out to millions of single strands distributed in miniature chambers, emulsions and other surfaces. A PCR reaction is then performed such that a positive (1) or negative (0) will result. The determination of how many copies were in the original sample can then be performed. Here technological improvements have allowed the partition and amplification in titrated emulsions of oil, which lowers the cost in comparison with other quantitative methods, such as qPCR. This approach of solid-phase PCR is also used in the template preparation for some of the next-generation sequencing methods.[15]

1.5.4. Sensitivity of the PCR

The exquisite sensitivity of the PCR system is also one of its main drawbacks since the very large degree of amplification may make the system vulnerable to contamination. Even a trace of foreign DNA or previously amplified patient DNA, even such as that present on laboratory bench surfaces, may be amplified to significant levels and may give misleading results. Hence, cleanliness is paramount when undertaking PCR, and dedicated equipment and, in some cases, dedicated laboratories are used. It is possible that amplified products may also contaminate the PCR although this may be overcome by UV irradiation to damage already amplified products so that they cannot be used as templates. A further solution is to incorporate uracil into the PCR and then treat the products with the enzyme uracil *n*-glycosylase (UNG) which degrades any PCR amplicons with incorporated uracil, rendering them useless as templates. This is especially important when analysing patient clinical samples. It is of course vital to include the correct negative and positive controls for each PCR to allow subsequent troubleshooting to be undertaken.

1.5.5. Quantitative or Real-time PCR

One important evolution of the PCR method has been the development of quantitative PCR (qPCR). With this method it is possible to determine initial amounts of starting template DNA and follow the reaction in real time. It also provides an early indication of any problems with the amplification process. Early quantitative PCR methods involved the comparison of a standard or control DNA template amplified with separate primers at the same time as the specific target DNA.[16] These types of quantification rely on the reaction being exponential and so any factors affecting this may also affect the result. Other methods have involved the incorporation of a radiolabel through the primers or nucleotides and its subsequent detection following purification of the amplicon.

The introduction of thermal cyclers that incorporate the ability to detect the accumulation of DNA through fluorescent dyes binding to the DNA has rapidly transformed this area.[17] One common DNA binding dye is SYBR green, this binds to dsDNA as it is amplified, albeit in a non-specific fashion, and may be detected in a qPCR thermal cycler. In order to quantitate unknown DNA templates a standard dilution is prepared using template DNA of known concentration. As the DNA accumulates during the early exponential phase of the reaction an arbitrary point is taken where each of the diluted DNA samples cross a line. This is termed the crossing threshold or Ct value. From the various Ct values a log graph is prepared from the standards and thus any unknown amplified at the same time can be deduced. Since SYBR green and similar DNA-binding dyes are non-specific, in order to determine if a correctly sized PCR product is present most qPCR cyclers have a built-in melting curve function. This increases the temperature of the amplicons in the cycler until the dsDNA is denatured. Each amplicon has a characteristic melt temperature and thus can be identified on the basis of theoretical and previous data. A refinement of this method, termed precision melt analysis, is able to differentiate between amplicons with one base difference and has value in mutation analysis. Subsequent DNA sequencing can provide a definitive confirmation.

In order to detect specific amplicons during PCR an oligonucleotide probe labelled with a fluorescent reporter and quencher molecule at either end can included in the reaction in the place of SYBR green. This method is termed the 5′ fluorogenic exonuclease detection system or more commonly the TaqMan system. When the oligonucleotide 'TaqMan' probe binds to the target sequence the 5′ exonuclease activity of *Taq* polymerase degrades and releases the reporter from the quencher during extension. A signal is thus generated which increases in direct proportion to the number of starting molecules. Thus the detection system is able to induce and detect fluorescence in real time as the PCR proceeds. Importantly, it provides confirmation that the correct DNA sequence has been amplified (Figure 1.15). Refinements of the PCR process are also in development and promise rapid amplification and reporting. This is mainly due to advances in miniaturising the amplification system and using microfluidics on a small chip. Efficiency, multiplex throughput and low reagent costs are advantages of these methods. Indeed, there are a number of further developments where PCR and other systems including isothermal cycling systems, such as loop-mediated isothermal amplification (LAMP), can be undertaken in various microfluidic platforms. At present, there is a move to provide a set of guidelines or information in the development and reporting of qPCR termed MIQE guidelines or minimum information for publication of quantitative real-time PCR

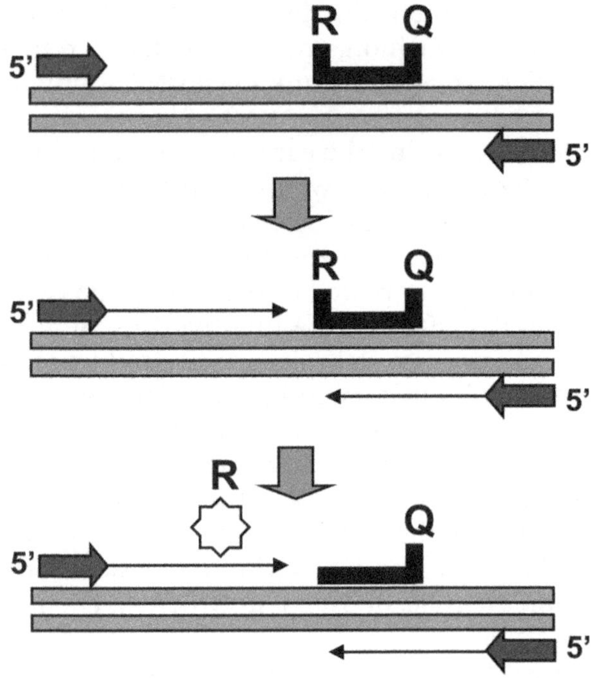

Figure 1.15 Representation of the 5′ nuclease assay or TaqMan approach of specific quantitative PCR where R is the reporter and Q is the quencher. Note: Taq polymerase extends from the 3′ end of the primer in the middle panel and cleaves the R from Q in the lower panel, producing a fluorescent signal. Reproduced from ref. 22 with permission from the Royal Society of Chemistry.

experiments.[18] Although not without its drawbacks PCR is a remarkable development that has changed the approach of many researchers to the analysis of nucleic acids and continues to have a profound effect on core genomic and genetic analysis.

1.5.6. Alternative Amplification Methods

Many traditional methods in molecular biology have now been superseded by PCR and the applications of the technique appear to be unlimited. The success of the PCR process has given impetus to the development of other amplification techniques that are based on either thermal cycling or non-thermal cycling (isothermal) methods. Indeed the development of isothermal systems, such as the LAMP DNA amplification system, obviates the need for a thermal cycler and has the advantage of being able to be used in non-laboratory settings.[19]

1.6. Analysing Genes *in Situ*

Gross chromosomal changes are often detectable by microscopic examination of the chromosomes within a karyotype. Single or restricted numbers of base substitutions, deletions, rearrangements or insertions are far less easily detectable but may induce similarly profound effects on normal cellular biochemistry. *In situ* hybridisation makes it possible to determine the chromosomal location of a particular gene fragment or gene mutation. This is undertaken by preparing a labelled DNA or RNA probe and applying this to a tissue or chromosomal preparation fixed to a microscope slide. Any probe that does not hybridise to complementary sequences is washed off and an image of the distribution or location of the bound probe is viewed with a detection system. Using tissue or cells fixed to slides it is also possible to carry out *in situ* PCR and qPCR. This is a highly sensitive technique where PCR is carried out directly on the tissue slide with the standard PCR reagents. Specially adapted thermal cycling machines are required to hold the slide preparations and allow the PCR to proceed. This allows the localisation and identification of, for example, single copies of intracellular viruses and, in the case of qPCR, the determination of initial concentrations of nucleic acid.

1.7. Analysing Genetic Mutations and Polymorphisms

There are several types of mutations that can occur in nucleic acids, either transiently or stably incorporated into the genome. During evolution, mutations may be inherited in one or both copies of a chromosome, resulting in polymorphisms within the population. Mutations may occur at any site within the genome; however, there are several instances whereby mutations occur in limited regions. This is particularly obvious in prokaryotes, where elements of the genome (termed hypervariable regions) undergo extensive mutations to generate large numbers of variants, by virtue of the high rate of replication of the organisms. Similar hypervariable sequences are generated in the normal antibody immune response in eukaryotes. Mutations may have several effects upon the structure and function of the genome. Some mutations may lead to undetectable effects upon normal cellular functions, termed conservative mutations. Examples of these are mutations that occur in intron sequences and therefore play no part in the final structure and function of the protein or its regulation. Alternatively, mutations may result in profound effects upon normal cell function, such as altered transcription rates, or on the sequence of mRNAs necessary for normal cellular processes.

Mutations occurring within exons may alter the amino acid composition of the encoded protein by causing amino acid substitution or by changing the reading frame used during translation. These point mutations were traditionally detected by Southern blotting or, if a convenient restriction site was available, by restriction fragment length polymorphism (RFLP). However, PCR has been used to great effect in mutation and polymorphism detection and numerous techniques have been developed over the last few years.[20] These involve both screening and confirmatory techniques. One such is single-strand conformational polymorphism (SSCP), this is a screening method and is able to identify that a mutation is within the amplified region even though the identity of the mutation cannot be determined (see Chapter 2). A further confirmatory PCR-based technique is allele-specific oligonucleotide PCR (ASO-PCR) where two competing primers and one general primer are used in two separate PCRs. One of the primers is directly complementary to the known point mutation whereas the other is a wild-type primer; that is, the primers are identical except for the terminal 3′ end base. Thus, if the DNA contains the point mutation only the primer with the complementary sequence will bind and be incorporated into the amplified DNA, whereas if the DNA is normal the wild-type primer is incorporated. The results of the PCR are analysed by agarose gel electrophoresis. A further modification of ASO-PCR has been developed where the primers are each labelled with a different fluorochrome. Since the primers are labelled differently a positive or negative result is produced directly without the need to examine the PCRs by gel electrophoresis.

Various modifications now allow more than one PCR to be carried out at a time (multiplex PCR), and hence the detection of more than one mutation is possible at the same time. Where the mutation is unknown it is also possible to use a PCR system with a gel-based detection method termed denaturing gradient gel electrophoresis (DGGE). In this technique a sample DNA heteroduplex containing a mutation is amplified by the PCR, which is also used to attach a GC-rich sequence (GC clamp) to one end of the heteroduplex. The mutated heteroduplex is identified by its altered melting properties through a polyacrylamide gel which contains a gradient of denaturant, such as urea. At a certain point in the gradient the heteroduplex

Table 1.3 Methods for detecting mutations in DNA samples. Reproduced from ref. 22 with permission from the Royal Society of Chemistry.

Technique	Basis of method	Main characteristics of detection
Southern blotting	Gel- or blot-based	Labelled probe hybridisation to DNA
Dot and slot blotting	Blot-based	Labelled probe hybridisation to DNA
Allele-specific oligo-PCR (ASO-PCR)	PCR-based	Oligonucleotide matching to DNA sample
Denaturing gradient gel electrophoresis (DGGE)	Gel- or-PCR-based	Melting temperature of DNA strands
Single-stranded conformation polymorphism (SSCP)	Gel- or PCR-based	Conformation difference of DNA strands
Multiplex ligation-dependent probe amplification (MLPA)	Gel- or PCR-based	Oligonucleotide matching to DNA sample or ligation
DNA sequencing	Sanger or next-generation sequencing	Nucleotide sequence analysis of DNA
DNA microchips	Glass-chip-based	Sample DNA hybridisation to oligo arrays

will denature relative to a perfectly matched homoduplex and thus may be identified. The GC clamp maintains the integrity of the end of the duplex on passage through the gel. The sensitivity of this and other mutation-detection methods has been substantially increased by the use of PCR, and further techniques used to detect known or unknown mutations are listed in Table 1.3. An extension of this principle is used in a number of detection methods employing denaturing high-performance liquid chromatography (dHPLC). Commonly known as wave technology, the detection of denatured single strands containing mismatches is rapid, allowing a high-throughput analysis of samples to be achieved.

Polymorphisms are particularly interesting elements of the human genome and as such may be used as the basis for differentiating between individuals. All humans carry repeats of sequences known as minisatellite DNA, of which the number of repeats varies between unrelated individuals. Hybridisation of probes which anneal to these sequences using Southern blotting provides the means to type and identify those individuals.

Multiplex ligation-dependent probe amplification (MLPA) is a PCR-based multiplex assay that allows a number of target sites, such as deletions, duplications, mutations or single-nucleotide polymorphisms (SNPs) to be amplified with a pair of primer-containing probes. The process involves a denaturation and hybridization stage, a ligation stage and a PCR amplification stage, followed by analysis of the products. The technique involves designing two adjacent hybridization probes that contain the fluorescently labelled forward and reverse primer sequence used the amplification stage. One of the probes is designed with a stuffer sequence that can be varied to suit the target. When binding to the target, the probes are first hybridized to the denatured sample DNA. In the next stage the two oligonucleotide probes are ligated with a ligase, only ligated probes can be used in the amplification stage. Oligonucleotide probes that are not ligated will only contain one primer sequence and, as a consequence, cannot be amplified and generate a signal.

An alternative labelling strategy used in karyotyping and gene localisation is fluorescent *in situ* hybridisation (FISH).[21] This method, sometimes termed chromosome painting, is based on *in situ* hybridisation but different gene probes are labelled with different fluorochromes, each specific for a particular chromosome. The advantage of this method is that separate gene regions may be identified and comparisons made within the same chromosome preparation. The technique is also likely to be highly useful in genome mapping for ordering DNA probes along a chromosomal segment.

References

1. J. D. Watson and F. H. Crick, *Nature*, 1953, **171**(4361), 964.
2. B. Alberts, A. Johnson, J. Lewis, M. Raff, K. Roberts and P. Walter, *Molecular Biology of the Cell*, Garland, 6th edn, 2014.
3. R. Rapley, *Principles and Techniques of Biochemistry and Molecular Biology*, 2018, pp. 73–176.
4. S.-B. Liang and L.-W. Fu, *Biotechnol. Adv.*, 2017, **35**(4), 443–449.
5. H. O. Smith and K. W. Wilcox, *J. Mol. Biol.*, 1970, **51**, 379.
6. E. M. Southern, *J. Mol. Biol.*, 1975, **98**, 503.
7. J. C. Alwine, D. J. Kemp and G. R. Stark, *Proc. Natl. Acad. Sci. U. S. A.*, 1977, **74**, 5350.
8. A. De Muro and R. Rapley, *Gene Probes*, Humana Press, Totowa, NY, 2002, vol. 179.
9. R. K. Saiki, S. Scharf, F. Faloona, K. B. Mullis, G. T. Horn, H. A. Erlich and N. Arnheim, *Science*, 1985, **230**, 1350.
10. R. K. Saiki, D. H. Gelfand, S. Stoffel, S. J. Scharf, R. Higuchi, G. T. Horn, K. B. Mullis and H. A. Erlich, *Science*, 1988, **239**, 487.
11. R. H. Don, P. T. Cox, B. J. Wainwright, K. Baker and J. S. Mattick, *Nucleic Acids Res.*, 1991, **19**, 4008.
12. F. M. Sang, X. Li and J. Liu, *Chin. J. Anal. Chem.*, 2017, **45**(11), 1745–1753.
13. *PCR Primer Design*, ed. A. Yuryev, Springer Protocols, 2015, vol. 1275.
14. S. Corbet, J. Bukh, A. Heinsen and A. Fomsgaard, *J. Clin. Microbiol.*, 2003, **41**, 1091.
15. R. Sanders, J. Huggett, C. A. Bushell, S. Cowen, D. J. Scott and C. A. Foy, *Anal. Chem.*, 2011, **83**(17), 6474.

16. L. Biesecker, W. Burke, L. Kohane, S. E. Plon and R. Zimmern, *Nat. Rev.*, 2012, **13**(1), 818.
17. R. Higuchi, C. Fockler, G. Dollinger and R. Watson, *Biotechnology*, 1993, **11**, 1026.
18. *PCR Technologies*, ed. T. Nolan and S. Bustin, CRC Press, 2013.
19. Y. P. Wong, S. Othman, Y. L. Lau, S. Radu and H. Y. Chee, *J. Appl. Microbiol.*, 2018, **124**(3), 626–643.
20. R. Rapley and B. D. M. Theophilus, *PCR Mutation Detection Protocols*, Humana Press, 2011.
21. C. Chenghua, W. Shu and L. Penning, *Front. Cell Dev. Biol.*, 2016, **4**, 89.
22. R. Rapley and N. Hupert, in *Molecular Biology and Biotechnology*, ed. R. Rapley and D. Whitehouse, Royal Society of Chemistry, Cambridge UK, 2014, vol. 6, ch. 1, pp. 1–39.

2 Introduction to Genomic Diagnostics

Gary Frewin and David Whitehouse*

School of Life and Medical Sciences, University of Hertfordshire, College Lane, Hatfield, Hertfordshire, AL10 9AB, UK
*E-mail: d.whitehouse@herts.ac.uk

2.1. Introduction

Ground-breaking discoveries in the 20th century include the double helix structure of DNA and the genetic code. This knowledge paved the way to the development of recombinant DNA technology and molecular biology that has since enabled the identification, mapping, isolation and sequencing of genes and genomes in numerous species. In the 50 years between the reporting of the DNA double helix in 1953 and the completion of the human genome sequence, dramatic progress in biotechnology has led to advances on many fronts, including molecular cloning, genome-wide mapping and mutation and polymorphism detection. In the last decade, the lowering cost and greater ease associated with nucleic acid sequencing has prompted proposals to sequence populations comprising thousands or even millions of genomes in the hunt for the genetic causes of disease. Some of the enabling molecular methodologies and their applications to diagnostics are addressed in this chapter.

Genomics and Clinical Diagnostics
Edited by David Whitehouse and Ralph Rapley

Published by the Royal Society of Chemistry, www.rsc.org

2.1.1. Genomics Background

'Genome' refers to the entire collection of hereditary material (DNA or RNA) in an organism, cell or virus. The eukaryote genome is subdivided into the nuclear genome and the much smaller mitochondrial genome. The human genome is some 3 billion base pairs and consists of 23 linear DNA molecules that are packaged into 23 chromosomes and is estimated to encode about 22 000 genes, somewhat less than the estimated 35 000 genes in the grape (*Vitis* sp.) genome.[1] In contrast, mitochondrial DNA consists of a single circular chromosome comprising of about 16 600 base pairs that encodes just 37 genes of which 13 are protein coding. Significant technological developments in DNA sequencing technology platforms and bioinformatics have accelerated genome sequencing and mapping and elevated genome analysis beyond research into the clinical domain. Large curated databases that catalogue pathogenic germ line and somatic variants (*e.g.* Catalogue of Somatic Mutations in Cancer (COSMIC), see Chapter 5) are being developed to aid clinical decision making. Genomics-based molecular diagnostics are now available for many genetic diseases, including cancers. In a growing number of pathologies molecular diagnostics is beginning inform the prescription of therapeutics and is expected to be a key component of the drive toward personalised medicine.

2.1.2. Genomes and Chromosomes

Cellular genomes (*i.e.* non-viral) vary tremendously in size and complexity. The smallest known from a bacterium contains about 600 000 DNA base pairs; in contrast, the eukaryote nuclear genome often consists of billions of base pairs (Table 2.1) In many eukaryotes, notably representatives of plants, fungi, invertebrates, fish, amphibians and reptiles, genomic analysis is complicated by the polyploid nature

Table 2.1 Approximate sizes of haploid genomes in various organisms.

Organism		Genome size
Bacteria	*Mycoplasma genitalium*	0.6 Mb
	Escherichia coli	4.6 Mb
Yeast	*Saccharomyces cerevisiae*	14 Mb
Nematode worm	*Caenorhabditis elegans*	100 Mb
Flowering plant	*Arabidopsis thalania*	115 Mb
Fruit fly	*Drosophila melanogaster*	165 Mb
Puffer fish	*Fugu rubripes rubripes*	400 Mb
Human	*Homo sapiens*	3300 Mb

of the genomes where nuclear DNA is packaged into several sets of homologous chromosomes. In humans and the majority of the animal kingdom, including 'model species' such as mice and fruit flies, DNA is packaged into two sets of homologous chromosomes *i.e.* the cells are said to be diploid. The human diploid genome consists of 23 pairs of chromosomes (22 autosomes and two sex chromosomes) that differ in size and staining properties, one set being inherited from each parent. When normal male cells undergoing metaphase are arrested, stained and examined under the light microscope, each pair of autosomes appears identical, but the sex chromosomes X and Y are distinct. In normal female cells, which have two X chromosomes, all 23 pairs of homologous chromosomes appear identical.

2.1.3. Genomic Analysis

In prokaryotes, genomic analysis is simplified because the genomes consist of single chromosomes and simple gene structures, *i.e.* the genes comprise uninterrupted arrays of amino-acid-coding DNA sequences that extend from the transcription start site to the end of the coding region and there is relatively little intergenic non-coding DNA. In contrast, the DNA sequences of most eukaryote genes are broken up into small segments of coding DNA called exons and non-coding intervening sequences called introns (see Chapter 1). Furthermore, eukaryotic genes are often separated by vast lengths of intergenic non-coding DNA as well as epigenetic modifications, or 'tags', which serve to regulate gene expression.

The first genome to be sequenced was that of the virus-like bacteriophage PhiX174. The single chromosome consists of 5386 nucleotides that encode 11 genes with no intergenic sequences.[2,3] Although the reporting of the PhiX174 sequence in 1977 preceded the publication the human genome sequence by some 25 years, Sanger's pivotal discovery of the 'chain termination' DNA sequencing method (see Chapter 3) was essential to the success of the Human Genome Project (HGP). 'Sanger sequencing' eventually led to the development of semi-automated high-throughput DNA sequencing strategies that, together with PCR and sophisticated bioinformatics tools, enabled the completion the draft human genome sequence in 2001.[4,5] In the 21st century, further advances in technologies, such as next-generation sequencing (NGS, see Chapter 3) and DNA microarray platforms, have enabled rapid, cost-effective sequencing and analysis of genomes, transcriptomes and epigenetic profiles. Genomic analysis of various disease populations has also accelerated the discovery of new clinically relevant biomarkers (see Chapters 15 and 16). Ongoing genetic changes may be monitored, for example by resequencing, throughout the life-span of a tumour, giving insight into the genomic instability that is the hallmark of many cancers. Third-generation sequencers are further decreasing the costs and turnaround-times while improving read lengths to millions of base-pairs per read on devices no larger than a USB drive,[6] enabling the possibility of rapid genome analysis in remote areas and hospital emergency rooms.

2.1.4. Gene Identification

Gene identification and mapping in eukaryotes is complicated because the majority of nuclear genomic DNA is devoid of genes. It has been estimated that less than 2% of human nuclear DNA comprises some 22 000 expressed genes.[1] This appears to be too few genes to constitute the genetic blueprint for complex organisms such as vertebrates; however, because of post-transcriptional processing the number of messenger RNA (mRNA) transcripts vastly exceeds the actual number of structural (protein-coding) genes. Thus, a relatively small number of genes produce a vast array of different mRNA transcripts that arise through various mechanisms, such as alternative promoter usage and alternative exon splicing.[7,8] Further heterogeneity is introduced by posttranslational modification of proteins during their maturation. Thus some 22 000 genes could give rise to approximately 1 000 000 proteins.

The 98% of the genome that is non-genic was sometimes referred to as "junk DNA" because it appeared to be non-functional. However, there is growing evidence that as much as 90% of the genomic DNA may be actively transcribed, giving rise to several classes of non-coding RNAs (ncRNA). Examples of well-known and important ncRNAs are transfer RNA (tRNA) and ribosomal RNA (rRNA), but the results of transcriptomic studies indicate there are many classes of non-coding RNA species. At one time it was assumed that the seemingly uncoordinated transcriptional activity was due to background noise. However, it is becoming clear that some classes of ncRNAs are functionally active in gene regulation and the transfer of genetic information. Thus, some ncRNAs could play key roles in molecular physiology. Thus, for example, abnormal expression of ncRNAs such as miRNAs (micro RNAs) and snRNAs (small non-coding RNAs), could yield insights into cancer biology and offer the potential for new approaches to diagnostics and therapeutics.[9]

2.1.5. Genetic Polymorphism

'Genetic polymorphism' is a term that relates to the presence of two or more discrete variants, or alleles, at a locus; in human genetics a gene locus is said to be polymorphic if at least 2% of the individuals in a population are heterozygous. 'Locus' means a specific location of a DNA sequence in a genome but need not be within a functional gene. Genetic polymorphisms arise from mutations that have increased in frequency in a population due the action of evolutionary forces such as natural selection, gene flow or genetic drift. The most common type of polymorphism arises from single-nucleotide substitutions (referred to as single-nucleotide polymorphisms or SNPs). The majority of SNPs are found in non-coding DNA and are assumed to be harmless, but when SNPs that occur in genes lead to amino acid substitutions, or changes in regulatory DNA, there is the potential for a loss or change of function, leading to deleterious effects. A notable example is the β-globin allele HbS, which is associated with blood disorders known as sickle cell disease (SCD). The HbS allele reaches polymorphic frequencies in some African populations and populations of African origin, as well as populations in India and the Arabian Peninsula. The SNP associated with HbS is a missense mutation in which A is replaced by T in codon 6 of the β-globin gene, leading to the replacement of glutamic acid (Glu) by valine (Val) in the β-globin chain. Persons who are homozygous for the Val allele suffer from SCD, which is caused by the polymerisation of haemoglobin, leading to erythrocytes becoming deformed and rigid under certain conditions.[10] Other classes of molecular variation that reach polymorphic frequencies are short insertions and deletions (indels) leading to copy number differences of tandem repeats: polymorphic minisatellite and microsatellite arrays have been important markers in many genome mapping projects. This type of copy number variation can also be associated with Mendelian genetic disease (see Section 2.2.2.2).

Polymorphism detection (see Section 2.3) has become indispensable to genome-wide analysis. Polymorphisms can be utilised as markers that, together with other types of markers, such as sequence-tagged sites (STSs) and expressed sequence tags (ESTs), help to assemble long segments of DNA sequence and map genes. For genetic mapping, panels of SNPs analysed using DNA microarrays have become the preferred method for genome-wide association studies (GWAS) that seek to identify susceptibility genes associated with complex (*i.e.* non-Mendelian) diseases and pharmacological interactions. For example, polymorphisms in several genes were revealed by a GWAS to influence an individual's response to the anticoagulant drug warfarin, a discovery that has begun to move into the clinical setting through the use of genetic tests.[11] Molecular diagnostics utilising, in most cases, hybridisation of a probe with target nucleic acid sequences have allowed the detection of polymorphisms for the rapid diagnosis of diseases at a speed and cost that are far superior to NGS.

2.2. Genome Analysis

2.2.1. Mapping Genomes and Genes

Genome and gene maps provide information on the location of genes and other markers that enable the user to navigate to specific features and regions of interest. The two main approaches for locating genic and other DNA fragments on chromosomes are genetic and physical mapping. Genetic or linkage mapping aims to determine the genetic distance between a pair of polymorphic markers, or a marker and a disease gene locus by estimating the recombination frequency between them. Linkage is the propensity for genes at two nearby loci to remain on the same chromatid at meiosis, *i.e.* they tend not to be separated by recombination. A linkage map is constructed by analysing recombination frequencies between several pairs of polymorphic marker loci. The genetic distance between the markers, which is determined by their recombination frequency, is measured in centimorgans (cM) rather than base pairs. For any pair of marker loci, the lower the recombination frequency, the shorter the genetic distance between the markers. Linkage mapping in humans is statistically complex and relies on the availability of DNA samples from several multi-generation pedigrees, with affected individuals if a Mendelian disease is being mapped.[12] In the pre-DNA era, linkage analysis in humans was a rather slow-moving process because of the dearth of suitable polymorphic markers. However, the introduction of DNA technology and the discovery of restriction fragment length polymorphisms (RFLPs) and then tandem repeat polymorphisms, such as TA repeats, augmented the use of linkage analysis, which became a major mapping tool in the HGP. Since less than 300 well-chosen genetic marker loci permit full coverage of the human genome, linkage mapping is an effective approach to discovering the approximate location of an unmapped gene or Mendelian disease. Where the lack of pedigree information prevents classical linkage mapping, other genetic mapping techniques can be employed such as population based non-random association studies (linkage disequilibrium), affected sib-pair analysis (ASP) and the transmission disequilibrium test (TDT).[12] In contrast to linkage- and association-based mapping, physical or genome mapping studies are used to determine the physical distances between sets of markers in base pairs or cytogenetic locations rather than recombination frequencies. Physical mapping is addressed in Section 2.4.1.

2.2.2. Tools for Genetic Mapping

2.2.2.1. RFLPs and Minisatellites

The main tools for genetic mapping are polymorphic genetic markers. The first DNA polymorphisms were detected using the Southern blotting technique for analysing genomic DNA digested with restriction enzymes.[13] RFLPs were revealed using a specific genomic or complementary DNA (cDNA) probe that hybridised with digested DNA from the locus of interest. Southern blotting involves the transfer of DNA fragments, generated by restriction enzyme digestion, from an agarose electrophoresis slab gel to an immobilising membrane, such as nitrocellulose or nylon. The immobilised DNA is hybridised with a locus-specific radioactive or biotin-labelled DNA probe that could be from a cDNA or genomic DNA clone or a PCR product (see Chapter 3 for a discussion of PCR). When the autoradiograph or fluorograph of the probed restriction fragments from a set of unrelated individuals is examined, person-to-person variation in the resulting band pattern reflects the presence of an RFLP.[14] The majority of RFLPs consist of two alleles (*i.e.* cut and uncut) and result from single nucleotide substitutions leading to alteration of a restriction site (Figure 2.1).

A second type of RFLP is the result of different lengths of the DNA fragments (variable number of tandem repeat polymorphisms, VNTR) rather than the presence or absence of restriction sites. VNTRs are extremely polymorphic blocks of repetitive DNA where each repeat is between 10 and 100 nucleotides (minisatellite DNA). Hypervariable VNTR bands were first detected on Southern blots by hybridisation with a labelled 'core' probe after digestion of human DNA digested with a restriction enzyme, often *Hin*F1, that cuts the DNA either side of the minisatellite arrays.[15] DNA core probes such as 33.6 and 33.15 that are derived from the human myoglobin gene sequence (*MB* on 22q13.1) recognise a common core sequence of about 10 to 15 base pairs that is shared between many different minisatellite loci. Such 'multilocus' probes revealed tremendous person-to-person variation

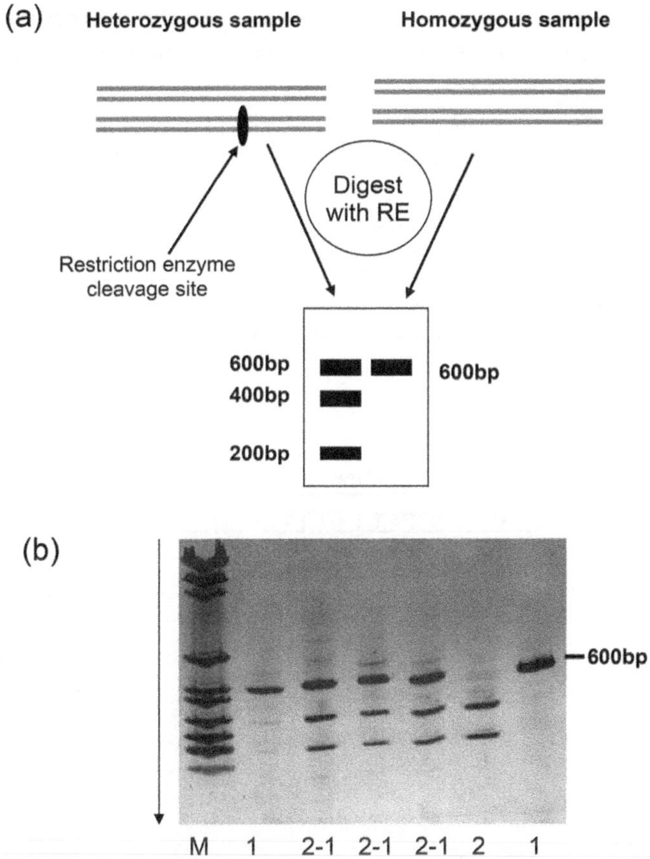

Figure 2.1 RFLP analysis of a 600 bp PCR product. (a) Schematic analysis of products from heterozygous and homozygous individuals digested with a restriction endonuclease (RE). Three bands arise from the heterozygous individual, one that represents the uncleaved fragment and the others, the cleaved fragments. (b) An *Sfa*I RFLP revealed in six individuals following digestion of a 600 bp PCR product.

in the complex patterns of bands that became known as DNA fingerprints.[16] DNA fingerprints have played a major role in forensic medicine where, with the exception of monozygotic twins, they can be used to identify individuals unambiguously and therefore match suspects and victims to crime scenes.

However, DNA fingerprints are not amenable to Mendelian segregation analysis or gene mapping because it is not possible to assign alleles to a specific locus due to the complexity of the electrophoretic band patterns. A considerable advance in this respect was the development of single-locus VNTR probes, which were isolated from specific cloned minisatellites and are able to detect individual minisatellite loci. Single-locus probes (SLPs) produce simple patterns in which heterozygous samples display two bands and homozygotes, a single band, thus they became widely used for gene mapping because the allelic bands could be readily traced through families (Figure 2.2).

The length of a single VNTR locus often exceeds several kilobases and thus VNTRs are generally analysed using Southern blotting techniques with SLPs rather than PCR. The average heterozygosity of polymorphic VNTR loci is about 70%; however, loci with the highest

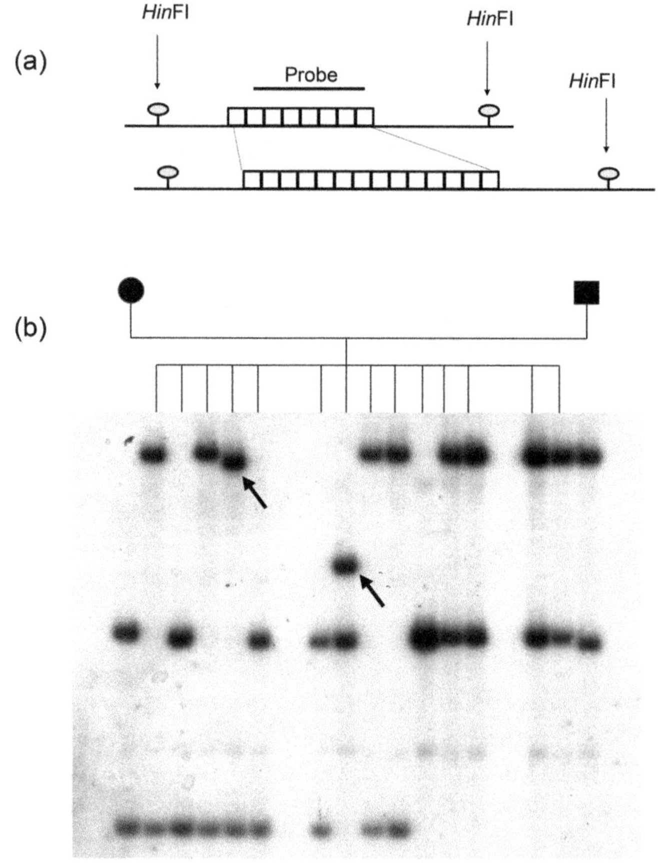

Figure 2.2 (a) Schematic illustration of a heterozygous VNTR locus. The invariant *Hinf*I sites reveal a length polymorphism resulting from the repeat number variation in the two minisatellite alleles. (b) Southern blot analysis of hypervariable minisatellite B6.7 using a single-locus probe on *Hinf*I digests of genomic DNA from a large family. The arrows indicate new mutations not found in either parent.

heterozygosity's tend to be unstable.[17] It has been estimated that there are between 15 000 and 20 000 VNTR loci in the human genome, but since they tend to cluster near the ends of chromosomes they are of somewhat limited use for gene mapping. Interestingly, the repeat blocks in many VNTR loci vary slightly within the sequences of the repeats, giving rise to repeat types along the length of the two alleles. These variants are detectable using PCR-based methods and are the basis of the minisatellite variant repeats (MVR) typing system, an elegant means of producing highly informative individual genetic band patterns that somewhat resemble barcodes.[18]

2.2.2.2. Microsatellites

Microsatellites (also referred to as short tandem repeats or STRs) have a much simpler structure than minisatellites, usually comprising between two and seven nucleotides per repeat; they are also relatively stable. Because the overall length of the tandem repeat is relatively small, microsatellites are amenable to PCR amplification. In general, STRs produce simple band patterns on electrophoresis gels; trinucleotide and tetranucleotide microsatellites tend to be preferred over dinucleotide repeat polymorphisms, which have been associated with a greater tendency to produce confusing secondary bands, sometimes called 'stutter' bands. STRs were the most widely used source of marker polymorphisms in the HGP, and they still form the basis of modern forensic DNA profiling, for example the Combined DNA Index System (CODIS) that was developed by the FBI. Although STRs tend to be less polymorphic than minisatellites, they are found scattered throughout the human genome and are thus a better mapping resource. The human genome contains over a million STR loci;[4] by the mid-1990s thousands of polymorphic STRs had been used to generate linkage maps of the human chromosomes.

Interestingly some of the triplet repeat microsatellite sequences have been found to play a key role in the aetiology of neurological genetic disorders such as Huntington's disease (HD, OMIM 143100), myotonic dystrophy (DM, OMIM 160900) and fragile-X syndrome (FMR1, OMIM 309550). In these cases, the onset and severity of the disorder is related to an increase in the number of triplet repeats that tend to expand from generation to generation, a phenomenon called 'anticipation' in medical genetics.[12] For example in HD, the onset of symptoms is associated with the increasing length of a CAG triplet repeat located within the Huntington gene (*HTT*). The CAG triplet encodes the amino acid glutamine, which gives rise to a poly-glutamine tract within the huntingtin protein. In healthy individuals the number of CAG repeats ranges between 6 and 26, whereas in patients with more than 40 repeats, full-blown disease is to be expected.[19]

Although the molecular mechanisms leading to microsatellite variation are not completely understood, it is thought that replication slippage is the most common mechanism leading to the gain or loss of one or more repeat units.[20] Other mutational mechanisms such as those involving unequal crossing-over or duplication events, have also been considered.[21] The tendency to mutate can also depend on the chromosomal environment of a particular locus and whether or not it is transcribed.[22–24] Microsatellite loci tend to exhibit higher mutation rates than single-nucleotide substitution loci.[24]

2.2.2.3. Single-nucleotide Polymorphisms (SNPs)

The most widespread type of genetic variation is the result of single nucleotide substitution. Commonly referred to as single nucleotide polymorphisms or SNPs, these variant sites are found scattered throughout the genomes of most species. A SNP is caused by a single nucleotide replacement, such as G**A**ATTC to G**T**ATTC, which occurs with a frequency of at least 1% in the population. SNP is a new name for an old phenomenon, namely the type of point mutation that can be detected by conventional RFLP analysis and indeed missense mutations that were detected in gene products (enzyme polymorphisms) using zone electrophoresis.[25] SNPs are abundant: there are an estimated 15 million SNPs in the human genome.[26] In a study to investigate genetic backgrounds in different populations, 1.6 million human SNPs were genotyped in 1184 individuals.[27] SNPs are not distributed evenly throughout the genome and their frequency can vary up to tenfold.[26,28] Unsurprisingly, because of selection pressure, SNPs are less abundant in coding regions than non-coding regions; other factors that affect their distribution are local recombination and mutation rates.[29] The majority of SNPs are located in repeat regions, which are difficult to analyse, even so SNPs are the best choice of polymorphic markers that offer coverage of the whole genome. It has been estimated that there are 5.3 million common SNPs, each with a frequency of 10–15%, that account for the bulk of DNA differences. Such SNPs occur on average once every 600 base pairs within the human genome in non-coding DNA,[30] thus, the level of heterozygosity in the human genome resulting from SNPs is extraordinarily high. Although most SNPs result from single-nucleotide base differences, many SNP detection methods can equally well detect small insertions and deletions.

Although SNPs have a use in classical gene mapping by linkage analysis, they have attracted great interest as marker polymorphisms for large-scale genome-wide association studies (GWAS) that attempt to discover associations between 'susceptibility' loci and complex traits, such as obesity or schizophrenia.[31] Most SNPs are 'anonymous' genetic markers used to narrow down the regions and identify variants that have phenotypic effects. On rare occasions the SNPs themselves may contribute to the phenotype. SNPs within genes, especially those in coding regions that lead to missense mutations, may well change the function, regulation or expression of a protein.[32] SNPs have also been discovered in the splice sites of genes, which can result in variant protein products with different exon arrays,[33] as well as in the controlling regions of genes, where they can affect gene expression and regulation.[34,35] Exonic single-nucleotide variants can also form part of a physician's diagnostic assessment. For instance, screening for germline missense mutations in the transcription factor gene *TP53* in autosomal dominant Li-Fraumeni kindreds, where some family members that are risk of developing cancers, has been carried out for several years. Germline mutations in *TP53* have also been associated with non-Li-Fraumeni familial cancers. Several exonic SNPs have been identified in *TP53*, a few of which result in amino acid substitution. Studies of such *TP53* missense SNPs in relation to cancer susceptibility are often inconclusive, though there is a suggestion that codon 72 [proline (Pro) or arginine (Arg)] is associated with early cancer diagnosis in persons carrying the Arg72 allele.[36] Most SNPs occur in non-coding sequences that are separated from functional genes by vast genomic distances and these are unlikely to have phenotypic significance.

The US National Centre for Biotechnology Information (NCBI) curates and maintains a SNP database (dbSNP) which contains sequence details of several million SNPs, together with short insertions and deletions, from a variety of species. The structural simplicity of SNPs compared with microsatellites renders them ideal for full-scale automation and unambiguous scoring of alleles. However, in contrast to microsatellite loci, because most SNPs represent single base changes or small insertions or deletions, they tend to be biallelic and thus their heterozygosity cannot exceed 0.5; microsatellite loci on the other hand tend to have multiple alleles with heterozygosities sometimes approaching 1.0. However, the low polymorphism information content (PIC value – a measure of the diversity of a genetic locus) of SNPs compared with microsatellites is offset by their relative abundance.

2.3. Mutation and Polymorphism Detection

For many years electrophoretic methods have prevailed for polymorphism typing and mutation detection. With the completion of the HGP there has been an intensifying requirement for faster and improved methods to discover and characterise polymorphic markers in order to map genomes, identify disease-related genes and provide rapid diagnosis for patients suspected of harbouring pathogenic genetic mutations.

Whereas DNA sequencing platforms, particularly next-generation sequencing, are now the prominent methods for mutation discovery (see Chapter 3),[37] a variety of alternative approaches were developed to discover and genotype SNPs and STR polymorphisms.[38] These methods can be categorised according to their platform technologies *e.g.* electrophoretic or non-electrophoretic as well as the molecular features they assess, such as the differential conformation of DNA strands or the melting properties of PCR products. In addition, many methods have been developed to genotype known SNPs, these often use allele-specific primers and probes and real-time quantitative PCR (qPCR) amplification (such as the 5′-nuclease assay, see Chapter 3) or microarray technology (Table 2.2). Following a brief introduction to high-throughput electrophoresis-based methods, a conformation-based approach (single-strand conformation polymorphism), a DNA-melting-based approach (denaturing high-performance liquid chromatography) and DNA microarray technologies are discussed.

A range of gel-based and capillary electrophoresis methods and instruments have been introduced that are suitable for large-scale high-throughput genotyping of STR and SNP loci. For instance, PCR products labeled with infrared fluorescent dyes can be separated by electrophoresis on thin denaturing gels (*e.g.* 0.25 mm) and results recorded and analysed on instruments such as the LI-COR 4300, which has a maximum capacity of 64 samples per gel and run times of between 1 and 3.5 hours depending on the resolution required. Gel-based methods have now largely been replaced by capillary electrophoresis (CE). CE was introduced as a workable technique in the 1990s and has since undergone many developments and sophistications. In contrast to gel-based methods, capillary electrophoresis can be fully automated as the products of a PCR reaction can be introduced robotically to an instrument with multiple capillaries, allowing large numbers of genotypes to be characterised in parallel and in a short time; therefore relatively

Table 2.2 Methods for discovering (D) and genotyping (G) SNPs and STR polymorphisms. SNPs (single nucleotide polymorphisms) can be assessed by the melting and conformational properties of double stranded (ds) and single stranded (ss) DNA respectively. STRs (short tandem repeats) are genotyped by analysis of fragment sizes. DHPLC, denaturing high-performance liquid chromatography; DGGE, denaturing gradient gel electrophoresis; qPCR; quantitative PCR; SSCP, single-strand conformation polymorphism; MALDI-TOF, matrix-assisted laser desorption/ionisation-time of flight; NGS, next-generation sequencing.

Type	Basis of detection	Technology	Application	Throughput
SNP	Melting of dsDNA	DHPLC	D, G	Medium
		DGGE	D	Low
		qPCR Melt-curve analysis[a]	D	Low
	Conformation of ssDNA	SSCP	D	Low
		Capillary electrophoresis SSCP	D	Medium
	Probe hybridisation	DNA microarrays	G, D	Very high
		MALDI-TOF	G, D	High
		Allele specific PCR[a]	G	Low
		5′ nuclease assay	G	Low
	Primer extension	Dideoxy sequencing[a]	D	Medium
		NGS (some platforms)[a]	D	Very high
STR	Fragment size analysis	Gel electrophoresis	G	Medium
		Capillary electrophoresis	G	High

[a]See Chapter 3 for details of DNA sequencing technologies.

few highly skilled personnel are needed. Once the polymorphic STRs are amplified using fluorophore-labelled PCR primers the PCR products are loaded into the capillaries and electrophoresed. They are identified by laser-stimulated fluorescence as they pass by a detector that records the electrophoretic mobilities. There are a number of CE instruments useful for genotyping (fragment) analysis with varying capacities. An example of a high-throughput system is the 48 capillary Genetic Analyzer 3730xL (Applied Biosystems), which has the capacity to electrophorese over 4000 samples a day (yielding more than 90 000 STR genotypes) and to operate unattended for 24 h. Dedicated software is used to process and determine the allele sizes and record the genotypes of the samples. SNP genotyping by CE is more complex since the allelic PCR products are normally the same length. There are a number of proprietary approaches to SNP genotyping, such as the TaqMan 5′-nuclease assay, and non-proprietary alternatives; some are based on mini-sequencing, others on ligation reactions, molecular beacons and allele-specific reagents (see Chapter 1).

2.3.1. Single Stranded Conformation Polymorphism (SSCP)

SSCP was first described in 1989,[39] and has been widely applied in genetic analysis both for the discovery and genotyping of SNPs.[40] The principle of SSCP analysis is to denature PCR products to make them single-stranded and then to separate the single strands by gel electrophoresis under non-denaturing conditions in which single-stranded fragments of DNA will adopt unique and specific conformations as they attempt to fold into the most stable structure. A mutated strand (allele) differing by just one nucleotide can adopt a different conformation to it is wild-type counterpart. The structural differences in the conformers can be identified by electrophoresis where a homozygous sample displays two bands and a heterozygote, four bands, one for each denatured strand (Figure 2.3). For mutation detection, the resolving power of the technique is improved by screening PCR fragments that are less than 300 bp in length.[41] Provided that PCR-SSCP analysis

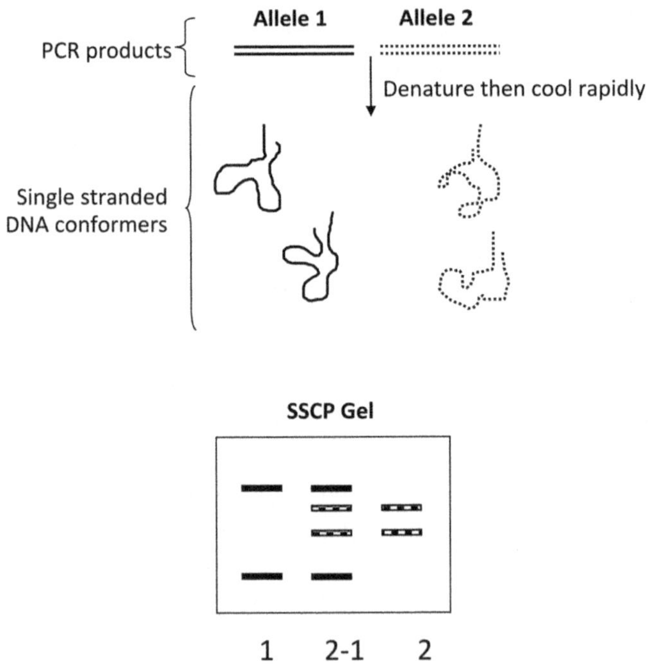

Figure 2.3 Principle of SSCP analysis. In the presence of heat and formamide the dsDNA PCR products from a heterozygote are denatured for several minutes to form single-stranded DNA (ssDNA). Immediately before loading onto a non-denaturing electrophoresis gel, the samples are cooled on ice to encourage the formation of ssDNA conformers. Each conformer has a unique electrophoretic mobility as shown on the gel diagram.

is conducted under appropriate conditions, for example by repeating each electrophoresis experiment at two running temperatures or by varying the concentration of denaturant (such as formamide) in the gel, it is an efficient method for discovering new alleles.[42–44] SSCP has been adapted for capillary electrophoresis; in this semi-automated format, PCR fluorophore-labelled primers generate products that are recorded and analysed by capillary electrophoresis and laser-induced fluorescence under non-denaturing conditions.[45] It is also amenable to be multiplexed, where several PCR products labelled with different fluorophores are analysed simultaneously in the same electrophoretic gel lane or capillary.[46]

2.3.2. Denaturing High-Performance Liquid Chromatography (DHPLC)

Denaturing high-performance liquid chromatography is a fast and reliable method for SNP detection and discovery and is an alternative to melt-curve analysis that can be performed following real time qPCR (see Chapter 3) or denaturing gradient gel electrophoresis (DGGE).[47–49] The DHPLC technique is based on the separation of homoduplexes and heteroduplexes formed between reference and mutant DNA molecules by ion-pair reversed-phase high-performance liquid chromatography under partially denaturing conditions. For DNA fragments of between about 100 and 1500 base pairs, DHPLC is capable of detecting all single-base substitutions as well as small insertions and deletions.

To perform the analysis, DNA fragments are amplified by PCR from both reference and test samples. The two PCR products are then mixed together and fully denatured at 95 °C before being slowly reannealed by reducing the temperature to about 25 °C. Three classes of duplexes are formed: homoduplexes of both the reference and test DNAs and heteroduplexes that form from the annealing of reference and mutant DNA strands. Any nucleotide sequence differences between the between the reference and test samples will produce mismatches in the heteroduplexes. Mismatched heteroduplexes and homoduplexes are then introduced to an ion-pair of the reversed-phase HPLC flow path where they bind to the HPLC capillary. The temperature is then increased to a threshold level, usually between 50 and 60 °C, whereupon the DNA duplexes become partially denatured. Initially, the duplexes are retained in the HPLC capillary through ionic binding interactions between the negatively charged phosphate backbone of the partially denatured DNA fragments and the beads

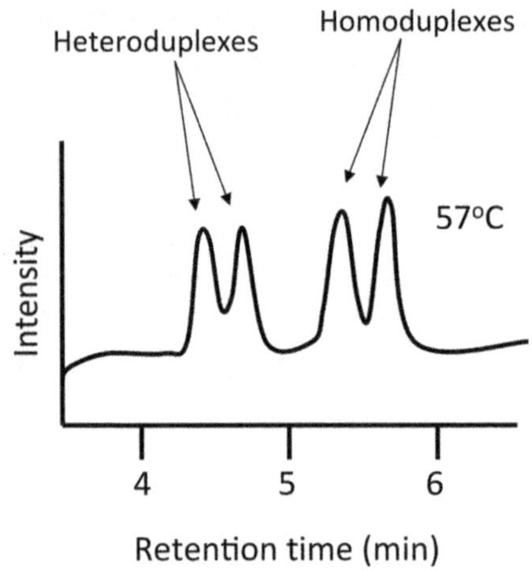

Figure 2.4 Schematic representation of DHPLC analysis of PCR products showing that the homoduplex and heteroduplex DNA species are differentially eluted when the column is held at the melting temperature T_m of the PCR fragment, in this case 57 °C. *n.b.* To detect homozygous point mutations, the mutant PCR product is mixed with reference DNA.

in the cartridge that are coated with the positively charged ion pair reagent triethylammonium acetate (TEAA). The column is then eluted with a TAEE-acetonitrile gradient under conditions in which the heteroduplexes with a mismatched base pair(s) elute before the more stable homoduplexes. Note that for SNP detection purposes, the two homoduplexes are indistinguishable. The eluted fragments then pass through the UV detector and the absorbance is measured and the data is analysed by computer. Two sets of peaks are observed for each duplex (Figure 2.4). The method is rapid, taking on average 7 minutes to analyse a genotype, and amenable to multiplexing through the use of laser-induced fluorescence detection that is analogous to high-throughput DNA sequencing strategies.[50] From a practical perspective the method requires the use of *Pfu* polymerase instead of *Taq* polymerase for generating PCR products. *Pfu* has proof-reading properties, which minimise the introduction of confusing PCR-induced mutations.

2.3.3. DNA Microarrays

The desire for rapid access to comprehensive genomic and gene expression data has stimulated the development of DNA microarrays, which, together with NGS and CE, platforms are key technologies for high-throughput genomic analysis. DNA microarrays provide a radically different approach to large-scale characterisation of polymorphic sites and other genomic features[51] and currently represent the extreme end of miniaturisation. A microarray consists of many thousands of DNA oligonucleotide probes, designed to encompass the features of interest, attached to a glass or silicon 'chip' that is about the size of a microscope coverslip, or to arrays of microspheres (beads). The probes can be applied to the solid supports either by printing or *via in situ* synthesis. They can then be hybridised with test amplicons from genomic DNA or cDNA, which have been labelled with one or two fluorescent dyes depending on the experimental protocol. Unhybridised material is washed away and the result is recorded using a confocal laser scanner. The data are collected, analysed and displayed using dedicated computer programs. The investment in equipment and skills needed to design and prepare microarrays is a disincentive for many research laboratories; this has encouraged some biotech companies to commercialise pre-prepared microarrays for a variety of applications (see https://www.labome.com) as well as the required instrumentation and software.

The first DNA microarrays were designed to measure mRNA transcript levels from thousands of genes in a single experiment.[52] Such functional genomics studies enable the physiological state of cells and overall gene expression pattern to be correlated. Gene expression arrays have many applications, including transcriptional profiling of human diseases, such as cancers, which promise to improve the understanding of neoplasia, improved diagnostics, analysis of response to drugs and identification of novel therapeutic targets.[53] Microarray data from functional genomics experiments can be accessed from the Genome Expression Omnibus (GEO) public database, which curates gene expression data from high-throughput platforms.

DNA microarrays are not limited to gene expression studies (Figure 2.5). The first genotyping arrays were devised to identify key mutations in highly variable medically important genes and genomes, such as the tumour suppressor gene *TP53* (OMIM 191170)[54] and the human immunodeficiency virus (HIV).[55] More recently genotyping chips have been designed to record SNP profiles on a 'whole genome' basis. Information from post-genomic projects such as the HapMap and 1000 genome initiatives to map haplotypes and characterise human variation in different population groups have been used to identify SNPs and provide coverage of expressed genes.[56] Commercial microarrays, such as the high density Affymetrix SNP 6.0, or Illumina's 1M bead array, simultaneously analyse many hundreds of thousands of human SNPs and other features, such as copy number variants in a genomic DNA sample. The

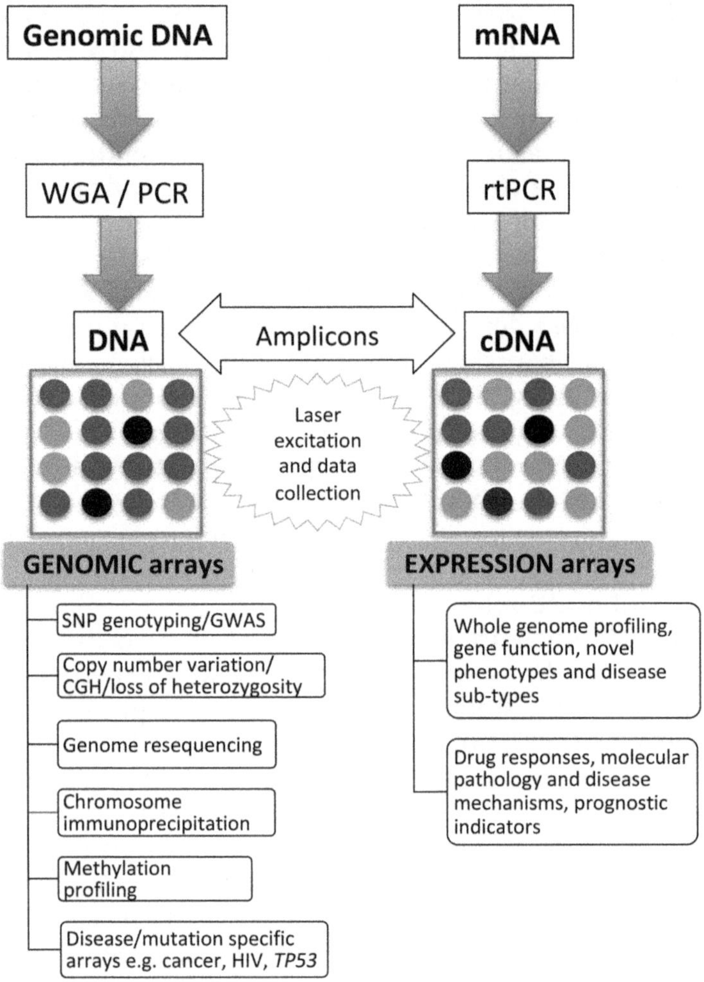

Figure 2.5 A summary of the major steps when using DNA microarrays for genomic and expression studies and some applications. For genomic microarrays, WGA (whole-genome amplification) can be achieved by PCR- or non-PCR-based routes. In expression studies, mRNA is converted to cDNA using reverse transcriptase followed by PCR (rtPCR). Fluorophore labels, such as cyanine dyes, are used to indicate the levels of hybridisation between the test (or target) DNA and the probes can be incorporated into the target DNA prior to hybridisation, or once hybridisation has occurred.

technology has proved instrumental for GWAS to investigate the genetics of complex diseases and conditions such as cardiovascular disease, cancers (dedicated cancer genome arrays are available), obesity, diabetes *etc.* and other applications, such as predicting responses to drugs and investigating the genetic structure of populations. Microarrays have also been or are being developed to monitor SNP profiles and copy number variation in economically and scientifically important species such as pigs,[57] cattle,[58] maize[59] and mice.[60]

The current estimate of around 10–15 million SNPs in the human genome indicates that around 10% have been covered by the currently available high-density arrays, which have been designed to include SNPs with minor alleles with frequencies not less than 0.025–0.05. Genomic mapping microarrays preferably include probes for tag-SNPs, which are markers for regions of high linkage disequilibrium, SNPs for autosomes, X and Y chromosomes, recombination hotspots *etc.* SNP arrays are designed for biallelic polymorphisms, thus three classes of genotype *e.g. AA*, *AB* and *BB* must be detected for each locus. Different DNA microarray platforms utilise different chemistries.[51] Following whole-genome 'one primer' PCR-based amplification and fluorescence labelling of the test DNA, Affymetrix SNP arrays work through differential hybridisation of labelled amplicons to 25 bp arrayed SNP-specific probes in a process that is somewhat analogous to Southern blot probing (longer probes are also included for detection of copy number variants). In contrast, the Illumina Infinium microsphere system recommends non-PCR amplification and utilises arrays of longer 50 bp probes: hybridisation of unlabelled test DNA followed by single-base extension with two-colour immunohistochemical staining are used to identify the genotypes. Although DNA microarray analysis provides a highly efficient GWAS platform for SNP genotyping,[51,61] it can also be used to detect other genomic features. For example, copy number variants that result from genomic events such as duplication, insertion or loss of sequences and loss of heterozygosity;[62] methylation profiles of CpGs including promoter regions, to study their epigenetic significance;[63] outcomes of chromosome immunoprecipitation (ChiP) experiments to identify features such as transcription factor binding sites[64] and histone modifications.[65]

An alternative technology platform for SNP detection is matrix-assisted laser desorption/ionisation-time of flight (MALDI-TOF) mass spectrometry (MS).[66] Whilst MS is limited by the number of SNPs that can be genotyped and is not capable of genome-wide analysis, it offers the advantage of exquisite precision and therefore can be used to confirm the results of SNP profiling by GWAS as well as for fine mapping. In the Sequenom MassARRAY iPLEX system up to 36 SNPs can be multiplexed by PCR and genotyped simultaneously, amplicons undergo single-base extensions to introduce large mass differences between alleles that enable unambiguous genotyping by medium-throughput (*i.e.* not whole-genome) MALDI-TOF MS. The MassARRAY platform has been recently been introduced for non-invasive tumour monitoring and multiplex mutation analysis of circulating tumour DNA (ctDNA) in plasma samples from colorectal cancer (CRC) patients.[67]

2.4. Genome Projects Background

The HGP was conceived in the mid-1980s, but launched in 1990. This led to an upsurge in DNA sequencing activity and an avalanche of sequence data. By 1995 the first cellular (as opposed to viral) genome of a bacterium had been completed and this was followed by a stream of sequenced genomes, from prokaryotes initially, and then eukaryotes. Two independent drafts of the human genome euchromatic DNA sequence were published simultaneously in 2001 as a result of the publicly funded International Human Genome Project and the privately funded initiative led by Celera Genomics.[4,5] This led to estimates that the human genome consists of between 30 500 and 35 500[4] and 26 000 to 38 000[5] expressed genes respectively. These are now considered to be overestimates, the actual number being closer to 22 000 genes.[1] Rapid and powerful advances in cloning and sequencing technologies and computational biology have transformed genome sequencing initiatives from massive long-term endeavours to relatively low cost, quick factory-style undertakings. By early 2018, genome sequences had been recorded for more than 36 000 species, including 5754 eukaryotes.[68] It is estimated that the cost of the HGP, which involved a massive international effort over more than 10 years was in the order of $500 million. In 2001 it was predicted that the cost of sequencing a human genome would, in the foreseeable future, be just $1000. As of January 2018, several companies have claimed to have fulfilled this prediction. Part of this challenge has included addressing not just the cost of sequencing, but also the associated costs of both equipment and interpretation of the genetic data to make it clinically meaningful. This was achieved by Veritas Genetics in 2015. From 2017, the $1000 goalposts have moved to $100 for a whole-genome sequence. In addition to reduced cost, other projects have attempted to enhance the portability and turn-around-time for sequencing. In 2016, Oxford Nanopore's MinION achieved WGS on a USB sized device with 11× coverage, taking 48 hours with alignment data returned in real-time on a laptop computer[69] at a cost of around $2750. Together, these advances are bringing human genomic analysis into the realms of personal medicine.

Following the success of the HGP in driving innovation towards cheaper sequencing and the desire to increase the clinical value of genetic data, more recent projects have aimed to sequence the genomes of specific populations of individuals. For example, the 100 000 Genome Project is a UK initiative proposing to sequence the genomes of 70 000 patients with rare and/or inherited diseases or common cancers. In cancer patients two genomes are sequenced, one from healthy cells and one from the tumour cells, while in those with rare diseases, such as neonatal diabetes, the genomes of two additional family members are also sequenced for comparison. In 2016, Genomics England announced that data acquired for the 100 000 Genome Project was being used for genetic diagnosis at Great Ormond Street Hospital, London.

While sequencing of both healthy and tumour DNA from a patient is expected to yield significant clinical insights, single-cell sequencing (SCS) is an approach that hopes to reveal even more of the genetic heterogeneity between cells. Traditional sequencing takes the DNA of thousands of cells and produces a physical map by averaging the sequence data. However, no two cells are genetically identical, and it is now possible to sequence the DNA of single cells despite the minimal volumes of starting material. It is hoped that such methods will resolve, for example, the complex clonal evolution and patterns of genetic instability that occur in tumour cells.

2.4.1. Mapping and Sequencing Strategies

An aim of the HGP was to produce four types of map: physical, linkage, DNA sequence and expressed gene. Physical and linkage maps provided essential anchor points and frameworks to align DNA sequences and assign genes. A high-resolution physical map based on the analysis of overlapping DNA clones represents the actual distance in DNA base pairs between genetic markers and other landmarks. But the ultimate physical map is the DNA sequence itself. Low-resolution physical maps were generated using techniques such as somatic cell hybridisation and fluorescence *in situ* hybridisation; these methods were also applicable for assigning genes to chromosomes and will be addressed in Section 2.5. More recently, other maps have been constructed, such as epigenetic maps containing information about DNA methylation sites and histone modifications.

There are two major approaches to genome sequencing: whole-genome shotgun sequencing (WGS) and the more labour intensive hierarchical shotgun sequencing (HS).

In simple organisms, such as bacteria and viruses, where the single chromosomes are haploid and very few repeat sequences occur, or indeed for sequencing individual human genes, WGS works well.[70,71] In contrast, for entire eukaryotic genomes, where in many cases repeat sequences abound, including the human genome (over 50% repeats) and there is considerable genetic variation within, it has been argued that HS offers considerable advantages over WGS sequencing and this was the approach adopted by the publicly funded International Human Genome Sequencing Consortium.[4]

2.4.1.1. Hierarchical Sequencing

This approach relies on the production of a set of large-insert clones (typically 100–200 kb each) that cover the entire genome.[4] Since the clones are all independent and ultimately positioned on a physical map in an order that represents each chromosome, repeated sequences are far less troublesome, leading to fewer gaps than encountered by the WGS approach. For many genome projects, high-capacity vectors, such as bacterial artificial chromosomes (BACs) are advantageously used for generating large-insert clones since they are less likely to rearrange their genetic material than alternatives, such as yeast artificial chromosomes (YACs). Long-range physical maps are generated from the production of 'contigs'. Contigs are sets of overlapping DNA fragments that have been obtained from independent clones and positioned relative to one another so that they form a contiguous array. To obtain contigs, genomic libraries must be prepared from high molecular weight DNA that has either been partially digested with restriction enzymes or randomly sheared. In contrast to complete digestion, partial digestion of DNA (or random shearing) will result in a set of overlapping fragments (Figure 2.6). Thus, partial digestion ensures that when each DNA fragment is cloned into a vector it has ends that will overlap with other clones. Thus, when the overlaps are identified and aligned the clones can be positioned, or ordered, so that a physical map is produced.

Large-insert clones are broken down further into sets of smaller overlapping subclones and sequenced using the shotgun sequencing approach. In order to position the overlapping ends into a contig representing the large insert clone, it is preferable to undertake DNA sequencing of both ends of the individual subclones (double-barrelled shotgun sequencing). Eventually the entire DNA sequence of the large-insert clone is obtained by computer-based alignment of individual sub-clone DNA end sequences. To minimise the overlaps

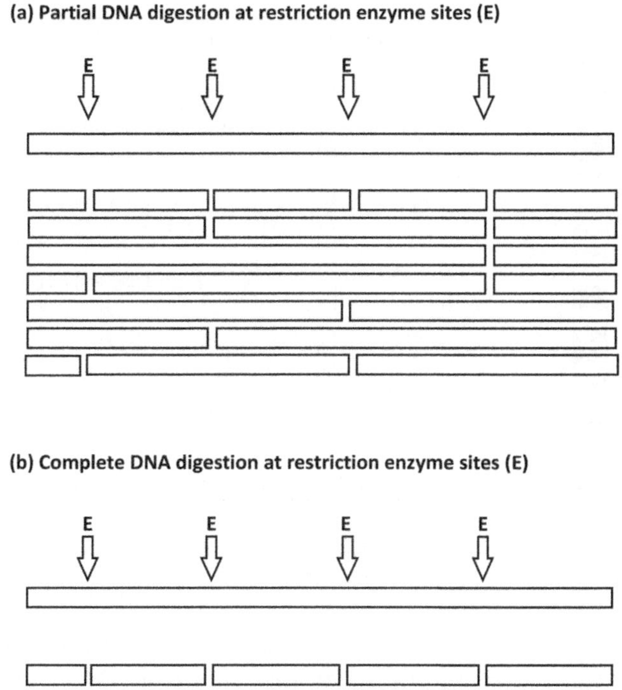

Figure 2.6 Comparison of partial and complete digestion of DNA molecules at restriction enzymes sites (E).

and identify the large-insert clones for further sequencing, *in silico* restriction enzyme mapping can be undertaken to produce a 'fingerprint clone contig'. In the HGP, fingerprint clone contigs were mapped to human chromosomal locations by each Chromosome Workgroup using resources such as panels of human radiation hybrids (RH), fluorescence *in situ* hybridisation (FISH) with human chromosomes and existing genetic maps. Radiation hybrids are panels of human-hamster cell hybrids formed by introducing X-ray-generated fragments of human chromosomes into hamster cells; the resulting panel of somatic hybrid cells that contain different regions of the human genome are typed for the presence or absence of DNA markers; the data are used to construct maps that are complementary to both recombination maps and physical maps based on contigs.[72]

In order to define a common way for all research laboratories to order clones and connect physical maps together, an arbitrary molecular technique based on the polymerase chain reaction has been developed to generate sequence-tagged sites (STS). These are small unique sequence of between 200 and 300 bp that are amplified by PCR.[73] The

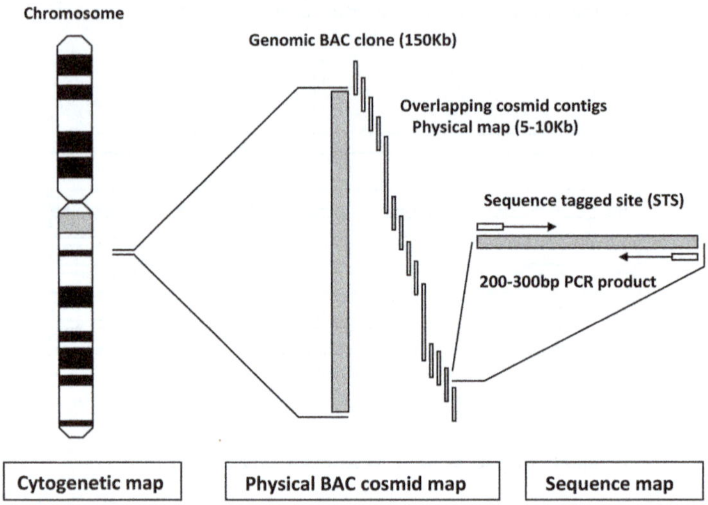

Figure 2.7 Scheme for the use of STS markers in the hierarchical physical mapping of a human chromosome using BAC clones.

uniqueness of the STS is defined by the PCR primers that flank the STS. If the PCR results in amplification, then the STS is present in the clone being tested. In this way defining STS markers that lie approximately 100 kb apart along a contig map allows the ordering of those contigs. Thus, all groups working with clones have publicly available defined landmarks with which to order clones produced in their DNA libraries (Figure 2.7). STSs may also be generated from polymorphic markers that may be traced through families along with other DNA markers and located on a genetic linkage map. Polymorphic STSs may thus serve as markers on both physical and genetic linkage maps for each chromosome and therefore provide a means for aligning the two types of map.

In addition to the human genome, the hierarchical sequencing approach has been used to sequence several genomes, including those of the yeast *Saccharomyces cerevisiae* and the nematode worm *Caenorhabditis elegans*.[74,75] High-quality BAC-clone-based physical genome maps in one species can be of great value to genome projects in other species where some conservation of genomic sequence and gene order might be expected. For example, the outputs of the HGP have provided anchors for ordering BAC clones generated in several other species, such as mice, rats and cattle, allowing simplification of clone alignments and physical map building as well as the generation of comparative maps in the respective species.[76–78] The International Human Genome Sequencing Consortium sequence was reported as finished in 2004 (Build 35) and contains 2.85 billion nucleotides interrupted by 341 gaps. It covers approximately 99% of the euchromatic genome.[79]

2.4.1.2. Whole-genome Shotgun Sequencing (WGS)

In contrast to the hierarchical BAC-by-BAC approach, which relies on the availability of genetic and physical maps for success, WGS is based on the strategy of sequencing a vast number of random genomic clones followed by a computer-based analysis of the DNA sequences, which identifies matching sequences in different clones. This enables the assembly of a chromosomal DNA sequence without resorting to other map resources. As with the HS approach, overlapping clones are required, but since the clones are destined for direct sequence analysis, only vectors that contained small to medium inserts are used. Thus, once the overlaps have been identified the entire sequence is assembled. WGS was the approach adopted by the privately funded human genome initiative.[5]

Although WGS remains somewhat controversial for sequencing complex genomes of 'higher' organisms because of the problems associated with repeat sequences and heterozygosity, it is a widely used approach. The number of complex genomes sequenced by WGS is increasing and includes the fruit fly *Drosophila*, mosquito (*Anopheles*), mouse, puffer fish, dog and grapevine.[80–85] However, in some cases, such as the silk worm genome project, the WGS method has resulted in many seemingly irresolvable gaps in the genome and so BAC-based hierarchical ordering of clones was used to close the gaps.[86] Advances in bioinformatics and computational analysis of WGS sequences indicate that the problems caused by repeat sequences could be overcome, thus, the approach can be expected to gain more ground in future genome projects.[87]

2.4.2. Gene Discovery and Clinical Validation

The ultimate goal of genome projects is to identify, map and characterise all the genes. Some well-established methods for mapping genes predated the DNA era including somatic cell hybridization and familial linkage analysis. Somatic cell hybridization was a forerunner of the radiation hybrid mapping panels that played a significant role in the HGP for long-range gene mapping. Somatic cell mapping panels are derived from human-rodent cell hybrids that are formed by fusing human cells with rodent cells. The hybrids tend to lose human chromosomes at random, which leads to the establishment of clones that retain different human chromosomes. This enables the assignment of a DNA probe, or an isozyme detected by its activity, to a particular human chromosome.[88] Both somatic cell hybridization and linkage analysis require knowledge of the gene to be mapped and thus cannot be used alone for gene discovery. However, both approaches can be used to provide low-resolution localisation of a gene: in the case of the somatic cell hybrids, to a particular cytogenetic location and for linkage analysis, to a linkage group, which, in some instances, may be associated with a chromosome.[12] A limitation of linkage analysis is that the gene being mapped and the marker genes must be heterozygous in key parent individuals; this was frequently a stumbling block before the availability of highly polymorphic DNA markers, such as microsatellites and SNPs.

Considerable progress in gene localisation and discovery has been made with PCR-based EST mapping. An EST or expressed sequence tag is a small PCR product that has been generated from a cDNA sequence, thus reflecting an expressed gene in the cell or tissue from which the mRNA was prepared. Public-domain ESTs are available from the dbEST database.[89] The latest release of dbEST contains more than 74 million ESTs from a large number of species. EST mapping used with large-insert genomic clones can provide information on genome organisation, including gene density as well as localisation.[90]

A further gene isolation system, which uses adapted vectors, termed exon trapping, or exon amplification, may be used to identify exon sequences. Exon trapping requires a specialised 'splicing' expression vector that will accept fragments of genomic DNA containing sequences that allow splicing reactions to take place. Following transfection of a eukaryotic cell line a transcript is produced that may be detected by using specific primers in a reverse transcription (RT)-PCR. This indicates the nature of the foreign DNA, which potentially contains novel exon sequences.[91]

2.5. Clinical Genomics

Mapping and characterising genes associated with disease are central to the development of clinical genomics. A pathogenic variant detected in a sample of patient DNA may provide information about the identity of the disease (diagnostic), the course of the disease (prognostic) and in some cases the response to specific treatments (pharmacogenetics). Molecular diagnostics is playing an increasing role in medicine and many techniques have been utilised to discover new or identify existing genetic variation in patient samples.

Where a pathogenic mutation has been characterised, PCR using allele-specific primers or sequencing methodologies can be used to assign a patient's genetic status and provide information for physicians and genetic counsellors. For example, the gene mutation that is associated with sickle cell disease in *HbS* homozygotes, is caused by a single A → T base substitution in codon 6 of the beta-globin gene. From the diagnostic perspective this leads to the loss of a *Dde*I restriction site, which can be readily detected using allele-specific PCR, allele-specific oligonucleotide hybridisation or PCR-RFLP (see above and also Chapter 3 for a discussion of PCR). Similar approaches can be used to type causative mutations (or tightly linked SNPs) in other Mendelian diseases.

When a genetic disease is suspected to arise from a hitherto uncharacterised mutation, strategies that do not rely on knowledge of the DNA sequence of the mutation are required to detect allelic status. SSCP, DGGE or heteroduplex analysis may be used for this purpose (Table 2.2). However, other approaches have sought to combine whole-exome sequencing (WES), where only protein coding genes (the exome) are sequenced, with bioinformatics tools to compare patient sequence data against a reference genome to identify causative genetic variants (see Section 2.4).

WES and WGS produce vast amounts of sequence information the analysis of which requires powerful computational tools. An overwhelming number of gene variants detected by WES and WGS, such as SNPs, will not be pathogenic. However, curated databases generated from large-scale sequencing projects can be used to filter out harmless SNPs and identify the potential pathogenic mutations required for validation and the development of molecular diagnostic tests. WGS potentially allows for the detection of all variants within a patient's genome, but the practical and cost implications of NGS technologies limit the approach to large-scale collaborative projects and service centres. With most known pathogenic mutations falling in exonic regions, WES, which sequences only a small portion of the genome, is a more affordable, and often more practical, approach, particularly in terms of the computational load of the bioinformatic analysis. While traditional gene-mapping approaches have revealed causative mutations for many Mendelian disorders, WES has begun to make significant contributions. For example, in a study of 250 patients carrying suspected genetic disorders where the underlying cause could not be inferred from the phenotype or other types of genetic testing, 25% of subjects were diagnosed.[92] 200 000–400 000 variants were identified in each patient, which were then filtered and compared against the Human Gene Mutation Database, an online repository of germline mutations implicated in disease held by the Institute of Medical Genetics in Cardiff, UK, to identify potentially deleterious mutations, approximately 400–700 per patient. Extensive clinical analysis was carried out to confirm the relevance of each mutated allele and eventually action a diagnosis.

This and other studies have demonstrated the value of WES in identifying pathogenic mutations that are resistant to linkage analysis. The ability to obtain a molecular diagnosis in this manner can significantly reduce the burden of disease cost from inappropriate care arising from delayed diagnosis or misdiagnosis, as well as the costs of unnecessary procedures often undertaken during the 'diagnostic odyssey' of patients with rare diseases, which can span many years, even decades.[93]

While WES has proved clinically valuable, 50–75% of patients do not receive a diagnosis. Some of these cases will be due to causative mutations being unclassified; the number of unclassified mutations will reduce as more gene-disease associations are reported. However, WES covers only the exonic regions of the genome and thus cannot address variants within the non-coding regulatory regions. WGS is an attractive alternative which could determine the role of non-genic regulatory regions of DNA in disease. In recent studies, WGS approaches have demonstrated detection rates of cancer-predisposing genetic variants in up to 21% of patients, as opposed to just 9% in patients screened by conventional methods.[94] Despite the many advantages of WES and WGS, the affordability of targeted mutation hotspot panels makes them the more attractive option for many clinics. The Ion Ampliseq Cancer Hotspot Panel contains primers targeted to 50 oncogenes and tumour suppressor genes and can detect more than 2800 mutations. Hotspot panels must be continuously updated as the number of clinically relevant cancer markers increases. Eventually, the cost trade-off may shift in favour of WES and/or WGS approaches.

2.5.1. Bioinformatics for Clinical Genetics

Parallel developments in bioinformatics have been essential for the success of the genome projects (see Chapter 5). In the closing stages of the Human Genome Project some institutions were generating up to 175 000 individual DNA sequence reads per day, information that needed to be analysed, submitted to databases, annotated and checked for quality. A multiplicity of DNA and protein sequences from different species is submitted daily by laboratories world-wide to databases such as GenBank (USA, DNA sequences), European Nucleotide Archive (Europe, DNA sequences), DDBJ (Japan, DNA sequences) and UniProt (protein sequences), which are all freely available. These massive primary sequence databases have stimulated the production of numerous bioinformatics tools as well as the emergence of many specialised genomic and protein databases dedicated to various subjects, such as a particular species group, biological features, such as disease markers, protein families, protein domains *etc.* The Nucleic Acids Research online Molecular Biology Database Collection is a public repository that lists more than 1500 databases.[95]

The torrent of data arriving from modern high-throughput sequencing efforts places a considerable burden on computational resources. The major rate-limiting step in biological discovery has thus shifted away from the laboratory to data analysis. WGS of a single human genome produces approximately 104 gigabytes of data, with terabytes of space required once backups are considered. It has been estimated that as many as 40 petabytes could be generated each year following the planning of population-scale genome projects, such as the Genomics England 100 000 genomes project and the proposal by AstraZeneca that aims to sequence as many as 2 million genomes. How best to deal with these data is the subject of energetic debate. With a predicted 2–40 exabytes of raw sequence data expected by 2025,[96] data storage is only part of the problem. Thus identifying and comparing variants within and between datasets can be expected to increase the necessary processing power exponentially, and much of this work is performed in cloud-based architectures.

Another important facet has been the development of algorithms to enable predictions of genes and gene features from otherwise anonymous genomic sequences. Programs such as Genescan at MIT that predict exons, genes, promoter regions, polyA tails and other features associated with expressed DNA have been used to indicate possible genic regions. Thus 'cloning' genes using database resources and dedicated algorithms is now a reality and biological computing has become an applied science that enables investigators to identify genes *in silico*. Interestingly *in silico* simulations may also assist in the construction of clinically relevant hypotheses that could then be tested in bacterial or animal model systems[97]

Integrating and annotating the expanding genomic data delivered by next-generation sequencing and DNA microarrays is a continuing challenge for bioinformaticians and database curators. The complete map of the human genome was published in 2003 but work on perfecting the assembly is ongoing with more than seven-hundred gaps at the date of writing.[98] Online access to genomic data with the ability to search, query and visualise large volumes of information has been an enormous boon to genetic research. However, many of the tools and services constructed over the last decade suffer from limited clinical uptake, often due to technical complexity requiring extensive specialist training. This 'first generation' library of tools has facilitated bioinformaticians to annotate the human genome but, as highlighted by a 2015 working party report from Genomics England and the NHS, there is a pressing need to develop a new iteration of tools and a user-interface (UI) design that will allow clinical reporting from genome-scale datasets to become routine.[99]

2.5.2. Clinical Bioinformatics Pipelines

The goal of clinical bioinformatics is to take the raw data derived from biological samples and arrive at clinically applicable knowledge. Methodologies and pipelines for bioinformatics and mutation detection are far from standardised, but generally three main steps can be defined: data preparation, analysis and clinical interpretation and reporting (Figure 2.8).

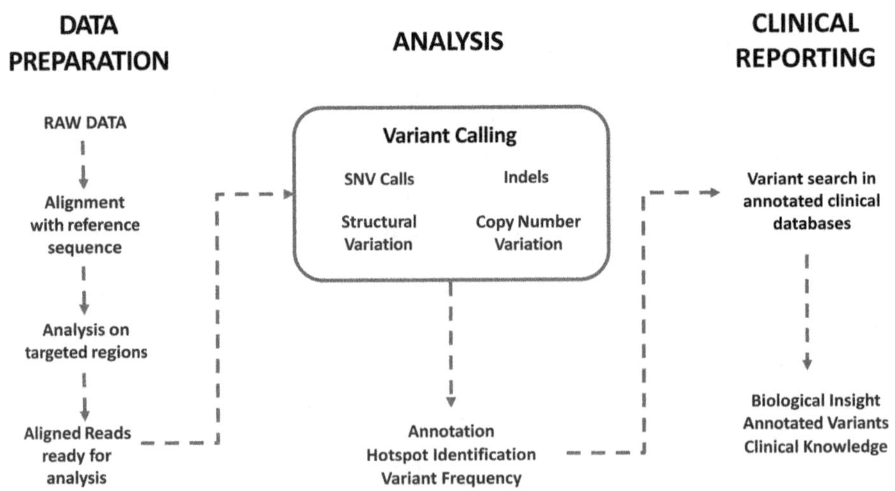

Figure 2.8 A generalised pipeline for genomic-based diagnostic bioinformatics. The scheme begins with raw DNA sequence reads derived from patient samples and results in an annotated mutation report to facilitate decision making at the clinical level.

Data preparation includes the steps required to convert the raw signals produced by the sequencing platforms into a text string of nucleotide bases. Each nucleotide in the string is accompanied by quality scores reflecting the confidence that the given base is correct, a quality score of 20 indicates a probability of 0.01 that the nucleotide read is erroneous. Different sequencing technologies generate different levels of accuracy, single-molecule sequencers such as nanopores or single molecule real time sequencing (SMRT) struggle with error rates as high a 5%. However, these may be corrected to within 99.99% accuracy using a combination of multiple reads, known as coverage, and algorithm design. The accuracy of the base-sequencing technology is critical to searching for rare genetic variants, which remains a problem even for the most accurate NGS technologies.[100] The final steps in the data preparation process is the generation of a digital file, most commonly in FASTQ and FASTA format, the latter being a simple text-based sequence of letters representing the four nucleotide bases. FASTQ contains the associated quality scores for each base.

Once the sequence file is prepared it is possible to apply a variety of computational tools. This part of the process will most often begin with aligning the sequence against the latest reference genome, or, in some cases, against an earlier read of the patient's own genetic data to observe temporal changes or differences between cell populations. Many tools such as BWA (Burrows–Wheeler Aligner) or BLAST[101] have been developed to search databases for matching sequences of nucleotides or amino acids and to score alignments within an acceptable timeframe. It is important that alignment algorithms can tolerate the intrinsic mismatching between samples and references, arising from, for example, SNPs and indels, but also from PCR artefacts and sequence gaps. Additional steps are often undertaken to further refine the previously assigned per-base quality scores and reduce the effect of experimentally introduced sequence artefacts and increase confidence in downstream variant calling. This can be achieved by comparing regions where SNPs are absent or very rare. It is important to understand the problems associated with some of the computational processes, particularly where biases have been reported as arising from algorithm design.[102,103]

To diagnose or discover the genetic basis of a disease phenotype it is essential that bioinformatics tools can also reliably identify all types of genetic variation in the region under study. This includes SNPs as well as indels, chromosomal inversions and translocations; the larger variations impose a significantly higher computational burden and complexity in algorithm design. The overall goal is to discover differences between the patient genetic data and the reference sequence. There are many bioinformatic tools and resources that go beyond simple SNP identification and can be used for variant calling between clinical and reference sequences, such as Clustal, Freebayes and GATK UnifiedGenotyper.

The next step is to construct a preliminary clinical assessment from the sequence analysis result, a task which requires considerable supporting evidence and validation by health practitioners. Currently, this is the least automated part of the clinical bioinformatics pipeline with clinical and diagnostic overlays being missing from most existing bioinformatics workflows. Efforts are underway to address this in order to reduce the increasing burden on clinicians who might otherwise have to acquire bioinformatics skills. One example is the output of Oxford Nanopore's 'What's in my Pot?' (WIMP) software, designed to provide a user-friendly feedback dashboard, which can indicate the presence of pathogenic nucleotide sequences, such as Ebola virus, in patient samples in a real time graphical interface.[104] Likewise, cBioPortal has become a valuable resource, bringing together large cancer datasets along with a constantly updated set of analytical tools and visual outputs to facilitate clinical decision making.

More commonly, online curated mutation databases such as ClinVar, OMIM (Online Mendelian Inheritance in Man), HGNC (the Human Genome Nomenclature database) and DoCM (Database of Curated Mutations) provide a means to manually search for the variants returned in patients' results and to connect with supporting literature in order to make a clinical assessment. OMIM (https://www.omim.org), catalogues clinical synopses, genomic features and comprehensive references for more than 18 000 human genetic traits (the majority of which are known at the molecular level) and the HGNC (https://www.genenames.org), assigns names to all known human genes and curates a searchable database. There are also dedicated knowledgebases, that enable investigators to undertake text searching (or mining) in addition to sequence searches, such as cancer and structure–activity relationship (canSAR), COSMIC and the International Agency for Research on Cancer (IARC) *TP53* Database; these resources serve various research and clinical communities.

2.6. Conclusion

In medicine a major challenge is to unravel the genetic basis of common diseases that together will affect the majority of individuals in all populations. The HGP has produced the genomic resources to tackle these multifactorial disorders, including cardiovascular disease, cancers, obesity, diabetes *etc.* Such complex traits do not exhibit Mendelian inheritance patterns. Instead, the actions of several genes, each with a small effect, and environmental influences modify the risk of disease. Various types of association studies are used in an attempt to define the genes that underlie the propensity for some individuals to develop such diseases. Population and family-based association studies, such as GWAS and affected sib pair analysis, make use of the gene maps, polymorphic marker resources and genome technologies in the attempt to pinpoint small regions of chromosomes associated with the disease phenotype. Chromosomal regions showing positive association with disease can then be resequenced[7] and combined with functional studies in an attempt to identify the causative mutations.[105] It is anticipated that advances on several fronts will lead to improved understanding of molecular pathology, discovery of new drug targets and better diagnostics. The translation of genomic-scale laboratory-based approaches to end users, such as hospitals and doctors' offices, require the development of fast, high-precision and low-cost technology platforms. Innovative instruments, such as Oxford Nanopore Technologies nanopore real time 'hand held' DNA sequencer,[106] and point-of-care diagnostic platforms, such as QuantuMDX's continuous flow nanowire biosensors,[107] should become available in the near future. The convenience of such miniaturized technologies might be expected to lead to the gradual replacement of current systems, such as the multiplex ligation-dependent probe amplification (MLPA) assays for routine diagnostic purposes.[108] In the next decade further improvements to genomic diagnostic technology platforms can be expected to emerge and their implementation in clinical settings promises to mitigate further the burden of disease in many populations.

References

1. M. Pertea and S. L. Salzberg, *Genome Biol.*, 2010, **11**, 206.
2. F. Sanger, S. Nicklen and A. R. Coulson, *Proc. Natl. Acad. Sci. U. S. A.*, 1977, **74**, 5463.
3. F. Sanger, G. M. Air, B. G. Barrell, N. L. Brown, A. R. Coulson, J. C. Fiddes, C. A. Hutchison III, P. M. Slocombe and M. Smith, *Nature*, 1977, **265**, 687.
4. International Human Genome Sequencing Consortium, *Nature*, 2001, **409**, 860.
5. J. C. Venter, M. D. Adams, E. W. Myers, P. W. Li, R. J. Mural, G. G. Sutton, H. O. Smith, M. Yandell, C. A. Evans and R. A. Holt, *et al.*, *Science*, 2001, **291**, 1304.
6. A. Edwards, A. R. Debbonaire, B. Sattler, L. A. Mur and A. J. Hodson, bioRxiv, 2016, , DOI: 10.1101/073965.
7. P. Carninci, *Trends Genet.*, 2006, **22**, 501.
8. P. Carninci, T. Kasukawa, S. Katayama, J. Gough, M. C. Frith, N. Maeda, R. Oyama, T. Ravasi, B. Lenhard and C. Wells, *et al.*, *Science*, 2005, **309**, 1559.

9. J. Sana, P. Faltejskova, M. Svoboda and O. Slaby, *J. Transl. Med.*, 2012, **10**, 103.
10. F. B. Piel, A. P. Patil, R. E. Howes, O. A. Nyangiri, P. W. Gething, T. N. Williams, D. J. Weatherall and S. I. Hay, *Nat. Commun.*, 2010, **1**, 104.
11. M. W. Moore, D. Babu and P. D. Cotter, *Curr. Top. Genet.*, 2012, **5**, 23.
12. *An Introduction to Genetic Analysis*, ed. A. J. F. Griffiths, S. R. Wessler, S. B. Carroll and J. Doebley Gelbart, W.H. Freeman and Company, New York, 10th edn, 2012.
13. E. M. Southern, *Trends Biochem. Sci.*, 2000, **25**, 585.
14. R. Leach, R. DeMars, S. Hasstedt and R. White, *Proc. Natl. Acad. Sci. U. S. A.*, 1986, **83**, 3909.
15. A. J. Jeffreys, V. Wilson and S. L. Thein, *Nature*, 1985, **314**, 67.
16. A. J. Jeffreys, V. Wilson and S. L. Thein, *Nature*, 1985, **316**, 76.
17. K. Tamaki, C. A. May, Y. E. Dubrova and A. J. Jeffreys, *Hum. Mol. Genet.*, 1999, **8**, 879.
18. K. Tamaki, C. H. Brenner and A. J. Jeffreys, *Forensic Sci. Int.*, 2000, **113**, 55.
19. R. A. C. Roos, *Orphanet J. Rare Dis.*, 2010, **5**, 40.
20. C. Schlötterer and D. Tautz, *Nucleic Acids Res.*, 1992, **20**, 211.
21. J. M. Hancock, in *Microsatellites: Evolution and Applications*, ed. D. Goldstein and C. Schlötterer, Oxford University Press, New York, 1999, pp. 1–9.
22. H. Ellegren, *Nat. Genet.*, 2000, **24**, 400.
23. J. D. Hawk, L. Stefanovic, J. C. Boyer, T. D. Petes and R. A. Farber, *Proc. Natl. Acad. Sci. U. S. A*, 2005, **102**, 8639.
24. T.-M. Kim, P. W. Laird and P. J. Park, *Cell*, 2013, **155**, 858.
25. H. Harris and D. A. Hopkinson, *Handbook of Enzyme Electrophoresis in Human Genetics (with Supplements)*, Oxford American Publishing Co., New York, l976.
26. The 1000 Genomes Project Consortium, *Nature*, 2010, **467**, 1061.
27. International HapMap 3 Consortium, *Nature*, 2010, **467**, 52.
28. International SNP Map Working Group, *Nature*, 2001, **409**, 928.
29. D. E. Reich, S. F. Schaffner, M. J. Daly, G. McVean, J. C. Mullikin, J. M. Higgins, D. J. Richter, E. S. Lander and D. Altshuler, *Nat. Genet.*, 2002, **32**, 135.
30. L. Kruglyak and D. A. Nickerson, *Nat. Genet.*, 2001, **27**, 234.
31. Y. F. Pei, L. Zhang, Y. Liu, J. Li, H. Shen, Y. Z. Liu, Q. Tian, H. He, S. Wu, S. Ran, Y. Han, R. Hai, Y. Lin, J. Zhu, X. Z. Zhu, C. J. Papasian and H.-W. Deng, *Hum. Mol. Genet.*, 2014, **23**, 820.
32. F. J. de Serres, *Environ. Health Perspect.*, 2003, **111**, 1851.
33. A. Woolfe, J. C. Mullikin and L. Elnitski, *Genome Biol.*, 2010, **11**, R20.
34. E. M. El-Omar, M. Carrington, W. H. Chow, K. E. McColl, J. H. Bream, H. A. Young, J. Herrera, J. Lissowska, C. C. Yuan and N. Rothman, *et al.*, *Nature*, 2000, **404**, 39.
35. A. Ligers, N. Teleshova, T. Masterman, W. X. Huang and J. Hillert, *Genes Immun.*, 2001, **2**, 45.
36. M. Olivier, M. Hollstein and P. Hainaut, *Cold Spring Harbor Perspect. Biol.*, 2010, **2**, a001008.
37. X. Wu, C. Ren, T. Joshi, T. Vuong, D. Xu and H. T. Nguyen, *BMC Genomics*, 2010, **11**, 469.
38. *Single Nucleotide Polymorphisms, Methods and Protocols*, ed. A. A. Komar, Humana Press, Totowa, New Jersey, 2009.
39. M. Orita, H. Iwahana, H. Kanazawa, K. Hayashi and T. Sekiya, *Proc. Natl. Acad. Sci. U. S. A.*, 1989, **86**, 2766.
40. T. Tozaki, N.-H. Choi-Miura, M. Taniyama, M. Kurosawa and M. Tomita, *BMC Med. Genet.*, 2002, **3**, 6.
41. K. Hayashi and D. W. Yandell, *Hum. Mutat.*, 1993, **2**, 338.
42. S. P. Yip, W. Putt, D. A. Hopkinson and D. B. Whitehouse, *Ann. Hum. Genet.*, 1999, **63**, 129.
43. W. Yan, H. Zhou, Y. Luo, J. Hu and J. G. H. Hickford, *PLoS One*, 2014, **9**, e88691.

44. S. P. Yip, D. A. Hopkinson and D. B. Whitehouse, *BioTechniques*, 1999, **27**, 20.
45. T. Tahira, A. Suzuki, Y. Kukita and K. Hayashi, *Methods Mol. Biol.*, 2003, **212**, 37.
46. W. Choi, G. W. Shin, H. S. Hwang, S. P. Pack, G. Y. Jung and G. Y. Jung, *Electrophoresis*, 2014, **35**, 1196.
47. L. Wang, S. Zhou, J. Xie, H. Gao, F. Wang, J. Zhou, Y. Wang and H. Wang, *J. Thorac. Dis.*, 2017, **9**, 5295.
48. R. L. Copeland, D. Beyene, V. Apprey, M. R. Daremipouran, T. J. Naab, O. O. Kassim and Y. M. Kanaan, *Cancer Genomics Proteomics*, 2017, **14**, 461.
49. P. Carbonell, M. C. Turpin, D. Torres-Moreno, I. Molina-Martínez, J. García-Solano, M. Perez-Guillermo and P. Conesa-Zamora, *J. Mol. Diagn.*, 2011, **13**, 467.
50. A. Premstaller, W. Xiao, H. Oberacher, M. O'Keefe, D. Stern, T. Willis, C. G. Huber and P. J. Oefner, *Genome Res.*, 2001, **11**, 1944.
51. T. LaFramboise, *Nucleic Acids Res.*, 2009, **37**, 4181.
52. V. Trevino, F. Falciani and H. A. Barrera-Saldaña, *Mol. Med.*, 2007, **13**, 527.
53. D. R. Rhodes, S. Kalyana-Sundaram, V. Mahavisno, R. Varambally, J. Yu, B. B. Briggs, T. R. Barrette, M. J. Anstet, C. Kincead-Beal and P. Kulkarni, *et al.*, *Neoplasia*, 2007, **9**, 166.
54. S. A. Ahrendt, S. Halachmi, J. T. Chow, L. Wu, N. Halachmi, S. C. Yang, S. Wehage, J. Jen and D. Sidransky, *Proc. Natl. Acad. Sci. U. S. A.*, 1999, **96**, 7382.
55. R. Gonzalez, B. Masquelier, H. Fleury, B. Lacroix, A. Troesch, G. Vernet and J. N. Telles, *J. Clin. Microbiol.*, 2004, **42**, 2907.
56. T. J. Hoffmann, M. N. Kvale, S. E. Hesselson, Y. Zhan, C. Aquino, Y. Cao, S. Cawley, E. Chung, S. Connell, J. Eshragh, M. Ewing, J. Gollub, M. Henderson, E. Hubbell, C. Iribarren, J. Kaufman, R. Z. Lao, Y. Lu, D. Ludwig, G. K. Mathauda, W. McGuire, G. Mei, S. Miles, M. M. Purdy, C. Quesenberry, D. Ranatunga, S. Rowell, M. Sadler, M. H. Shapero, L. Shen, T. R. Shenoy, D. Smethurst, S. K. Van den Eeden, L. Walter, E. Wan, R. Wearley, T. Webster, C. C. Wen, Li Weng, R. A. Whitmer, A. Williams, S. C. Wong, C. Zau, A. Finn, C. Schaefer, P.-Y. Kwok and N. Risch, *Genomics*, 2013, **98**, 79.
57. W. Luo, S. Chen, D. Cheng, L. Wang, Y. Li, X. Ma, X. Song, X. Liu, W. Li, J. Liang, H. Yan, K. Zhao, C. Wang, L. Wang and L. Zhang, *Int. J. Biol. Sci.*, 2012, **8**, 870.
58. M. Gautier, D. Laloë and K. Moazami-Goudarzi, *PLoS One*, 2010, **5**, e13038.
59. R. A. Swanson-Wagner, S. R. Eichten, S. Kumari, P. Tiffin, J. C. Stein, D. Ware and N. M. Springer, *Genome Res.*, 2010, **20**, 1689.
60. Y. Kayashima, H. Tomita, S. Zhilicheva, S. Kim, H. S. Kim, B. J. Bennett and N. Maeda, *PLoS One*, 2014, **9**, e88274.
61. K. J. Lindquist, E. Jorgenson, T. J. Hoffmann and J. S. Witte, *Genet. Epidemiol.*, 2013, **37**, 383.
62. C. R. Coughlin II, G. H. Scharer and T. H. Shaikh, *Genome Med.*, 2012, **4**, 80.
63. J. Liu, Z. Zhang, M. Bando, T. Itoh, M. A. Deardorff, J. R. Li, D. Clark, M. Kaur, K. Tatsuro, A. D. Kline, C. Chang, H. Vega, L. G. Jackson, N. B. Spinner, K. Shirahige and I. D. Krantz, *Nucleic Acids Res.*, 2010, **38**, 5657.
64. G. M. Euskirchen, J. S. Rozowsky, C.-L. Wei, W. H. Lee, Z. D. Zhang, S. Hartman, O. Emanuelsson, V. Stolc, S. Weissman, M. B. Gerstein, Y. Ruan and M. Snyder, *Genome Res.*, 2007, **17**, 898.
65. J. A. McCann, E. M. Muro, C. Palmer, G. Palidwor, C. J. Porter, M. A. Andrade-Navarro and M. A. Rudnicki, *BMC Genomics*, 2007, **8**, 322.
66. I. G. Gut, *Hum. Mutat.*, 2004, **23**, 437.
67. S.-J. Shin, S.-M. Chun, T.-I. Kim, Y. J. Kim, H.-J. Choi, S. J. Jang and J. Kim, *PLoS One*, 2017, **12**, e0176340.
68. NCBI, Genome List, 2018, https://www.ncbi.nlm.nih.gov/genome/browse/#!/overview.
69. H. Lu, F. Giordano and Z. Ning, *Genomics, Proteomics Bioinf.*, 2016, **14**, 265.

70. R. C. Gardner, A. J. Howarth, P. Hahn, M. Brown-Luedi, R. J. Shepherd and J. Messing, *Nucleic Acids Res.*, 1981, **9**, 2871.
71. S. L. Chissoe, A. Bodenteich, Y. F. Wang, Y. P. Wang, D. Burian, S. W. Clifton, J. Crabtree, A. Freeman, K. Iyer and L. Jian, *et al.*, *Genomics*, 1995, **27**, 67.
72. G. Gyapay, K. Schmitt, C. Fizames, H. Jones, N. Vega-Czarny, D. Spillett, D. Muselet, J. F. Prud'homme, C. Dib and C. Auffray, *et al.*, *Hum. Mol. Genet.*, 1996, **5**, 339.
73. E. A. Stewart, K. B. McKusick, A. Aggarwal, E. Bajorek, S. Brady, A. Chu, N. Fang, D. Hadley, M. Harris and S. Hussain, *et al.*, *Genome Res.*, 1997, **7**, 422.
74. H. W. Mewes, K. Albermann, M. Bahr, D. Frishman, A. Gleissner, J. Hani, K. Heumann, K. Kleine, A. Maierl and S. G. Oliver, *et al.*, *Nature*, 1997, **387**, 7.
75. *C. elegans* Sequencing Consortium, *Science*, 1998, **282**, 2012.
76. S. G. Gregory, M. Sekhon, J. Schein, S. Zhao, K. Osoegawa, C. E. Scott, R. S. Evans, P. W. Burridge, T. V. Cox and C. A. Fox, *et al.*, *Nature*, 2002, **418**, 743.
77. M. Krzywinski, J. Wallis, C. Gosele, I. Bosdet, R. Chiu, T. Graves, O. Hummel, D. Layman, C. Mathewson and N. Wye, *et al.*, *Genome Res.*, 2004, **14**, 766.
78. International Bovine BAC Mapping Consortium, *Genome Biol.*, 2007, **8**, R165.
79. International Human Genome Sequencing Consortium, *Nature*, 2004, **431**, 931.
80. M. D. Adams, S. E. Celniker, R. A. Holt, C. A. Evans, J. D. Gocayne, P. G. Amanatides, S. E. Scherer, P. W. Li, R. A. Hoskins and R. F. Galle, *et al.*, *Science*, 2000, **287**, 2185.
81. R. A. Holt, G. M. Subramanian, A. Halpern, G. G. Sutton, R. Charlab, D. R. Nusskern, P. Wincker, A. G. Clark, J. M. C. Ribeiro and R. Wides, *et al.*, *Science*, 2002, **298**, 129.
82. Mouse Genome Sequencing Consortium, *Nature*, 2002, **420**, 520.
83. S. Aparicio, J. Chapman, E. Stupka, N. Putnam, J. M. Chia, P. Dehal, A. Christoffels, S. Rash, S. Hoon and A. Smit, *et al.*, *Science*, 2002, **297**, 1301.
84. E. F. Kirkness, V. Bafna, A. L. Halpern, S. Levy, K. Remington, D. B. Rusch, A. L. Delcher, M. Pop, W. Wang and C. M. Fraser, *et al.*, *Science*, 2003, **301**, 1898.
85. R. Velasco, A. Zharkikh, M. Troggio, D. A. Cartwright, A. Cestaro, D. Pruss, M. Pindo, L. M. FitzGerald, S. Vezzulli, J. Reid, G. Malacarne, D. Iliev, G. Coppola, B. Wardell, D. Micheletti, T. Macalma, M. Facci, J. T. Mitchell, M. Perazzolli, G. Eldredge, P. Gatto, R. Oyzerski, M. Moretto, N. Gutin, M. Stefanini, Y. Chen, C. Segala, C. Davenport, L. Demattè, A. Mraz, J. Battilana, K. Stormo, F. Costa, Q. Tao, A. Si-Ammour, T. Harkins, A. Lackey, C. Perbost, B. Taillon, A. Stella, V. Solovyev, J. A. Fawcett, L. Sterck, K. Vandepoele, S. M. Grando, S. Toppo, C. Moser, J. Lanchbury, R. Bogden, M. Skolnick, V. Sgaramella, S. K. Bhatnagar, P. Fontana, A. Gutin, Y. Van de Peer, F. Salamini and R. Viola, *PLoS One*, 2007, **2**, e1326.
86. S. Zhan, J. Huang, Q. Guo, Y. Zhao, W. Li, X. Miao, M. R. Goldsmith, M. Li and Y. Huang, *BMC Genomics*, 2009, **10**, 389.
87. S. Istrail, G. G. Sutton, L. Florea, A. L. Halpern, C. M. Mobarry, R. Lippert, B. Walenz, H. Shatkay, I. Dew, J. R. Miller, M. J. Flanigan, N. J. Edwards, R. Bolanos, D. Fasulo, B. V. Halldorsson, S. Hannenhalli, R. Turner, S. Yooseph, F. Lu, D. R. Nusskern, B. Chris Shue, X. Holly Zheng, F. Zhong, A. L. Delcher, D. H. Huson, S. A. Kravitz, L. Mouchard, K. Reinert, K. A. Remington, A. G. Clark, M. S. Waterman, E. E. Eichler, M. D. Adams, M. W. Hunkapiller, E. W. Myers and J. C. Venter, *Proc. Natl. Acad. Sci. U. S. A.*, 2004, **101**, 1916.
88. H. N. Bu-Ghanim, C. M. Casimir, S. Povey and A. W. Segal, *Genomics*, 1990, **8**, 56.
89. M. S. Boguski, T. M. Lowe and C. M. Tolstoshev, *Nat. Genet.*, 1993, **4**, 332.
90. L. L. Qi, B. Echalier, S. Chao, G. R. Lazo, G. E. Butler, O. D. Anderson, E. D. Akhunov, J. Dvořák, A. M. Linkiewicz, A. Ratnasiri, J. Dubcovsky, C. E. Bermudez-Kandianis, R. A. Greene, R. Kantety, C. M. La Rota, J. D. Munkvold, S. F. Sorrells, M. E. Sorrells, M. Dilbirligi, D. Sidhu, M. Erayman, H. S. Randhawa, D. Sandhu, S. N. Bondareva, K. S. Gill, A. A. Mahmoud, X.-F. Ma, Miftahudin, J. P. Gustafson, E. J. Conley, V. Nduati, J. L. Gonzalez-Hernandez, J. A. Anderson,

J. H. Peng, N. L. V. Lapitan, K. G. Hossain, V. Kalavacharla, S. F. Kianian, M. S. Pathan, D. S. Zhang, H. T. Nguyen, D.-W. Choi, R. D. Fenton, T. J. Close, P. E. McGuire, C. O. Qualset and B. S. Gill, *Genetics*, 2004, **168**, 701.

91. G. M. Duyk, S. W. Kim, R. M. Myers and D. R. Cox, *Proc. Natl. Acad. Sci. U. S. A.*, 1990, **87**, 8995.
92. Y. Yang, D. M. Muzny, J. G. Reid, M. N. Bainbridge, A. Willis, P. A. Ward, A. Braxton, J. Beuten, F. Xia, Z. Niu, M. Hardison, R. Person, R. Bekheirnia, M. S. Leduc, A. Kirby, P. Pham, J. Scull, M. Wang, Y. Ding, S. E. Plon, J. R. Lupski, A. L. Beaudet, R. A. Gibbs and C. M. Eng, *N. Engl. J. Med.*, 2013, **369**, 1502.
93. K. D. Christensen, D. Dukhovny, U. Siebert and R. C. Green, *J. Pers. Med.*, 2015, **5**, 470.
94. S. B. Foley, J. J. Rios, V. E. Mgbemena, L. S. Robinson, H. L. Hampel, A. E. Toland, L. Durham and T. S. Ross, *EBioMedicine*, 2015, **2**, 74.
95. X. M. Fernández-Suárez and M. Y. Galperin, *Nucleic Acids Res.*, 2013, **41**, D1.
96. Z. D. Stephens, S. Y. Lee, F. Faghri, R. H. Campbell, C. Zhai, M. J. Efron, R. Iyer, M. C. Schatz, S. Sinha and G. E. Robinson, *PLoS Biol.*, 2015, **13**, e1002195.
97. R. A. Studer, B. H. Dessailly and C. A. Orengo, *Biochem. J.*, 2013, **449**, 581.
98. GRC, Human Genome Assembly GRCh38.p12-Genome Reference Consortium, https://www.ncbi.nlm.nih.gov/grc/human/data, accessed 21 May 2018.
99. Clinical Bioinformatics Task Group, Developing Clinical Bioinformatics Training in the NHS-a timeline for action, 2016, https://www.genomicseducation.hee.nhs.uk/images/publications/Developing_NHS_Clinical_Bioinformatics_Training.pdf.
100. T. Le Van, M. Van Leeuwen, A. C. Fierro, D. De Maeyer, J. Van Den Eynden, L. Verbeke, L. De Raedt, K. Marchal and S. Nijssen, *Bioinformatics*, 2016, **32**, i445.
101. S. F. Altschul, T. L. Madden, A. A. Schaffer, J. Zhang, Z. Zhang, W. Miller and D. J. Lipman, *Nucleic Acids Res.*, 1997, **25**, 3389.
102. R. Luthra, H. Chen, S. Roy-Chowdhuri and R. R. Singh, *Cancers*, 2015, **7**, 2023.
103. J. Xuan, Y. Yu, T. Qing, L. Guo and L. Shi, *Cancer Lett.*, 2013, **340**, 284.
104. S. Juul, F. Izquierdo, A. Hurst, X. Dai and A. Wright, bioRxiv Prepr, 2015, http://biorxiv.org/content/biorxiv/early/2015/11/06/030742.full.pdf.
105. W. Bodmer and C. Bonilla, *Nat. Genet.*, 2008, **40**, 695.
106. E. Pennisi, *Science*, 2014, **343**, 829.
107. J. O'Halloran, *Mol. Cytogenet.*, 2014, **21**(7(Suppl. 1)), I27.
108. S. Sengstake, N. Bablishvili, A. Schuitema, N. Bzekalava, E. Abadia, J. de Beer, N. Tadumadze, M. Akhalaia, K. Tuin, N. Tukvadze, R. Aspindzelashvili, E. Bachiyska, S. Panaiotov, C. Sola, D. van Soolingen, P. Klatser, R. Anthony and I. Bergval, *BMC Genomics*, 2014, **15**, 572.

3 Molecular Cloning and DNA Sequencing

Ralph Rapley

University of Hertfordshire, School of Life and Medical Sciences, College Lane, Hatfield, AL10 9AB, UK
*E-mail: r.rapley@herts.ac.uk

3.1. Introduction

The many important advances in the techniques of molecular biology and biotechnology have provided the means for, and accelerated, the identification, analysis and manipulation of nucleic acids. Many of these have been through the isolation and characterisation of numerous DNA-manipulating enzymes, such as DNA polymerase, DNA ligase and reverse transcriptase, for example. However, perhaps one of the most important was the isolation and application of enzymes that enabled the reliable and reproducible digestion of DNA. These enzymes, termed restriction endonucleases, not only provided a turning point for the analysis of DNA but also provided the means to develop recombinant DNA technology as well as key methods for gene library production, molecular cloning, DNA sequencing and also gene editing.

Genomics and Clinical Diagnostics
Edited by David Whitehouse and Ralph Rapley

Published by the Royal Society of Chemistry, www.rsc.org

3.1.1. The Molecular Cloning Process

Molecular cloning provides a convenient way of dividing the very large segments of DNA associated with genomes into smaller pieces that are suited to detailed analysis.[1] Cloning involves the amplification of specific segments of DNA, leading to the accumulation of large amounts of identical DNA fragments that can be used for various purposes. In many ways the process is analogous to amplification of DNA by PCR. Classical molecular cloning employs bespoke DNA cloning vectors, which are often derived from circular DNA molecules called bacterial plasmids and bacteriophages, such recombinant vectors replicate efficiently in specialised strains of bacterial host cells, resulting in the production of multiple copies of the recombinant DNA molecule that was inserted into the vector. Before a DNA fragment can be inserted into a cloning vector, it must be prepared so that the precise DNA sequences of its ends are compatible with the ends of the vector molecule DNA sequence. There are ways to produce recombinant molecules independent of this process but cell-based cloning still remains the method of choice for most situations. Cell-based cloning is achieved using commercially available enzymes that cleave double stranded DNA at specific sequences; these are known as restriction enzymes or restriction endonucleases (see Chapter 1). If both the vector and the fragments of 'foreign' DNA are cut with the same restriction enzyme, such as *Eco*RI which cuts at the sequence 5′GAATTC3′, it is a relatively straightforward procedure to join, or ligate, the two molecules because their matching ends have small overhangs with complementary base pairs.[2] The resulting molecule is a recombinant construct which is capable of growing in the host cell, such as *Escherichia coli*. In general, there are two types of libraries, those that represent the DNA from a cell, termed a genomic library, or the representation of messenger RNA (mRNA) from a cell at a given time, termed a complementary DNA (cDNA) library.

3.1.2. Types of Cloning Vectors

Modern cloning vectors are engineered so that they contain 'multiple cloning sites' or 'polylinkers'. These are segments of DNA that consist of several unique restriction enzyme recognition sites for the most widely available restriction enzymes.[3] Thus, if a circular vector, such as pUC18, is cut with any one of the polylinker-specific restriction enzymes, it will become linearized but unfragmented (Figure 3.1). Under the correct conditions the linear vector is able to ligate to a fragment of DNA, which has been generated using the same restriction

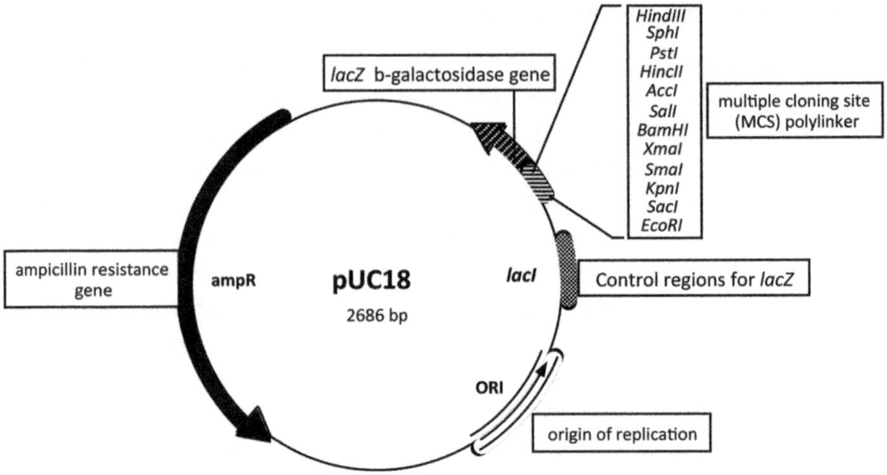

Figure 3.1 Map indicating the important features of the plasmid cloning vector pUC18. Reproduced from ref. 21 with permission from the Royal Society of Chemistry.

enzyme. Upon successful ligation, using the enzyme DNA ligase, the vector molecule containing the recombinant insert DNA is re-circularised. Other necessary features of cloning vectors include an origin of DNA replication and one or more selectable marker genes, typically these are antibiotic-resistance genes.

3.1.3. Cloning PCR Products

While PCR has, to some extent, replaced cloning as a method for the generation of large quantities of a desired DNA fragment there is, in certain circumstances, still a requirement for the cloning of PCR-amplified DNA. For example certain techniques, such as *in vitro* protein synthesis, are best achieved with the DNA fragment inserted into an appropriate plasmid or phage cloning vector. Cloning methods for PCR follow closely the cloning of DNA fragments derived from the conventional manipulation of DNA. The techniques with which this may be achieved are traditionally through one of two ways, blunt-ended or cohesive-ended cloning. Certain thermostable DNA polymerases such as *Taq* DNA polymerase and *Tth* DNA polymerase give rise to PCR products having a 3′ overhanging A residue.[4] It is therefore possible to clone the PCR product into specially constructed dT vectors, this is termed dA:dT cloning. This makes use of the fact that the terminal additions of A residues may be successfully ligated to vectors prepared with T residue overhangs to allow ligation of the PCR

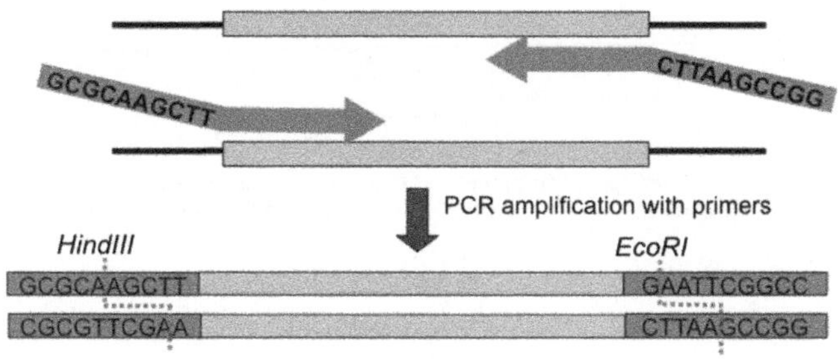

Figure 3.2 Restriction primer PCR. Application of restriction sites to the 5′ end of the primers used in PCR allowing digestion and ligation to cloning vectors.

product. The reaction is catalysed by DNA ligase as in conventional ligation reactions. Furthermore TA-TOPO cloning is very similar and employs a precut vector with 5′-(C/T)CCTT-3′ at the linear ends but with a linked topoisomerase enzyme that carries out the ligation of the PCR products to the vector.

It is also possible to carry out cohesive-ended cloning with PCR products. In this case oligonucleotide primers are designed with a restriction endonuclease site incorporated into them. Since the complementarity of the primers needs to be absolute at the 3′ end the 5′ end of the primer is usually the region for the location of the restriction site. This needs to be designed with care, since the efficiency of digestion with certain restriction endonuclease decreases if extra nucleotides, not involved in recognition, are absent at the 5′ end. In this case the digestion and ligation reactions are the same as those undertaken for conventional reactions (Figure 3.2). An adaptation of this principal is used in the in fusion cloning system, where homologous sequences in an amplified PCR product are joined with complementary sequences in a cloning vector.

3.1.4. Isothermal Assembly – Gibson Assembly Cloning System

One further key method of efficiently cloning DNA that obviates the need for restriction enzymes is termed the Gibson assembly. This is a three-enzyme-based single-step technique allowing assembly of multiple DNA fragments in an isothermal single-tube reaction. Here an exonuclease is used to create single-stranded 3′ overhangs that allow annealing of the DNA fragments at the overlap region. A DNA polymerase is then used to fill in the gaps within each annealed fragment. Finally, a DNA ligase seals the nicks in the assembled DNA.

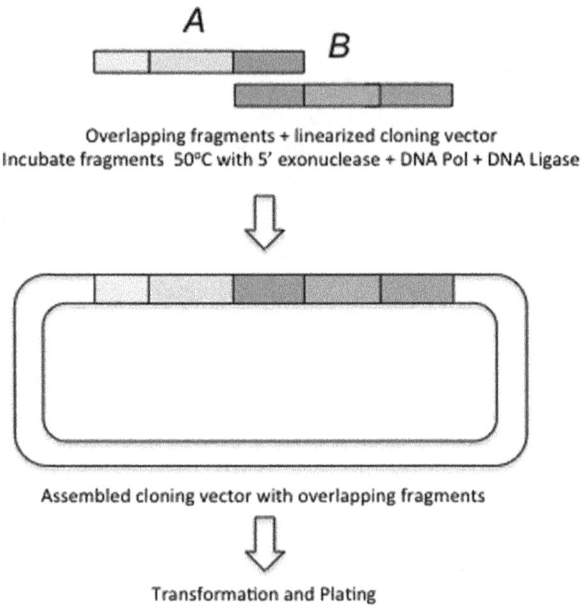

Figure 3.3 The Gibson assembly method of cloning without the need for restriction endonuclease digestion and ligation.

A key feature of the method is the design of the primers, requiring overlapping sequences with complementarity of 20–40 bases for the assembly of the construct (Figure 3.3). Up to six fragments of up to 12 kb can be assembled. Compared with conventional cloning involving restriction enzymes, where joining two fragments can be time-consuming and may involve many manipulation steps, the Gibson assembly is very efficient and provides DNA joining which is seamless.[5] A number of bioinformatics resources are also available to assist in the planning of the sequences required for the process, such as Benchling or NEBuilder.

A number of alternative cloning methods are also available. One, termed Gene Art, Seamless Cloning, omits the initial ligation step but allows assembly of up to four inserts and a linearized vector of up to 40 kb in a 30 min room-temperature reaction, the ligation being carried out in *E. coli*. Other non-traditional cloning systems are also available including those based on Type IIs restriction enzymes (GoldenGate system), ligation-independent cloning (LIC), and for some applications biotechnology companies, such as Synthetic Genomics Inc., can synthesise error-free gene sequences of up to 2 Mb in size.[6]

Table 3.1 Characteristics of cloning vectors. The main types of cloning vectors with various insert capacities and host cells. Vectors are mainly based on plasmids or viruses or have been constructed artificially. Reproduced from ref. 21 with permission from the Royal Society of Chemistry.[a]

Vector	Host cell	Vector structure	Insert range (kb)
M13	*E. coli*	Circular virus	1–4
Plasmid	*E. coli*	Circular plasmid	1–5
Phage I	*E. coli*	Linear virus	2–25
Cosmids	*E. coli*	Circular plasmid	35–45
BACs	*E. coli*	Circular plasmid	50–500
YACs	*Saccharomyces cerevisiae*	Linear chromosome	100–2000

[a]BAC, bacterial artificial chromosome; YAC, yeast artificial chromosome.

3.1.5. Aspects of Cloning Vectors

Many types of vector and host cells are employed in cloning experiments. The choice of vector depends on the size of the insert to be cloned and the purpose of the experiment (Table 3.1). Plasmids and lambda bacteriophage vectors are used for cloning small inserts of up to a few kb of DNA, cosmid vectors on the other hand can replicate efficiently with up to about 50 kb of DNA insert; bacterial artificial chromosomes (BACs) and yeast artificial chromosomes (YACs) are capable of replicating significantly larger inserts of up to about 500 kb of DNA (BACs) and more than 1000 kb of DNA (YACs). The last two vectors have been vital tools in the eukaryotic genome projects, where vast lengths of DNA had to be mapped and arranged into a series of overlapping clones, or contigs. A number of studies have indicated that whereas YACs were capable of holding more than 1 megabase of insert DNA they tended to become unstable, causing the inserts to rearrange, whereas BACs were found to be far more stable and have become the mainstay for many genome projects.[7]

3.1.6. Consideration of DNA Libraries

'DNA library' is the term used to describe a collection of recombinant clones, or DNA molecules, generated from a specific source of DNA. There are two main types of DNA library, which are very distinct in their origin and purpose. DNA from a nucleated cell, whatever the tissue source from a specific organism is used to make a 'genomic' library. The idea of a general genomic library is to produce a set of clones that contain enough DNA fragments so that the entire genome of the organism is represented. Variations of genomic DNA libraries, such as chromosome-specific libraries that were prepared from chromosomes sorted by flow cytometry, were employed in the Human Genome Project in an attempt to shorten the path between the starting DNA and generation of the genome map. The second type is the cDNA library that is made from mRNA that has been reverse transcribed by the enzyme reverse transcriptase. This is essentially the same process as for reverse transcription (RT)-PCR for analysis of RNA (see Chapter 1). Reverse transcriptase produces complementary DNA (or cDNA) fragments which are then cloned into a vector. Therefore, unlike a genomic DNA, library, a cDNA library is representative of the expressed genes in a particular cell or tissue type at a particular time. The choice of the particular type of gene library depends on a number of factors, the most important being the final application of any DNA fragment derived from the library.[8] If the ultimate aim is to understand the control of protein production for a particular gene or the analysis of its architecture, then genomic libraries must be used (Figure 3.4).

However, if the goal is the production of new or modified proteins, or the determination of the tissue-specific expression and timing patterns, cDNA libraries are more appropriate. The main consideration in the construction of genomic or cDNA libraries is therefore the nucleic acid starting material. Since the genome of an organism is fixed, chromosomal DNA may be isolated from almost any cell type in order to prepare genomic libraries. In contrast, however, cDNA libraries only represent the mRNA being produced from a specific cell type at a particular time. Thus, it is important to consider carefully the cell or tissue type from which the mRNA is to be derived in the construction of cDNA libraries. A comparison of genomic and cDNA libraries is shown in Figure 3.5.

3.1.7. Production of Recombinant Constructs

The recombinant vector molecule must be introduced into its 'matching' host cell in order to replicate and produce multiple copies. The process by which DNA is introduced into the host cell is known as bacterial transformation. Since 'naked' vector DNA is hydrophilic and the bacterial cell wall is normally impermeable to such molecules, the host cell must be made 'competent' by treatment with calcium chloride in early log phase of growth. This causes the cell to become permeable to chloride ions. When competent cells are mixed with DNA and heat shocked at 42 °C, the swollen cells are able to take up the naked DNA molecules. It is thought that only a single DNA molecule

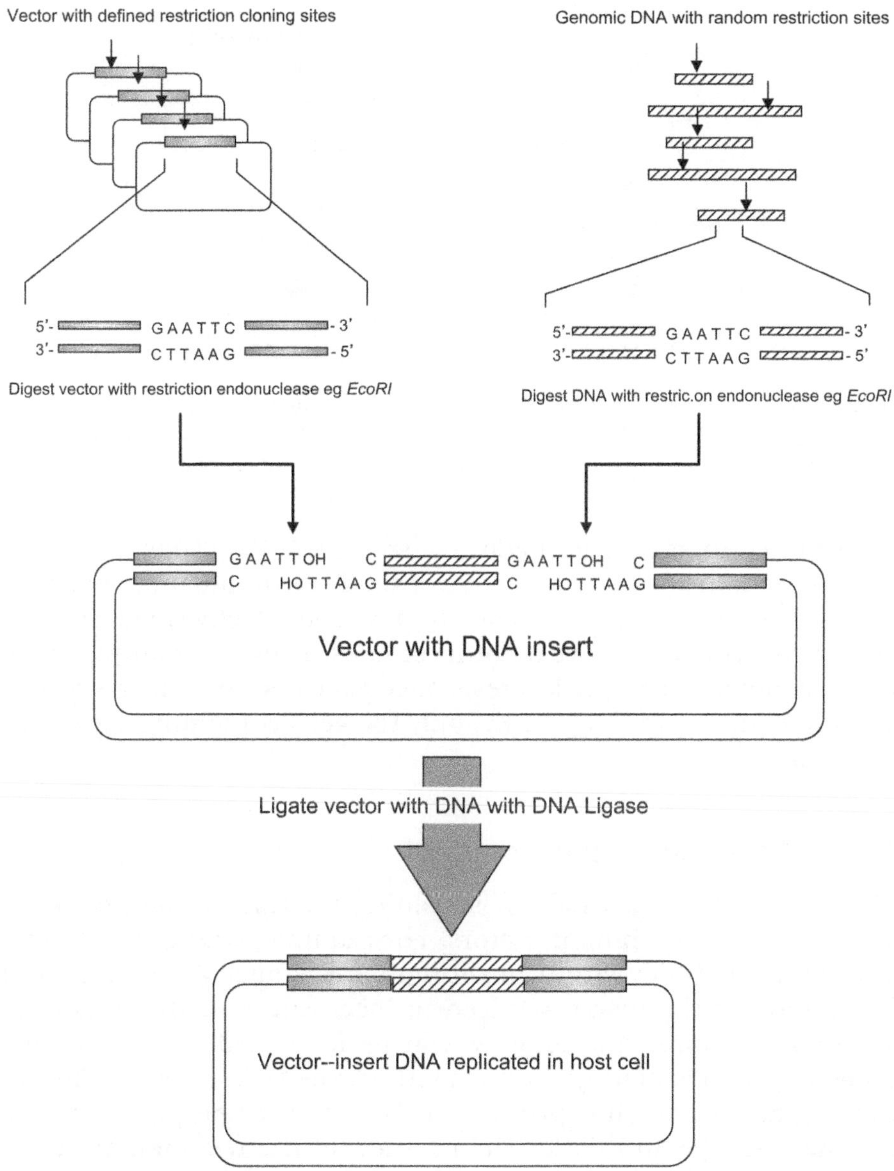

Figure 3.4 Overview of the restriction endonuclease digestion and DNA ligation process in DNA cloning. Reproduced from ref. 21 with permission from the Royal Society of Chemistry.

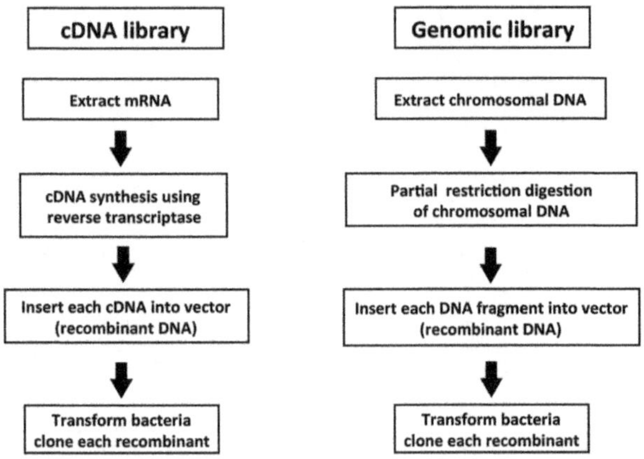

Figure 3.5 Comparison of the general steps involved in the construction of genomic and cDNA libraries. Reproduced from ref. 21 with permission from the Royal Society of Chemistry.

is permitted to enter any single cell, such as *E. coli*. Thus individual colonies of transformed bacteria grow single recombinant vector molecules. The bacterial cells are usually grown on selective media so that only transformants survive to form colonies. Thus, for example, if the vector contains an ampicillin resistance gene and the cells are grown on ampicillin-containing media, only the cells containing the vector will form colonies.[9]

3.1.8. Screening Gene Libraries

There are several methods of screening for transformed colonies that contain recombinant vectors. For example, where the cloning site in the vector lies within an antibiotic-resistance gene, successful integration of the insert will lead to inactivation of the resistance gene and recombinant colonies can be identified by a technique known as replica plating. In this method, the pattern of colonies in the original Petri dish is printed onto a nutrient agar plate containing the selective antibiotic. The position of the recombinant colonies, *i.e.* those that fail to grow on the selective antibiotic, is noted so that they can be picked from the master plate (Figure 3.6). In other vectors, a method called blue/white selection can be performed. In this technique, the MCS is part of a gene in its own right and codes for a portion of a polypeptide called β-galactosidase. When the pUC plasmid has been used to transform the host cell *E. coli* the gene

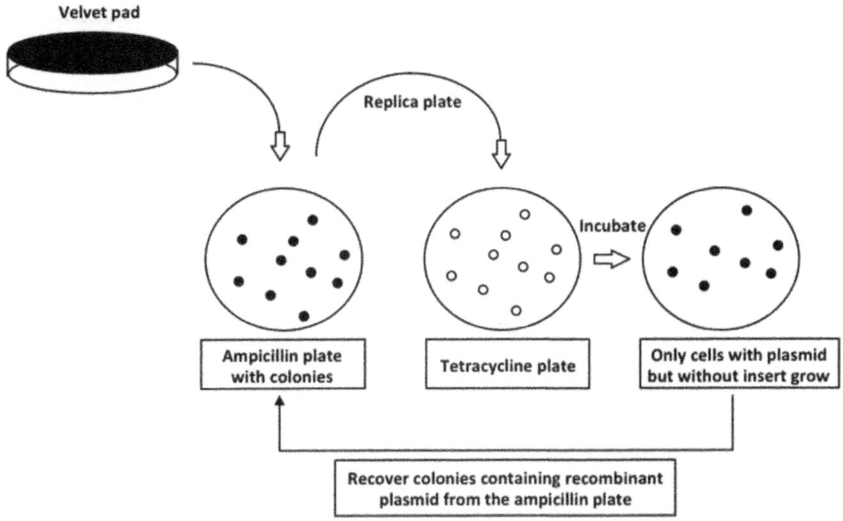

Figure 3.6 Replica plating to detect recombinant plasmids. A sterile velvet pad is pressed on to the surface of an agar plate, picking up some cells from each colony growing on that plate. The pad is then pressed on to a fresh agar plate containing the selective antibiotic, thus inoculating it with cells in a pattern identical with that of the original colonies. Clones of cells that fail to grow on the second plate (owing to the loss of antibiotic resistance) can be recovered from their corresponding colonies on the first plate. Reproduced from ref. 21 with permission from the Royal Society of Chemistry.

may be switched on by adding the inducer IPTG (isopropyl-β-D-thiogalactopyranoside). Its presence causes the enzyme β-galactosidase to be produced. The functional enzyme is able to hydrolyse a colourless substance called X-gal (5-bromo-4-chloro-3-indolyl-β-galactopyranoside) into a blue insoluble material (5,5′-dibromo-4,4′-dichloro indigo). However, if the gene is disrupted by the insertion of a foreign fragment of DNA, a non-functional enzyme results, which is unable to carry out hydrolysis of X-gal. Thus, a colony bearing a recombinant pUC plasmid may be easily detected since it is white or colourless in the presence of X-gal, whereas a colony bearing an intact non-recombinant pUC plasmid will be blue since its gene is fully functional and not disrupted. This elegant system, termed blue/white selection, allows the initial identification of recombinants to be undertaken very quickly and has been included in a number of subsequent vector systems (Figure 3.7). This selection method and insertional inactivation of antibiotic resistance genes do not, however, provide any information on the character of the DNA insert, just the status of the vector.

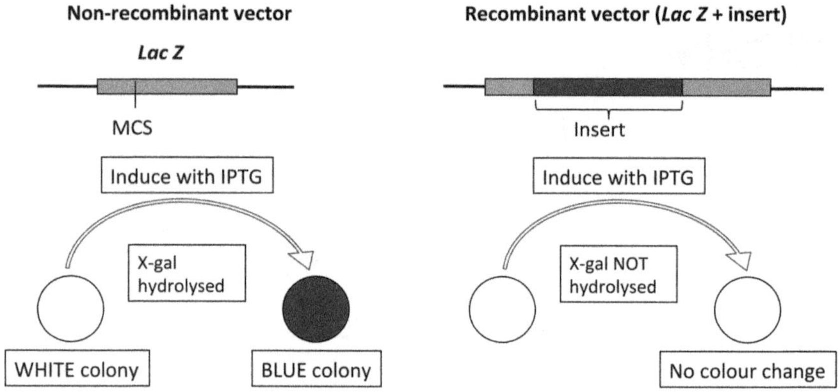

Figure 3.7 Principle of blue/white selection for the detection of recombinant vectors. In the presence of the inducer IPTG the β-galactosidase (*LacZ*) gene is transcribed. The recombinant DNA insert disrupts the expression of the *LacZ* gene which encompasses the multiple cloning site (MCS), hence the substrate X-gal is not hydrolysed and the recombinant colonies remain white. Reproduced from ref. 21 with permission from the Royal Society of Chemistry.

3.1.9. Identification of Inserts in Clones

Bacterial clones containing the required insert can be identified by hybridisation with specifically labelled or enzyme-labelled cDNA or genomic DNA probes, by PCR or, alternatively, by the immunodetection of protein products (using specialised expression vectors which allow a cloned foreign cDNA to be transcribed to express its protein product). Both approaches are technically straightforward. Both involve the transfer of bacterial colonies from a master agar plate onto carefully orientated nitrocellulose or nylon membranes. The cells are then lysed and the DNA (or protein) from the lysed colonies is immobilised on the membrane, which is used for the probing step. Recombinant colonies can be detected as spots on X-ray film, either by autoradiography or enzyme-generated chemiluminescence. The spots on the X-ray film can then be aligned with the agar master plate, allowing the correct colonies to be picked (Figure 3.8). For protein detection in expression vectors, antibody probes are employed, in a manner analogous to an enzyme-linked immunosorbent assay (ELISA) test. The antibody probe is conjugated to an enzyme, such as horseradish peroxidase or alkaline phosphatase. It is the action of the bound enzyme on its chemiluminescence or chromogenic substrate that reveals the position of the recombinant colonies.

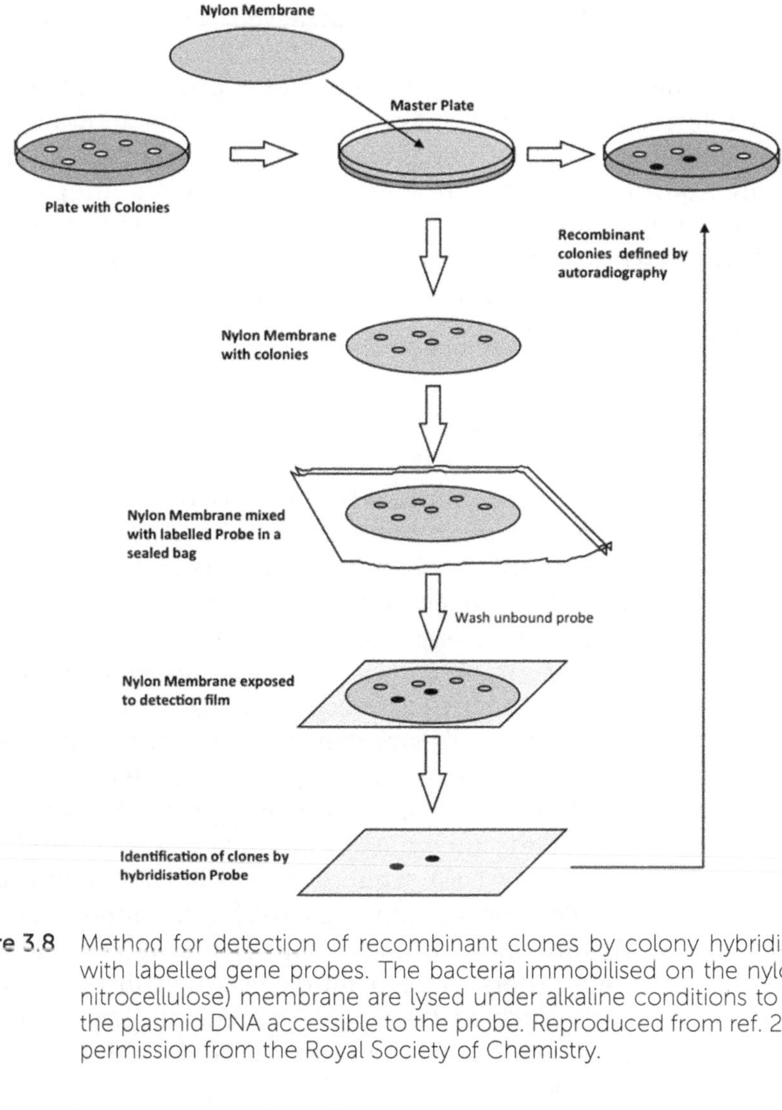

Figure 3.8 Method for detection of recombinant clones by colony hybridisation with labelled gene probes. The bacteria immobilised on the nylon (or nitrocellulose) membrane are lysed under alkaline conditions to make the plasmid DNA accessible to the probe. Reproduced from ref. 21 with permission from the Royal Society of Chemistry.

3.1.10. PCR Screening of Gene Libraries

In many cases it is possible to use PCR to screen cDNA or genomic libraries constructed in plasmids or bacteriophage vectors. This is usually undertaken with primers which anneal to the vector rather than the foreign DNA insert. The size of an amplified product may be used to characterise the cloned DNA and subsequent restriction mapping is then carried out. The main advantage of PCR over traditional hybridisation-based screening is the rapidity of the technique, as PCR screening may be undertaken in 3–4 h, whereas it may be several days before detection by hybridisation is achieved. The PCR screening technique gives an indication of the size of the cloned insert rather than the sequence of the insert; however, PCR primers that are specific for a foreign DNA insert may also be used. This allows a more rigorous characterisation of clones from cDNA and genomic libraries. There are now huge numbers of identified clones containing human genes, gene fragments and other elements, such as gene promoters, in laboratories worldwide. This, combined with bioinformatics annotation, provides a huge resource for the study of genes for medical and biotechnological analysis.

3.2. DNA Sequencing

The determination of the linear order or sequence of nucleotide bases along a length of DNA is one of the central techniques in molecular biology and has played the key role in numerous genome mapping and sequencing projects. Two techniques were developed in the 1970s for efficient DNA sequencing, one based on an enzymatic method, frequently termed Sanger sequencing, after its developer Fred Sanger, and a chemical method, termed Maxam and Gilbert sequencing, named for the same reason. For standard DNA sequence analysis the Sanger method and its variants were by far the most effective method and have been the mainstay of DNA sequencing for many years and numerous commercial kits and strategies have been developed for its use.[10] However, the field of DNA sequencing has been revolutionised by so called next-generation sequencing (NGS) or second-generation sequencing (SGS), with methodology such as Illumina sequencing by synthesis, among others. Indeed more recently third-generation sequencing methods, such as Nanopore sequencing, have also and continue to be developed, providing many exciting alternatives for DNA sequence analysis.

3.2.1. Sanger 'Chain Termination' Sequencing

Sanger sequencing is an enzymatic method that requires single-stranded DNA as the template. Traditionally this demanded that the DNA fragment of interest be inserted and cloned into the specialised bacteriophage vector termed M13, which is naturally single-stranded. Although M13 is still available for use, the advent of PCR has provided the means not only to amplify a region of any genome or cDNA for which primer sequences are available, but also to very quickly generate the corresponding nucleotide sequence.

This has led to an explosion in the accumulation of DNA sequence information and has provided much impetus for many areas of genomic analysis.

The Sanger method is simple and elegant and in many ways mimics the natural ability of DNA polymerase to extend a growing nucleotide chain based on an existing template. Initially the DNA to be sequenced is allowed to hybridise with an oligonucleotide primer, which is complementary to a sequence adjacent to the 3′ side of DNA. The oligonucleotide will then act as a primer for synthesis of a strand of DNA, catalysed by DNA polymerase. Either one of the deoxynucleotide triphosphates (dNTPs) required for DNA synthesis is labelled (or all four in some methods) with a fluorescent molecule or, alternatively, the primer may be labelled to provide the means of detection following the completion of the method.

3.2.2. Dideoxynucleotide Chain Terminators

The interesting aspect of the Sanger method is the use of dideoxy chain terminators (Figure 3.9). Here the reaction mixture is divided into four aliquots, representing the four dNTPs A, C, G and T. Using the adenine (A) tube as an example, in addition to all of the dNTPs being present in the mix, an analogue of dATP is added

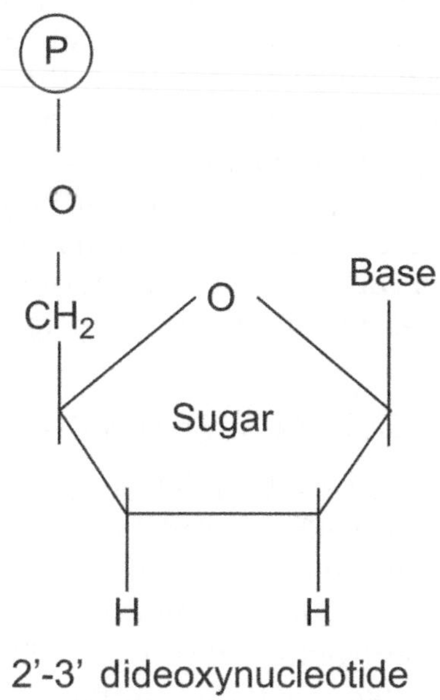

Figure 3.9 Chemical structure of a dideoxynucleotide analogue showing lack of oxygen at the 2′ and 3′ position in the ribose ring. Reproduced from ref. 21 with permission from the Royal Society of Chemistry.

[2′3′-dideoxyadenosine triphosphate (ddATP)], which is similar to A except that it has no 3′ hydroxyl group. Since a 5′ to 3′ phosphodiester linkage cannot be formed without a 3′-hydroxyl group, the presence of the ddATP will terminate the growing chain. The situation for tube C is identical except that ddCTP is added; similarly the G and T tubes contain ddGTP and ddTTP respectively.

Since the incorporation of a ddNTP rather than a dNTP is a random event, the reaction will produce new molecules varying widely in length, but all terminating at the same type of base. Thus, four sets of DNA sequence are generated, each terminating at a different type of base, but all having a common 5′ end (the primer). In the original method the four labelled and chain-terminated samples are then denatured by heating and loaded next to each other on a polyacrylamide gel for electrophoretic separation. The electrophoresis was performed at approximately 70 °C in the presence of urea, to prevent renaturation of the DNA, since even partial renaturation alters the rates of migration of DNA fragments. Very thin, long electrophoresis gels were used for maximum resolution over a wide range of fragment lengths. After electrophoresis, the positions of labelled DNA bands on the gel were determined by autoradiography. Since every band in the lane from the ddATP sample must contain molecules which terminate at adenine, and that those in the ddCTP terminate at cytosine, *etc.*, it is possible to read the sequence of the newly synthesised strand from the autoradiogram, provided that the gel can resolve differences in length equal to a single nucleotide, hence, the ability to detect and characterize features in sequences, such as point mutations. Under ideal conditions, sequences up to about 300–500 bases in length can be read from one gel (Figure 3.10). Although the principle has remained essentially the same, the reactions, detection and analysis methodology used to identify the chain-terminated fragments has become automated and thus much more efficient.[11]

3.2.3. Sequencing Double-Stranded DNA

It is also possible to undertake nucleotide sequencing from double-stranded molecules, such as plasmid cloning vectors and PCR amplicons, directly. The double-stranded DNA must be denatured prior to annealing with primer. In the case of plasmids, an alkaline denaturation step is sufficient. However, for PCR amplicons this is more problematic. Unlike plasmids, amplicons are short and reanneal rapidly. Denaturants such as formamide or dimethylsulphoxide (DMSO) have been used to prevent the reannealing of PCR strands

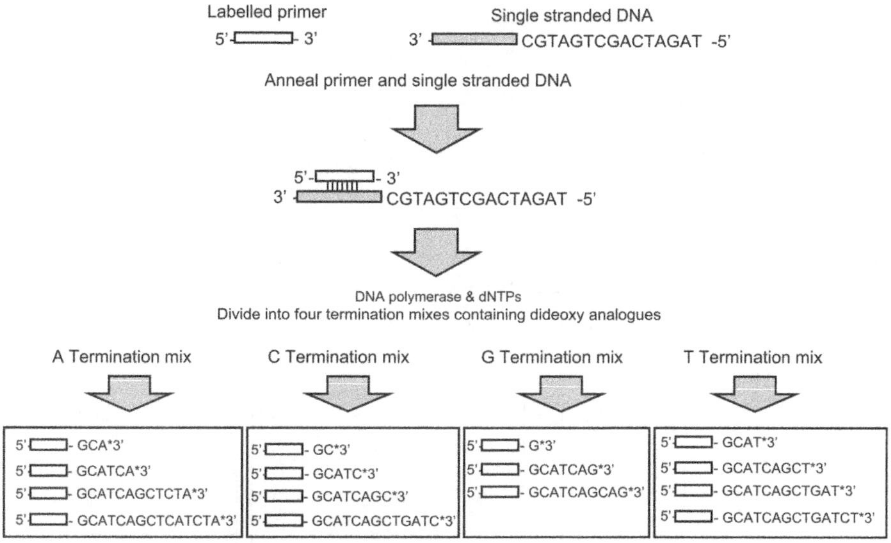

Figure 3.10 Principal of Sanger 'chain termination sequencing'. The initial stage involves annealing of a primer to the DNA sequence. The reaction is separated into four equal aliquots where each tube contains a chain terminator [*e.g.* dideoxyadenine (ddA)]. This terminates the growing chain and provides differing lengths but all ending in the same base. Once these are separated by electrophoresis and visualised the sequence may be determined. Reproduced from ref. 21 with permission from the Royal Society of Chemistry.

following their separation. Another strategy is to bias the amplification towards one strand by using a primer ratio of 100 : 1, which also overcomes this problem to a certain extent.

It is possible to physically separate and retain one PCR product strand by incorporating a molecule such as biotin into one of the primers. Following PCR, the strand that contains the biotinylated primer may be removed by affinity chromatography with streptavidin-coated magnetic beads, leaving the complementary PCR strand. This magnetic affinity purification provides single-stranded DNA derived from the PCR amplicon and although somewhat time-consuming, it does provide high-quality single-stranded DNA for sequencing.

3.2.4. Automated Dye-Terminator DNA Sequencing

Advances in fluorescent-labelling chemistry provided the impetus for the development of high-throughput automated sequencing techniques. Most systems involve the use of dNTPs labelled with different fluorochromes (often referred to as dye terminators). The advantage

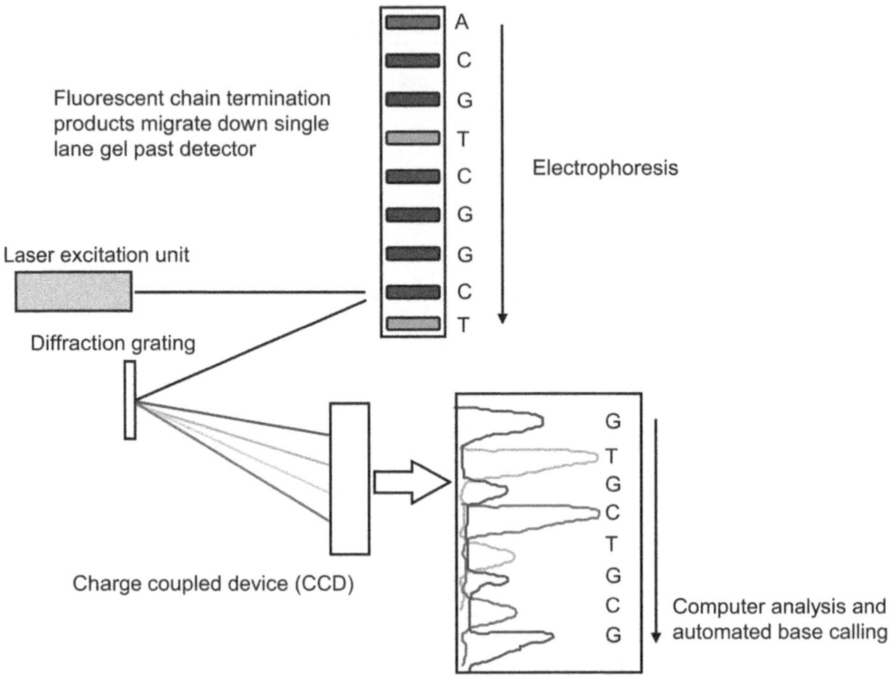

Figure 3.11 Dye terminator sequencing. Here the terminated products are labelled with different fluorophores that are separated by capillary electrophoresis and pass through a CCD, which detects each fluorophore and hence the sequence of the DNA bases. Reproduced from ref. 21 with permission from the Royal Society of Chemistry.

of this modification is that since a different label is incorporated with each ddNTP it is unnecessary to perform four separate reactions. The products are analysed using liquid polymers in thin capillary tubes, obviating the need to pour sequencing gels and requiring little manual operation. This substantially reduces the electrophoresis run times and allows a higher throughput to be achieved. Each product with a base-specific dye, such as fluorescein or tetramethyl-rhodamine, is excited by a laser and the dye then emits light at its characteristic wavelength. A diffraction grating separates the emissions which are detected by a charge coupled device (CCD) and the resulting sequence is analysed by a computer (Figure 3.11). The advantages of the techniques include real time detection of the sequence and the lengths of sequence that may be analysed are between 500 and 1000 bases. Capillary electrophoresis is now a standard method for the detection of Sanger sequencing products. A number of large-scale sequencing facilities are fully automated, allowing the rapid acquisition of sequence data. Automated sequencing for genome projects was previously based on sequencing using instruments such as the ABI PRISM 3700 DNA Analyzer. This could be formatted to produce simultaneous reads in 384-well cycle sequencing reaction plates. The nucleotide sequences could then be downloaded automatically to databases and manipulated using a variety of bioinformatics resources. The first human genome to be sequenced was completed in 2001, using chain termination chemistry. This task took 13 years and cost approximately $3 billion to complete. The current cost using new approaches, detailed below, has decreased to such an extent that sequencing a human genome now costs less than $1000.

3.2.5. Next-Generation Sequencing

Progress in sequencing methods and further developments in automation have led to new ways of sequencing, termed NGS or next-generation sequencing.[12] This collection of methods, also referred to as second-generation sequencing (SGS), is based on different platforms and offers rapid, high-throughput analysis that can determine the sequence of large numbers of different DNA strands at one time. Currently, the main NGS systems involve template preparation with some form of enrichment by PCR, usually followed by sequencing by synthesis. Various platforms have been developed to undertake these processes. They include Illumina/Solexa sequencing, Roche 454 sequencing and ABI's SOLiD sequencing. Ion torrent, Proton/PGM and Pacific Biosciences real time sequencing system are also among other systems that have been developed. NGS has numerous applications in many fields including whole-genome sequencing (WGS), RNA-sequencing (RNA-Seq), chromatin immunoprecipitation (ChIP) sequencing, exome sequencing, microbiome sequencing, ribonomics and epigenomics. One general feature of these systems is that they perform the methods in parallel, termed 'massively parallel sequencing' so that large amounts of sequence data are produced simultaneously.[13] The length of the sequence reads is usually short (100–150 bases) and thus the fragments have to be aligned together by bioinformatics software following the sequencing. Each fragment that is sequenced is also uniquely tagged with a short identifying barcode in the initial sample preparation stage. The management, processing and analysis of enormous amounts of data in the reconstruction of large fragments of DNA derived from shorter reads requires powerful computer hardware processing and the development and availability of large-capacity data centres. In addition, an efficient laboratory information management system (LIMS) is required to handle the sample labelling and data analysis. At present

Table 3.2 Selected examples of sequencing methods. The table highlights the main types of DNA sequencing techniques, the principles by which they work and other aspects of their operations.

Method	Principle	Length (bp)	Reads	Run time	Accuracy
Sanger	Chain termination	400–900	1	3 hours	99.9%
Illumina	Sequencing by synthesis	50–250	3 million	Days	99.9%
Roche 454	Pyrosequencing	700	1 million	24 hours	99.9%
Ion Torrent	Ion semiconductor	400	80 million	2 hours	98%
SOLiD	Sequencing by ligation	50 + 35/50	1.4 billion	7 days	99.9%
Pacific Bio	Single-molecule real time	8000	400 Mb	2 hours	87% raw read
Nanopore	DNA transport *via* porins	1 Mb	1	Variable	97%

the set up and capital costs for NGS systems are high in comparison to traditional Sanger sequencing although this is decreasing as uptake of NGS rises. Table 3.2 includes a list of a number of current NGS platforms.

3.2.6. Second-Generation Sequencing Platforms

The most commonly used NGS sequencing platform has been developed by Illumina (Solexa). Here a range of automated sequencing systems termed Hi-Seq, Mi-Seq and Next-Seq, which generally differ in their capacity, have been developed, the dye-terminator chemistry used for DNA sequencing is similar. This process is termed sequencing by synthesis (SBS) and in some ways is similar to Sanger sequencing but in a way that's scaled up or massively parallel.

Initially genomic DNA is fragmented into shorter fragments and adapters are ligated to each end of the DNA. The DNA fragments have identification tags added and are then immobilised onto a slide or flow cell containing a lawn of primers. The DNA fragments hybridise at one end to one of the immobilised primers and then bend over and bind to a complementary primer also bound on the slide surface. The bridge formed is amplified by PCR (termed bridge PCR) and produces DNA colonies or clusters in each channel of the flow cell. The sequencing part of the technology involves the use of reversible dye terminators. Following denaturation one strand is removed and sequencing primers, polymerase and nucleotides are added. Each nucleotide is tagged with a different fluorophore and incorporates a 3′ block or reversible terminator. The complementary bases are flowed across the cell and are incorporated by a DNA polymerase. A laser is used to

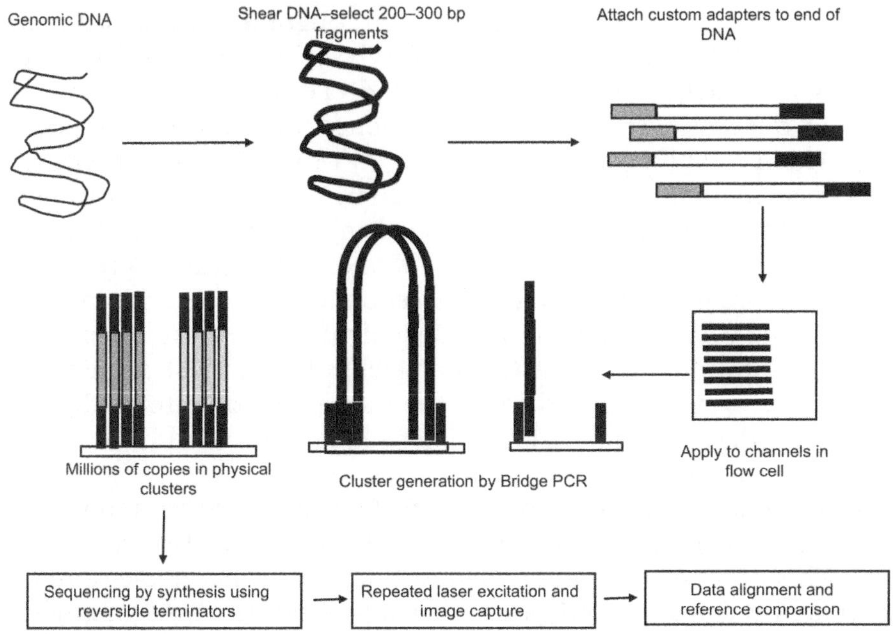

Figure 3.12 Representation of Illumina sequencing by synthesis method. Preparation and enrichment of the sequences are performed initially by sheering and adding adapters, employing bridge PCR. Sequencing involves reversible dye terminators followed by excitation and image capture. Bioinformatic analysis is then needed to align the sequenced fragments. Reproduced from ref. 21 with permission from the Royal Society of Chemistry.

excite the fluorescence, thus identifying the base, which is recorded in an image and takes place in parallel at each cluster. Following this the fluorophore is cleaved, the block removed and the second cycle begins. All the information is processed to build up the sequence in each of the clusters (Figure 3.12). The sequences produced are relatively short (100–150 bases) but numerous and have to be assembled using specialised bioinformatics software.

The 454 platform utilizes the pyrosequencing format and involves direct analysis of DNA fragments and allows the rapid sequencing of entire genomes by the 'shotgun' approach.[14] First, genomic DNA is randomly sheared and ligated to linker sequences that permit individual molecules captured on the surface of a bead to be amplified while isolated within an emulsion droplet by emulsion PCR (emPCR). Each bead is designed to contain one DNA fragment. A very large collection of such beads is arrayed in the 1.6 million wells of a fibre-optic slide.

Each PCR template is hybridised to an oligonucleotide and incubated with DNA polymerase, ATP sulfurylase, luciferase and apyrase. During the reaction the first of the four dNTPs is added and if incorporated releases pyrophosphate (PPi), hence the name 'pyrosequencing'. The ATP sulfurylase converts the PPi to ATP which drives the luciferase-mediated conversion of luciferin to oxyluciferin to generate light. Apyrase degrades the resulting component dNTPs and ATP. This is followed by another round of dNTP addition. A resulting pyrogram provides an output of the sequence. The method provides short reads very quickly and is especially useful for the determination of mutations or SNPs. Current technology of 454 sequencing allows one giga base or a third of the human genome to be derived in a single day, in comparison, Sanger sequencing would take over a year to complete.

An alternative NGS platform developed by Applied Biosystems is the SOLiD (Sequencing by Oligonucleotide Ligation and Detection) system. This process is similar to 454 technology in that it uses emPCR to generate clonal amplification of single-stranded DNA fragments (ligated to adapters) attached to beads. The beads are then attached to a glass surface at a high density. A universal sequencing primer is hybridized to the adapter sequence on the beads containing the template. DNA ligase and a set of four fluorescently labelled di-base probes are added to the reaction and compete for ligation to the sequencing primer. Specificity of the di-base probes is achieved by interrogating every first and second base in each ligation reaction. Hybridisation, ligation, fluorescence detection and cleavage are performed for a predetermined number of cycles, after which the extension product is removed and the template is reset with a primer offset by one base for a second round of ligation cycles.

In general, second-generation sequencing approaches produce large amounts of short-length sequence data that has to be assembled and compiled by sophisticated techniques and analysed by specialist bioinformatics software. They have, however, revolutionised they way in which DNA sequence information has been produced on a massive scale. A comparison of selected sequencing methods is presented in Table 3.2.

3.2.7. Third-Generation Sequencing Platforms

Further approaches that are gaining increasing use are the so-called third-generation sequencing techniques.[15] These are potentially advantageous methods as they obviate the need for rigorous initial

DNA preparation and clonal amplification before sequencing. One interesting format is used by Pacific Biosciences with their single-molecule real-time sequencing (SMRT) platform. The method utilises phospholinked nucleotides, where a fluorescent dye is attached through the phosphate group in the molecule, each of the four nucleotides having a different fluorescent dye. Thus, during incorporation of the nucleotide into a growing chain by a DNA polymerase the phosphate is cleaved while a detector identifies which of the four nucleotides has been incorporated. The technology is based around a laser-light-illuminated nanophotonic containment cylinder, termed a zero-mode waveguide, within which the reaction takes place. Remarkably these cylinders are smaller than the wavelength of light and a sophisticated CCD detects flashes of light when a nucleotide is incorporated and records this as a movie file.

A number of developments in DNA sequencing have been based on tiny semiconductor or biological 'nanopores' through which DNA can be passed and sequence deduced.[16] The technique is based on a transmembrane protein 'pore', such as alpha haemolysin spanning a lipid bilayer. A voltage is placed across the membrane, thus an ion current is produced and ions move through the pore. To sequence a fragment of DNA the sample DNA strand moves through the pore and as different bases pass they alter the current and thus the linear DNA sequence can be identified. One such example is a USB powered device termed the MinION, developed by Oxford Nanopore. This is an ultraportable device, similar in size to a mobile phone, allowing its use in the field. The advantage over NGS methods is that sample preparation is simplified, as no PCR enrichment or labelling is required and the data processing is less complex.

3.3. Microarrays and DNA Microchips

One firmly established area under rapid development in molecular biology is the use of microarrays or DNA microchips. These provide a radically different approach to current laboratory molecular biology research strategies in that large-scale analysis and quantification of genes and gene expression is possible simultaneously. A microarray consists of an ordered arrangement of potentially hundreds of thousands of DNA sequences, such as oligonucleotides or cDNAs, deposited onto a solid surface. The solid support may be either glass or silicon and currently the arrays are synthesised on or off the chip. They require complex fabrication methods similar to those used in producing computer microchips. Most commercial productions employ robotic ultrafine microarray deposition instruments, which dispense volumes in the picolitre range. Alternatively on-chip fabrication, as used by Affymetrix, builds up layers of nucleotides using a process borrowed from the computer industry termed photolithography. Here wafer-thin masks with holes allow photoactivation of specific dNTPs, which are linked together at specific regions on the chip. The whole process allows layers of oligonucleotides to be built up, with each nucleotide at each position being defined by computer.

3.3.1. Types of Microarrays

The arrays themselves may represent a variety of nucleic acid materials. This may be mRNA produced in a particular cell type, termed cDNA expression arrays, or may alternatively represent coding and regulatory regions of a particular gene or group of genes.[17] A number of arrays are now available that may determine mutations in DNA, mRNA transcript levels or other polymorphisms, such as SNPs. Sample DNA is placed on the array and any unhybridised DNA washed off. The array is then analysed and scanned for patterns of hybridisation by detection of fluorescence signals. Any mutations or genetic polymorphisms in relevant genes may be rapidly analysed by computer interpretation of the resulting hybridisation pattern and mutation, transcript level or polymorphism defined. Indeed the collation and manipulation of data from microarrays presents as big a problem as fabricating the chips in the first place. Microarrays have been developed for the detection of various genetic mutations, including in the cystic fibrosis CFTR gene (cystic fibrosis transmembrane regulator), the breast cancer gene BRCA1 and for use in the study of the human immunodeficiency virus (HIV).

At present, microarrays require DNA to be highly purified, which limits their applicability. However, as DNA purification becomes automated and microarray technology develops it is not difficult to envisage numerous laboratory tests on a single DNA microchip. This could not only be used for analysing single genes but large numbers of genes or DNA representing microorganisms, viruses, *etc.* Since the potential for quantitation of gene transcription exists, expression arrays could also be used in defining a particular disease status. This technique may be very significant since it will allow large amounts of sequence information to be gathered very rapidly and assist in many fields of molecular biology, especially in large genome sequencing projects or in so-called resequencing projects, where gene regions such as those containing potentially important polymorphisms require analysis in a number of samples.

3.3.2. SNP Microarrays

One current application of microarray technology is the generation of a catalogue of SNPs across the human genome. Estimates indicate that there are approximately 85 million SNPs and, importantly, some may point to the development of certain diseases. For example, an SNP in the F5 gene is implicated in a thrombophilia, Factor V Leiden, a genetic blood clotting disorder. SNP analysis is therefore clearly a candidate for microarray analysis and genome-wide SNP arrays enable the simultaneous analysis of nearly 1 million SNPs on one gene chip. In order to simplify the problem of the vast numbers of SNPs that need to be analysed the HapMap project currently analyses SNPs that are inherited as a block, and, in theory, as few as 500 000 SNPs will be required to genotype an individual.

3.3.3. Tissue Microarrays

An extension of microarray technology may also be used to analyse tissue sections. This process, termed tissue microarrays (TMA), uses tissue cores or biopsies from conventional paraffin-embedded tissues. Thousands of tissue cores are sliced and placed on a solid support, such as glass, where they may all be subjected to the same immunohistochemical staining process or analysis with gene probes using *in situ* hybridisation. As with DNA microarrays, many samples may be analysed simultaneously, less tissue is required and greater standardisation is possible.

3.4. Gene Modulation and Editing

The advent of advanced technologies, such as molecular cloning and DNA sequencing, has allowed the development of new area of gene modulation and genome editing. These new and exciting areas hold great potential for altering expression of genes or correcting single-base mutations. With appropriate ethical and regulatory consideration they may provide a new era in molecular biology and biotechnology.

3.4.1. Modulating Gene Expression by RNAi

There are a number of ways of experimentally changing the expression of genes. Traditionally, methods have focussed on altering the levels of mRNA by manipulation of promoter sequences or levels of accessory proteins involved in control of expression.[18] In addition, post-mRNA production methods have also been employed, such as antisense RNA, where a nucleic acid sequence complementary to an expressed mRNA is delivered into the cell. This antisense sequence binds to the mRNA and prevents its translation. A development of this theme and a process that is found in a variety of normal cellular processes is termed RNA interference (RNAi) and uses microRNA. Here a number of techniques have been developed that allow the modulation of gene expression in certain cells. This type of cellular-based gene expression modulation will no doubt be extended to many organisms in the next few years.

3.4.2. Endonuclease-Based Genome Editing

There are a number of genome editing systems now available, based on engineered endonucleases that allow the creation of gene knockouts and knockins. These systems, rather than modulating the expression of a gene, alter the genes themselves and are essentially engineered restriction endonucleases produced to modify specific DNA sequences. One system is based on zinc finger proteins that are artificially fused to restriction enzyme *Fok*I. Another similar system is termed TALENs or transcription activator like effector nucleases. In this case a transcription activator like a DNA binding domain (TAL) is fused with a DNA cleavage domain to form a hybrid enzyme. The DNA binding domain of the hybrid molecule provides specificity, which, importantly, can be designed for any particular DNA sequence. The constructs can be delivered to the pronucleus or cytoplasm for protein synthesis, after which double strand breaks are induced at the target DNA sequence. This is able to stimulate one of two endogenous cellular mechanisms: non-homologous end joining (NHEJ) or homology-directed repair (HDR). Although elegant, these systems require extensive engineering of the protein.

3.4.3. CRISPR/Cas9-Based Genome Editing

One major gene editing system is CRISPR (clustered regularly interspaced short palindromic repeats) involving RNA-guided engineered nucleases.[19] The system was first identified as an adaptive immunity pathway in prokaryotes, providing resistance to bacterial phage and analogous to RNA interference processes found in eukaryotes. The system has been modified and adapted for the engineering of genomes and, unlike TALENs, doesn't require new nucleases to be designed for every genomic target.

Essentially CRISPR may be thought of as a programmable restriction enzyme system. It utilises double stranded sequence-specific breaks and repair using HDR *via* homologous recombination. The availability of vast amounts of sequence information has allowed the development of the system into an essential molecular biology technique.

The system consists of two parts, a short synthetic guide RNA and a non-specific double stranded endonuclease termed CRISPR-associated 9 (Cas9). The guide RNA has a specific 20-nucleotide spacer/targeting sequence (CRISPR crRNA), which determines the genomic target to be identified and modified, linked to a *trans*-activating (tracrRNA) sequence necessary for recruitment and stability of Cas9 nuclease. The guide (gRNA) sequence and Cas9 nuclease elements are constructed into a plasmid expression vector, which can be used to transfect cells under study. Transcription of the gRNA and Cas9 results in crRNA that binds to the DNA to be altered, which is held in the Cas9 nuclease with the aid of the tracrRNA. Indeed it is the fact that the target sequence can be reprogrammed simply by changing the 20 nucleotides in the crRNA that makes the system so elegant. DNA is then digested by the Cas9 nuclease at specific points on both DNA strands. A donor DNA sequence may be included with a desired feature such as a base change or insertion or deletion (Figure 3.13).

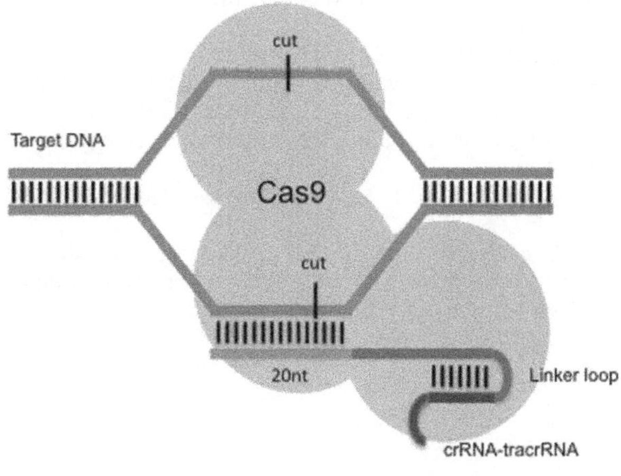

Figure 3.13 CRISPR/Cas9 gene editing system, where the nuclease is targeted to DNA by a linked to a *trans*-activating (tracRNA) required by recruitment of the Cas9 nuclease specific 20-nucleotide guide RNA. Following DNA digestion by Cas9 at specific points repair by homologous recombination then takes place and a new sequence is incorporated.

The cell then employs homologous recombination to repair the break and incorporate the new DNA sequence. It is also possible to program Cas9 with multiple guide RNAs to allow multiplex site-specific editing for designing large deletions, inversions and translocations.[20] Further refinements of the editing system have allowed the targeting of protein domains for transcriptional regulation and epigenetic modification.

References

1. J. D. Watson, *Recombinant DNA: Genes and Genomes: A Short Course*, W.H. Freeman, San Francisco, 2007.
2. H. O. Smith and K. W. Wilcox, *J. Mol. Biol.*, 1970, **51**, 379.
3. N. Casali and A. Preston, *E. coli Plasmid Vectors. Methods in Molecular Biology*, 2015, vol. 235, p. 23.
4. T. A. Holton and M. W. Graham, *Nucleic Acids Res.*, 1991, **19**(5), 1156.
5. D. G. Gibson, *Methods Enzymol.*, 2011, **498**, 349–361.
6. B. Sands and R. Brent, *Curr. Protoc. Mol. Biol.*, 2016, **113**, 3.26.1–3.26.20.
7. M. O'Connor, M. Peifer and W. Bender, *Science*, 1989, **244**(4910), 1307–1312.
8. R. Rapley and J. M. Walker, *Molecular Biomethods Handbook*, Humana Press, NY, 2007.
9. D. W. Russell and J. Sambrook, *Molecular Cloning: A Laboratory Manual*, Cold Spring Harbor Laboratory, N.Y, 2001.
10. F. Sanger, S. Nicklen and A. R. Coulson, *Proc. Natl. Acad. Sci. U. S. A.*, 1977, **74**, 5463–5467.
11. E. Pettersson, J. Lundeberg and A. Ahmadian, *Genomics*, 2009, **93**(2), 105–111.
12. M. L. Metzker, *Nat. Rev. Genet.*, 2010, **11**, 31–46.
13. S. E. Levy and R. M. Myers Annual, *Review of Genomics and Human Genetics 2016*, 2016, vol. 17, pp. 95–115.
14. S. Goodwin, J. D. McPherson and R. McCombie, *Nature Reviews Genetics*, 2016, vol. 17, pp. 333–351.
15. C. Bleidorn, *Syst. Biodivers.*, 2016, **14**(1), 1–8.
16. N. J. Loman and M. Watson, *Nat. Methods*, 2015, **12**(4), 303–304.
17. R. Bumgarner, *Curr. Protoc. Mol. Biol.*, 2013, **0**(22), Unit-22.1.
18. M. L. Babbin and J. J. Rossi, *Annual Review of Pharmacology and Toxicology*, 2016, vol. 56, pp. 103–122.
19. D. M. Thurtle-Schmidt and T. W. Lo, *Biochem. Mol. Biol. Educ.*, 2018, **46**(2), 195–205.
20. J. G. Doench *Nature Reviews Genetics*, 2017, vol. 19, pp. 67–80.
21. R. Rapley and D. Whitehouse, *Molecular Biology and Biotechnology*, Royal Society of Chemistry, Cambridge, 6th edn, 2014.

4 Microbial Proteomics

Hershna Patel and David B. Whitehouse*

School of Life and Medical Sciences, University of Hertfordshire, College Lane, Hatfield, Hertfordshire, AL10 9AB, United Kingdom
*E-mail: d.whitehouse@herts.ac.uk

4.1. Introduction

Proteomics is the large-scale and high-throughput study of proteins and their functions.[1] The term 'proteome' was first used in 1995 and derives from PROTEins expressed in a genOME.[2] A proteome is the population of all proteins expressed by the genome in a specific cell type, tissue or biological fluid at a particular time and under specific conditions. 'Proteomics' was coined in 1997 to introduce uniformity of terminology with 'genomics', the study of the genome.[3] Microbial proteomics investigations enable the construction of detailed cellular and biochemical maps in specific microorganisms that are beyond the reach of genomics and transcriptomics. For instance, proteomics can be used to identify precisely the expression and cellular locations of specific proteins at defined times that could not be accurately predicted from mRNA levels, to identify posttranslational modifications that could not be ascertained from gene sequences and to characterise many aspects of protein function and protein interactions within an organism. In addition, the identification, classification and analysis of proteins from microbial pathogens, along with genomics studies are integral to clinical research and diagnostics. The implementation

Genomics and Clinical Diagnostics
Edited by David Whitehouse and Ralph Rapley

Published by the Royal Society of Chemistry, www.rsc.org

of various proteomic techniques in medical microbiology laboratories has greatly improved the ability to make accurate and timely diagnoses of infectious diseases that pose a significant burden to public health.[4] Any tissue, cell type or extracellular fluid is amenable to proteomic analysis.[5] In contrast to the genome, which comprises the set of all genes in a species, the proteome in different cells, tissues and fluids is dynamic and constantly changing in response to metabolic requirements, developmental stage and environmental stresses.[6] The total number of human proteins far exceeds the number of structural genes; it is estimated that the human genome comprises 19 000–20 000 structural genes with some 180 000 exons, which occupy about 1% of the genome.[7] Through the action of alternative splicing (it is estimated that 74–99% of human multi-exon genes are alternatively spliced), mutation and other molecular events some 150 000 mRNA transcripts could be produced from 20 000 genes,[8] giving rise to the same number of primary protein products. The number of protein species is further increased through the action of events such as heteromer formation and posttranslational modifications, such as glycosylation and phosphorylation. Therefore, in principle, 20 000 structural genes could give rise to a million or more different proteins.[9] In contrast to the human (eukaryotic) genome, the genes of prokaryotes are intronless and the genomes are thus considerably less complex and encode fewer proteins.[10]

4.1.1. Outline of Microbial Proteomics Workflow

A typical aim of a straightforward proteomics investigation is to isolate and characterise a single protein from a complex mixture of proteins in a biological sample. In clinical settings, microbiological specimens for analysis may include whole-cell lysates, colony cultures, blood cultures and other biological fluids. The first step in the proteomics workflow is the extraction of proteins from the biological material of interest. This is followed by their separation, normally by gel electrophoresis or liquid chromatography, to isolate a single protein from the mixture. A frequently used method is two-dimensional polyacrylamide gel electrophoresis (2DGE), where proteins are first separated by their isoelectric points and then by their molecular masses.[11] Proteins of interest are cut from the gel, digested with a proteinase and the resulting peptides analysed by mass spectrometry. Protein identification for bacteria, viruses and protozoa is achieved by matching the precise masses of the experimental peptide fragments with those from known proteins recorded in sequence databases. If peptide matching (sometimes referred to as peptide mass fingerprinting) is

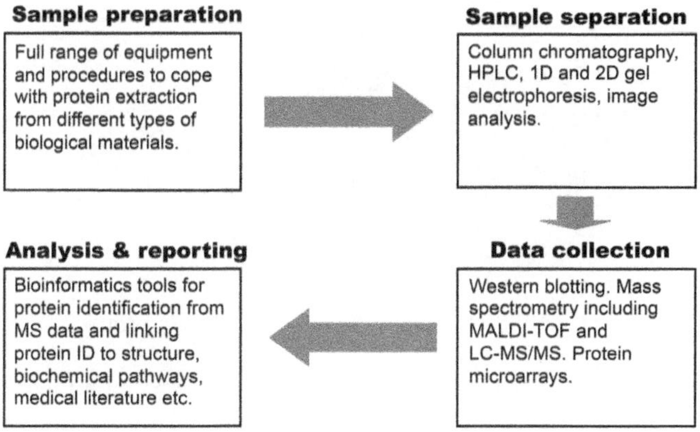

Figure 4.1 A typical proteomics laboratory workflow. Reproduced from ref. 88 with permission from the Royal Society of Chemistry.

unsuccessful, a peptide can be further fragmented and analysed to allow the *de novo* identification of amino acid sequences that can be used to search sequence databases. Various bacteria, such as species of the genera *Escherichia*, *Staphylococcus*, *Pseudomonas* and *Vibrio*, can be analysed on the basis of their protein profiles obtained through mass spectrometry approaches,[12] clinical samples are routinely investigated in this way to identify and differentiate pathogens or to confirm a diagnosis. A generalised proteomics workflow is shown in Figure 4.1. Proteomics depends on sophisticated instrumentation, capable of processing multiple samples with a very high degree of precision, and powerful bioinformatics tools. However, its roots can be traced to traditional protein biochemistry and the separation methods that were employed to identify specific proteins in complex biological mixtures.[13] There are many applications for microbial proteomics, ranging from the identification and characterisation of proteins and their interactions with the host, to probing metabolic pathways and the assessment of protein expression profiles in disease. Following a brief introduction to basic protein biochemistry, this chapter describes each stage of the proteomics workflow and some applications of proteomics in medical microbiology.

4.2. Protein Biochemistry

Proteins are complex biomolecules of varying size that are composed of linear arrays of amino acids. Each amino acid contains a central (alpha) carbon atom to which an amino group ($-NH_2$) and a carboxyl group ($-COOH$) are attached (Figure 4.2). Also, attached to the alpha

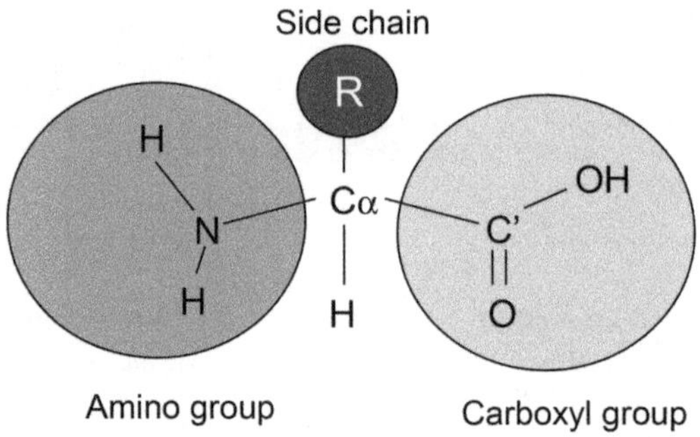

Figure 4.2 The general structure of an amino acid. Reproduced from ref. 88 with permission from the Royal Society of Chemistry.

Table 4.1 The functions of some major groups of proteins.

Function	Examples
Energy metabolism	Enzymes *e.g.* transferases, isomerases
Cellular communication	Peptide hormones, receptor proteins
Defence	Cytokines *e.g.* interferons, interleukins
Adaptive immunity	Antibodies, T-cell receptors
Transport	Haemoglobin, transferrin
Structural framework	Collagens, elastin
Maintaining osmotic pressure of blood	Albumin
Antioxidants	Albumin, superoxide dismutases

carbon atom are a hydrogen atom and a side group (R-group). The R-group varies between different amino acids, giving each its own unique chemical properties. For example, some amino acids are positively or negatively charged while others are hydrophilic or hydrophobic. The R-groups are important in determining the three-dimensional structure and function of proteins. Twenty amino acids are encoded in DNA, these make up the vast majority of proteins in living things; the number of protein structures and hence functions is enormous.[14] An overview of some of the most important functions of proteins is shown in Table 4.1.

Many proteins undergo post-translational modifications and form complexes with other types of molecules. For example, they may combine with carbohydrates to form glycoproteins (glycosylation) or lipids to form lipoproteins[15] or nucleic acids to form ribonucleoproteins. Glycosylation is a common protein modification seen in pathogenic organisms: many bacterial glycoproteins are associated with virulence factors. Structural features, such as pili and flagella, are glycosylated, as well as other surface and membrane proteins, which could be probed as targets for proteomics-based identification.[16] Glycosylation is also an essential process for the synthesis of peptidoglycan, a polymer comprising polysaccharide and amino acids found in bacterial cell walls.

4.2.1. Protein Structure

Amino acids are linked together by peptide bonds. The peptide bond is formed by a condensation reaction when a carboxyl group of one amino acid reacts with the amino group of another and a water molecule is released (Figure 4.3). A peptide bond formed between two amino acids gives rise to a dipeptide and the joining together of three amino acids creates a tripeptide *etc.* Historically, peptide chains with less than 20 amino acids were referred to as 'oligopeptides', 'polypeptides' were associated with 20–100 amino acids and proteins with more than 100 amino acids.[17] Protein structure is organized into four categories: primary, secondary, tertiary and quaternary structures.

4.2.1.1. Primary Structure

The primary structure is the linear sequence of amino acids in the protein. One of the first proteins for which the primary structure was determined was the hormone insulin (51 amino acids). Proteins made up of approximately the same ratios of amino acids but arranged in

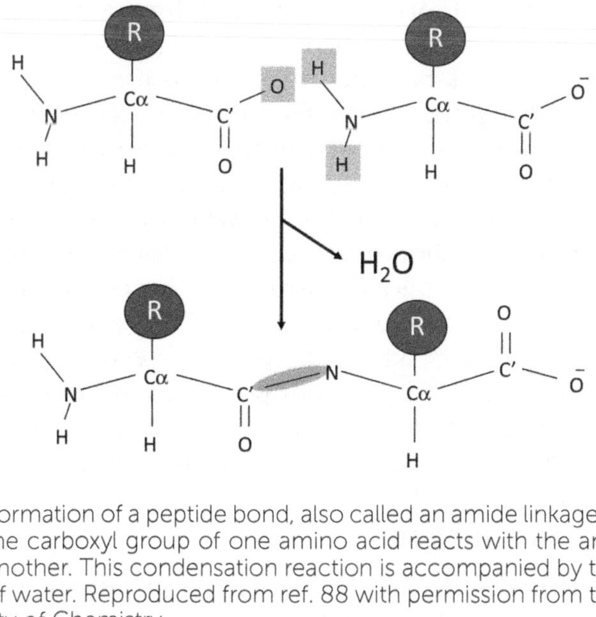

Figure 4.3 Formation of a peptide bond, also called an amide linkage, occurs when the carboxyl group of one amino acid reacts with the amino group of another. This condensation reaction is accompanied by the elimination of water. Reproduced from ref. 88 with permission from the Royal Society of Chemistry.

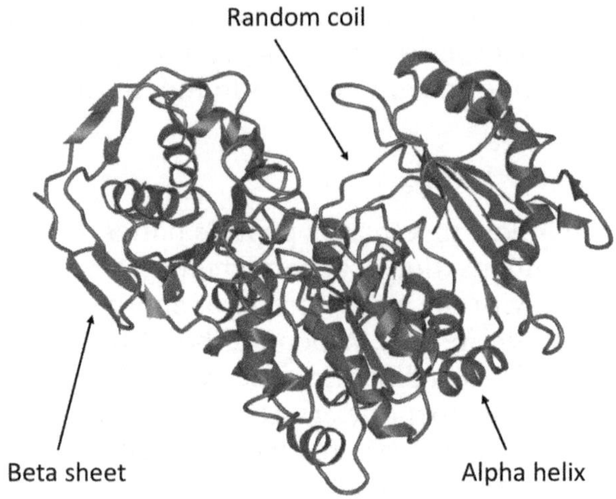

Figure 4.4 Secondary structures of proteins: examples of α-helix, β-sheet and random coil in a 3D ribbon model of the phosphoglucomutase (PGM1) enzyme, EC 5.4.2.2. Image generated from PDB ID: 1C4G (S. Baranidharan, W. J. Ray Jr. DOI: 10.2210/pdb1C4G/pdb).

different orders have different secondary and tertiary structures and perform different functions.

4.2.1.2. *Secondary Structure*

Primary structures can form unstructured domains or fold and twist into more complex secondary structures (Figure 4.4). The two main types of regular secondary structure elements are α-helix and β-sheet, which are joined together by loops and turns. An α-helix is formed as a result of a right-handed twist of the polypeptide chain along its axis. The C=O group of one amino acid and the N–H group of another are held together by hydrogen bonds, which are made every fourth amino acid. Hydrogen bonds are essential for stabilising α-helixes. The less common β-sheet is a flat structure formed by hydrogen bonding between adjacent strands. Some proteins have only α-helixes or β-sheets, whereas others have both types of secondary structure; in general proteins with high levels of β-sheets tend to be more stable than those without. A primary sequence that does not form a regular structure is referred to as unstructured or random coil.

4.2.1.3. *Tertiary Structure*

The tertiary structure is the three-dimensional shape of the folded polypeptide chain (*e.g.* Figure 4.4) and is stabilised by different types of chemical bonds, including covalent bonds, such as disulphide bridges that form between adjacent cysteine residues, ionic and hydrogen bonds, as well as hydrophobic interactions. Because many of the bonds stabilising the tertiary structure are relatively weak, the majority of proteins are liable to denature at high temperatures or when exposed to denaturants, such as urea and guanidinium thiocyanate, or extremes of pH.

Proteins are divided into two main classes based on their 3D structure: globular and fibrous. The majority of proteins, such as the enzymes, are globular and have irregular shapes, tend to be soluble in aqueous solvents and are most amenable to proteomic analysis. Fibrous proteins are tough and mainly insoluble molecules found only in animals. Examples of fibrous proteins include collagen, elastin, keratin and actin.

4.2.1.4. Quaternary Structure

Proteins comprising more than one polypeptide chain are said to have a quaternary structure, an example is the haemoglobin heterotetramer, which consists of four globin polypeptide chains (in adults, two α-chains and two β-chains).

Since proteins are the main targets for antibiotics and antiviral drugs, knowledge of the microbial proteome and constituent protein structures is of great advantage in the field of drug discovery and development. Expression-based proteomic studies can thus be used to identify and evaluate novel proteins that may be of therapeutic interest.

4.3. Sample Preparation

Success in proteomics experiments depends on appropriate sample preparation.[18] The challenges are different to those encountered with nucleic acid research since, as noted above, proteins represent a vast diversity of structures, including fibrous and globular proteins, protein complexes and those with posttranslational modifications. Consequently, a proteomics laboratory requires a full range of procedures to cope with a huge variety of biological materials and protein structures.

4.3.1. Cell Lysis and Protein Extraction

Biochemical analysis of microorganisms often requires the protein or proteins of interest to be purified. In microbiology the starting point is frequently a bacterial culture. There are several approaches to lyse bacterial cells, including enzymatic digestion to remove the bacterial cell wall, sonication (using an ultrasonic probe to produce the acoustic signals), osmotic shock, mechanical disruption (homogenisation) and exposure to chaotropes and detergents capable of breaking plasma membranes. The resulting homogenate contains all the macromolecules, including proteins and nucleic acids, as well as organelles, such as ribosomes, pili and flagella.[19]

4.3.1.1. Centrifugation

A crude bacterial homogenate may be centrifuged at low-speed to produce a clear extract containing soluble proteins that is suitable for electrophoresis or liquid chromatography without further treatment. Examples of protein fractions obtained from Gram-negative bacteria include the outer membrane, cytoplasmic membrane, periplasmic and cytosolic fractions.[20] For eukaryotic cells, centrifugation may also be used to fractionate the extract: at a low speed, large and dense components such as nuclei, fragments of membrane, *etc.* are pelleted at the bottom of the tube, as the speed is increased, smaller sub-cellular components are pelleted (see Figure 4.5). Differential ultracentrifugation is a widely used procedure for separating organelles, membrane

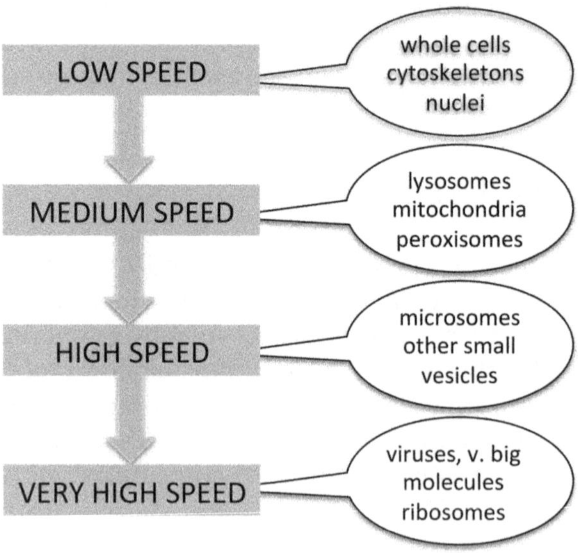

Figure 4.5 Differential centrifugation: repeated centrifugation with increasing relative centrifugal forces enables fractionation of subcellular components according to size and density. Reproduced from ref. 88 with permission from the Royal Society of Chemistry.

fragments, ribosomes and large molecules. Centrifugation in a sucrose gradient is useful for sub cellular fractionation of ribosomes, mitochondria and other organelles: a shallow density gradient, which increases in concentration toward the bottom of the tube, serves to stabilise the cell components and prevent convective mixing. The sub-cellular particles form bands as they migrate through the gradient at different rates.[21] An alternative method is buoyant density centrifugation, usually done in a caesium chloride solution. During ultracentrifugation the caesium chloride solution forms a density gradient along which the components move until they reach the region equal to their own density. Components with the highest buoyant density accumulate towards the bottom of the tube, and those with the lowest buoyant density towards the top. Buoyant density gradient centrifugation, also known as equilibrium centrifugation is a highly efficient fractionation method.[22] On completion of centrifugation, the fractions can be collected through a hole pierced in the base of the tube.

4.3.1.2. Removing Contaminants

Protein extracts often contain substances, such as lipids, salts and detergents, that could interfere with downstream operations,[23] such as the resolution of bands or spots on electrophoresis gels. There are several techniques to remove contaminants and 'clean up' samples, such as gel filtration (desalting), dialysis and precipitation. Gel filtration and dialysis remove contaminants by size exclusion and dilution, whereas precipitation, which involves lowering the solubility of proteins by the addition of salts, such as ammonium sulphate, or organic compounds, such as TCA, enables the precipitated proteins to be isolated from soluble contaminants.[24] In contrast to gel filtration and dialysis, precipitation allows the concentration of protein samples and is thus especially useful when a relatively dilute source of proteins, such as body fluids, is used; it also limits the number of proteins lost in a sample.[25]

4.3.1.3. Protein Assays

Following the 'clean up' stage, it is essential to determine the total protein concentration in the sample; it is important to load the appropriate concentration of protein to ensure good 2DGE images *etc.*[26] The most extensively used protein assays are the reagent-based colourimetric assays, such as the Bradford and Lowry methods, in which the sample is mixed with an aqueous reagent that results in a colour change. The intensity of the colour is measured by spectrophotometry and a standard curve of known protein concentrations against their absorbance values makes it possible to determine the concentration of proteins in an unknown sample.

The Bradford assay involves a chemical reaction between Coomassie Brilliant Blue G-250 and several amino acid residues, primarily arginine and, to a lesser extent, lysine, tyrosine, histidine, tryptophan and phenylalanine. The reaction leads to a shift in absorbance of the dye from 465 nm to 595 nm. The Bradford assay is a straightforward method and can be used with reducing agents to stabilise proteins. However, some detergents, such as sodium dodecyl sulphate (SDS), may compromise the results.[27] The Lowry assay involves two chemical reactions with the Folin phenol reagent (phosphor-molybdic-phosphotungstic acid) and copper sulphate. First, copper reacts with peptide bonds in an alkaline solution followed by reduction of the Folin reagent. The reactions produce a blue colour and the absorbance is read at 750 nm.[28] A drawback of this method is the possibility of interference by some reagents, such as barbital, chloride, citrate, TRIS (tris(hydroxymethyl)aminomethane) and EDTA (ethylenediaminetetraacetic acid). The Lowry assay has been modified for use with samples containing reducing agents and detergents; this is known as RC DC (reducing agent compatible (RC), detergent compatible (DC)) protein assay. Both the Bradford and Lowry assays have advantages and disadvantages and that need to be considered.

4.3.2. Protein Separation

The protein extract goes through a separation step to produce single proteins suitable for mass spectrometry. Techniques for separation of single proteins from complex mixtures include column chromatography, which is often used to part-purify a sample, and gel electrophoresis.

4.3.2.1. Column Chromatography

Complex mixtures of proteins can be fractionated according to their ionic properties (ion exchange), relative sizes (size exclusion or gel filtration) or their affinity with a ligand (affinity chromatography). The proteins are separated as they pass through a column packed with the appropriate porous stationary phase. Small volumes of the column eluate are then collected using a fraction collector and tested for the

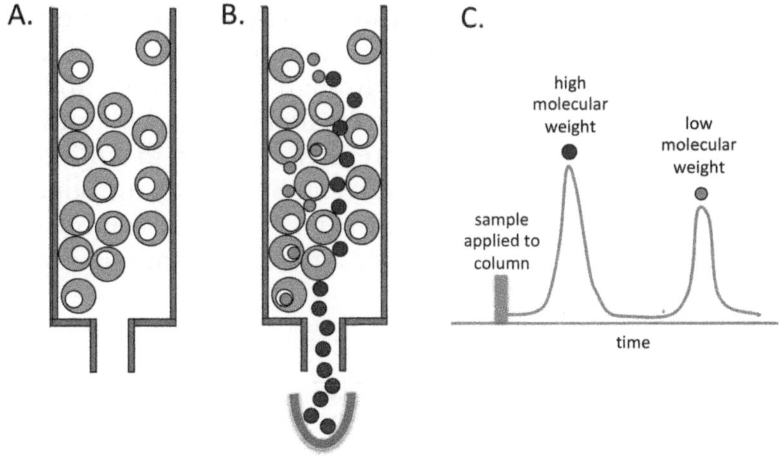

Figure 4.6 Gel filtration, or size exclusion column chromatography: (A) The column is packed with beads of cross-linked dextran (*e.g.* Sephadex ™) that consist of a network of pores large enough to trap small proteins but too small for large proteins, which are therefore excluded. (B) When the extract is applied, the large proteins are eluted first and smaller proteins are retarded. (C) The eluted fractions can be monitored by UV spectrophotometry. Reproduced from ref. 88 with permission from the Royal Society of Chemistry.

presence of protein. Figure 4.6 illustrates size exclusion (gel filtration) chromatography. The matrix for ion-exchange chromatography consists of either negatively (*e.g.* carboxymethyl groups) or positively (*e.g.* diethylaminoethyl groups) charged beads. Proteins are thus separated according to their charge. Positively charged proteins bind to negatively charged beads and negatively charged proteins bind to positively charged beads. Proteins are eluted from the column by gradually changing the buffer pH or ionic strength. In gel filtration chromatography (Figure 4.6), proteins are separated according to their size using a matrix of porous beads. Small proteins enter the pores and therefore migrate through a column at a slower rate than large proteins that cannot enter the pores. Large proteins are thus eluted first. In affinity chromatography, the matrix is covalently linked to a ligand that is specific for the protein of interest (*e.g.* an antibody). Proteins are separated according to their specific binding properties. Unbound proteins are washed out of the column, whereas proteins that bind to the ligand are retained and eluted by changing ionic strength or pH of the buffer.[29]

From the perspective of isolating a single protein species or peptide, the most useful format is high-performance liquid chromatography

(HPLC). HPLC offers more separation strategies than traditional column chromatography and has several additional advantages, including speed (the samples are forced through the columns under great pressure), the requirement for much smaller starting volumes, higher sensitivity of detection and greater resolving power. In addition, once the detector records a protein peak, it can be directly introduced into an ioniser and then a mass spectrometer (MS).[30] In contrast to gel-based methods, HPLC is suitable for automating sample separation and hence data collection.

4.4. Electrophoretic Methods

Since the early work on proteomics, electrophoresis has been the mainstay for isolating and preparing proteins for mass spectrometry. Polyacrylamide gel electrophoresis is the most widely used separation technique in proteomics laboratories. Gel-based techniques can be grouped into three categories, (1) zone electrophoresis, where separation of proteins is based on their net charge at a certain buffer pH and the sieving effect of the gel, on completion of electrophoresis the gel is stained to reveal the position of the separated proteins, (2) gel isoelectric focusing (IEF), in which native, or denatured, proteins migrate through a pH gradient generated by polyampholytes until they reach their isoelectric point, (where their net electrostatic charge is zero), and (3) SDS gel electrophoresis, where proteins are denatured, coated with sodium dodecyl sulphate and separated according to their molecular masses.[31] The most widely used method in proteomics is two-dimensional gel electrophoresis (2DGE), which combines IEF and SDS. In the first dimension, the proteins are separated according to their isoelectric points on an IEF gel strip, which is then applied to an SDS slab gel, in which the proteins are separated according to their molecular masses in the second dimension.[32] 2DGE is also a common technique in the study of bacterial pathogens to analyse and differentiate the cellular protein composition in a sample.[33]

4.4.1. Polyacrylamide Gel Isoelectric Focusing (IEF)

The IEF method for separating proteins in their native state is conducted in a gel that contains a stable pH gradient,[34] which allows proteins to be separated according to their isoelectric points. In contrast

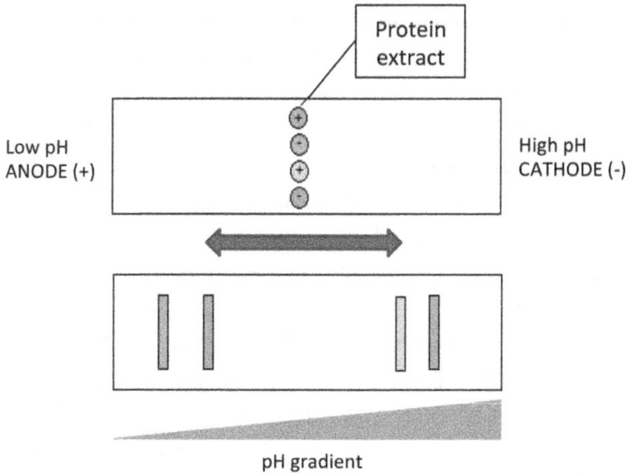

Figure 4.7 Polyacrylamide gel isoelectric focusing (IEF). The proteins in a complex mixture separate along the pH gradient according to their isoelectric points (pI). Reproduced from ref. 88 with permission from the Royal Society of Chemistry.

to zone electrophoresis where the proteins tend to diffuse as soon as they are applied to the gel, IEF is a 'concentration method', which means the operator is not constrained to apply the protein extracts in narrow zones. As IEF proceeds, all proteins carrying a net charge will migrate along the pH gradient, acidic proteins will migrate towards the anode and basic proteins towards the cathode (Figure 4.7). As individual proteins approach the pH value of their isoelectric points, they begin to lose their net charges. On reaching their isoelectric points their net charges will be zero. Once focused into narrow bands, any diffusion away from their pI (isoelectric point) will restore a charge on the proteins, which will immediately 'refocus' to their pI, thus maintaining the tight resolution of the protein bands. In the first version, gel IEF, the pH gradient was established using mixtures of soluble polyampholytes such as Ampholines® and Pharmalytes®. Polyampholytes are mixtures of small amphoteric molecules that are weak acids or weak bases that differ in their acid dissociation constants (pK_a); they resemble oligomers of amino acids.[35] In an IEF gel the polyampholytes migrate towards the electrodes according to their charges. Commercial polyampholytes were available in a wide range of pH values ranging from pH 3.5–10 to sub pH unit ranges such as pH 4.2–4.9. IEF is typically conducted in thin polyacrylamide sheets using horizontal flatbed equipment with efficient cooling. A disadvantage of soluble polyampholytes is that the pH gradient will only retain its linear shape for a limited period of time, after which cathodal drift will cause the gradient to stretch in the acidic region of the gel.[36] For this reason the use of immobilised pH gradients (IPG) has gained a foothold and is preferred for IEF procedures in proteomics studies; IPGs eliminate gradient instability whilst providing a high buffering capacity. IPGs are formed by making a gel with acrylamide monomers that are derivatised with buffering molecules that have carboxylic acid groups or tertiary amines, which differ considerably in acid dissociation constants. Thus, the pH gradient, which is covalently attached to the gel matrix, is prepared by mixing two Immobilines® such as pH 4.0 and pH 6.0 using a gradient maker so that their ratios in the gel will produce a stable linear pH gradient between pH 4.0 and pH 6.0.

4.4.2. SDS-PAGE

Sodium dodecyl sulphate polyacrylamide gel electrophoresis (SDS-PAGE), which was introduced in the late 1960s, separates proteins according to their molecular mass rather than their intrinsic charge characteristics, hence it is a useful method for estimating the molecular masses of unknown proteins.[37] SDS is an anionic detergent with chaotropic properties that binds to proteins, causing them to lose their native conformations. In the absence of SDS, proteins migrate at a rate determined by their intrinsic electrostatic charge and shape. Prior to electrophoresis, the proteins should be heat denatured and the charge on their R-groups masked through the binding of SDS. Studies have indicated that the number of SDS molecules that bind a protein is about half of the number of amino acids and that SDS binds to all proteins in a uniform manner.[38] The high levels of bound SDS completely overwhelm the intrinsic electrostatic charges, causing all of the proteins to become negatively charged; thus, they migrate towards the anode. The mobility differences observed are due principally to the sieving effect of the gel: smaller proteins will migrate fastest. Because of their constant charge : mass ratio, the electrophoretic mobility of SDS treated proteins is a function of their masses, thereby enabling the molecular mass of unknown proteins to be estimated. In most experiments, sets of marker proteins of known mass are electrophoresed together in one lane and used to produce a calibration graph of distance against log molecular weight (M_w) on which the distance migrated by the unknown protein can be plotted, revealing its apparent molecular weight.

Successful separation of proteins depends on the acrylamide and bisacrylamide concentration in the gel; this is dictated by the molecular masses of proteins to be resolved. For instance, proteins with high molecular mass should be separated in gels with lower acrylamide concentrations and *vice versa*. Proteins with a molecular mass of 60–200 kDa require 5% gels; proteins with a molecular mass of 16–70 kDa require 10% gels; proteins with a molecular mass of 12–45 kDa require 15% gels. Provided a constant ratio of acrylamide to bisacrylamide is used, as the total acrylamide concentration increases the size of the pores decreases and smaller molecules are separated more readily.[39]

In a typical SDS-PAGE experiment the gel consists of two phases: a stacking gel and a resolving gel. The stacking gel is a narrow strip with a low concentration of acrylamide (often 3%), which allows the SDS-treated proteins to concentrate in a tight band by the process of isotachophoresis before they enter the resolving gel.[40] The stacking gel usually has a pH of 6.8 (Tris–HCl), whereas the resolving gel has a pH of 8.8 (Tris–HCl). The tank, or running buffer, which connects the gels to the electrodes has a pH of 8.3 (Tris–glycine). Isotachophoresis is well understood; under the influence of an electric current, glycine ions in the running buffer migrate into the stacking gel, where the pH is reduced to 6.8, during which the negatively charged glycine loses its net charge. This decreases the migration rate of glycine in the electric field. In contrast, the chloride ions (from Tris–HCl) travel at a higher speed in the electric field. The difference in the mobility of ions results in the formation of two distinct fronts: the chloride front and the glycine front. The samples become trapped between the chloride and glycine fronts and they begin to concentrate into tight bands. When the protein samples enter the resolving gel (pH 8.8), the glycine becomes negatively charged again and starts migrating through the gel at a greater speed; leaving the SDS-coated proteins to electrophorese at their own rates.[41]

SDS-PAGE has many applications in protein chemistry, proteomics and molecular microbiology. For example, a migration pattern of protein bands specific to a bacterial strain can be obtained, the steps in a protein purification effort can be monitored, protein structure can be interrogated and even allelic variation of polymorphic proteins can be assessed, as in the case of some human mucin proteins.[42] However, some proteins are too alike in size and mobility and may not be separated on a gel in one dimension. Thus the resolving power of SDS-PAGE alone is limited. Hence 2DGE can be used to combine high-resolution charge separation using IEF in the first dimension

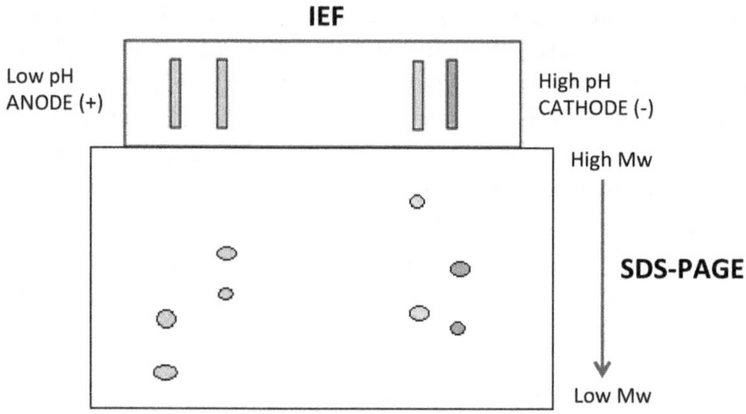

Figure 4.8 Schematic representation of two-dimensional electrophoresis (2DGE). The proteins are separated in the first dimension according to their isoelectric points and in the second dimension according to their molecular weights. Reproduced from ref. 88 with permission from the Royal Society of Chemistry.

and molecular mass separation using SDS-PAGE in the second dimension (Figure 4.8). The 2DGE procedure commences with the IEF step, which takes place in gel strips containing immobilised pH gradients (IPG).

4.4.2.1. The First Dimension: IPG Strip Selection

Dehydrated IPG strips are commercially available (*e.g.* ReadyStrip, BioRad) and have pH gradients that range from broad, *e.g.* pH 3–10, to narrow, such as pH 3.5–4.5. Typically, a broad pH range should be chosen in order to obtain an overview of proteins in the sample. It is then possible to focus in on a region of interest by narrowing the pH range. Narrow-pH-range strips provide better separation of proteins and may reveal greater numbers of spots in comparison to broad-pH-range strips. IPG strips are available in several sizes *e.g.* 24 cm, 18 cm, 13 cm, 11 cm and 7 cm. The choice depends on the size of the SDS gel to be used in the 2D separation. Generally, longer IPG strips provide greater separation and loading capacity for larger samples.

4.4.2.2. Sample Loading Methods

There are two methods of sample application: the protein sample can either be mixed with a rehydration buffer or it can be applied to the IPG strip (already rehydrated) through sample wells. The rehydration loading method is easier and it allows loading and separation of greater amounts of proteins.

4.4.2.3. IPG Strip Rehydration

The selection of the most suitable rehydration buffer is generally dictated by the solubility of the proteins in the sample. Typically a rehydration solution includes urea, detergent, such as CHAPS (3-((3-cholamidopropyl) dimethylammonio)-1-propanesulfonate), a reducing agent, such as β-mercaptoethanol or dithiothreitol (DTT), carrier ampholyte mixtures (Pharmalytes®) and an optional dye. Urea is used to solubilise and denature proteins, whereas CHAPS prevents protein aggregation. The reducing agent is included to cleave disulphide bonds and the carrier ampholyte mixtures enhance protein separation.[43] The amount of the sample applied to a rehydration buffer depends on the method chosen to visualize the results. For instance, the recommended amount of proteins for silver staining is 5–20 µg, whereas for Coomassie Blue staining, 50–100 µg of the sample is used. The proteins will not be visible if too little protein is loaded or be overstained if too much is loaded.

4.4.2.4. IPG Strip Equilibration

Following IEF, the IPG strip has to be equilibrated. This involves incubation in the SDS buffer followed by reduction and oxidation steps. A reducing agent, such as DTT, breaks disulphide bonds of proteins, this is followed by incubation with an oxidising agent, such as iodoacetamide, which alkylates reactive sulfhydryl groups and thus prevents the generation of oxidation artefacts during the SDS-PAGE step. The equilibration solution also contains urea and glycerol, which reduce electroendosmosis and improve transfer of proteins from the first to the second dimension.[44] Pre-prepared IPG strips of the desired pH range are typically used for the first dimension of 2DGE. Several strips can be focused side by side using a flatbed IEF apparatus, allowing for replicates or different samples to be focused simultaneously.

4.4.3. The Second Dimension: SDS-PAGE

Following IEF and the second equilibration step, the IPG strips are carefully removed and washed in SDS gel tank buffer (*e.g.* Tris–HCl) to remove any equilibration solution. The top of an SDS-PAGE gel is then rinsed with water to remove any acrylamide and the IPG strip carefully placed along the gel so that there is uniform contact and no air bubbles. At this stage Mw markers are placed in a well next to the IPG strip. Finally molten agarose is used to seal the top of the gel and strip. Once solidified, the gel cassette is placed in a gel tank, buffer added and the proteins are electrophoresed at a constant voltage. After the extract has been separated by 2DGE, the gel has to be stained to reveal the positions of the protein spots. Several different detection methods are available. The method chosen should be sensitive enough to detect the proteins of interest in the sample being analysed. It should also be compatible with mass spectrometry (MS), and wherever possible be non-toxic and environmentally friendly.[45] The most commonly used stains are Coomassie Blue, fluorescent and silver stains. All have advantages and disadvantages. Coomassie Blue staining is easy to perform, cheap and compatible with MS, but it has a low sensitivity of detection; fluorescence-based stains have the highest sensitivity of detection and most are suitable for use with MS, but they are relatively expensive and need to be visualised using a fluorescence imager; silver staining is more difficult to perform and shows higher sensitivity of detection than Coomassie Blue. However, MS analysis of silver-stained proteins is more problematic than the other staining methods, although several strategies are available to prepare silver stained proteins for matrix assisted laser desorption/ionisation-time of flight (MALDI-TOF) MS.

4.4.4. Specific Protein Detection: Western Blotting

Where specific antibodies are available, protein detection can be facilitated by western blotting without proceeding to mass spectrometry. Proteins are transferred electrophoretically from the gel to a nitrocellulose or polyvinylidene fluoride (PVDF) membrane. Before immunodetection using a specific antibody, the membrane is incubated with a 'blocking' solution (containing agents such as bovine serum albumin (BSA), milk or Tween 20) to prevent non-specific binding of antibody. The membrane is then incubated with a primary antibody that is specific for the protein under investigation. The primary antibody may be labelled with an enzyme, such as horseradish peroxidase (HRP) or a fluorescent dye such as rhodamine or Cy3 and used for 'direct detection'.[46] However indirect detection is generally more successful and widely used. In this case, the primary antibody is not labelled. Once the antibody, for example, a mouse-anti-human protein IgG, has reacted with the antigen, unbound antibody is washed away and the membrane is incubated with a second antibody, for example a rabbit-anti-mouse IgG, conjugated to an enzyme or fluorophore, which binds to the primary antibody. The protein spots are visualised following incubation with the enzyme substrate, leading to the production of an insoluble coloured product, or a chemiluminescence signal that can be recorded with a CCD camera. Horseradish peroxidase (HRP) and alkaline phosphatase (ALP) are both used as enzyme conjugate labels for western blot detection. Alternatively, a fluorescence imager can be used to record fluorescent labels. In each case, clearly visible bands (if a 1D gel is used) or spots (on a 2D gel) are revealed that corresponds to the position of the specific protein.[47] Western blot assays are commonly used to detect bacterial and viral proteins.

4.5. Mass Spectrometry: Identification of Bacterial Targets

Mass spectrometry is an important method for identifying proteins and has revolutionised the process of pathogen identification and strain typing in recent years. The technique enables rapid and reliable diagnosis of infectious disease through detection of specific proteins, such as ribosomal proteins or potent toxins produced by highly pathogenic bacteria. Following trypsin digestion of a protein recovered from a 2D gel, the masses of the resulting tryptic peptides are determined and compared with those generated *in silico* from protein sequence databases.

4.5.1. Matrix Assisted Laser Desorption/Ionisation-Time of Flight (MALDI-TOF)

A variety of established mass spectrometry methods are used to identify bacterial targets, but perhaps the most widely used is matrix-assisted laser desorption/ionization-time of flight (MALDI-TOF) MS. The fully automated technique is used alongside and complements conventional phenotypic or molecular methods for direct microbial identification. Examples of infectious agents commonly encountered in microbiology laboratories which are identified by MALDI-TOF MS include strains of species of the genera *Escherichia*, *Shigella* and *Salmonella*, which are associated with intestinal diseases, as well as strains of species of the genera *Staphylococcus*, *Streptococcus* and *Pseudomonas*. Bacteria which are difficult to culture, such as Mycobacteria, and species of the genera *Bartonella* and *Legionella* can also be detected by this method. The reproducibility, speed and sensitivity of analysis make MALDI-TOF MS ideal for routine use in clinical laboratories.[48]

To conduct a MALDI-TOF analysis a polyacrylamide plug containing the protein spot to be identified is cut out of the 2D gel and digested with trypsin to yield peptides for mass spectrometric analysis. Trypsin cuts the protein sequence after the basic amino acids lysine (K) and arginine (R), except when the residue is followed by a proline (P). Peptide fragments of 6–20 residues are optimal for MALDI-TOF. The peptides are then mixed with an organic compound, such as alpha-cyano-4-hydroxycinnamic acid, *i.e.* the matrix substance, and spots of matrix plus peptides are then dried at room temperature on the surface of the metal MALDI plate.[49] Alternatively, a pure bacterial culture colony obtained from a clinical isolate can be applied directly to the MALDI plate prior to matrix overlay.[48] Specimens such as urine, blood cultures and cerebrospinal fluid can also be used.[48] The plate is inserted into the mass spectrometer where a pulsed laser beam excites the crystallised matrix molecules that absorb most of the laser light energy, thereby protecting peptide mixtures from fragmentation. This leads to matrix desorption and ionization of the microbial peptides by proton transfer from the matrix. Hence sample ions that are produced in the ion source have a single positive charge (although ions carrying charges of 2+ and 3+ can also be generated) and travel into the gas phase (Figure 4.9).

The ionised peptides are accelerated in an electric field and travel through a flight tube to the detector (Figure 4.10). The time taken to

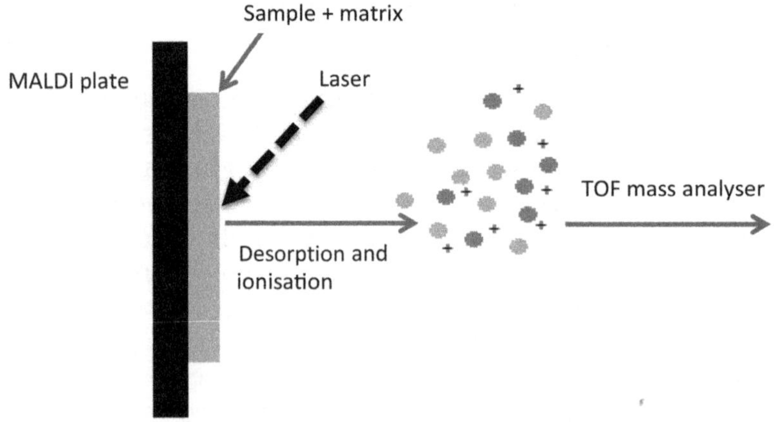

Figure 4.9 Matrix-assisted laser desorption and ionisation (MALDI). Reproduced from ref. 88 with permission from the Royal Society of Chemistry.

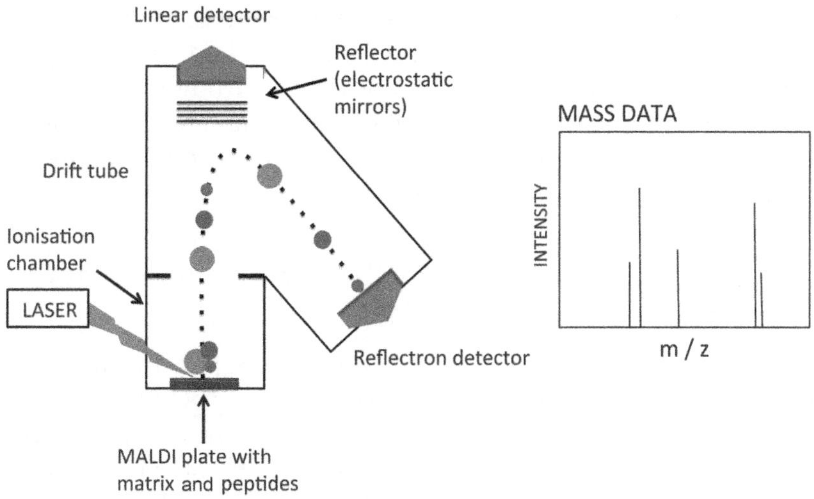

Figure 4.10 Schematic representation of MALDI time of flight (TOF) mass spectrometry. Reproduced from ref. 88 with permission from the Royal Society of Chemistry.

reach the detector depends on the mass to charge (m:z) ratio of the peptide ion, since z (the charge) is almost always 1, the m:z ratio can be taken as the mass of the peptide. Ions with larger m:z ratios travel more slowly through the flight tube than smaller ions and thus arrive at the detector behind the smaller ions. The MS gathers the mass data and generates a spectrum of ion intensity plotted against m:z ratio. The identification of proteins through the determination of peptide masses, which are searched against reference database of known proteins, is referred to as peptide-mass fingerprinting. Advantageously, in MALDI-TOF MS the matrix absorbs most of the laser energy, thus, ionisation conditions are relatively gentle, which means that peptides are not fragmented and high-molecular-weight species, even whole proteins, can be analysed. The results of bacterial identification are available within hours and the MALDI-TOF device is often connected with the laboratory information management system (LIMS) enabling immediate updating of patient records. However, because of the requirement for mixing the analyte with the matrix, the throughput is largely determined by the size of the MALDI plate. The ability to identify specific species by peptide-mass fingerprinting is limited by the spectrum of bacterial species available in the reference database.[50]

A useful modification of the MALDI method is surface enhanced laser desorption ionisation-time of flight (SELDI-TOF) MS technique that utilises a specifically modified surface protein chip to which the sample is applied. Proteins of interest in the sample are captured and bound by the chip on the basis of their physical and chemical properties. Any unbound proteins are then removed through washing steps, purifying the sample to facilitate the MS analysis. The desorption and ionisation stages proceed and a mass spectrum is generated based on the peptide mass to charge ratio.[51]

4.5.2. Liquid Chromatography Coupled Tandem Mass Spectrometry (LC-MS/MS)

In contrast to MALDI-TOF MS and SELDI-TOF MS, liquid chromatography coupled tandem mass spectrometry (LC-MS/MS) allows for 'in-line' processing of samples and is therefore amenable to automation and high-throughput operations. LC-MS/MS can be used to identify various molecules from complex biological samples, as well as to determine the amino acid sequences of peptides that are refractory to peptide mass fingerprinting.[52] *De novo* sequencing is indispensable if the genome of the organism being studied has not yet been sequenced. In addition, LC-MS/MS is used to identify and characterise posttranslational modifications by analysing the sequence of the peptide and looking for additional chemical groups such as carbohydrates (glycosylation) or phosphates (phosphorylation).

The instrumentation for LC-MS/MS consists of three elements: an HPLC, an electrospray ionisation source and a mass analyser. In a typical experiment, the protein is subjected to enzymatic cleavage to generate a mixture of peptides. The peptide fragments are then separated using HPLC and transferred to a fine capillary tube that is subjected to a high voltage, *i.e.* the electrospray ionisation source. This results in the formation of highly charged droplets consisting of both sample and a solvent. As the droplets exit the capillary, the solvent starts to evaporate so the droplets gradually decrease in size. Finally, highly charged analyte molecules are ejected from the droplets and travel into the mass analyser (Figure 4.11). The LC-MS/MS mass analyser consists of four parallel metal rods (a quadrupole) to which both constant and oscillating electric fields are applied. Peptide ions, which were produced in the ion source, migrate along the middle of the quadrupole and are separated according to their $m:z$ ratio. Only ions of particular $m:z$ values reach the detector. The separation of peptides with various $m:z$ ratios is possible by altering the electric field in the quadrupoles.[53] The system typically uses of three quadrupoles (Figure 4.12). In the first (Q1), a single specific ion (the precursor ion) is

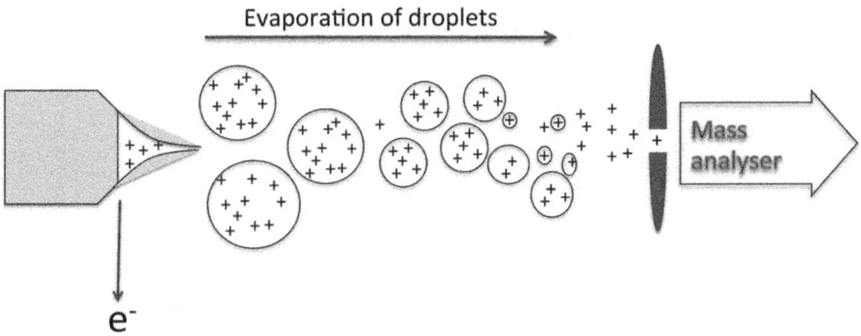

Figure 4.11 Electrospray ionisation. The large initial droplets in the aerosol evaporate and reduce in size until all the solvent has evaporated, leaving the charged analytes to form a charged gas phase before they enter the mass analyser. Reproduced from ref. 88 with permission from the Royal Society of Chemistry.

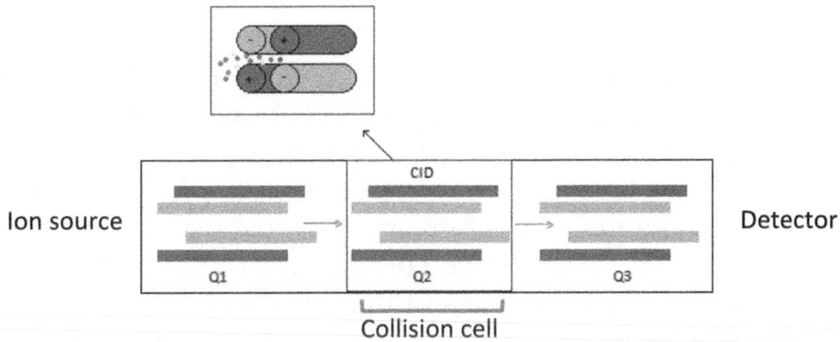

Figure 4.12 Tandem quadrupole system. Q1 and Q3 are mass spectrometers and Q2 is a collision cell.[87]

selected according to its $m:z$ ratio and travels to the second quadrupole (Q2). In Q2 the precursor ion collides with a collision gas (*e.g.* argon) in the process called collision-induced dissociation (CID) and is further fragmented. This fragmentation procedure breaks the peptide bonds, producing various fragments that differ by a single amino acid, low energy conditions during CID mean that the peptide ions fragment in predictable patterns.[52] The mass analysis of fragmented peptides takes place in the third quadrupole (Q3) to produce the mass spectrum, which is used to deduce the *de novo* amino acid sequence of the peptide or characterise posttranslational modifications. The mass spectrum is a graphical representation of the relationship between relative abundance of ions and their $m:z$ ratios. All the peaks that are presented in the mass spectrum shown in Figure 4.13 refer to an

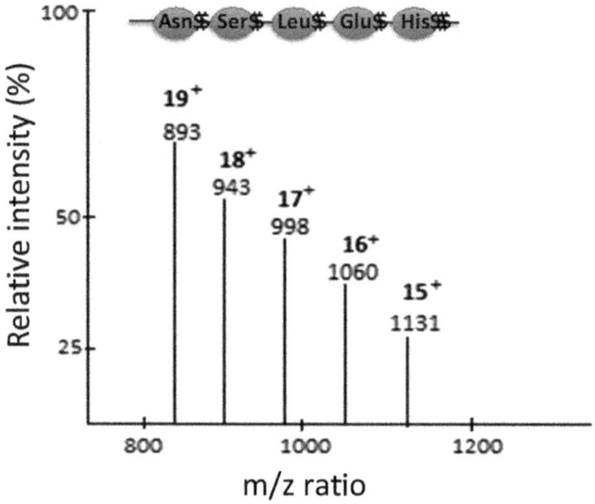

Figure 4.13 Example of an ESI mass spectrum.[87]

individual peptide carrying multiple charges. The number of charges on the protein depends on the peptide's molecular weight. By knowing the charge of each peak and its $m:z$ ratio it is possible to determine the mass of the molecule. For example, a peptide with an $m:z$ ratio of 1131 carries 15 charges. Therefore, $m = 15 \times 1131$ so the mass equals 16965 daltons (Da). 15 must be subtracted to obtain the peptide's mass, which is 16950 Da in this example. The mass of the remaining peaks can be determined in the same way. An average value for all peaks is calculated to measure the masses of peptides accurately, the more accurate the mass of the peptides the fewer the number of hits in the database. The *de novo* amino acid sequence is determined by analysing the differences in mass of several small peptides. Posttranslational modifications can also be detected by identifying an increase in the mass of a particular amino acid.

4.6. Bioinformatics Tools for Protein Identification

Mass spectrometry experiments generate mass spectra of tryptic peptides (*e.g.* from MALDI-TOF MS), mass data that can be used for *de novo* peptide sequencing or that represents posttranslational modifications (MS/MS). Following the determination of the peptide mass fingerprint or *de novo* peptide sequence, bioinformatics tools are used to identify the protein of interest. Proteins can be identified by matching a list of experimental tryptic peptide masses (the mass fingerprint) to the list of all peptide masses generated by simulated tryptic cleavage in the database.[54] The aim of database search is to find a profile match. In the real world, tryptic digestion is not always perfect and may result in partial digestion of proteins due to the failure of trypsin to cut after Lys and Arg residues, especially where there is an adjacent Pro. In addition, peptides produced by MALDI-MS are highly charged, increasing their relative masses. For these reasons it is important that the database allows for some degree of freedom in matching the fragmentation patterns. The probability of successful identification can be improved by inputting as much information as possible to the search engine. For instance, in addition to the mass of the peptide and its sequence, the molecular weight, isoelectric point and the species from which the protein was obtained can aid identification. This approach is the best suited to species whose genomes have been fully sequenced and mapped.[55]

Dedicated reference databases of microbial proteomes have been developed for final identification and classification based on mass spectra of known organisms. In the case of MALDI-TOF MS, devices often use integrated commercial manufacturer reference databases such as VITEK MS *in vitro* diagnostic (bioMérieux Inc.), which automatically compares the sample mass spectrum generated, resulting in rapid microbial identification of clinically relevant species.

SpectraBank (http://www.spectrabank.org) is an open-access database, which contains MALDI-TOF mass spectra of various bacterial species and can be used to compare spectra obtained from a sample for identification.

Mascot (http://www.matrixscience.com) is a widely used suite of programs dedicated to the analysis of MS data and can be used for identification, quantification and further characterisation of proteins using mass data generated by different MS platforms.

Expasy (Expert Protein Analysis System) is an open access Swiss Institute of Bioinformatics suite of tools for the analysis of protein sequences and structures as well as MS and 2DGE data analysis. Expasy offers a range of programs for peptide analysis. For example, the AACompIdent program is capable of identifying the protein of interest on the basis of its amino acid composition by comparing the experimental data with known amino acid compositions held on the UniProtKB database. Information such as the name of the species or keywords can narrow the database search results. TagIdent identifies the protein by comparing an unknown sequence tag (up to six residues) with proteins held on UniProtKB database (SWISS-PROT, PIR and TrEMBL). It can generate a list of candidate proteins that match the isoelectric point and molecular weight of an unknown protein. MultiIdent uses peptide mass fingerprinting data together with isoelectric point, molecular weight, amino acid composition, sequence tag and species name to identify the protein. Additionally, the HAMAP database (https://hamap.expasy.org/proteomes.html) within Expasy provides a useful resource to search for manually curated microbial proteomes.

ProFound (prowl.rockefeller.edu) is also a valuable tool, which searches protein sequence databases utilizing peptide mass fingerprinting data. It takes advantage of a Bayesian algorithm that ranks the protein sequences found in the database according to their probability of generating the peptide mass fingerprints.

Where it is not possible to match peptides in the protein sequence databases, the *de novo* amino acid sequences obtained by MS/MS can be analysed using tblastn (see Chapter 5) to search for homologues in DNA sequence databases such as GenBank. In order to further the analysis of structure and function of a given protein, laboratory studies are required. The Worldwide Protein Data Bank[56] (wwPDB) is also a useful resource to obtain structural information for experimentally determined protein structures.

4.7. Expression Proteomics

2DGE has become a widely used technique for the analysis of protein expression patterns, primarily due to its relative simplicity. There are, however, drawbacks to this method, such as poor reproducibility between replicates and low sensitivity of detection using Coomassie Blue. Furthermore, 2DGE is capable of visualising only proteins with molecular weights between 10 and 100 kDa and it generally fails to separate membrane proteins because of the formation of protein aggregates during IEF.[57] Gel-based methods circumventing some of these problems have been developed, such as difference gel electrophoresis (DIGE).

4.7.1. Difference Gel Electrophoresis (DIGE)

The DIGE method involves labelling of protein samples with fluorescent dyes before 2DGE. Frequently used fluorophores include Cy2, Cy3 or Cy5; these cyanine dyes carry an *N*-hydroxysuccinimidyl ester reactive group that forms covalent bonds with the ε-amino side group of lysine residues.[1] A maximum of three protein samples can be analysed in a DIGE experiment; treated and control samples are labelled with fluorescent dyes that have different excitation wavelengths. They are then adjusted to equalise total protein concentration, mixed together and separated by 2DGE and the individual protein spots are detected using a fluorescence scanner. If there are no differences in individual protein concentrations between the samples, excitation of the 'multiplexed' spots will yield images of intermediate colour. Any difference in expression between the samples will result in the appearance or disappearance, or gradation, of one of the fluorescent dyes in the merged image. Thus, DIGE is a straightforward method to compare specific protein levels from two or three samples in a single gel[58] that ensures improved sensitivity of detection and reproducibility in comparison with the standard 2DGE approaches.[59] The analysis of protein expression patterns is carried out using computer programs dedicated to 2D-DIGE analysis such as De-Cyder.[60] There are some limitations, however, for example 2D-DIGE is not suitable for extremely acidic or basic proteins, or hydrophobic proteins or those smaller than about 15 kDa. In addition, the fluorescent dyes are expensive compared with silver or Coomassie blue staining and there is requirement for lysine residues.[58]

4.7.2. Isotope Coded Affinity Tags (ICAT)

A useful non-gel-based method for protein expression studies is ICAT (isotope coded affinity tags). The ICAT method is capable of a greater degree of sensitivity and reproducibility than DIGE.[1] It quantifies differences in protein levels in two samples by MS.[61] ICAT is a chemical isotope labelling method that uses the ICAT reagent, which consists of three components, a reactive group specific for thiol groups *e.g.* on cysteine residues, an isotope-coded light or heavy linker and a biotin moiety for affinity chromatography.[62] As with DIGE, two samples are compared simultaneously. For ICAT, the samples are labelled with either light or heavy ICAT reagents. The two samples are then mixed together and digested with trypsin; next, a streptavidin affinity column is used to isolate tryptic peptides with the coded tags from unlabelled peptides. ICAT-labelled peptides are analysed using LC-ESI-tandem mass spectrometry[63] and the resulting mass data are analysed by searching a database such as proICAT.[64] ICAT is a powerful method because peptides with low expression levels can still be detected on 2D gels.[65] The limitations of this approach include the inability to detect acidic proteins, proteins that lack cysteine residues and posttranslational modifications that are not on cysteine-containing tryptic peptides.[66]

4.7.3. Protein Microarrays

An exciting development in protein science in the era of functional genomics has been the introduction of protein microarrays that have the potential for use in high-throughput studies of protein interactions and multiplexing for diagnostic applications. Protein microarrays allow screening of thousands of proteins in a single experiment and are analogous to DNA microarrays (see Chapters 2 and 9). The types of analysis amenable to microarrays include protein–protein, protein–DNA/RNA and protein–small molecule interactions as well as enzyme reactions, such as phosphorylation.[67] A solid support medium with arrays of specific proteins or ligands capable of binding to their complementary molecules is incubated with an extract containing the molecules of interest. A characteristic signal pattern is produced when the query molecules that are fluorescently labelled react with proteins on the array or an enzyme substrate produces a fluorescent or coloured product. The intensity of the signal is directly proportional to the amount or activity of the molecule captured on the microarray.[68] Fluorescent spots are recorded and the resulting patterns are interpreted.

There are two main types of protein microarrays: the forward-phase arrays (FPA) and reverse-phase arrays (RPA). In FPA, capture molecules, usually antibodies, are fixed on a solid surface. Each microscopic spot comprises one type of antibody. After the appropriate incubation time, only the specific analyte from the mixture of other molecules binds the antibody. The protein interaction is detected following the addition of a second labelled antibody that binds to the specific analyte. In RPA, unknown protein mixtures are immobilized on the solid phase and a fluorescently labelled specific antibody is applied. Interactions are detected only for complementary molecules.[68] The FPA is conceptually more similar to DNA microarrays, whereas the RPA is more similar to western blots.

In contrast to DNA microarrays, which have reached a high level of sophistication, protein microarray technology continues to pose many challenges. Unlike oligonucleotide probes, the behaviour of proteins arrayed on a solid phase is not nearly so predictable. For instance, proteins tend to cross-react with non-specific molecules, some classes of proteins may become unstable and change their physicochemical properties and the experimental conditions may affect protein conformation and hence its activity or binding affinity. There are also some biological challenges: many genes produce several protein variants from differentially spliced mRNAs, such variants may differ in their antigenicity and biological activity; posttranslational modifications can add an additional layer of complexity. Discriminating between various isoforms of a single protein is a major challenge for protein microarray technology. In theory, it is possible to produce specific protein binders for each query molecule, but this would be very expensive. Furthermore, in contrast to DNA microarray technology, it is not possible to amplify low-abundance proteins.[69]

Despite obvious problems associated with the analysis of proteins, microarrays have several advantages. They offer the potential advantage of being scalable, flexible, automatable and easy to perform.[69] Furthermore, microarray technology allows the operator to control environmental parameters such as pH, temperature and concentration of the labelled probe.[70] In medical microbiology, protein microarrays can be used to investigate key protein interactions with other biomolecules and for disease-state biomarker discovery, as shown by the development of a microarray for the respiratory pathogen *Mycobacterium tuberculosis*.[71]

4.8. Practical Applications

Proteomic studies can provide fundamental information on cellular and molecular physiology in health and disease by assessing changes in protein expression in normal and abnormal samples. Gel-based methods, such as 2DGE, continue to be of great value for detailed separation and visualisation of microbial proteomes under different conditions, which enables profiling and downstream analysis of specific proteins. Furthermore, the application of mass-spectrometry-based technologies in medical microbiology extends beyond high-resolution pathogen identification. The MALDI-TOF MS method has the capability to differentiate various bacterial strains for antibiotic resistance/susceptibility profiling.[72] An example is the detection of antibiotic resistance in Enterobacteriaceae such as *Escherichia coli* to β-lactams, a broad spectrum class of antibiotics. The bacterial sample is mixed with the antibiotic of interest; if the bacterium is capable of breaking down the antibiotic by hydrolysis through production of the enzyme β-lactamase, a hydrolysed form of the antibiotic will be formed. The difference in molecular weight of hydrolysed and non-hydrolysed β-lactams will produce different mass spectra, which can indicate antibiotic resistance.[73] Similarly, the LC-MS/MS method can be utilised to investigate β-lactamase activity in bacteria by analysing the differences between the antibiotic and its metabolite profile resulting from β-lactam hydrolysis, as shown for the β-lactam ampicillin.[74] LC-MS/MS has also shown many applications in microbial peptide sequencing, targeted protein identification and quantitation for various bacteria of medical importance.[75] An example is the detection and typing of the flagella protein H antigen in the common food-borne pathogen *Salmonella enterica*.[76]

The applicability of MS in clinical situations can therefore lead to improved patient treatment; assist in the management of an outbreak and also aid with epidemiological surveillance of infectious diseases on a large scale. The speed, sensitivity and reproducibility of mass-spectrometry-based techniques[75] also make them an attractive choice compared with traditional methods for microbial identification. Examples of proteins from major bacterial pathogens which have been identified using MS techniques are given in Table 4.2.

Proteomic studies could pave the way to the identification of novel antimicrobial drug targets which, in turn, could accelerate new drug development.[84] The discovery of drug targets is dependent on the analysis of the structure and function of a given protein. By understanding the 3D structure of a protein, small molecules can be designed, for example, to inhibit the active site of an enzyme. Signal transduction pathways and cellular processes facilitated by enzymes could thus be blocked. This is an important approach, which is often exploited in the drug-development process and it is expected that the power of proteomics could contribute to the development of both conventional and personalized drugs. For example, knowledge of the influenza A proteome has enabled constituent proteins implicated in the virus life cycle to be evaluated as potential drug targets. Typically, an *in silico* approach is used to target regions of structural or functional importance from structural data or molecular models. Molecular docking or virtual screening techniques can be applied to determine the quality of fit of the drug molecule to various sites in the protein in order to alter or disable its functionality, as shown for the influenza A PB2 protein.[85]

The identification of alterations in the structure and function of the protein, as well as changes in protein expression patterns and protein–protein interactions are important for diagnosis and prognosis of infectious disease. The high-throughput identification and quantification of proteins in healthy and diseased tissues and fluids promises to improve *in vitro* diagnosis on a broad front.

Table 4.2 Examples of microbial proteins/targets detected using MS-based techniques for major bacterial pathogens including *Escherichia*, *Salmonella*, *Clostridium*, *Staphylococcus*, *Streptococcus* and *Bacillus*.

Species	Target Protein	ID Method	Notes	Reference
Escherichia coli	Flagella H-antigen	LC-MS/MS, trypsin digested	Flagella isolated from pure culture. H-antigen typed from various *E. coli* reference strains and clinical isolates.	77
Shiga-toxin-producing *Escherichia coli* (STEC)	Shiga toxin 2 (Stx2)	MALDI-TOF/TOF MS and MS/MS, undigested	Stx2 subtypes identified in unfractionated cell lysates from 26 reference and environmental STEC strains.	78
Salmonella enterica	Subspecies-specific biomarker peptides	LC separation, MALDI-TOF/TOF MS, trypsin digested	Biomarkers identified for: *S. enterica* subsp. *arizonae S. enterica* subsp. *enterica S. enterica* subsp. *houtenae*	79
Clostridium difficile	TcdA and TcdB toxins	1D SDS-PAGE – LC-MS/MS trypsin digested; and gel free, LC-MS/MS, digested	Toxins identified in purified samples and commercial *C. difficile* culture filtrate.	80
Staphylococcus aureus	Delta-toxin	MALDI-TOF MS	Whole-cell analysis, toxin detected in over 100 clinical *S. aureus* isolates	81
Group A *Streptococcus* (GAS)	Various ribosomal proteins (biomarker peptides)	MALDI-TOF MS	Whole-cell analysis biomarkers detected in *S. pyogenes* strain ATCC 700294, plus eight other GAS clinical isolates.	82
Bacillus anthracis	Anthrax lethal factor toxin cleaved peptides	MALDI-TOF MS, undigested	Sample preparation required for purification and peptide hydrolysis of substrate. Toxin activity detected in clinical samples.	83

4.9. Summary and Conclusions

Proteomics is the high-throughput and automated study of protein expression, structure and function; it offers unparalleled opportunities to delve into protein transactions and the cellular mechanisms they underpin. Proteomics techniques are currently employed in clinical laboratories to investigate microbial pathogens through their protein profiles and assist with early diagnosis. Compared with genomics and transcriptomics, proteomics addresses biology directly and can provide insights into dynamic aspects of pathology and cell physiology. A proteomics experiment frequently sets out to identify all the proteins expressed in a cell type, tissue or fluid under defined conditions. The first step in the workflow, sample preparation, is crucial to the success of the experiment. Thereafter, separation of the sample by electrophoresis or liquid chromatography enables the isolation of specific proteins that can be identified by immunodetection or, more frequently, mass spectrometry. Thus, a typical proteomic workflow might include separation by 2DGE, gel staining and image analysis followed by spot excision and tryptic digestion. The masses of the tryptic peptides are measured by MALDI-TOF mass spectrometry and the resulting peptide mass spectra searched in reference databases to find sufficient matches to identify the protein. In recent years, the MALDI-TOF MS method in particular has shown great success in the area of microbial identification.[50] If protein identification is unsuccessful using MALDI, peptide ions can be selected and fragmented using MS/MS to produce *de novo* amino acid sequences. The aim of proteome profiling and identification is to catalogue all the proteins in a particular sample and optionally analyse posttranslational modifications. In contrast, expression proteomics aims to identify proteins that are expressed at different levels in test samples and controls. The test samples may be from individuals affected by a specific pathology or from experiments to monitor the effect of pharmaceuticals *etc.* Whilst standard 2DGE can be used for expression proteomics, there are drawbacks, such as poor reproducibility and the need for multiple replicates of the gels. Two very effective methods have been devised to overcome these drawbacks: gel-based DIGE and liquid-chromatography-based ICAT, both are widely used for expression studies. A third category of proteomics, sometimes called functional proteomics, is being developed to assess protein–protein and protein–ligand interactions. Interaction proteomics is currently based on the relatively new protein microarray technology. It is anticipated that protein microarrays, particularly in the reverse-phase format, will provide valuable insights into many aspects of pathology and cell physiology through the understanding of the networks of protein interactions that underline cellular responses. Integrating computational tools and bioinformatics approaches in proteomic investigations can also improve understanding of protein biology.

The application of proteomics technologies is yet to be fully realised, though there are indications that there will be marked clinical benefits. Through expression proteomic studies it is hoped that new biomarkers, which are defined as 'measurable characteristics that reflect physiological, pharmacological or disease processes' will be discovered and employed to improve clinical outcomes for many types of pathology.[86] These could include diagnostic and prognostic markers, as well as those that could predict responses to drug therapy *etc.* Overall, there is great potential for proteomics approaches in the study of microorganisms, and from the current stage in the cycle of value creation; it is likely that proteomics technologies will have significantly increased uses in clinical diagnostics in the near future.

Bioinformatics sources cited in the text

Expasy. (2010). Protein Analysis System. Available from www.expasy.org.

Mascot. (2013). Identification and characterisation with MS/MS data. Available from: www.matrixscience.com.

Nijmegen Proteomics Facility. (2013). Maldi-TOF MS. Available from: www.proteomicsnijmegen.nl

ProFound. (2012). Protein identification and characterisation database. Available from: prowl.rockefeller.edu.

SpectraBank. (2012). SpectraBank: An open access tool for rapid microbial identification by MALDI-TOF MS fingerprinting. Available from: http://www.spectrabank.org

References

1. K. Chandramouli and P. Y. Qian, *Hum. Genomics Proteomics*, 2009, **1**, 1.
2. V. C. Wasinger, S. J. Cordwell, A. Cerpa-Polijak, J. X. Yan, A. A. Gooley, M. R. Wilkins, M. W. Duncan, R. Harris, K. L. Williams and I. Humphery-Smith, *Electrophoresis*, 1995, **16**, 1090.
3. M. R. Wilkins, K. L. Williams, R. D. Appel and D. F. Hochstrasser, *Proteome Research: New Frontiers in Functional Genomics*, Berlin Heidelberg, Springer Verlag, 1997.
4. F. Del Chierico, A. Petrucca, P. Vernocchi, G. Bracaglia, E. Fiscarelli, P. Bernaschi, M. Muraca, A. Urbani and L. Putignani, *J. Proteomics*, 2014, **97**, 69.
5. M. A. Alaoui-Jamali and Y. J. Xu, *J. Zhejiang Univ., Sci., B*, 2006, **7**, 411.
6. S. J. Fey and P. M. Larsen, *Curr. Opin. Chem. Biol.*, 2001, **5**, 26.
7. D. W. Nebert, G. Zhang and E. S. Vesell, *Annu. Rev. Pharmacol. Toxicol.*, 2013, **53**, 355.
8. J. M. Johnson, J. Castle, P. Garrett-Engele, Z. Kan, P. M. Loerch, C. D. Armour, R. Santos, E. E. Schadt, R. Stoughton and D. D. Shoemaker, *Science*, 2003, **302**, 2141.

9. J. Godovac-Zimmermann and L. R. Brown, *Mass Spectrom. Rev.*, 2001, **20**, 1.
10. E. V. Koonin and Y. I. Wolf, *Nucleic Acids Res.*, 2008, **36**, 6688.
11. R. L. Gundry, M. Y. White, C. I. Murray, L. A. Kane, Q. Fu, B. A. Stanley and J. E. Van Eyk, *Curr. Protoc. Mol. Biol.*, 2009, **10.25**, 1.
12. Y.-P. Ho and P. M. Reddy, *Clin. Chem.*, 2010, **56**, 525.
13. S. D. Patterson and R. H. Aebersold, *Nat. Genet.*, 2003, **33**, 311.
14. M. Boyle and K. Senior, *Human Biology*, Collins, London, 2008.
15. W. K. Purves, D. Sadava, G. H. Orians and H. C. Heller, *Life: The Science of Biology*, Gordonsville: W.H. Freeman and Company, 2004.
16. P. G. Hitchen and A. Dell, *Microbiology*, 2006, **152**, 1575.
17. F. H. Martini and J. L. Nath, *Fundamentals of Anatomy and Physiology*, Pearson Education, Inc, San Francisco, 2009.
18. V. Thongboonkerd, *Proteomics of Human Body Fluids: Principles, Methods, and Applications*, Humana Press, New York, 2007.
19. B. Alberts, A. Johnson, J. Lewis, M. Raff, K. Roberts and P. Walter, *Molecular Biology of the Cell*, Garland Science, New York. 2002.
20. M. Thein, G. Sauer, N. Paramasivam, I. Grin and D. Linke, *J. Proteome Res.*, 2010, **9**, 6135.
21. J. L. Cole, J. W. Lary, T. P. Moody and T. M. Laue, *Methods Cell Biol.*, 2008, **84**, 143.
22. B. Alberts, D. Bray, K. Hopkin, A. Johnson, J. Lewis, M. Raff, K. Roberts and P. Walter, *Essential Cell Biology*, Garland Science, New York, 2010.
23. L. Jiang, L. He and M. Fountoulakis, *J. Chromatogr. A*, 2004, **1023**, 317.
24. A. Bodzon-Kulakowska, A. Bierczynska-Krzysik, T. Dylag, A. Drabik, P. Suder, M. Noga, J. Jarzebinska and J. Silberring, *J. Chromatogr. B.*, 2006, **849**, 1.
25. D. Martins de Souza, B. M. Oliveira, E. Castro-Dias, F. V. Winck, R. S. Horiuchi, P. A. Baldasso, H. T. Caetano, N. K. D. Pires, S. Marangoni and J. C. Novello, *Briefings Funct. Genomics Proteomics*, 2008, **7**, 312.
26. T. Berkelman, *Mol. Biol.*, 2008, **424**, 43.
27. M. M. Bradford, *Anal. Biochem.*, 1976, **72**, 248.
28. O. H. Lowry, N. J. Rosebrough, A. L. Farr and R. J. Randell, *J. Biol. Chem.*, 1951, **193**, 265.
29. W. W. Ward and G. Swiatek, *Curr. Anal. Chem.*, 2009, **5**, 1.
30. J. J. Pitt, *Clin. Biochem. Rev.*, 2009, **30**, 19.
31. F. Chevalier, *Proteome Sci.*, 2010, **8**, 1.
32. A. Gorg, W. Weiss and M. J. Dunn, *Proteomics*, 2004, **4**, 3665.
33. S. O. T. Curreem, R. M. Watt, S. K. P. Lau and P. C. Y. Woo, *Protein Cell*, 2012, **3**, 346.
34. H. J. Isaaq and T. D. Veenstra, *BioTechniques*, 2008, **44**, 697.
35. R. J. Braun, N. Kinkl, M. Beer and M. Ueffing, *Anal. Bioanal. Chem.*, 2007, **389**, 1033.
36. T. Rabilloud and C. Lelong, *J. Proteomics*, 2011, **74**, 1829.
37. A. L. Shapiro, E. Viñuela and J. V. Maizel, *Biochem. Biophys. Res. Commun.*, 1967, **28**, 815.
38. P. G. Righetti, A. V. Stoyanov and M. Y. Zhukov, *The Proteome Revisited: Theory and Practice of All Relevant Electrophoretic Steps*, Elsevier Science, Amsterdam, 2001.
39. S. R. Gallagher, *Curr. Protoc. Mol. Biol.*, 2006, **1**, 10.2.1.
40. D. M. Hawcroft, *Electrophoresis*, Oxford University Press, New York, 1997.
41. Science Squared, *How SDS-PAGE Works*, Available from: http://bitesizebio.com/articles/how-sds-page-works/, 2013.
42. D. M. Swallow, S. Gendler, B. Griffiths, G. Corney, J. Taylor-Papadimitriou and M. E. Bramwell, *Nature*, 1987, **328**, 82.
43. T. Berkelman and T. Stenstedt, *2-D Electrophoresis: Principles and Methods*, Amersham Biosciences AB, Sweden, Uppsala, 2001.
44. A. Gorg, O. Drews, C. Luck, F. Weiland and W. Weiss, *Electrophoresis*, 2009, **30**, 1.
45. R. Westermeier and R. Marouga, *Biosci. Rep.*, 2005, **25**, 19.

46. T. Mahmood and P. C. Yang, *N. Am. J. Med. Sci.*, 2012, **4**, 429.
47. C. Moore, *Introduction to Western Blotting*, MorphoSys, Oxford, 2009.
48. Public Health England (2016), Matrix-assisted laser desorption/ionisation - time of flight mass spectrometry (MALDI-TOF MS) test, *UK Standards for Microbiology Investigations*, TP 40 Issue 1, https://www.gov.uk/guidance/uk-standards-for-microbiology-investigations-smi-quality-and-consistency-in-clinical-laboratories, (accessed March 2018).
49. J. Gobom, M. Schuerenberg, M. Mueller, D. Theiss, H. Lehrach and E. Nordhoff, *Anal. Chem.*, 2001, **73**, 434.
50. A. Bizzini and G. Greub, *Clin. Microbiol. Infect.*, 2010, **16**, 1614.
51. D. Emerson, L. Agulto, H. Liu and L. Liu, *Bioscience*, 2008, **58**, 925.
52. C. Delahunty and J. R. Yates, *Methods*, 2005, **35**, 248.
53. W. Li, J. Zhang and F. L. S. Tse, *Handbook of LC-MS Bioanalysis: Best Practices, Experimental Protocols and Regulations*, John Wiley & Sons, Inc., New Jersey, 2013.
54. R. Aebersold and M. Mann, *Nature*, 2003, **422**, 198.
55. J. Xiong, *Essential Bioinformatics*, Cambridge University Press, Cambridge, 2006.
56. H. Berman, K. Henrick and H. Nakamura, *Nat. Struct. Mol. Biol.*, 2003, **10**, 980.
57. F. Delom and E. Chevet, *Proteome Sci.*, 2006, **4**, 1.
58. G. Van den Bergh and L. Arckens, *Expert Rev. Proteomics*, 2005, **2**, 243.
59. G. Zhou, H. Li, D. DeCamp, S. Chen, H. Shu, Y. Gong, M. Flaig, J. W. Gillespie, N. Hu, P. R. Taylor, M. R. Emmert-Buck, L. A. Liotta, E. F. Petricoin III and Y. Zhao, *Mol. Cell. Proteomics*, 2002, **1**, 117.
60. R. Marouga, S. David and E. Hawkins, *Anal. Bioanal. Chem.*, 2005, **382**, 669.
61. A. S. Haqqani, J. F. Kelly and D. B. Stanimirovic, *Methods Mol. Biol.*, 2008, **439**, 225.
62. L. Monteoliva and J. P. Albar, *Briefings Funct. Genomics Proteomics*, 2004, **3**, 220.
63. M. Sethuraman, M. E. McComb, H. Huang, S. Huang, T. Heibeck, C. E. Costello and R. A. Cohen, *J. Proteome Res.*, 2004, **3**, 1228.
64. X. J. Li, H. Zhang, J. A. Ranish and R. Aebersold, *Anal. Chem.*, 2003, **75**, 6648.
65. F. Schmidt, S. Donahoe, K. Hagens, J. Mattow, U. E. Schaible, S. H. Kaufmann, R. Aebersold and P. R. Jungblut, *Mol. Cell. Proteomics*, 2004, **3**, 24.
66. H. Zhou, J. A. Ranish, J. D. Watts and R. Aebersold, *Nat. Biotechnol.*, 2002, **20**, 512.
67. A. J. Nijdam, M. R. Zianni, E. E. Herderick, M. M.-C. Cheng, J. R. Prosperi, F. A. Robertson, E. F. Petricoin III, L. A. Liotta and M. Ferrari, *J. Proteome Res.*, 2009, **8**, 1247.
68. L. A. Liotta, V. Espina, A. I. Mehta, V. Calvert, K. Rosenblatt, D. Geho, P. J. Munson, L. Young, J. Wulfkuhle and E. F. Petricoin III, *Cancer Cell*, 2003, **3**, 317.
69. P. Cutler, *Proteomics*, 2003, **3**, 3.
70. G. MacBeath, *Nat. Biotechnol.*, 2001, **19**, 828.
71. J. Deng, L. Bi, L. Zhou, S. Guo, J. Fleming, H. Jiang, Y. Zhou, J. Gu, Q. Zhong, Z. Wang, Z. Liu, R. Deng, J. Gao, T. Chen, W. Li, J. Wang, X. Wang, H. Li, F. Ge, G. Zhu, H. Zhang, J. Gu, F. Wu, Z. Zhang, D. Wang, H. Hang, Y. Li, L. Cheng, X. He, S. Tao and X.-E. Zhang, *Cell Rep.*, 2014, **9**, 2317.
72. J. Hrabák, E. Chudácková and R. Walková, *Clin. Microbiol. Rev.*, 2013, **26**, 103.
73. J. S. Jung, C. Popp, K. Sparbier, C. Lange, M. Kostrzewa and S. Schubert, *J. Clin. Microbiol.*, 2014, **52**, 924.
74. A. Grundt, P. Findeisen, T. Miethke, E. Jäger, P. Ahmad-Nejad and M. Neumaier, *J. Clin. Microbiol.*, 2012, **50**, 1727.
75. K. Cheng, H. Chui, L. Domish, D. Hernandez and G. Wang, *Proteomics: Clin. Appl.*, 2016, **10**, 346.
76. K. Cheng, A. Sloan, J. Meakin, S. McCorrister, M. Jerome, G. Westmacott, M. Drebot, C. Nadon, J. D. Knox and G. Wang, *J. Clin. Microbiol.*, 2014, **52**, 2189.
77. K. Cheng, M. Drebot, J. McCrea, L. Peterson, D. Lee, S. McCorrister, R. Nickel, A. Gerbasi, A. Sloan, D. Janella, G. Van Domselaar, D. Beniac, T. Booth, L. Chui, H. Tabor, G. Westmacott, M. Gilmour and G. Wang, *PLoS One*, 2013, **8**, e57339.

78. C. K. Fagerquist, W. J. Zaragoza, O. Sultan, N. Woo, B. Quiñones, M. B. Cooley and R. E. Mandrell, *Appl. Environ. Microbiol.*, 2014, **80**, 2928.
79. M.-T. Gekenidis, P. Studer, S. Wüthrich, R. Brunisholz and D. Drissner, *Appl. Environ. Microbiol.*, 2014, **80**, 4234.
80. H. Moura, R. R. Terilli, A. R. Woolfitt, Y. M. Williamson, G. Wagner, T. A. Blake, M. I. Solano and J. R. Barr, *Int. J. Proteomics*, 2013, **2013**, 1.
81. J. Gagnaire, O. Dauwalder, S. Boisset, D. Khau, A.-M. Freydière, F. Ader, M. Bes, G. Lina, A. Tristan, M.-E. Reverdy, A. Marchand, T. Geissmann, Y. Benito, G. Durand, J.-P. Charrier, J. Etienne, M. Welker, A. Van Belkum and F. Vandenesch, *PLoS One*, 2012, 7, e40660.
82. H. Moura, A. R. Woolfitt, M. G. Carvalho, A. Pavlopoulos, L. M. Teixeira, G. A. Satten and J. R. Barr, *FEMS Immunol. Med. Microbiol.*, 2008, **53**, 333.
83. M. Gallegos-Candela, A. E. Boyer, A. R. Woolfitt, J. Brumlow, R. C. Lins, C. P. Quinn, A. R. Hoffmaster, G. Meister and J. R. Barr, *Anal. Biochem.*, 2018, **543**, 97.
84. M. Abhilash, *Internet J. Genomics Proteomics*, 2008, **4**, 1.
85. H. Patel and A. Kukol, *Virology*, 2017, **509**, 112.
86. P. C. Guest, M. G. Gottschalk and S. Bahn, *Genome Med.*, 2013, **5**, 17.
87. C. S. Ho, C. W. K. Lam, M. H. M. Chan, R. C. K. Cheung, L. K. Law, L. C. W. Lit, K. F. Ng, M. W. M. Suen and H. L. Tai, *Clin. Biochem. Rev.*, 2003, **24**, 3.
88. R. Rapley and D. Whitehouse, *Molecular Biology and Biotechnology*, Royal Society of Chemistry, Cambridge, 6, 2014.

5 Introductory Bioinformatics

David Whitehouse* and Ralph Rapley

School of Life and Medical Sciences, University of Hertfordshire, College Lane, Hatfield, Hertfordshire, AL10 9AB, United Kingdom
*E-mail: d.whitehouse@herts.ac.uk

5.1. Introduction

Bioinformatics is a major branch of computational biology that has developed in parallel with advances in molecular biology and biochemistry. For several decades increasing amounts of biological data have been produced and deposited in computerised databases. The maintenance of such databases and the utilisation of software tools to curate and interrogate the data form the basis of bioinformatics. Bioinformatics aims to increase our knowledge and understanding of biology and facilitate scientific discovery, ranging from sequencing genomes, mapping genes and relating genotypes to phenotypes to the structure and function of proteins and small molecules. Bioinformatics encompasses an array of multidisciplinary approaches and is pivotal to all the areas of bioscience that are associated with massive datasets, such as the Human Genome Project (HGP) sequencing endeavour and the current NHS England 100 000 genomes project. Since completion of the HGP in 2003 bioinformatics has paved the way to analysing genes and gene products in numerous organisms where the genomes have been fully sequenced. This chapter commences with an introduction to databases and

Genomics and Clinical Diagnostics
Edited by David Whitehouse and Ralph Rapley

Published by the Royal Society of Chemistry, www.rsc.org

some of the software tools used in basic research and medical genetics. There follows a discussion of clinically relevant databases and some practical applications of bioinformatics in the laboratory, such as PCR primer design.

5.2. Databases

The development of computerised databases that store, organise and permit the analysis of biological information is crucial to the understanding of many biological processes.[1] Raw biological and chemical data are collected and organised into structured and curated databases that enable scientists to access and exploit the information for both basic and applied projects. In molecular biology, curated databases containing massive datasets of nucleic and amino acid sequence provide unparalleled opportunities for discovery and for characterizing genomes, genes and proteins in health and disease.

Many biological databases have been developed for projects such as genome sequencing, gene mapping, protein structure display and analysis, drug design, proteomic analysis and a range of other biomedical scenarios.[2] Biological databases can be broadly classified as primary, secondary and specialised, depending on the nature of their contents (Figure 5.1). Primary databases contain raw sequence or structural data, examples include GenBank and the Protein Data Bank (PDB). Secondary databases contain information

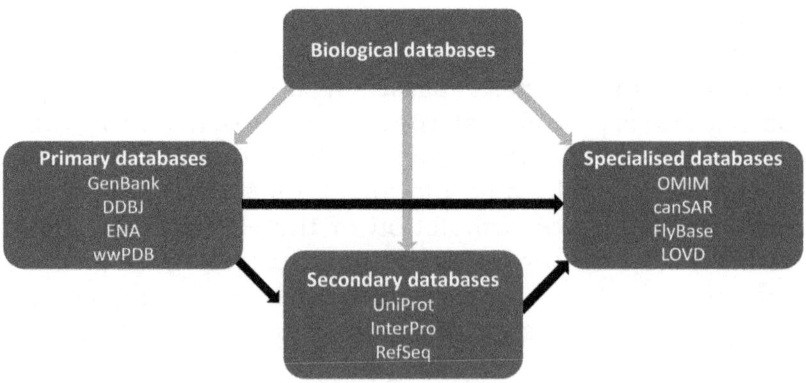

Figure 5.1 An overview diagram of the classes of databases referred to in the text. Information in the primary sequence databases is used to create secondary databases. Specialised databases utilize varying degrees of information from primary and secondary databases along with extensive curation.

derived from the primary databases. These are more sophisticated and require computational processing of raw data. Examples of secondary databases include UniProt and Pfam. These protein resource databases provide a range of tools including, for example, a means for allocating amino acid sequences to protein families based on conserved features. Specialised databases are archival resources related to specific research areas, genetic diseases or organisms such as LOVD (human disease genes), HIV Databases or FlyBase (*Drosophila* genetics). Examples of databases with their respective URLs are presented in Table 5.1.

Table 5.1 Some bioinformatic resources referred to in the text that are freely available over the world wide web.

Resource	About	URL
BLAST	A comprehensive suite of programs to find regions of similarity between nucleotide or protein sequences	https://blast.ncbi.nlm.nih.gov/Blast.cgi
GenBank	Genetic sequence database: annotated collection of all publically available DNA sequences	https://www.ncbi.nlm.nih.gov/genbank/
UniProt	Comprehensive resource of protein sequence and functional information	https://www.uniprot.org
RefSeq	Non-redundant annotated set of sequences, including genomic DNA, transcripts and proteins	https://www.ncbi.nlm.nih.gov/refseq/
wwPDB	Repository of information about the 3D structures of proteins, nucleic acids and complex assemblies	http://www.wwpdb.org
InterPro	Database and tools to classify protein sequences into families and predict the presence of important domains and sites	http://www.ebi.ac.uk/interpro/search/sequence-search
Webcutter2	Restriction enzyme recognition site mapping tool	http://www.firstmarket.com/cutter/cut2.html
GeneFisher2	Interactive program for designing degenerate primers	https://bibiserv.cebitec.uni-bielefeld.de/sadr2/primerdesign/genefisher2/index.html
LOVD	Comprehensive database of human disease genes and variants	https://databases.lovd.nl/shared/diseases
FlyBase	*Drosophila melanogaster* genomics resource	http://flybase.org
HIV databases	Comprehensive data on HIV genetic sequences and epitopes plus analytical tools	https://www.hiv.lanl.gov/content/index
CanSAR	An integrated knowledge-base to enable translational research in cancer	https://cansar.icr.ac.uk
COSMIC	Catalogue of somatic mutations in cancer	https://cancer.sanger.ac.uk/cosmic
OMIM	Authoritative text-based compendium of human genes and genetic phenotypes that is updated daily	https://www.omim.org

5.2.1. Nucleotide Sequence Databases

There are three main publically available nucleotide sequence databases on the world wide web: GenBank, European Nucleotide Archive (ENA) and DNA Data Bank of Japan (DDBJ). Nucleotide sequence data generated throughout the world are submitted to one of these public repositories. The three databases closely collaborate and share information on a regular basis; together they constitute the International Nucleotide Sequence Database Collaboration (INSC). This means that the same nucleotide sequence is available to a user regardless of the database used. Despite the fact that these three databases contain identical raw sequence data, the output is presented in slightly different formats. The format of GenBank records is described in this chapter.

5.2.1.1. GenBank

Along with the ENA and DDBJ, GenBank provides open access to annotated nucleotide sequences. The number of sequences continues to increase; by April 2018 more than 8×10^8 sequences had been deposited. The continued growth of the INSC databases is mainly due to the expansion of whole-genome shotgun sequencing (WGS, see Chapter 2) data submissions.

GenBank, accessed at the National Centre for Biotechnology Information (NCBI), stores sequences that are obtained from species representing the major taxonomic groupings.[3] Due to the enormous volume of data, GenBank files are separated into "organismal" and "functional" divisions according to the origin and properties of the sequence. Examples of "sequence" divisions include primates (PRI), other mammals (MAM), plant and fungi (PLN), viral (VRL), bacterial (BCT), high-throughput genomic (HTG) and expressed sequence tags (EST). There are currently 16 such divisions. Many secondary and specialised biological databases contain links to the relevant records in GenBank, which itself contains important basic information for each sequence record.

A GenBank record has a flat file format and is divided into three sections: header, features and the nucleotide sequence; wherever the record is a coding sequence (CDS) *i.e.* a cDNA, the translated amino acid sequence is also shown. The header section consists of a number of different subsections. The first line in the header section is the LOCUS number, which is a unique identifier of the sequence. There is also information on sequence length, molecule type, GenBank division and publication date. DEFINITION refers to a brief description of the sequence. The next line contains the ACCESSION field, which provides a code number for a sequence record. VERSION and GENE INDEX (GI) present the number of sequence revisions. If the sequence data has been changed then the accession number remains the same, but version and GI values are modified. The KEYWORDS field lists words and phrases associated with the sequence. SOURCE indicates the origin of the organism from which the data were derived. There is information on publications related to a given sequence in the REFERENCE subsection. The features section contains the SOURCE field that determines sequence length, scientific name of the source of the organism and the taxonomy identification number. GenBank records for cDNA sequence also contain the CDS field, which refers to the boundaries of the gene's coding sequence. This subsection also contains information on the start and stop codons and an amino acid translation. The final section is ORIGIN. Although this is generally left blank in current sequence records, in the older records the ORIGIN points to the start of the sequence as determined experimentally. The nucleotide sequence itself begins on the line immediately below ORIGIN.

5.2.1.2. RefSeq

The reference sequence (RefSeq) database at NCBI holds comprehensive information for a particular gene, such as DNA, RNA transcripts (including alternatively spliced transcripts from the same gene), non-coding RNA, variations, conserved domains and protein sequences as well as bibliographic information, aliases, cross-ref erences *etc.* RefSeq is a non-redundant database constructed from GenBank archives. Individual elements of information and data are all provided through a single record with a unique accession number. This information has been developed for a number of model organisms and is a frequent starting point for many projects and investigations. In contrast to GenBank, RefSeq is limited to storing information from well-characterised organisms and consequently contains far fewer records. Because the genetic and functional annotations contained within the database are curated manually and linked together, they provide a rich source of informative data. This is one of the unique features of RefSeq in comparison with databases such as GenBank, where the information may contain redundant sequences.

5.2.2. Bioinformatics Data Formats

In a standard GenBank record, the nucleotide sequence is displayed in GenBank format; each row starts with the nucleotide number and the sequence itself is displayed in six blocks of ten nucleotides. Whilst visually helpful, this format cannot be used by data processing software; it must first be converted to a simple text format. There are many alternative formats, but the most useful and widely used is the FASTA format. A FASTA file contains some brief information about the ID of the sequence and the sequence itself in simple text format. The FASTA format is suitable for use with a wide range of software tools and databases. The first line of a FASTA file always starts with a ">" followed by the sequence identifier *e.g.* the gene name. Comments can also be made on the first line or on different lines, provided they are preceded by ";". These start-of-line symbols (> and ;) are ignored by most software. The actual sequence is written in single letter code starting on the next line. Conveniently, GenBank nucleotide sequence records can be instantly converted to FASTA format by clicking a link beneath the record title at the top of the page. In next-generation sequencing (NGS; see Chapter 3), FASTQ file formats are used by some sequencing systems, such as the Illumina method. Developed by the Sanger Institute, the FASTQ format consists of four lines. The first line is a description (similar to FASTA format), the second is the raw sequence, the third line is also a sequence identifier and the final line is a quality value for the raw sequence that starts in the second line. There are a number of different file formats used for NGS at various stages of the data processing which are either text based, such as FASTA, FASTQ, SAM and VCF, or binary-based, such as BAM and BCF (see Section 5.7.2.).

5.3. Bioinformatics Resources: BLAST, PubMed and OMIM

Three key bioinformatics resources for medical biotechnology are the program suite BLAST and the specialised text archives PubMed and OMIM. All three are freely available and accessible *via* the NCBI website. BLAST and PubMed interact directly with other NCBI resources.

(1) Basic Local Alignment Search Tool (BLAST) is an important bioinformatic tool that allows nucleotide and amino acid databases to be searched in order to identify sequences that match a sequence of interest.[4] The sequence of interest (the query sequence) is compared

with all sequences (subject sequences) in the selected database and the statistical significance of the best matches is calculated. BLAST can be used to identify members of gene families, investigate evolutionary relationships and potentially investigate functional relationships. There are five main BLAST programs that can be selected according to the purpose of the investigation and the type of query sequence to be analysed:

- blastn compares a nucleotide sequence to nucleotide database,
- blastp compares a protein sequence to protein database,
- blastx compares translated nucleotide sequence to protein database,
- tblastn compares protein sequence to translated nucleotide database,
- tblastx compares translated nucleotide sequence to translated nucleotide database.

Once the appropriate BLAST program has been selected, a page is returned that allows the query sequence, in FASTA format, to be pasted into the query window and the program is run.

BLAST outputs both graphical and text representations of the results (Figure 5.2). The graphical view includes colour coded horizontal lines that correspond to the number of sequence matches in the search and their quality (red: most related – high quality; green and blue: moderately related; black: unrelated – low quality). The length of the coloured line indicates the region of similarity between sequences. Each individual line is linked to a 'summary' pop-up window that shows the name of the matching sequence, the alignment (bit) score and the accession number *etc.* Below the graphical view is a table of entries matching the query sequence. Entries are sorted according to expect (*E*)-values in declining order. The alignment output is arranged into header, statistics and alignment subsections. The header subsection provides the name of the sequence and its accession (ID) number. The statistics subsection contains the range (the extent of the sequence alignment), bit score, *E*-value, identity, gap existence and strand (plus/plus or plus/minus). The bit score is a statistical measure of the sequence similarity, where higher numbers correspond to higher similarity. The *E*-value describes the number of hits that can be expected by chance when searching a database of a particular size. It is therefore a probability measure of the statistical significance of the match, the lower the *E*-value the more significant the result. The alignment subsection presents the aligned sequences and usefully permits visual inspection of the data. There is also a link to the NCBI Genome Data Viewer (GDV), a browser that

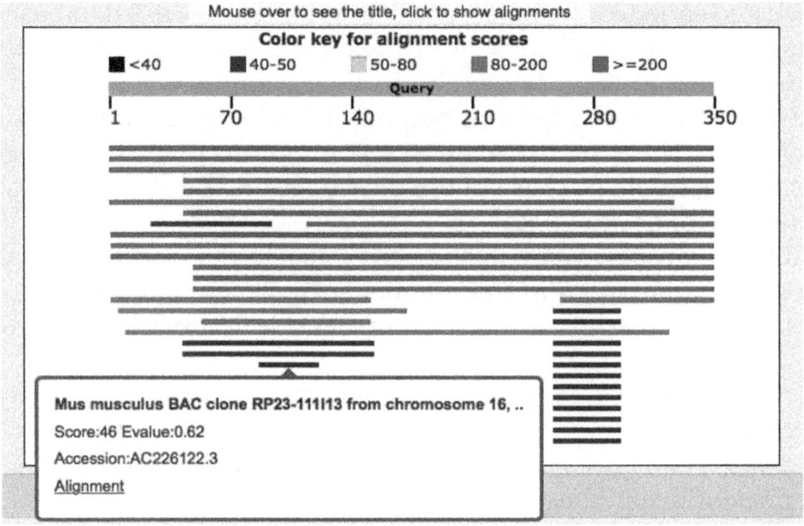

Figure 5.2 Graphical view of BLAST scores. A genomic DNA query sequence matching *Homo sapiens* chromosome 1, GRCh38.p12 primary assembly NCBI reference sequence: NC_000001.11–phosphoglucomutase 1 was used to search the non-redundant database using blastn (nucleotide blast). The output shown is a summary of the sequences in the database that aligned to the query sequence. The horizontal bars are colour coded to indicate the extent and the quality of the alignments where red bars are the most similar (best matches) and black are the least. Selecting individual horizontal bars displays a text box containing information about the alignment and a link to the alignment itself in the BLAST alignment output.

enables the scientist to explore and analyse the sequence of interest in the context of more than 580 eukaryotic genome assemblies from the NCBI Reference Sequence collection. Additionally a set of specialised programs can be accessed from the BLAST homepage, such as Primer-BLAST, which is an aid to PCR primer design and CD-search, which looks for conserved domains in protein and nucleotide sequences.

(2) PubMed, is a free search engine accessible from the NCBI website, and one of the most powerful general text resources for biomedical research. Currently it contains more than 28 million references from peer-reviewed biomedical literature. PubMed takes advantage of the MeSH (Medical Subject Headings) system, which is effectively a thesaurus that helps to search for relevant literature efficiently and accurately. PubMed is constantly being updated by MEDLINE, peer-reviewed scientific journals and online books. It is an ideal literature resource for any scientist entering a

new field of biomedicine, or to keep abreast with the current literature. PubMed results can be set to display abstracts or just titles. In some cases, a results page links directly to publically accessible full-text journal article. A related, but independently managed publically available searchable text resource is PubMed Central (PMC, at NCBI), which allows the user free access to nearly five million full-text articles. In addition to specific search results, both PubMed and PMC display links to similar articles and a search history. PMC also has links to some molecular resources.

(3) OMIM (Online Mendelian Inheritance in Man) is a long-established highly specialised and comprehensive text database that contains information about all known human Mendelian disorders, hence, it is an indispensable resource for clinical geneticists and medical genetics researchers. The online version, which gradually replaced Victor McKusick's authoritative tome "Mendelian Inheritance in Man", first appeared on the internet in 1987. More than 15 000 genes are catalogued in OMIM, which can be searched by key words, such as gene names or disorders, or by advanced search criteria, including creation date, genomic region or chromosome number. Each full-text and referenced report contains a description of the disorder and/or gene, along with clinical features emphasising the correlation of genotype and phenotype. Allelic variants are listed along with cytogenetic locations and genomic mapping information. OMIM (maintained by Johns Hopkins University) is updated daily and contains an array of external links to several information resources including GenBank records.

5.4. Protein Databases: UniProt and wwPDB

The flow of information between nucleotide and protein databases is essential for understanding of biological processes in health and disease, such as signalling and metabolic pathways, drug design and analysing the results of proteomics experiments (see Chapter 4), as well as protein structure and evolution. The Human Genome Project has led to a multitude of translated amino acid sequences that are publically available on dedicated searchable protein databases, such as UniProt and Pfam. These 'primary sequence' databases are complemented by structural databases, such as the Protein Data Bank (PDB), that contain experimentally obtained structures from studies on purified proteins using X-ray crystallography or nuclear magnetic resonance (NMR).

5.4.1. UniProt

UniProt (the Universal Protein resource) is a major central repository of protein data and information which is jointly managed by the European Bioinformatics Institute (EBI), the Protein Information Resource (PIR) and the Swiss Institute of Bioinformatics (SIB).[5] The resource was created in 2002 with the integration of SWISS-PROT, TrEMBL (Translated European Molecular Biology Laboratory Nucleotide Sequence Database) and PIR. UniProt is a publically accessible resource that incorporates numerous features and external links for each sequence record. In addition to the protein sequence records that include genetic variants and isoforms, UniProt provides comprehensive information on protein function, interactions, secondary structure, metabolic pathways *etc.* The UniProt homepage displays ten main UniProt resources: UniProtKB, three further databases: UniRef, UniParc and Proteomes and "Supporting data" consisting of literature, taxonomy, subcellular locations, cross references, diseases and keywords.

UniProtKB (UniProt Knowledgebase), the primary source of protein sequence information is divided into two sections: SWISS-PROT, which contains high-quality, manually annotated sequences that are reviewed by expert curators before they are entered on the database, and TrEMBL, which contains sequences that are annotated automatically and not reviewed. In May 2018 there were 557 275 SWISS-PROT records and 114 759 640 TrEMBL records. A text search (such as a protein or gene name) or a peptide search of UniProtKB returns a table of results that indicates whether a protein sequence is from SWISS-PROT (curated – gold icon) or TrEMBL (automatically annotated – blue icon) along with its entry number and name, protein name, gene name, organism and length in amino acids. Selecting the entry number opens the UniProtKB record page, which contains tabs to tools, including BLAST, for conducting sequence-based searches along the lines described above, FASTA, which converts UniProt sequences to the FASTA format, and Align.

Selecting Align opens Clustal Omega,[6] an EBI application for creating and analysing multiple sequence alignments (MSAs) between two or more amino acid sequences in the case of the UniProt version. (Clustal Omega accessed directly from the EBI tools website can align protein, DNA or RNA sequences). Analysis of protein MSAs is fundamental to the study of evolutionary relationships in protein families and the identification of functionally important motifs. Phylogenetic analysis and homology searching within protein superfamilies, when combined with laboratory findings, has helped to identify hitherto uncharacterised genes that share consensus structural features and functional motifs, such as co-factor binding and active sites, such as within the phosphoglucomutase superfamily.[7] Sequences to be aligned can be obtained directly from UniProtKB by entering the UniProt identifiers into the query window or by entering single-letter-code protein sequences in FASTA format. When the program has finished running the MSA is displayed. The identifier of each sequence in the alignment is in the left margin and beneath the bottom row of amino acids there are symbols that indicate the degree of similarity for the amino acids in each column: an asterisk (*) indicates 'identity' that is all the residues in the column are fully conserved; a colon (:) indicates that some residues are different but all are highly conserved in that they share similar properties; a period (.) indicates weakly conserved residues with more divergent properties. Absence of a symbol indicates lack of similarity. Gaps created by sequences of different lengths are indicated by hyphens in the row with a missing residue(s) (Figure 5.3). Clustal Omega provides a comprehensive list of colour-coded annotations for a range of features, such as glycosylation sites, active site, leader sequence, propeptide, secondary structure features *etc.* When selected, the relevant amino acids or groups of amino acids on the alignments are highlighted in the appropriate colour. Similar colour-coded highlighting can be applied to indicate amino acid properties, such as charge status, hydrophobicity, aliphatic residues, polar residues *etc.* Optionally, a phylogenetic tree for the alignment can be displayed. Clustal Omega, which can align up to 4000 amino acid (or nucleotide) sequences, is reported to have an accuracy equivalent to ClustalW2, also available from the EBI website.

On each UniProt record page, in addition to the sequence, copious information is presented on the protein's function, posttranslational modifications, domain structure, binding sites, subcellular location, similarities to other protein sequences, protein variants associated with disease *etc.* In addition, records contain multiple links to other biological databases, such as OMIM (Online Mendelian Inheritance in Man; see above), DrugBank (a drug and drug target database), 3D structure databases where the structure has been solved and also to specialised databases such as COSMIC (Catalogue of Somatic Mutations in Cancer; see Section 5.5), where appropriate. Selecting the UniProtKB "Feature Viewer" opens a new page with a 'selectable' horizontal bar diagram of the protein that visually indicates the position and details of molecular features, such as disulphide bridges, glycosylation sites, leader sequence and propeptide, as well as genetic

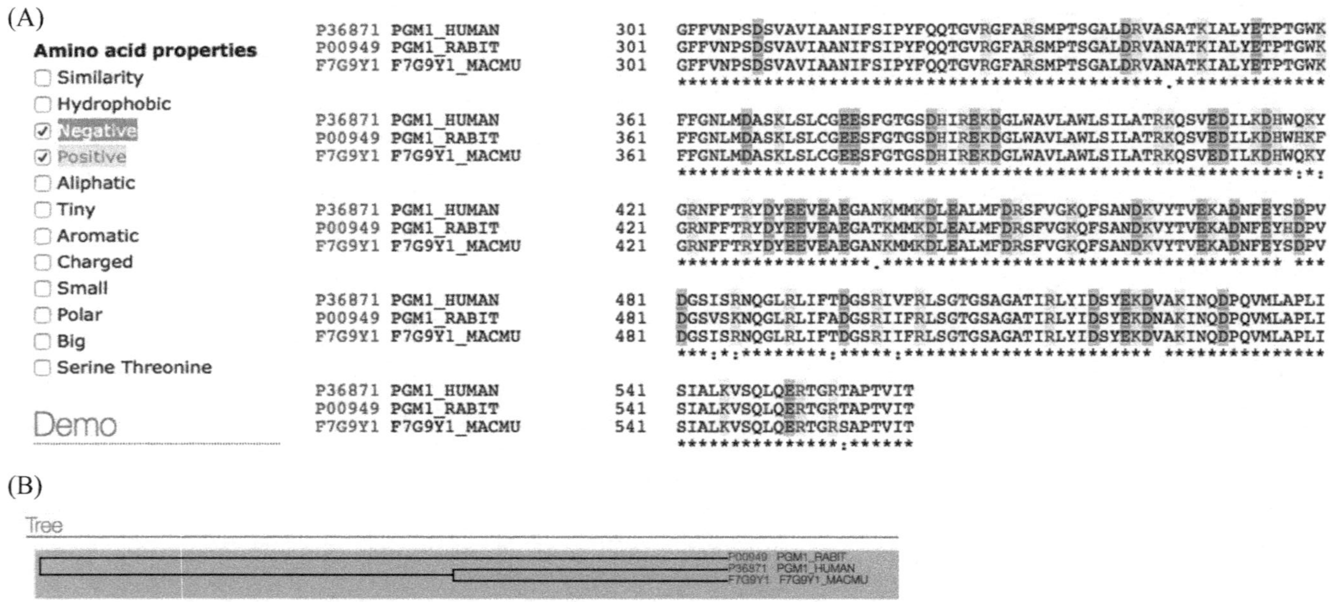

Figure 5.3 Output from Clustal Omega accessed using the Align program subset of the UniProt integrated database. (A) An extract of a simple multiple sequence alignment. Three phosphoglucomutase 1 amino acid sequences (Human, Rabbit and Macaque) were aligned: the fourth row indicates the quality of the alignment where an asterisk indicates that all the residues in the column above are identical; a colon indicates that some residues are different but all are highly conserved. An absence of a symbol indicates lack of similarity. The alignment has been modified to display basic (R and K) and acidic (D and E) residues. An evolutionary tree diagram is also displayed (B).

variants that are colour-coded to indicate their biological and/or pathological consequence. Two of the subsidiary databases UniRef, which provides non-redundant clustered sets of protein sequences from the UniProtKB and selected UniParc records, and Proteome, which provides protein sequences from fully characterised genomes, increase the speed of searches for particular classes of sequences; UniParc contains sequences of proteins that are no longer represented in UniProtKb along with information on their history. Most usefully each one of the ten UniProt databases (*i.e.* including the six "Supporting data" resources: literature, taxonomy, subcellular locations, cross references, diseases and keywords) can act as entry points for text searches to rapidly focus in on features of interest. The majority of scientific data in UniProtKB is collected from direct submissions of amino acid and nucleotide sequences, publications and review articles.

5.4.2. Structural Databases: World Wide Protein Data Bank (wwPDB)

Up to this point all databases and bioinformatic applications and tools have been concerned with storing, mining and analysing massive datasets of linear DNA or amino acid sequences, the latter typically obtained by translating the former. In contrast wwPDB was set up in 2003 to provide a single publically available database archive of three-dimensional structures of macromolecules, including proteins and nucleic acids. The data files in wwPDB represent the outputs of X-ray crystallography and nuclear magnetic resonance (NMR) studies of purified macromolecules and molecular complexes. Highly sophisticated software enables the 'flat file' data to be rendered as three-dimensional models, which can be labelled, displayed in different formats, such as ribbon diagrams or space filling models, and manipulated. There are four members of wwPDB: Protein Data Bank in Europe (PDBe), Protein Data Bank of Japan (PDBj), Research Collaboratory for Structural Bioinformatics Protein Data Bank (RCSB PDB) and Biological Magnetic Resonance Bank (BMRB). These four databanks host the same PDB archive, but as with the INSC, each member presents the output in slightly different formats with their own views of the primary coordinate data, as well as tools and resources.[2] Chen *et al.* have reported that in 2015, there were 113 971 biological macromolecular structures in the wwPDB database including 37 049 distinct protein sequences, 30 099 structures of human sequences and 8096 nucleic-acid-containing structures.[2]

5.5. Specialised Databases: Disease-Related

There are a number of specialised curated database repositories of nucleotide and amino acid sequences derived from clinical samples with specific pathologies. Two examples are COSMIC and cBioPortal.

COSMIC (Catalogue of Somatic Mutations in Cancer) takes and integrates data from scientific publications, metadata and targeted gene-screening panels.[10] Data is also derived from large scale-genome screens, other databases, such as TCGA (The Cancer Genome Atlas) and ICGC (International Cancer Genome Consortium), as well as the cancer genome project at the Sanger Institute in Cambridge, UK (Figure 5.4). These databases are becoming very important in the new era of clinical genomics, where rapid accumulation of clinical patient DNA can be analysed and compared with normal or reference sequences and disease sequences. This provides annotated information for 265 tumour suppressor genes and 276 oncogenes regarding copy number variations (CNV) and tissue distribution.

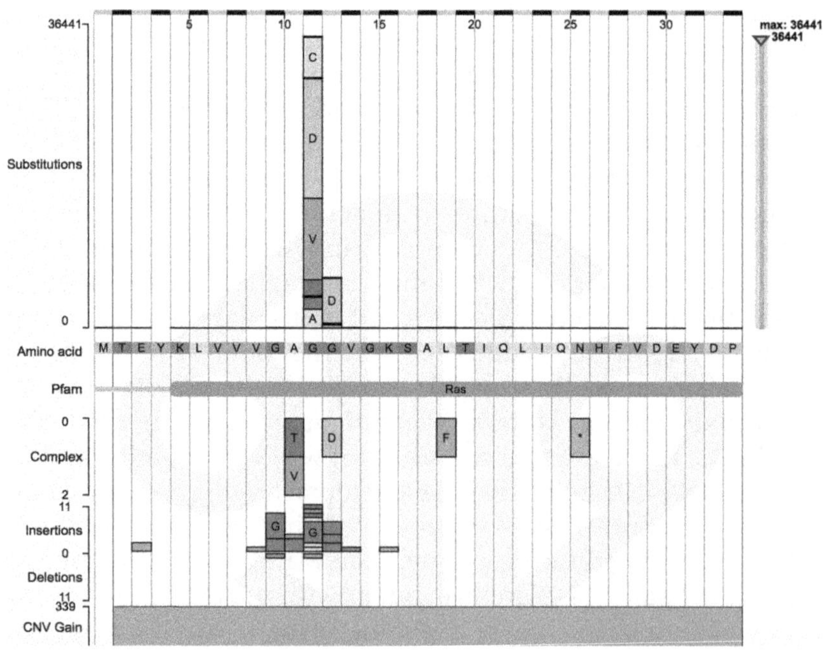

Figure 5.4 Output from a typical search using the biomarker KRAS in the COSMIC database search system. The graphical representation shows the gene expanded to the region of the major mutations between codon 12 and codon 61. Amino acids and nucleotides are shown and hovering a mouse over the region where the mutations appear gives information regarding the mutation.

cBioPortal is another important resource for cancer genomic information, again, much of it derived from TCGA.[11] Information on cancer types, clinical data, copy number data, mRNA and miRNA expression data as well as other data types are available. My Cancer Genome further provides detail on disease, cancer genes, mutations and variants and related clinical trial data and potential therapeutic interventions. Further databases have been developed and are evolving to provide information on, for example, micro RNA (miRNA) expression, methylation status and cancer cell line data for certain cancer types.

In addition to TCGA and COSMIC, and in a similar manner to My Cancer Genome, further databases have been developed at various institutions worldwide to evaluate the clinical significance of gene mutations and integrate therapeutic options. TARGET (Tumor Alterations Relevant for Genomics-Driven Therapy) developed at the Broad Institute and PCT (Personalised Cancer Therapy) hosted by the MD Anderson Cancer Center in Texas, USA attempt to draw information from various sources to provide an insight into genomic alterations with respect to current conventional treatments and more personalized therapeutic options. This includes the problems of sensitivity and resistance to certain drug treatments in certain tumour types. Importantly, the development of databases of information for prognostics, based on biomarker profiling and clinical outcomes is also a rapidly evolving area that will be of major benefit in the future and further datasets relating to this can be found in Gene Expression Omnibus (GEO) and the International Cancer Genome Consortium (ICGC).

Further drug discovery and drug target resources have also been developed over the last few years in an approach to integrate genomic, chemical, pharmacological, structural and clinical trials information among others types of data. One such resource is CanSAR (cancer and structure–activity relationship), which takes data from sources such as COSMIC and TCGA as well as protein data from UniProt,[5] 3D data from PDB and expression analysis information and drug and clinical trials information.[8] Using this integrated resource it is possible to assess whether a protein of interest is druggable, that is, where its activity can be modulated, through structural analysis of binding areas, cavities and physicochemical data using data from PDB with a small chemical molecule or drug. The selected data are used in an algorithm which indicates the likely drugability of the various structural regions of the protein under investigation.

In addition to databases of information regarding mutations and clinical data there are also specific resources relating to disease and even genes themselves. Typical examples are METABRIC (Molecular Taxonomy of Breast Cancer International Consortium) which contains large datasets of clinical information on breast cancers and outcomes in addition to corresponding CNV, SNP (single-nucleotide polymorphisms), CNA (copy number aberrations) and somatic mutations. There are also a number of gene-specific databases relating to oncogenes and tumour suppressor genes, and one of the most comprehensive is the IARC (International Agency for Research on Cancer) p53 database. This is a comprehensive database containing information on all aspects of the p53 tumour suppressor gene, including genotype and phenotype information, germline and sporadic mutations, functional and structural data, as well as useful information on cell culture and mouse models in which the gene has been investigated.

5.6. Secondary and Specialised Databases: Other Applications

Advances in database management and the development of bespoke software have led to bioinformatic utilities that address a range of specialised disciplines. For instance, amino acid and nucleotide sequence databases and tools have been developed to identify diverse members of protein families and superfamilies through protein signatures, such as conserved functional and structural domains, motifs and other features. InterPro is a publically available curated database which consists of annotated protein sequences and aims to predict important sites and domains in query protein and nucleic acid sequences. The InterPro resource integrates data and predictive models from many protein signature databases, each with somewhat differing properties, such as Pfam, PROSITE, PRINTS and PANTHER.[9] The resource offers functional annotation, sometimes on a residue-by-residue basis, which allows users to identify and characterise their query sequence and predict the presence of potential functional and structural sequences. There are numerous specialised annotated databases including those that aim to infer evolutionary relationships of species with fully sequenced genomes or proteomes, such as HOGENOM, eggNOG and OMA. Dedicated databases have been developed to address a multitude of diverse topics, including 2-D gel analysis, post translational modifications, metabolic pathways and the reactome, polymorphisms and mutations, proteomics, ontology *etc.*[2]

5.7. Laboratory Bioinformatics: Probe and Primer Design

Bioinformatic tools play an important role in the research laboratory. For example, oligonucleotide primer design for PCR (Chapter 1). The specificity of PCR lies in the design of the two oligonucleotide primers. These have to not only be complementary to sequences flanking the target DNA but also must not be self-complementary or bind each other to form dimers, since both prevent DNA amplification. They also have to be matched in their GC content and have similar annealing temperatures. The increasing use of bioinformatics resources, such as Primer3 and NCBI Pick Primers, in the design of primers makes the design and the selection of reaction conditions much more straightforward. These resources allow the sequences to be amplified, primer length, product size, GC content, *etc.* to be input and, following analysis, provide a choice of matched primer sequences. Indeed, the initial selection and design of primers without the aid of bioinformatics would now be unnecessarily time-consuming. There are numerous bioinformatics resources for designing primers for use in PCR. Most allow some degree of flexibility and are completely automated, although there may be times when manual design is needed or a combination of both is required.

5.7.1. PCR Primer Design Considerations

It is of crucial importance to carefully design PCR primers in order for PCR to be effective. The location of primers defines the target DNA sequence to be amplified. The amplification of PCR product is dependent on the annealing temperature of the primers. The two primers used in PCR should be specific (complementary) to the target DNA and have similar annealing temperatures. In addition, it is important to ensure that primers are not complementary with each other and will not bind to each other forming 'primer dimers' or hairpins. There are a number of automated primer design resources that are freely available on the web, such as Primer3, Primer-Blast and Genefisher among others. It is usually possible to enter the accession number representing the sequence, although raw sequence information and those in the FASTA format are usually acceptable. Primer3 output provides sets of primer combinations and the first set also indicates where the primers bind to the target sequence in the graphic with the sequence. Pick Primers at NCBI uses a similar Primer 3 algorithm.

It is possible to check the primers amplify the correct sequence and generate a theoretical PCR by using *in silico* PCR. A number of programs are available for this and the most widely used one is at UCSC. It is possible to identify potential restriction sites in the amplicon using the Webcutter resource. This allows the identification of any restriction endonuclease target sites that could be used to confirm the identity of the amplicon in the laboratory following a successful amplification. If an endonuclease target site can be identified the resulting theoretical products can be determined using standard gel electrophoresis and staining. Comparison of the Webcutter data and gel electrophoresis size data may be used to confirm a correctly amplified PCR.

5.7.2. NGS and Bioinformatic Workflow Systems

The generation of large data sets, such as those produced with next-generation sequencing methods, demands complex algorithms and programs, very powerful computer processing, and high-capacity data centres.[13] In many cases the amount of data produced by NGS systems is very large and can exceed gigabytes of information for each run. One of the challenges therefore lies in the computation and storage of the data. Most manufacturers of NGS systems have their own proprietary software and workflows, such as the Illumina BaseSpace Sequence Hub, however, a number of publically available resources have also been developed to address and support the handling and analysis of NGS data, including GenePattern, BioQueue and Galaxy.[12] The latter is a commonly used system maintained by the Galaxy project in the US at Johns Hopkins University and deals with the management, integration and assembly of large genomic data sets from NGS experiments. Written in Java and Python and running on the Linux operating system it allows an accessible means of data analysis for those with little knowledge of coding, through a graphical user interface (Figure 5.5). It is also open-access, allowing developers to provide further enhancements and updates as necessary. A number of specific resources for analysing biological information may be obtained by accessing the Galaxy ToolShed. There are over 4000 available tools and it is possible to build a sequence of tools into a pipeline. This can be achieved by building pipelines onto a so-called workflow canvas using a drag-and-drop interface built into Galaxy. A large number of these workflows have now been generated, shared with other users and also published to provide a wide range of accessible resources ranging from genomics (*e.g.* MGEscan) to metabolomics (*e.g.* Galaxy-M).

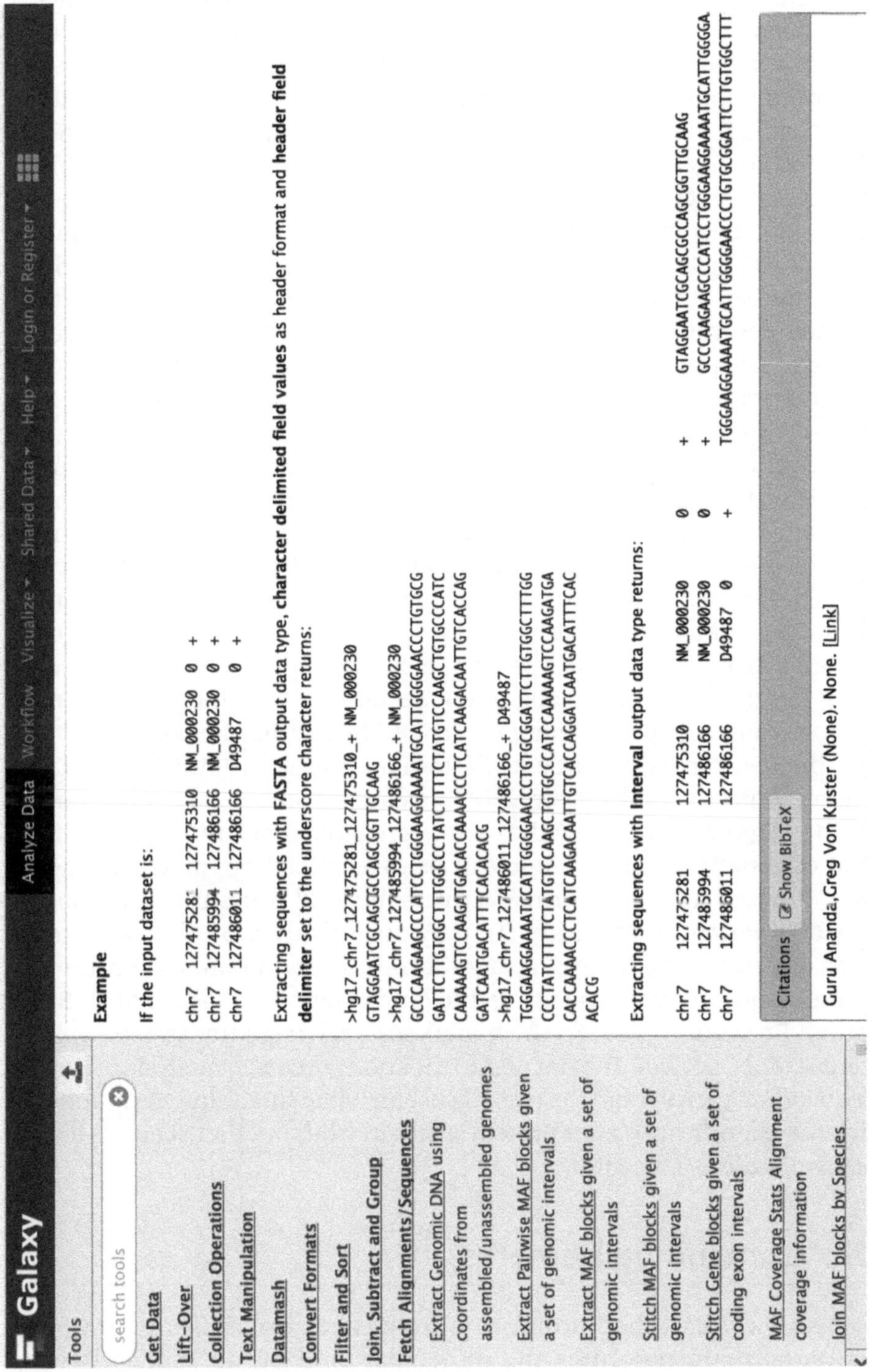

Figure 5.5 A typical screen from the Galaxy interface showing genomic data analysis in the FASTA output data type format with a sequence from chromosome 17.

NGS analysis is used with increasing frequency for clinical analysis of patient samples. This is especially true for initiatives such as the Genomic England 100 000 genomes project, of which one arm is aimed at examining patient DNA for particular features or structural variants (variant calling) in biomarker genes, such as point mutations, inversions, indels and translocations in particular cancers. In order to achieve this it is necessary to compare those patient DNA sequences to a human reference sequence, such as the Genome Reference Consortium human genome (GRChg38). Thus, in a typical clinical bioinformatics NGS workflow a number of tools and filters are used to produce the FASTQ data sequence files from the sequence from the machine. In a number of cases numerous samples are analysed at the same time and therefore have to be de-multiplexed to individual sequences. A number of manufacturers have their own proprietary workflow and software for all aspects of an NGS pipeline. In the primary analysis phase the raw sequence has to be cleaned or trimmed to remove artifacts, such as adapter sequences, using a program such as Trimmomatic.[14] In the secondary analysis phase the sequence is aligned to a reference genome with a further program such as BWA or Bowtie2 which provides a sequence alignment map (SAM) output file. The SAM file is a text-based format, which contains a number of data fields including a header section including the reference sequence, library information and an alignment section of 11 columns of alignment position, mapping quality *etc.* The SAM file may be converted to a compressed binary alignment file format, termed a BAM file. GATK (Genome Analysis Tool Kit) or SAMTools are resources that then compare the aligned sequence with the reference sequence for the identification of genetic variants. These can be determined by variant calling, which employs a further format termed the VCF (variant call format) that stores variant information. Tertiary analysis then involves annotation to determine the biological significance of any variants, validation, interpretation and reporting using Integrative Genomics Viewer or UCSC Genome Browser. RNA-Seq, which analyses and maps transcripts, uses different tools, such as TopHat and STAR and exome-Seq analysis of protein encoding genes, this is especially useful when analysing mutations, which may significantly alter protein structure and function, it also is less costly and utilises less data.

5.8. Concluding Remarks

It is clearly evident that bioinformatics has become fully integrated into mainstream genomics and molecular biology. Indeed, without computerised and annotated databases, sophisticated software and widespread public access to the bioinformatics resources, progress with the Human Genome Project and subsequent developments would have held back by many years. A strong message to anyone starting to use bioinformatics in the research laboratory is that the database resources and tools necessary to analyse DNA and amino acid sequences are freely available and generally simple to use (Table 5.1).

On a broader level, as a result of developments in next-generation sequencing and projects such as the Genomics England 100 000 Genomes initiative, vast amounts of human variation data are being amassed and a proportion of these might be expected to be of clinical significance. There is the opportunity to achieve a deep understanding of the significance of the genetic variants in complexes diseases, such as cancers, heart disease and diabetes. Pathological variants will offer new avenues to therapeutics, but both pathological and associated variants will contribute diagnostic biomarkers. Arrays of variant biomarkers could be used as entry points for personalised medicine, where markers might be expected to assist with diagnosis and prognosis as well as determining treatment options. It may be expected that bioinformatics will develop further into computational systems biology to yield insights into intracellular pathways and control networks. The relevance of human molecular variation could then be assessed for its pathological, therapeutic and diagnostic significance.

References

1. P. Hogeweg, *PLoS Comput. Biol.*, 2011, 7, e1002021.
2. C. Chen, H. Huang and C. H. Wu, *Methods Mol. Biol.*, 2017, **1558**, 3.
3. D. A. Benson, I. Karsch-Mizrachi, D. J. Lipman, J. Ostell and D. L. Wheeler, *Nucleic Acids Res.*, 2004, **32**, D23.
4. S. F. Altschul, T. L. Madden, A. A. Schäffer, J. Zhang, Z. Zhang, W. Miller and D. J. Lipman, *Nucleic Acids Res.*, 1997, **25**, 3389.
5. The UniProt Consortium, *Nucleic Acids Res.*, 2017, **45**, D158.
6. F. Sievers and D. G. Higgins, *Methods Mol. Biol.*, 2014, **1079**, 105.
7. D. B. Whitehouse, J. Tomkins, J. U. Lovegrove, D. A. Hopkinson and W. O. McMillan, *Mol. Biol. Evol.*, 1998, **15**, 456.
8. J. E. Tym, C. Mitsopoulos, E. A. Coker, P. Razaz, A. C. Schierz, A. A. Antolin and B. Al-Lazikani, *Nucleic Acids Res.*, 2016, **44**, D938.
9. R. D. Finn, T. K. Attwood, P. C. Babbitt, A. Bateman, P. Bork, A. J. Bridge, H.-Y. Chang, Z. Dosztányi, S. El-Gebali, M. Fraser, J. Gough, D. Haft, G. L. Holliday, H. Huang, X. Huang, I. Letunic, R. Lopez, S. Lu, A. Marchler-Bauer, H. Mi, J. Mistry, D. A. Natale, M. Necci, G. Nuka, C. A. Orengo, Y. Park, S. Pesseat, D. Piovesan, S. C. Potter, N. D. Rawlings, N. Redaschi, L. Richardson, C. Rivoire, A. Sangrador-Vegas, C. Sigrist, I. Sillitoe, B. Smithers, S. Squizzato, G. Sutton, N. Thanki, P. D. Thomas, S. C. E. Tosatto, C. H. Wu, I. Xenarios, L.-S. Yeh, S.-Y. Young and A. L. Mitchell, *Nucleic Acids Res.*, 2017, **45**, D190.

10. S. A. Forbes, D. Beare, H. Boutselakis, S. Bamford, N. Bindal, J. Tate, C. G. Cole, S. Ward, E. Dawson, L. Ponting, R. Stefancsik, B. Harsha, C. Yin, K. Mingming Jia, H. Jubb, Z. Sondka, S. Thompson, T. De and P. J. Campbell, *Nucleic Acids Res.*, 2017, **45**, D777.
11. E. Cerami, J. Gao, U. Dogrusoz, B. E. Gross, S. O. Sumer, B. A. Aksoy, A. Jacobsen, C. J. Byrne, M. L. Heuer, E. Larsson, Y. Antipin, B. Reva, A. P. Goldberg, C. Sander and N. Schultz, *Cancer Discovery*, 2012, **2**, 401.
12. C. Sloggett, N. Goonasekera and E. Afgan, *Bioinformatics*, 2013, **29**, 1685.
13. J. Li, A. M. N. Batcha, B. Gruning and U. R. Mansmann, *Cancer Inf.*, 2015, **14**, 87.
14. A. M. Bolger, M. Lohse and B. Usadel, *Bioinformatics*, 2014, **30**, 2114.

6 Genomics and Diagnostics in Acute Myeloid Leukaemia

Ahad F. Al Seraihi*[a], Ana Rio-Machin[a], Kiran Tawana[a], Sarah Charrot[a,c], Doriana Di Bella[a], Csaba Bödör[b], Tom Butler[c], Timothy Farren[c], Marianne Grantham[c], and Jude Fitzgibbon*[a]

[a]Centre for Haemato-Oncology, Barts Cancer Institute, Queen Mary University of London, London, UK; [b]MTA-SE Lendulet Molecular Oncohematology Research Group, 1st Department of Pathology and Experimental Cancer Research, Semmelweis University, Budapest, Hungary; [c]Specialist Integrated Haematological Malignancy Diagnostic Service (SIHMDS), Pathology, Royal London Hospital, Barts Health NHS Trust, London, UK
*E-mail: j.fitzgibbon@qmul.ac.uk, a.f.h.alseraihi@qmul.ac.uk

6.1. Introduction

6.1.1. Molecular Diagnostics of Cancers and the Genomic Revolution

Cancer is a disease of the genome where each tumour has its own key set of genetic changes. Unlocking these changes in patients' tumour cells forms the basis of molecular diagnostics. Over the last 10 years, our understanding of cancer biology has evolved, with a more realistic grasp of its multilayered complexity, owing to developments in molecular and genomics-based technologies. One field that has benefitted from such advancements is that of haematological

Genomics and Clinical Diagnostics
Edited by David Whitehouse and Ralph Rapley

Published by the Royal Society of Chemistry, www.rsc.org

malignancies. These are diseases of the blood, bone marrow and lymph nodes with a myriad of clinical manifestations and poor survival outcomes and where improvements in the diagnosis, prognosis and treatment of patients must parallel a wider introduction of molecular diagnostics.

The journey commenced with cytogenetics as one of the cornerstones of molecular diagnosis in haematological cancers; karyotyping or fluorescence *in situ* hybridisation (FISH) have been "gold standard" since the 1970's, allowing for the discovery of recurring chromosomal translocations (*e.g.* the Philadelphia chromosome) in leukaemias.[1] However, following the completion of the Human Genome Project in 2003,[2,3] the opportunities to define patient molecular profiles have intensified immensely with the introduction of next-generation sequencing (NGS)[4,5] (Figure 6.1). These "state of the art" technologies have transformed our understanding of the underlying genetics of these diseases, opening the door to a new era of genomics research and an inflation in the discovery of new genes and pathways through global initiatives, such as the International Cancer Genome Consortium (ICGC)[6] and The Cancer Genome Atlas (TCGA).[7] Indeed, one could argue that a further revolution has already started where genomic information and emerging new algorithms are set to change the way we diagnose and manage these patients. However, the challenge remains as to how best to exploit this new knowledge and the balance between discovery and translation into routine clinical practice.

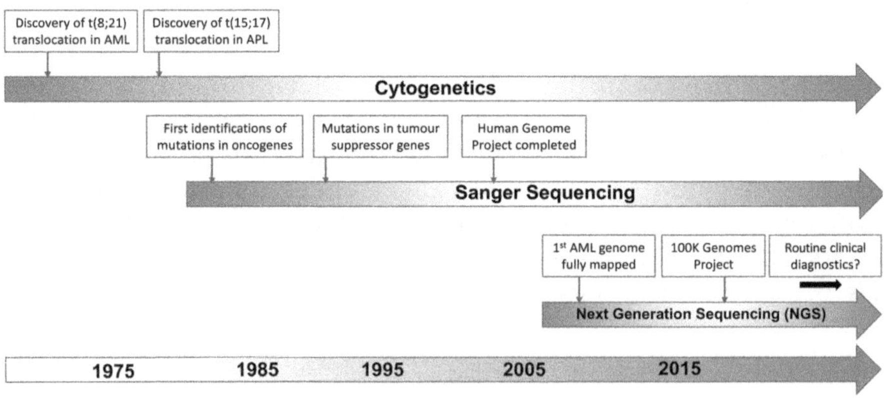

Figure 6.1 Timeline of genetic discoveries and diagnostic tests in haematological malignancies. AML, acute myeloid leukaemia; APL, acute promyelocytic leukaemia. Reproduced from ref. 34 with permission from Elsevier, Copyright 2015.

In this chapter, we discuss the influence of genomic technologies on our understanding of the molecular topography of these tumours, focusing on acute myeloid leukaemia (AML) as the first cancer genome to be fully sequenced[8] and how our new knowledge is set to advance the field of molecular diagnostics and guide patient care going forward.

6.1.2. Acute Myeloid Leukaemia

AML is the most common form of acute leukaemia in adults, with approximately 3000 new cases each year in the United Kingdom (UK) and a median age of disease onset approaching 63 years old.[9] It is a clinically diverse and genetically heterogeneous stem cell malignancy characterised by defective haematopoiesis and premature mortality in many patients. Whilst AML usually occurs *de novo* or secondary to a precursor myeloid malignancy such as myelodysplasia (MDS) or an exposure to DNA-damaging agents, such as chemotherapy (sporadic AML), there are rare occurrences of familial cases (<5%) where disease manifestations run in at least two members of the same family (familial AML). Altogether, these diseases are characterised by an uncontrolled proliferation of immature myeloid progenitor cells (myeloblasts) within the bone marrow (BM) and the inability of these cells to fully differentiate into mature myeloid cells. Patients predominantly present with symptoms suggestive of BM failure; namely fatigue or breathlessness (due to anaemia), recurrent or severe infections (due to neutropenia) or a bleeding propensity (due to thrombocytopenia).[10]

In Figure 6.2, we provide a schematic outline of the current AML diagnostic pathway routinely followed at our institution in the UK: a suspicion of AML is primarily made on the basis of an abnormal full blood count followed by examination of a blood film to assess for the presence and percentage of leukaemic blasts. The diagnosis is typically confirmed on a BM aspirate and/or a trephine biopsy looking at cell numbers, morphology and BM architecture. Furthermore, immunophenotyping (immunohistochemical and flow cytometric), cytogenetic and molecular genetic analyses are conducted in parallel, after which tests results are integrated and optimal treatment options are considered on a case-by-case basis as part of a multidisciplinary meeting. Samples are also taken to identify a leukaemia-specific phenotype or a molecular marker for the assessment of minimal residual disease (MRD), which involves detecting the persistence of AML post-remission and represents an important prognostic marker for relapse.[11]

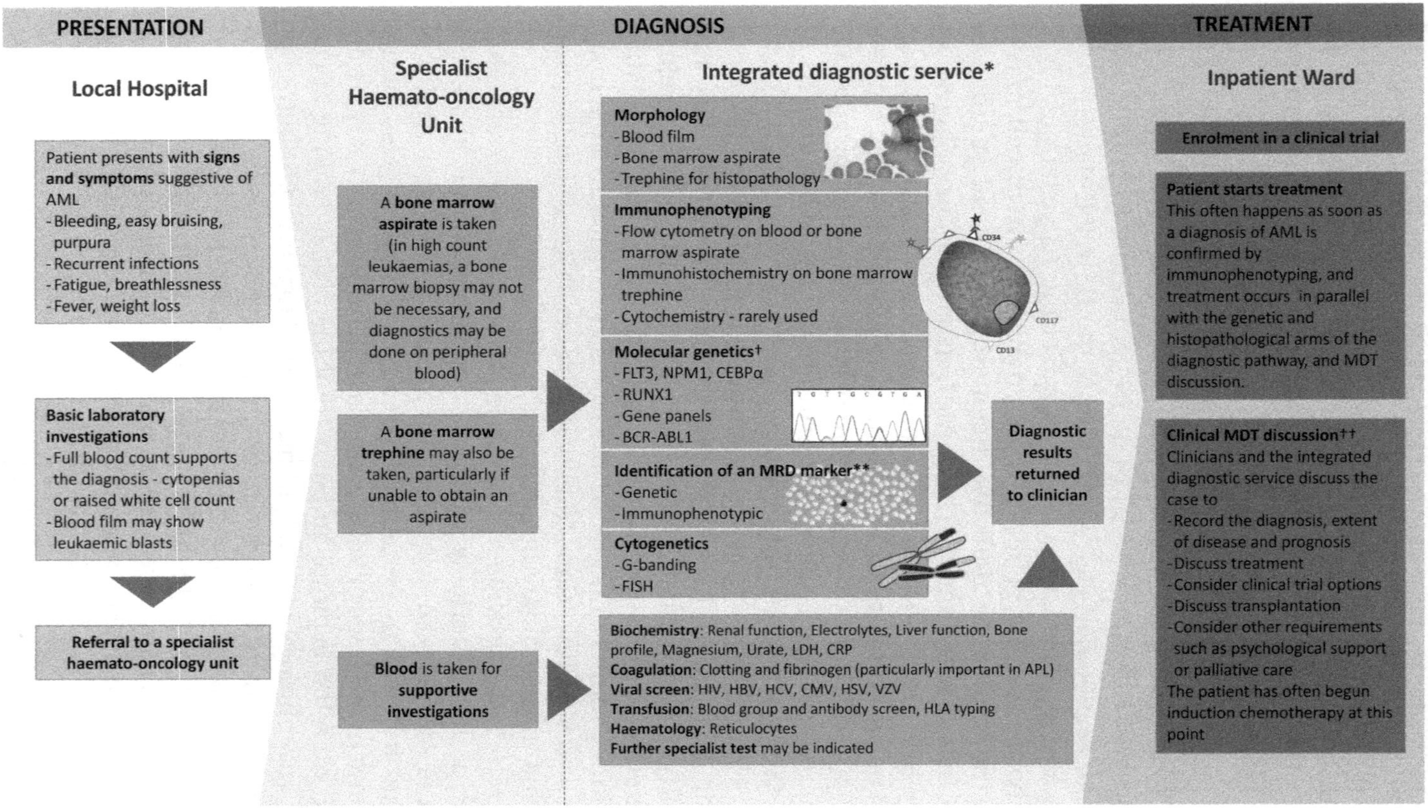

Figure 6.2 The diagnostic pathway for AML. *Morphology of the blood film and bone marrow aspirate, as well as flow cytometry results, can be available within 4 hours, and a rapid diagnosis can be returned to the clinician whilst awaiting further results. **Identification of an MRD marker is not required for diagnosis, but is important in assessing the depth of response to treatment. †*FLT3*, *NPM1* and CEBPα are routinely available through most diagnostic services. AML with mutated *RUNX1* is a provisional entity

In this chapter, we follow how the journey of a typical AML patient visiting the clinic is changing based on the introduction of new NGS technologies and how our knowledge of the molecular landscape of AML has evolved since the first AML genome was sequenced in 2008 (ref. 8) (Figure 6.3).

6.2. Molecular Landscape of AML in the Pre-NGS Era

6.2.1. Cytogenetic and Molecular Classification of AML

Leukaemias are genetically simple and the relatively easy access to malignant cells has made them the forerunner for molecular-based discoveries. Forty years ago, the pioneering work of Janet Rowley and others uncovered the importance of somatic chromosomal abnormalities including balanced translocations (*e.g.* t(8;21), t(9;22) and t(15;17)) giving rise to in-frame chimeric fusion and target genes encoding haematopoietic transcription factors (*e.g. RARA*, *RUNX1*), epigenetic regulators (*e.g. MLL* (*KMT2A*)) and components of the nuclear pore complex (*e.g. NUP98*).[12–14]

These molecular discoveries paved the way towards the introduction of disease classification systems, the most widely known schema is the World Health Organization (WHO) classification, providing an important framework for diagnosis, patient risk stratification and therapy decision making. In 2001, the WHO published the 'Classification of Tumours of the Hematopoietic and Lymphoid Tissues'[15] which superseded its predecessors (*e.g.* the French–American–British (FAB) classification) by redefining the AML BM blast threshold to 20%.

---

into the 2016 WHO criteria, and this test is likely to become more widely available. [††]The members of a haemato-oncology MDT include: haemato-oncologists, haematopathologists, clinical nurse specialists, palliative care, radiologists, clinical oncologists, members of the clinical transplant team, microbiologists, pharmacists, dieticians, and the MDT co-ordinator. AML, acute myeloid leukaemia; LDH, lactate dehydrogenase; CRP, C-reactive protein; HIV, human immunodeficiency virus; HBV, hepatitis B virus; HCV, hepatitis C virus; CMV, cytomegalovirus; HSV, herpes simplex virus; VZV, varicella zoster virus; HLA; human leukocyte antigen; APL, acute promyelocytic leukaemia; FISH, fluorescence *in situ* hybridisation; *FLT3*, Fms-related tyrosine kinase 3; *NPM1* nucleophosmin 1; CEBPα, CCAAT/enhancer-binding protein alpha; MDT, multi-disciplinary team; MRD, minimal residual disease; *RUNX1*, Runt-related transcription factor 1; BCR–ABL1, breakpoint cluster region–Abelson murine leukemia viral oncogene homolog 1 fusion gene.

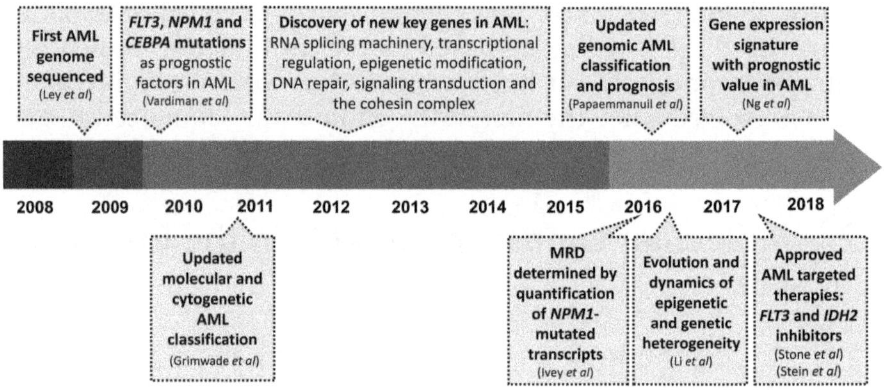

Figure 6.3 Milestones of key genetic breakthroughs in AML made over the past 10 years.

Further investigations of the molecular aberrations in AML prompted the revised WHO classification in 2008 (ref. 16 and 17) and its current incarnation in 2016 (ref. 18) (Table 6.1). This integrated framework encompasses morphological, chromosomal and molecular findings and remains part and parcel of all multidisciplinary AML meetings today.

6.2.1.1. Cytogenetic Risk Classification

Recurrent chromosomal rearrangements remain the most established prognostic markers in AML, dividing patients into three outcome groups according to favourable, intermediate and poor risk cytogenetic abnormalities.[10,19,20] The favourable risk group includes t(8;21) translocations, inv(16)/t(16;16) and the t(15;17) *PML–RARA* fusion gene, whilst the poor risk group represents approximately 17% of patients with AML and includes those with monosomies or deletions of chromosomes 5 and 7, chromosome 3q or 11q abnormalities or with complex karyotypes (four or more unrelated abnormalities).[20] The intermediate risk category represents approximately 50% of AML cases, including all normal karyotype AML (NK-AML) and those not otherwise classified as favourable or poor risk[20] (Table 6.2). Clinical outcomes in intermediate-risk patients are heterogeneous, with long-term survival estimated at 35–40% in young adults.[21,22] This has prompted the quest for additional prognostic markers to further stratify the disease and offer a more accurate prognosis for patients in this group. Collectively, whilst cytogenetic risk group rarely influences the choice of induction therapy

Table 6.1 The 2016 WHO Classification of AML and related neoplasms, myeloid neoplasms with germline predisposition and AML of ambiguous lineage.[18]

(1) AML and Related Neoplasms	
AML with recurrent genetic abnormalities	- AML with t(8;21)(q22;q22); *RUNX1-RUNX1T1* - AML with inv(16)(p13.1q22) or t(16;16)(p13.1;q22); *CBFB-MYH11* - APL with t(15;17)(q22;q12); *PML-RARA* - AML with t(9;11)(p22;q23); *MLLT3-MLL* - AML with t(6;9)(p23;924); *DEK-NUP214* - AML with inv(3)(q21q26.2) or t(3;3)(q21;q26.2); *RPN-EVI1* - AML (megakaryoblastic) with t(1;22)(p13;q13); *RBM15-MKL1* - Provisional entity: AML with *BCR-ABL1* - AML with mutated *NPM1* - AML with mutated *CEBPA* - Provisional entity: AML with mutated *RUNX1*
AML with MDS-related changes	
Therapy-related myeloid neoplasms	
AML, not otherwise specified	- AML with minimal differentiation - AML without maturation - AML with maturation - Acute myelomonocytic leukaemia - Acute monoblastic/monocytic leukaemia - Acute erythroid leukaemia i) pure erythroid leukaemia ii) Erythroleukaemia - Acute megakaryoblastic leukaemia - Acute basophilic leukaemia - Acute panmyelosis with myelofibrosis
Myeloid sarcoma	
Myeloid proliferations related to down syndrome	- Transient abnormal myelopoiesis - Myeloid leukaemia associated with down syndrome
Blastic/Plasmacytoid dendritic cell neoplasm	
(2) Myeloid neoplasms with germline predisposition	
(3) Acute leukaemias of ambiguous lineage	
	- Acute undifferentiated leukaemia - Mixed phenotype acute leukaemia with t(9;22)(q34.1;q11.2); *BCR-ABL1* - Mixed-phenotype acute leukaemia with t(v;11q23.3); *KMT2A* rearranged - Mixed phenotype acute leukaemia, B/myeloid, NOS - Mixed phenotype acute leukaemia, T/myeloid, NOS - Provisional entity: Natural killer (NK) cell lymphoblastic leukaemia/lymphoma

Table 6.2 The 2017 European LeukemiaNet (ELN) prognostication of AML risk groups according to cytogenetic and molecular genetic abnormalities.[12]

Risk category	Cytogenetic or molecular genetic abnormality
Favourable	t(8,21)(q22;q22); *RUNX1–RUNX1T1*
	inv(16)(p13.1;q22) or t(16;16)(p13.1;q22); *CBFB–MYH11*
	t(15;17)(q24;q21) *PM–RARA*
	Mutated *NPM1* without *FLT3–ITD* (normal karyotype)
	Biallelic mutated *CEBPA* (normal karyotype)
Intermediate	Mutated *NPM1* and *FLT3–ITD* (normal karyotype)
	Wildtype *NPM1* and *FLT3–ITD* (normal karyotype)
	Wildtype *NPM1* without *FLT3–ITD* (normal karyotype)
	Normal karyotype not classified as favourable.
	t(9;11)(p22;q23); *MLLT3–KMT2A*
	Cytogenetic abnormalities not classified as favourable or poor
Poor	inv(3)(q21q26.2) or t(3;3)(q21;q26.2);*GATA2–MECOM* (*EVI1*)
	t(6;9)(p23;q34); *DEK–NUP214*
	t(v;11)(v;q23); KMT2A rearranged
	t(9;22)(q34.1;q11.2); *BCR–ABL1*
	−5 or del(5q); −7; −17/abn(17p)
	Complex karyotype (>3), monosomal karyotype
	Wild type *NPM1* and *FLT3–ITD*
	Mutated *RUNX1*
	Mutated *ASXL1*
	Mutated *TP53*

(as treatment usually begins before these results are available), it has proven powerful for informing optimal consolidation strategies, in particular, whether a patient should undergo allogeneic haematopoietic stem cell transplantation (HSCT).[23] The current degree of uncertainty around patients with intermediate cytogenetic risk means that a proportion of patients who would have been cured by chemotherapy alone will go on to have a transplant with its associated morbidity, and improved prognostic scoring is therefore warranted in order to target this potentially toxic therapy to only those patients who really need it.

6.2.1.2. *Identification of Molecular Markers with a Prognostic Impact*

Over the last 20 years, research efforts have focused on deciphering the molecular basis of AML (particularly NK-AML), leading to the discovery of recurrent mutations in three genes: *FLT3*, *NPM1* and *CEBPA*, all of which were identified before the introduction of NGS technologies and are considered important prognostic markers that define survival outcomes[16,24] (Table 6.2). For example, while

FLT3 mutations modify risk from intermediate to poor and lead to reduced overall survival (OS), *NPM1* and biallelic *CEBPA* mutations have been linked with favourable prognosis and long term OS approaching 60%.[16,25] These genes are now recognised as an integral component of the AML risk stratification system, leading to their incorporation into the revised WHO classification as confirmed disease entities[12,18] and subsequently, into routine clinical testing.

6.2.2. Classical Treatment Regimens in AML

Therapeutic strategies in AML have not significantly changed in over 40 years. Generally, the standard approach to treatment consists of two cycles of intensive induction chemotherapy to induce remission, most commonly 7 days of cytarabine and 3 days of daunorubicin (so-called 7 + 3 regimen), followed by a consolidation regimen depending on the patient's prognostic factors (*e.g.* age, co-morbidities, disease status following induction and genetic risk).[26,27] In the UK, the majority of patients are enrolled in national clinical trials, which may include randomisation to receive investigational therapies considered to be of potential benefit from earlier phase I and II trials. Furthermore, advances in donor selection techniques (human leukocyte antigen (HLA) typing) combined with improvements in the management of graft *vs.* host disease (GvHD) and widespread adoption of reduced-intensity conditioning protocols have led to reductions in transplant related mortality.[28] However, despite progress in therapeutics and supportive care, AML continues to pose a major clinical burden as many cases (especially elderly patients over 60 years of age with five-year survival estimated at 14%) experience short remission periods and ultimately relapse and develop resistance to standard chemotherapy, fully warranting the development of new therapeutic strategies[29] (discussed in section 6.4.3). Moreover, elderly patients with significant co-morbidities are often not able to withstand standard intensive chemotherapy regimens due to high treatment-related mortality, leaving them with a negligible chance of cure, and effective treatment options with reduced toxicity are therefore urgently needed. This can only be afforded by improving the diagnostic and prognostic tools at our disposal and increasing our understanding of the biology underlying this disease, uncovering new pathways and investigating ways in which these pathways can be therapeutically exploited.

6.3. Recent Technical Advancements in NGS-based Genomic Profiling

In addition to routine cytogenetics, the emergence of NGS "multi-omics" technologies in the early 2000's has revolutionised our ability to perform both comprehensive and unbiased investigation of the genome at a very high resolution. This approach has unravelled deeper layers of complexity and led to the discovery of multiple novel genetic lesions and pathways (discussed in section 6.4.1). Indeed, unbiased investigation of the entire set of genetic information (whole-genome sequencing (WGS)) or the coding region (whole-exome sequencing (WES)) can now be performed with relative efficiency and accuracy, whilst targeted deep sequencing and multi-gene sequencing panels offer rapid analysis of multiple known genetic loci sequenced simultaneously with increased sensitivity for detection of subclonal mutations. With the increasing use of these "tour de force" technologies in the research and diagnostics settings, we are seeing a paradigm shift in the molecular investigation of AML from that of a single-gene approach towards the detailed characterisation of tumour architecture, co-operating molecular lesions and clonal evolution. Table 6.3 compares the most popular NGS technologies on the basis of their sequencing depth, gene coverage and other capabilities.

Table 6.3 Characteristics of three main NGS technologies[81] Reproduced from ref. 81 with permission from John Wiley and Sons, © 2015 WILEY PERIODICALS, INC.

	WGS	WES	Targeted deep sequencing
Description	Coverage of the entire genome	Coverage of coding regions	Deep sequencing of specific genes or regions of interest
Relative costs (compared with WES)	≈2.5×	1	Depends on the size of gene panel
Sequencing depth	10–65×	10–500×	100–10 000×
Gene coverage	>97.5% (at 60× depth)	97.5% (at 154× depth)	100%
Identification of new disease genes	YES	YES	NO
CNV-calling	ALL	More than three exons	Depends on the size of gene panel
Non-coding variants	YES	NO	NO
Incidental findings	YES	YES	NO

6.3.1. Multi-gene Sequencing Panels

Recently, genomic technologies have advanced, allowing the adoption of multi-gene testing panels for quick and reliable assessment of multiple genes in multiple samples simultaneously (so-called targeted resequencing) using a tagged amplicon approach. An example of such tests is Oncotype DX, which examines a panel of 21 genes, including *BRCA1* and *BRCA2*, where mutations in either of these genes increase the risk of developing breast and ovarian cancers.[30] In AML, the current diagnostic work-up include screening for mutations in *FLT3*, *NPM1*, *CEBPA* and *RUNX1*, however the list of disease genes will probably expand following the identification of new mutations with prognostic and therapeutic impact[31] or a role in familial disease (discussed in section 6.4.1). These gene panels typically include all exons or selected commonly mutated regions belonging to 10–400 disease genes; the number and choice of which is at the discretion of the host diagnostic laboratory. The analytical sensitivity for these tests reflects the high sequencing depth and can be as low as 1% variant allele frequency (VAF) therefore superseding traditional Sanger Sequencing methods with VAF of approximately 20% (calculated as shown in eqn (1)). The list of variants from raw sequencing data is filtered to exclude non-pathogenic variants or polymorphisms. Determining the pathogenicity of a particular variant requires employing fixed criteria taking into account: (1) variant frequency in healthy populations (ExAC or GnomAD databases); and (2) prediction of the biochemical function of variants by virtue of *in silico* tools such as (cBioPortal, OMIM, SIFT, Provean, PolyPhen-2, MutationTaster, GERP and CADD among others).[32]

$$\text{VAF} = \text{number of variant allele reads/total (variant + reference allele) reads} \quad (6.1)$$

6.3.2. Beyond Sequencing the Genome

What is important to keep in mind is that the aforementioned technologies profile a single, albeit important, layer of AML complexity, however, they do not provide information regarding genome-wide epigenetic modifications or alterations in gene expression. Therefore, assays to define the transcriptome (*e.g.* RNA-sequencing) or the methylome (*e.g.* whole-genome bisulphite sequencing) are employed to fully characterise the genetic and epigenetic repertoires associated with AML pathogenesis, although, at present, these analyses are firmly embedded within a discovery and research setting only.

6.3.3. Bioinformatics and Data Interpretation

Nowadays, with thousands of cancer genomes being sequenced and analysed worldwide, large consortia have formed public databases, enabling sharing of data and advancement of research. In particular, the ICGC and TCGA research networks are providing comprehensive knowledge data banks for coding variants from over 10 000 cancers with the Catalogue of Somatic Mutations in Cancer database (COSMIC) being regularly updated with novel variants identified from these extensive sequencing projects.[33] To analyse this exponential growth of NGS data, several technical and bioinformatics infrastructures have been developed for the assembly, annotation and comparison of genomes to comprehensively detail genetic variants. The future interpretation of this information will involve in-depth functional and clinical-pathological correlations and perhaps most importantly, the selection of molecular targets for novel therapeutic intervention (discussed in section 6.4.3).

With its high throughput, sensitivity and information capabilities, the future of NGS as a clinical tool is compelling and will probably lead to the phasing out of existing standards such as Sanger sequencing.[34] In the following section of this chapter, we present three different examples of how NGS technologies can shape the current diagnosis and management of AML patients by:

1. Informing disease classification and prognostic stratification.
2. Detecting minimal residual disease (MRD).
3. Identifying novel therapeutic targets.

6.4. Molecular Diagnostics Applications in the Post-NGS Era

6.4.1. Informing Disease Classification and Prognostic Stratification

6.4.1.1. Defining the Spectrum of Recurrent Driver Mutations in AML

Aside from the earliest genes detected in AML (*FLT3*, *NPM1* and *CEBPA*) (discussed in section 6.2.1), recent advances in NGS-based technologies have pushed the boundaries of unbiased AML genomic discoveries even further by capturing the enormous genetic heterogeneity. This has led to the characterisation of a plethora of genes and pathways known to have a role in haematopoiesis, with the potential to inform classification and selection of therapies. Notably, AML and MDS were shown to share molecular lesions occurring due to deregulation in pathways involving: RNA splicing machinery, transcriptional regulation, epigenetic modification, DNA repair, signal transduction and the cohesin complex[35–40] (Figure 6.3). Indeed, there are now over 100 genes known to be recurrently mutated in AML albeit many occur at low frequencies (<5%). In Figure 6.4, we show a list of the top 25 mutated genes ranging in frequency from 1% to 30%, the incidence of which varies across different cytogenetic groups, and the prognostic significance of these molecular lesions is outlined in Table 6.4.

Perhaps one of the key findings from these NGS studies was the identification of recurrent actionable mutations affecting components of the epigenetic machinery, resulting in the co-ordinated disruption of transcriptional control and chromatin remodelling. Prior to the introduction of NGS, mixed-lineage leukaemia gene partial tandem duplications (*MLL*-PTDs) were the sole recurrent epigenetic lesions and cumulatively, additional epigenetic lesions have been identified. These are enriched in genes encoding regulators of DNA methylation (*DNMT3A*, *TET2*, *IDH1*, *IDH2*) and post-translational histone modifications (*ASXL1*, *EZH2*), the majority of which are loss-of-function aberrations, further demonstrating the complexity of the (epi)genetic make-up of AML.[40–43]

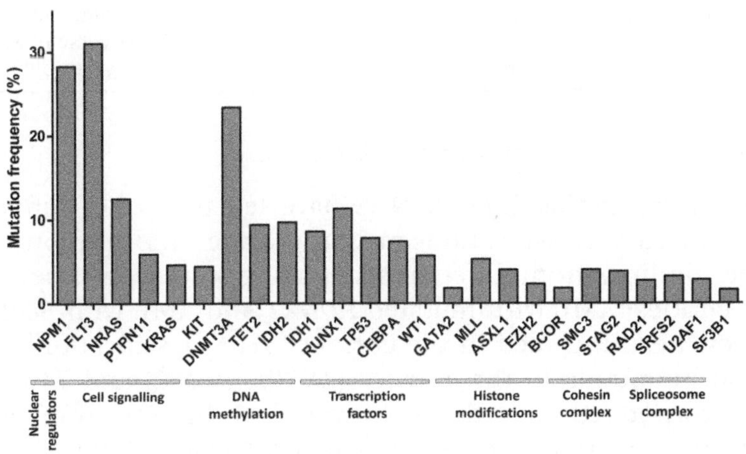

Figure 6.4 Mutational landscape of AML. Bar graph depicting the frequencies of 25 mutated genes in AML (Data extracted from ref. 39 and 44). Each gene is assigned to one of seven distinct functional categories.

Table 6.4 Key pathways and genes recurrently mutated in AML together with their frequencies and predicted effects on clinical outcome[82,83] (Data derived from ref. 39, 44, 82 and 83).

Functional category		Gene	Mutation frequency in AML (%)	Prognostic effect
Nuclear regulators		*NPM1*	28.3	Favourable in *NPM1*-mutant without *FLT3*–ITD
Cell signalling		*FLT3*	31	Adverse
		NRAS	12.5	*Conflicting data*
		PTPN11	5.9	Adverse in *NPM1:DNMT3A* mutant
		KRAS	4.6	*Conflicting data*
		KIT	4.4	Adverse in CBF–AML
Transcription factors		*RUNX1*	11.3	Adverse
		TP53	7.8	Adverse
		CEBPA	7.4	Double-mutant: Favourable
		WT1	5.7	*Conflicting data*
		GATA2	1.8	Favourable
Epigenetic modifiers	DNA methylation	*DNMT3A*	23.4	Adverse
		TET2	9.4	Adverse
		IDH2	9.7	*Conflicting data*
		IDH1	8.6	*Conflicting data*
	Histone modifications	*MLL*	5.3	Adverse
		ASXL1	4	Adverse
		EZH2	2.3	Adverse in MDS
		BCOR	1.8	Adverse in MDS
Cohesin complex		*SMC3*	4	*Conflicting data*
		STAG2	3.8	*Conflicting data*
		RAD21	2.7	*Conflicting data*
Spliceosome complex		*SRFS2*	3.2	Adverse in MDS
		U2AF1	2.8	Adverse in MDS
		SF3B1	1.6	Favourable in MDS

6.4.1.2. *Improved Genomic Classification*

In past years, applications of NGS have led to further refinement of AML disease classification and prognostic stratification. Taken together findings from different sequencing studies, two new provisional entities: AML with mutated *RUNX1* and AML with *BCR–ABL1* have now been included in the current WHO classification[18] (Table 6.1) and mutations in three genes (*ASXL1*, *RUNX1* and *TP53*) have been incorporated into the risk stratification algorithm as new molecular markers that confer a poor prognosis[18] (Table 6.2). This is just the beginning; a recent study by Papaemmanuil *et al.* defined the driver genomic landscape further in more than 1500 patients with AML.[39] This comprehensive analysis revealed four new AML molecular subgroups, including *NPM1* emerging as a separate entity, rarely found in isolation and for which its prognostic effect is largely dependent on concurrent mutation(s). Remaining subgroups include AML with mutated chromatin or spliceosome genes, AML with *TP53* mutations, chromosomal aneuploidy or both and AML with $IDH2^{R172}$ mutations.[39] By implementing these classification schemes, patients with AML can be segregated into several distinct subgroups on the basis of their underlying genetic abnormalities, allowing for more stratified approach to treatment and management.

6.4.1.3. Patterns of Mutational Co-occurrence Modifying Clinical Outcomes

It is worth noting that not all of these mutations occur independently and non-random molecular associations often take place and have the potential to shape prognosis.[39,44] For example, extensive analyses from the TCGA consortium sequencing approximately 200 sporadic AML cases reported an average of 13 coding mutations per genome, of which recurrent mutations in *DNMT3A*, *IDH1/2*, *RUNX1* and *CEBPA* were found to be mutually exclusive with transcription factor fusion genes, indicating that the former were putative initiating lesions.[44] In contrast, molecular co-operation is frequently noted between *FLT3–ITD* and *NPM1* mutations,[40,45] CBF translocations and *KIT* mutations[46] and between *TP53* mutations and chromosomal aneuploidies, predicting a very poor outcome.[47] Recent studies also reported somatic *GATA2* mutations to co-occur with biallelic *CEBPA* mutations in approximately 40% of *CEBPA*-mutated cases and were associated with a favourable prognosis.[48–50] These studies shed light on the notion of predestination (canalisation), where the sequential order of mutation acquisition is essential in governing tumour initiation and evolution (discussed in 6.4.1.5).

6.4.1.4. Germline Mutational Repertoire of Inherited Leukaemias

Heritable myeloid malignancies, while rare, are now regarded as a separate disease category in the most recent WHO classification[18] (Table 6.1). Studying familial AML provides valuable insights into the pathogenesis of sporadic diseases and recently, advancements in NGS technologies have dramatically accelerated efforts to elucidate the genomic landscape of these inherited tumours. AML predisposition syndromes have thus far been associated with germline mutations in 14 disease genes with variable clinical manifestations, illustrated in Figure 6.5. The most well-established syndromes are: familial

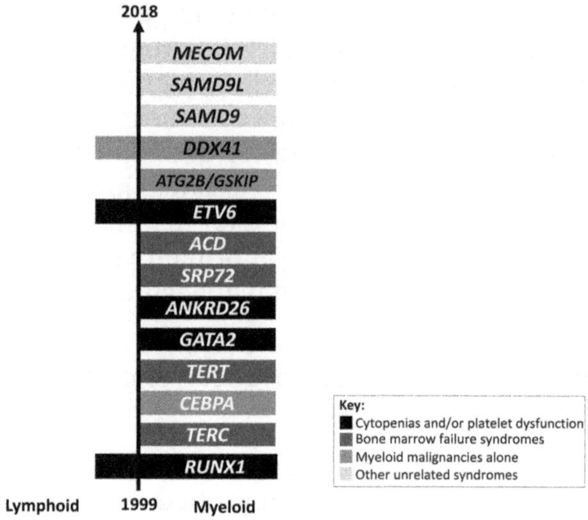

Figure 6.5 Mutational landscape in familial MDS/AML. The order of the 14 germline mutations is shown based on the year they were first discovered. These genes display diversity in phenotypes across the lymphoid and myeloid malignancies and can be sub-divided into four groups according to clinical phenotype: myeloid malignancies alone, cytopenias and/or platelet dysfunction, bone marrow failure syndromes and other unrelated syndromes.[79] This research was originally published in *Blood*, K. Tawana and J. Fitzgibbon, Inherited *DDX41* mutations: 11 genes and counting, 2016, 127(8), 960–961, ©the American Society of Hematology.

platelet disorder with propensity to AML (FPD/AML) due to mutations in *RUNX1*; familial AML with mutated *CEBPA*; familial MDS/AML with mutated *GATA2* and the inherited bone marrow failure syndromes (IBMFS) caused by mutations in key telomere regulators (*TERC* and *TERT*).[51–53] In light of these findings, these disease loci have now been integrated into dedicated diagnostic screening panels (in the UK this is located at the West Midlands Regional Genetics Labs in Birmingham) so that patients with a presumed familial involvement can be genetically tested for surveillance and exclusion of asymptomatic "silent" mutation carriers from the bone marrow transplant scenario. Moreover, familial leukaemia provides an unrivalled model with which to investigate both the stepwise evolution of disease and to discover as yet novel disease causing alleles. Indeed, ongoing investigations are now focused on identifying novel inherited disease genes and expanding the current mutational repertoire, exemplified by the recent discovery of germline variants in *DDX41*, *SAMD9* and *SAMD9L*, segregating in families with MDS/AML and/or other accessory phenotypes (*e.g.* neurological disorders).[54,55]

Collectively, as gene testing methods continue to improve, routinely screening these disease genes is becoming more feasible and, indeed, targeted myeloid gene panels are already available in diagnostic settings. These findings hold considerable promise that not only does testing for mutations in these genes inform prognosis but that it can also guide future therapeutic decisions.[56]

6.4.1.5. Insights on Clonal Evolution

An important revelation emerging in AML and cancer as a whole is that seldom does a biopsy represent a homogeneous population of tumour cells but rather, it is composed of a mosaic of multiple genomes, reflecting the expansion of subclones that confer a selective advantage and compete with progenitor clones for survival.[57,58] Indeed, this concept was first introduced by Nowell in 1976 who postulated that cancer cells sequentially accumulate genetic and epigenetic alterations over time, some of which provide a survival advantage and subsequently lead to a "Darwinian" natural selection that promotes clonal expansion and tumour progression[59,60] (Figure 6.6).

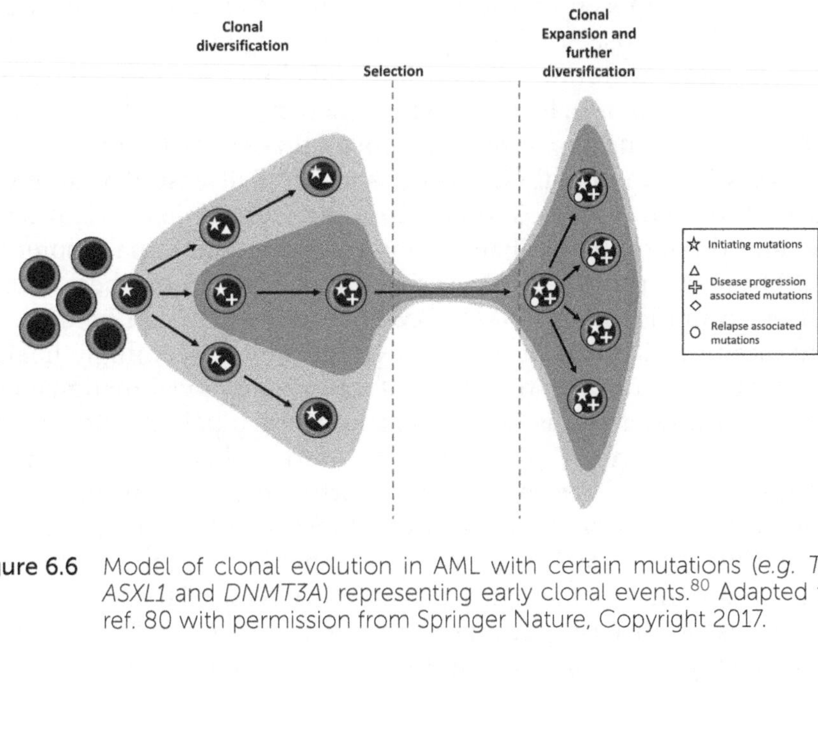

Figure 6.6 Model of clonal evolution in AML with certain mutations (*e.g. TET2, ASXL1* and *DNMT3A*) representing early clonal events.[80] Adapted from ref. 80 with permission from Springer Nature, Copyright 2017.

Forty years later, and following the publication of the first AML genome by Ley *et al.* in 2008,[8] subsequent NGS studies have led to the development of novel computational algorithms with which to study AML clonal architecture and evolution by analysing the mutant allele fraction (or VAF) of clonal cell populations.[61] These studies revealed that mutational patterns seem to follow specific temporally ordered trajectories from which evolutionary phylogenetic trees can be inferred. For example, mutations in genes involved in epigenetic regulation such as *ASXL1*, *DNMT3A*, *IDH1/2* and *TET2* represent initiating clonal events (*i.e.* the trunk of the tree) with seemingly high VAFs. In contrast, mutations involving *FLT3* or *NPM1* represent secondary genetic events that occur later in leukaemogenesis (*i.e.* the branches of the tree) and therefore are likely to have lower VAF values.[23]

These findings support the notion that stepwise acquisition of genetic events is required to fashion disease development. For examples, initiating mutations (*e.g. DNMT3A*, *TET2* or *IDH1/2*) can follow different routes of transformation depending on the secondary mutational event: *NPM1* mutations would lead to AML, *JAK2* mutations to myeloproliferative neoplasms and *SF3B1* mutations to MDS with ring sideroblasts.[39] Another example is a study by Ding *et al.* who employed NGS to trace the clonal evolution of AML by comparing the mutational reservoir between paired diagnostic and relapse patient samples and proposed two scenarios of disease relapse: a linear pattern, whereby relapse is driven by the evolution of the founding clone, or a branching evolution, where fitter subclones acquire additional mutations and expand, leading to disease recurrence.[62,63]

It has also become apparent that some of these somatic clones might not always lead to clonal expansion and overt disease development. Instead, they can remain dormant in a pre-leukaemic or a quiescent state, and the term clonal haematopoiesis of indeterminate potential (CHIP) has been proposed to describe the phenomenon where certain recurrent mutations (*e.g. ASXL1*, *DNMT3A* and *TET2*) can be detected in the peripheral blood of elderly (over 65 years), seemingly healthy individuals without prior diagnosis of a haematological malignancy.[64] This raises the question as to whether these individuals should be monitored to detect signs of progression to overt malignancy when these dormant clones have reawakened and started to proliferate aberrantly.

Nevertheless, a key challenge from these studies remains in discriminating driver from innocuous passenger or bystander mutations and further contextualising the prognostic effects of these molecular aberrations on disease pathogenesis. This is still a field very much in its infancy and work is now focusing on characterising tumour architecture and complexity in greater detail, by employing xenograft models of purified haematopoietic stem cell (HSC) subsets and in-depth genetic profiling techniques, such as single-cell sequencing, with a view to dissecting the precise molecular lesions governing sensitivity *vs.* resistance to chemotherapy. Such findings will ultimately guide the future design and implementation of targeted therapies to minimise the risk of relapse, the leading cause of mortality in AML.

6.4.2. Minimal Residual Disease (MRD) Detection

Another emerging area of potential risk-adapted therapy is the monitoring of persistent AML-associated mutations after therapy (also known as MRD). The conventional definition of remission post therapy involves the achievement of specific haematological parameters (indicating restoration of normal haematopoiesis) and the identification of <5% BM blasts.[65] Assays for the detection of MRD at levels not otherwise identified by standard morphology include real-time quantitative polymerase chain reaction (RT-qPCR) for detecting fusion gene transcripts and molecular variants and multiparameter flow cytometry (MFC) for the detection of leukaemia-associated immunophenotypes (LAIPs). Newer NGS advances, such as single-cell sequencing, offer the possibility of capturing disease heterogeneity at the single-cell level, therefore allowing for more accurate assessment of MRD.

Currently, the clinical implementation of MRD analysis is largely limited to acute promyelocytic leukaemia (APL) and monitoring of t(15;17) *PML–RARA* transcripts.[63] In AML, however, MRD positivity of molecular abnormalities (*e.g. NPM1*) or LAIPs has been associated with relapse and adverse outcomes,[20,66,67] suggesting its potential utility as a prognostic marker for relapse. Discussions are often polarised regarding the advantages of MRD analysis and are currently the subject of investigation in the UK NCRI AML17/AML19 trials, but it seems likely that this information will be a useful adjunct to select patients who will benefit from more aggressive therapy (*e.g.* allogeneic HSCT) in first remission (CR1)[11] and in the absence of less effective therapies to target this residual population of leukaemia initiating cells (LIC).

6.4.3. Novel Therapeutic Targets and the Promise of Precision Medicine

Perhaps one of the most important implementations of NGS-based profiling directly relating to patient care is the identification of a specific pathway or mutation in AML that could define response (or a lack thereof) to targeted therapies. The most successful example has been the use of all-trans retinoic acid (ATRA) in *PML–RARA* positive cases, changing this disease from one with a dire prognosis to one with the highest cure rate of all AML subtypes[68,69] and it is a key research aim to replicate the success of this agent in other types of AML. Indeed, giant strides are being made in unravelling the biological consequences of mutations discovered through large AML sequencing projects, opening up new opportunities for therapeutic intervention. Recently approved targeted therapies include *FLT3* tyrosine kinase inhibitors (*e.g.* midostaurin) in poor-risk AML cases with *FLT3–ITD* mutations and *IDH2* inhibitors (*e.g.* Enasidenib or Idhifa) in *IDH2*-mutant AMLs.[70–73] Other promising examples include drugs that influence epigenetic regulation of gene expression *e.g.* histone deacetylase (HDAC) inhibitors or hypomethylating agents (*e.g.* 5-azacytidine) in *TET2*/*DNMT3A*/*ASXL1*-mutant AMLs and DOT1L inhibitors in *MLL*-rearranged leukaemias, demonstrating clinical utility in ongoing clinical trials.[74–77] Ultimately, it is hoped that the future clinical management of AML patients will move away from the classic 'one size fits all' formula and towards a more stratified, patient- and disease-targeted approach. This long-term mission might finally become a reality through initiatives such as the Beat AML® Master Trial, a collaborative effort led by the Leukemia & Lymphoma Society (LLS) where each AML patient enrolled on the trial can be screened for actionable mutations and then assigned to a targeted therapy based on his or her underlying genetic profile.[78] While these initiatives are by no means a final solution, they offer a substantive opportunity for precision medicine.

6.5. Conclusions and Future Perspective

We have a new-found ability through NGS, to interrogate the cancer genome and this is providing an intricate understanding of cancer complexity and the challenges now faced in order to improve on the existing standards of care. The early inroads in diseases like AML are encouraging and there have been tangible successes in molecular classification and risk-stratification. While we still await a step change in how AML is treated and measurable improvements in patient outcome, it is no longer a matter of 'if' but instead a case of when these advances are likely to be delivered.

Acknowledgement

The authors would like to thank members of the Barts Health NHS Trust in London, in particular, haematologists, oncologists and scientists, for their contributions to improving care and outcome for AML patients. The authors would also like to acknowledge support from Bloodwise Programme grant (14032) awarded to J. F. The authors declare no conflicts of interest.

References

1. P. C. Nowell and D. A. Hungerford, Chromosome studies on normal and leukemic human leukocytes, *J. Natl. Cancer Inst.*, 1960, **25**, 85–109.
2. F. S. Collins, M. Morgan and A. Patrinos, The Human Genome Project: lessons from large-scale biology, *Science*, 2003, **300**(5617), 286–290.
3. International Human Genome Sequencing C, Finishing the euchromatic sequence of the human genome, *Nature*, 2004, **431**(7011), 931–945.
4. D. A. Martinez and M. A. Nelson, The next generation becomes the now generation, *PLoS Genet.*, 2010, **6**(4), e1000906.
5. E. Braggio, J. B. Egan, R. Fonseca and A. K. Stewart, Lessons from next-generation sequencing analysis in hematological malignancies, *Blood Cancer J.*, 2013, **3**, e127.
6. http://icgc.org/. (last accessed January 2018).
7. https://cancergenome.nih.gov. (last accessed January 2018).
8. T. J. Ley, E. R. Mardis, L. Ding, B. Fulton, M. D. McLellan and K. Chen, *et al.*, DNA sequencing of a cytogenetically normal acute myeloid leukaemia genome, *Nature*, 2008, **456**(7218), 66–72.
9. http://www.cancerresearchuk.org/about-cancer/acute-myeloid-leukaemia-aml/about-acute-myeloid-leukaemia. (last accessed January 2018).
10. D. Grimwade, The changing paradigm of prognostic factors in acute myeloid leukaemia, *Best Pract. Res., Clin. Haematol.*, 2012, **25**(4), 419–425.
11. D. Grimwade and S. D. Freeman, Defining minimal residual disease in acute myeloid leukemia: which platforms are ready for "prime time"? *Hematology*, 2014, **2014**(1), 222–233.
12. H. Dohner, E. Estey, D. Grimwade, S. Amadori, F. R. Appelbaum and T. Buchner, *et al.*, Diagnosis and management of AML in adults: 2017 ELN recommendations from an international expert panel, *Blood*, 2017, **129**(4), 424–447.
13. J. D. Rowley, Letter: A new consistent chromosomal abnormality in chronic myelogenous leukaemia identified by quinacrine fluorescence and Giemsa staining, *Nature*, 1973, **243**(5405), 290–293.
14. J. D. Rowley, Chromosome changes in acute leukaemia, *Br. J. Haematol.*, 1980, **44**(3), 339–346.
15. E. S. Jaffe, N. L. Harris, H. Stein and J. W. Vardiman, *Pathology and Genetics of Tumours of Haematopoietic and Lymphoid Tissues*. WHO press, Lyon, France, 2001, vol. 3.
16. J. W. Vardiman, J. Thiele, D. A. Arber, R. D. Brunning, M. J. Borowitz and A. Porwit, *et al.*, The 2008 revision of the World Health Organization (WHO) classification of myeloid neoplasms and acute leukemia: rationale and important changes, *Blood*, 2009, **114**(5), 937–951.
17. S. H. Swerdlow, E. Campo, N. L. Harris, E. S. Jaffe, S. A. Pileri, H. Stein, J. Thiele and J. W. Vardiman, *WHO Classification of Tumours of Haematopoietic and Lymphoid Tissues*, WHO press, Lyon, France, 2008, vol. 2.

18. D. A. Arber, A. Orazi, R. Hasserjian, J. Thiele, M. J. Borowitz and M. M. Le Beau, *et al.*, The 2016 revision to the World Health Organization classification of myeloid neoplasms and acute leukemia, *Blood*, 2016, **127**(20), 2391–2405.
19. D. Grimwade, H. Walker, F. Oliver, K. Wheatley, C. Harrison and G. Harrison, *et al.*, The importance of diagnostic cytogenetics on outcome in AML: analysis of 1,612 patients entered into the MRC AML 10 trial. The Medical Research Council Adult and Children's Leukaemia Working Parties, *Blood*, 1998, **92**(7), 2322–2333.
20. D. Grimwade, R. K. Hills, A. V. Moorman, H. Walker, S. Chatters and A. H. Goldstone, *et al.*, Refinement of cytogenetic classification in acute myeloid leukemia: determination of prognostic significance of rare recurring chromosomal abnormalities among 5876 younger adult patients treated in the United Kingdom Medical Research Council trials, *Blood*, 2010, **116**(3), 354–365.
21. D. A. Breems, W. L. Van Putten, P. C. Huijgens, G. J. Ossenkoppele, G. E. Verhoef and L. F. Verdonck, *et al.*, Prognostic index for adult patients with acute myeloid leukemia in first relapse, *J. Clin. Oncol.*, 2005, **23**(9), 1969–1978.
22. V. Rockova, S. Abbas, B. J. Wouters, C. A. Erpelinck, H. B. Beverloo and R. Delwel, *et al.*, Risk stratification of intermediate-risk acute myeloid leukemia: integrative analysis of a multitude of gene mutation and gene expression markers, *Blood*, 2011, **118**(4), 1069–1076.
23. L. Bullinger, K. Dohner and H. Dohner, Genomics of Acute Myeloid Leukemia Diagnosis and Pathways, *J. Clin. Oncol.*, 2017, **35**(9), 934–946.
24. R. F. Schlenk, K. Dohner, J. Krauter, S. Frohling, A. Corbacioglu and L. Bullinger, *et al.*, Mutations and treatment outcome in cytogenetically normal acute myeloid leukemia, *N. Engl. J. Med.*, 2008, **358**(18), 1909–1918.
25. A. Dufour, F. Schneider, K. H. Metzeler, E. Hoster, S. Schneider and E. Zellmeier, *et al.*, Acute myeloid leukemia with biallelic CEBPA gene mutations and normal karyotype represents a distinct genetic entity associated with a favorable clinical outcome, *J. Clin. Oncol.*, 2010, **28**(4), 570–577.
26. J. W. Yates, H. J. Wallace Jr, R. R. Ellison and J. F. Holland, Cytosine arabinoside (NSC-63878) and daunorubicin (NSC-83142) therapy in acute nonlymphocytic leukemia, *Cancer Chemother. Rep.*, 1973, **57**(4), 485–488.
27. K. Dohner and P. Paschka, Intermediate-risk acute myeloid leukemia therapy: current and future, *Hematology*, 2014, **2014**(1), 34–43.
28. J. T. Horan, B. R. Logan, M. A. Agovi-Johnson, H. M. Lazarus, A. A. Bacigalupo and K. K. Ballen, *et al.*, Reducing the risk for transplantation-related mortality after allogeneic hematopoietic cell transplantation: how much progress has been made? *J. Clin. Oncol.*, 2011, **29**(7), 805–813.
29. T. Buchner, W. E. Berdel, C. Haferlach, T. Haferlach, S. Schnittger and C. Muller-Tidow, *et al.*, Age-related risk profile and chemotherapy dose response in acute myeloid leukemia: a study by the German Acute Myeloid Leukemia Cooperative Group, *J. Clin. Oncol.*, 2009, **27**(1), 61–69.
30. D. F. Easton, P. D. Pharoah, A. C. Antoniou, M. Tischkowitz, S. V. Tavtigian and K. L. Nathanson, *et al.*, Gene-panel sequencing and the prediction of breast-cancer risk, *N. Engl. J. Med.*, 2015, **372**(23), 2243–2257.
31. J. P. Patel, M. Gonen, M. E. Figueroa, H. Fernandez, Z. Sun and J. Racevskis, *et al.*, Prognostic relevance of integrated genetic profiling in acute myeloid leukemia, *N. Engl. J. Med.*, 2012, **366**(12), 1079–1089.
32. F. P. R. Yang, *Next-Generation Sequencing Multi-Gene Mutation Panels in Myeloid Malignancies*, 2016, vol. 13.
33. http://cancer.sanger.ac.uk/cosmic. (last accessed January 2018).
34. J. S. Black, M. Salto-Tellez, K. I. Mills and M. A. Catherwood, The impact of next generation sequencing technologies on haematological research – A review, *Pathogenesis*, 2015, **2**(1–2), 9–16.

35. R. Bejar, K. Stevenson, O. Abdel-Wahab, N. Galili, B. Nilsson and G. Garcia-Manero, *et al.*, Clinical effect of point mutations in myelodysplastic syndromes, *N. Engl. J. Med.*, 2011, **364**(26), 2496–2506.
36. M. Cazzola, M. Rossi and L. Malcovati, Associazione Italiana per la Ricerca sul Cancro Gruppo Italiano Malattie M. Biologic and clinical significance of somatic mutations of SF3B1 in myeloid and lymphoid neoplasms, *Blood*, 2013, **121**(2), 260–269.
37. V. Grossmann, S. Schnittger, A. Kohlmann, C. Eder, A. Roller and F. Dicker, *et al.*, A novel hierarchical prognostic model of AML solely based on molecular mutations, *Blood*, 2012, **120**(15), 2963–2972.
38. G. Marcucci, T. Haferlach and H. Dohner, Molecular genetics of adult acute myeloid leukemia: prognostic and therapeutic implications, *J. Clin. Oncol.*, 2011, **29**(5), 475–486.
39. E. Papaemmanuil, M. Gerstung, L. Bullinger, V. I. Gaidzik, P. Paschka and N. D. Roberts, *et al.*, Genomic Classification and Prognosis in Acute Myeloid Leukemia, *N. Engl. J. Med.*, 2016, **374**(23), 2209–2221.
40. D. Grimwade, A. Ivey and B. J. Huntly, Molecular landscape of acute myeloid leukemia in younger adults and its clinical relevance, *Blood*, 2016, **127**(1), 29–41.
41. T. J. Ley, L. Ding, M. J. Walter, M. D. McLellan, T. Lamprecht and D. E. Larson, *et al.*, DNMT3A mutations in acute myeloid leukemia, *N. Engl. J. Med.*, 2010, **363**(25), 2424–2433.
42. P. Ntziachristos, O. Abdel-Wahab and I. Aifantis, Emerging concepts of epigenetic dysregulation in hematological malignancies, *Nat. Immunol.*, 2016, **17**(9), 1016–1024.
43. F. Delhommeau, S. Dupont, V. Della Valle, C. James, S. Trannoy and A. Masse, *et al.*, Mutation in TET2 in myeloid cancers, *N. Engl. J. Med.*, 2009, **360**(22), 2289–2301.
44. Cancer Genome Atlas Research N, Genomic and epigenomic landscapes of adult *de novo* acute myeloid leukemia, *N. Engl. J. Med.*, 2013, **368**(22), 2059–2074.
45. R. E. Gale, C. Green, C. Allen, A. J. Mead, A. K. Burnett and R. K. Hills, *et al.*, The impact of FLT3 internal tandem duplication mutant level, number, size, and interaction with NPM1 mutations in a large cohort of young adult patients with acute myeloid leukemia, *Blood*, 2008, **111**(5), 2776–2784.
46. P. Paschka, G. Marcucci, A. S. Ruppert, K. Mrozek, H. Chen and R. A. Kittles, *et al.*, Adverse prognostic significance of KIT mutations in adult acute myeloid leukemia with inv(16) and t(8;21): a Cancer and Leukemia Group B Study, *J. Clin. Oncol.*, 2006, **24**(24), 3904–3911.
47. K. H. Metzeler, T. Herold, M. Rothenberg-Thurley, S. Amler, M. C. Sauerland and D. Gorlich, *et al.*, Spectrum and prognostic relevance of driver gene mutations in acute myeloid leukemia, *Blood*, 2016, **128**(5), 686–698.
48. A. Fasan, C. Eder, C. Haferlach, V. Grossmann, A. Kohlmann and F. Dicker, *et al.*, GATA2 mutations are frequent in intermediate-risk karyotype AML with biallelic CEBPA mutations and are associated with favorable prognosis, *Leukemia*, 2013, **27**(2), 482–485.
49. C. L. Green, K. Tawana, R. K. Hills, C. Bodor, J. Fitzgibbon and S. Inglott, *et al.*, GATA2 mutations in sporadic and familial acute myeloid leukaemia patients with CEBPA mutations, *Br. J. Haematol.*, 2013, **161**(5), 701–705.
50. P. A. Greif, A. Dufour, N. P. Konstandin, B. Ksienzyk, E. Zellmeier and B. Tizazu, *et al.*, GATA2 zinc finger 1 mutations associated with biallelic CEBPA mutations define a unique genetic entity of acute myeloid leukemia, *Blood*, 2012, **120**(2), 395–403.
51. C. Owen, M. Barnett and J. Fitzgibbon, Familial myelodysplasia and acute myeloid leukaemia–a review, *Br. J. Haematol.*, 2008, **140**(2), 123–132.

52. E. M. Nickels, J. Soodalter, J. E. Churpek and L. A. Godley, Recognizing familial myeloid leukemia in adults, *Ther. Adv. Hematol.*, 2013, **4**(4), 254–269.
53. A. H. West, L. A. Godley and J. E. Churpek, Familial myelodysplastic syndrome/acute leukemia syndromes: a review and utility for translational investigations, *Ann. N. Y. Acad. Sci.*, 2014, **1310**, 111–118.
54. C. Polprasert, I. Schulze, M. A. Sekeres, H. Makishima, B. Przychodzen and N. Hosono, *et al.*, Inherited and Somatic Defects in DDX41 in Myeloid Neoplasms, *Cancer Cell*, 2015, **27**(5), 658–670.
55. B. Tesi, J. Davidsson, M. Voss, E. Rahikkala, T. D. Holmes and S. C. C. Chiang, *et al.*, Gain-of-function SAMD9L mutations cause a syndrome of cytopenia, immunodeficiency, MDS, and neurological symptoms, *Blood*, 2017, **129**(16), 2266–2279.
56. S. Yohe, Molecular Genetic Markers in Acute Myeloid Leukemia, *J. Clin. Med.*, 2015, **4**(3), 460–478.
57. T. A. Graubert and E. R. Mardis, Genomics of acute myeloid leukemia, *Cancer J.*, 2011, **17**(6), 487–491.
58. X. Zhang, D. Lv, Y. Zhang, Q. Liu and Z. Li, Clonal evolution of acute myeloid leukemia highlighted by latest genome sequencing studies, *Oncotarget*, 2016, **7**(36), 58586–58594.
59. P. C. Nowell, The clonal evolution of tumor cell populations, *Science*, 1976, **194**(4260), 23–28.
60. L. R. Yates and P. J. Campbell, Evolution of the cancer genome, *Nat. Rev. Genet.*, 2012, **13**(11), 795–806.
61. J. S. Welch, T. J. Ley, D. C. Link, C. A. Miller, D. E. Larson and D. C. Koboldt, *et al.*, The origin and evolution of mutations in acute myeloid leukemia, *Cell*, 2012, **150**(2), 264–278.
62. L. Ding, T. J. Ley, D. E. Larson, C. A. Miller, D. C. Koboldt and J. S. Welch, *et al.*, Clonal evolution in relapsed acute myeloid leukaemia revealed by whole-genome sequencing, *Nature*, 2012, **481**(7382), 506–510.
63. K. Tawana, C. Bödör, J. Cavenagh, M. Jenner and J. Fitzgibbon, The Molecular Pathogenesis of Acute Myeloid Leukemia, CML, *Leuk. Lymphoma*, 2013, **21**(3), 67–77.
64. P. S. Becker, Clonal Hematopoiesis: The Seeds of Leukemia or Innocuous Bystander? *The Hematologist*, 2016, **13**(1), 7.
65. G. Ossenkoppele and G. J. Schuurhuis, MRD in AML: does it already guide therapy decision-making? *Hematology*, 2016, **2016**(1), 356–365.
66. S. Schnittger, W. Kern, C. Tschulik, T. Weiss, F. Dicker and B. Falini, *et al.*, Minimal residual disease levels assessed by NPM1 mutation-specific RQ-PCR provide important prognostic information in AML, *Blood*, 2009, **114**(11), 2220–2231.
67. A. Ivey, R. K. Hills, M. A. Simpson, J. V. Jovanovic, A. Gilkes and A. Grech, *et al.*, Assessment of Minimal Residual Disease in Standard-Risk AML, *N. Engl. J. Med.*, 2016, **374**(5), 422–433.
68. S. Castaigne, C. Chomienne, M. T. Daniel, P. Ballerini, R. Berger and P. Fenaux, *et al.*, All-trans retinoic acid as a differentiation therapy for acute promyelocytic leukemia. I. Clinical results, *Blood*, 1990, **76**(9), 1704–1709.
69. A. Kanamaru, Y. Takemoto, M. Tanimoto, H. Murakami, N. Asou and T. Kobayashi, *et al.*, All-trans retinoic acid for the treatment of newly diagnosed acute promyelocytic leukemia. Japan Adult Leukemia Study Group, *Blood*, 1995, **85**(5), 1202–1206.
70. R. M. Stone, S. J. Mandrekar, B. L. Sanford, K. Laumann, S. Geyer and C. D. Bloomfield, *et al.*, Midostaurin plus Chemotherapy for Acute Myeloid Leukemia with a FLT3 Mutation, *N. Engl. J. Med.*, 2017, **377**(5), 454–464.
71. J. S. Garcia and R. M. Stone, The Development of FLT3 Inhibitors in Acute Myeloid Leukemia, *Hematology/Oncology Clinics of North America*, 2017, **31**(4), 663–680.

72. E. S. Kim, Enasidenib: First Global Approval, *Drugs*, 2017, **77**(15), 1705–1711.
73. E. M. Stein, C. D. DiNardo, D. A. Pollyea, A. T. Fathi, G. J. Roboz and J. K. Altman, *et al.*, Enasidenib in mutant IDH2 relapsed or refractory acute myeloid leukemia, *Blood*, 2017, **130**(6), 722–731.
74. H. Dombret, J. F. Seymour, A. Butrym, A. Wierzbowska, D. Selleslag and J. H. Jang, *et al.*, International phase 3 study of azacitidine *vs.* conventional care regimens in older patients with newly diagnosed AML with >30% blasts, *Blood*, 2015, **126**(3), 291–299.
75. M. Cruijsen, M. Lubbert, P. Wijermans and G. Huls, Clinical Results of Hypomethylating Agents in AML Treatment, *J. Clin. Med.*, 2014, **4**(1), 1–17.
76. S. R. Daigle, E. J. Olhava, C. A. Therkelsen, A. Basavapathruni, L. Jin and P. A. Boriack-Sjodin, *et al.*, Potent inhibition of DOT1L as treatment of MLL-fusion leukemia, *Blood*, 2013, **122**(6), 1017–1025.
77. E. M. Stein and M. S. Tallman, Mixed lineage rearranged leukaemia: pathogenesis and targeting DOT1L, *Curr. Opin. Hematol.*, 2015, **22**(2), 92–96.
78. https://www.lls.org/beat-aml. (last accessed January 2018).
79. K. Tawana and J. Fitzgibbon, Inherited DDX41 mutations: 11 genes and counting, *Blood*, 2016, **127**(8), 960–961.
80. A. A. Ferrando and C. Lopez-Otin, Clonal evolution in leukemia, *Nat. Med.*, 2017, **23**(10), 1135–1145.
81. Y. Sun, C. A. Ruivenkamp, M. J. Hoffer, T. Vrijenhoek, M. Kriek and C. J. van Asperen, *et al.*, Next-generation diagnostics: gene panel, exome, or whole genome? *Hum. Mutat.*, 2015, **36**(6), 648–655.
82. V. Visconte, A. Tabarroki, H. J. Rogers, E. Hasrouni, F. Traina and H. Makishima, *et al.*, SF3B1 mutations are infrequently found in non-myelodysplastic bone marrow failure syndromes and mast cell diseases but, if present, are associated with the ring sideroblast phenotype, *Haematologica*, 2013, **98**(9), e105–107.
83. F. Thol, R. Bollin, M. Gehlhaar, C. Walter, M. Dugas and K. J. Suchanek, *et al.*, Mutations in the cohesin complex in acute myeloid leukemia: clinical and prognostic implications, *Blood*, 2014, **123**(6), 914–920.

7 Comparative Genomics and the Improved Diagnosis of Trypanosomiasis and Visceral Leishmaniasis†

Tapan Bhattacharyya*[a], Tegwen Marlais[a], Bruno Hinckel[b,c], David Mateus[a], Hai Le[a], Niamh Murphy[a], Noah Fongwen[a], Matthew Yeo[a], Pascal Mertens[b] and Michael Miles*[a]

[a]Faculty of Infectious & Tropical Diseases, London School of Hygiene & Tropical Medicine, London, UK; [b]Coris BioConcept, Gembloux, Belgium; [c]University of Antwerp, Belgium
*E-mail: tapan.bhattacharyya@lshtm.ac.uk; michael.miles@lshtm.ac.uk

7.1. The Trypanosomatid Parasites

Trypanosomes and leishmanias (the trypanosomatids) cause devastating human and animal diseases, particularly in Africa, Latin America and Asia. An understanding of the diversity of *Trypanosoma* and *Leishmania* species and the genetics of natural populations is fundamental to the development of diagnostics and their deployment in endemic regions.

†Electronic supplementary information (ESI) available: Colour versions of Figures 7.3 and 7.5. See DOI: 10.1039/9781782628217.

Genomics and Clinical Diagnostics
Edited by David Whitehouse and Ralph Rapley

Published by the Royal Society of Chemistry, www.rsc.org

Furthermore, the distribution of these neglected tropical diseases (NTDs) is changing, with emergence of new genotypes, discovery of new species, urbanisation and their wider dissemination with ecological changes and the migration of human populations. There are no vaccines for human leishmaniasis or trypanosomiasis.

In this volume (Chapter 12), Polley describes technical approaches to the development of DNA-based diagnostic assays and summarises the current status of DNA-detection assays for a wide range of parasitic infections. Here, we focus on trypanosomiasis and leishmaniasis and demonstrate how comparative genetic and genomic analyses have informed and will continue to empower diagnostics discovery, and have led to the development of low-cost, point-of-care, rapid diagnostic tests, with both clinical and research applications. We also highlight the still underexploited potential of a systematic combined 'omics' approach to both the identification of candidate vaccines and production of prototype diagnostics.

7.1.1. Chagas Disease

Chagas disease, named after its discoverer, Carlos Chagas, is caused by the protozoan parasite *Trypanosoma cruzi*. Transmission is by contamination from blood sucking triatomine bug species, which infest houses in Latin America. The bugs defecate and urinate while or shortly after taking a blood meal. Infective forms of *T. cruzi* in the faeces can establish infection by crossing mucous membranes, the bite wound or abraded skin. Infection may also be acquired by blood or organ donation, congenitally or orally from food contaminated by triatomines or from uncooked meat of zoonotic mammalian reservoir hosts. The acute phase of infection may be fatal in 5–10% of cases, particularly in children and young adults. Prevalence of infection has decreased significantly due to vector control initiatives. However, it is estimated that around six million individuals in Latin America still carry *T. cruzi* infection.[1] Without successful treatment, *T. cruzi* infection is usually lifelong, and is opportunistic, potentially re-emerging in immunocompromised carriers, such as patients with HIV/AIDS. In the chronic phase, up to 30% of those infected suffer chagasic cardiomyopathy and electrocardiogram (ECG) abnormalities, some of whom may also develop aperistalsis and enlargement of hollow organs such as the oesophagus (megaoesophagus) or colon (megacolon). Only two toxic drugs of unreliable efficacy are currently available for treatment, benznidazole or nifurtimox, and severe dermatological side effects often interrupt treatment of adults. It is presumed that treatment of chronic cases may limit disease progression but this is not clear from clinical trials. The acute phase of *T. cruzi* infection often goes unrecognised, although in sporadic cases that have the conjunctiva as the portal of entry there may be characteristic unilateral ocular oedema and associated lymphadenopathy (Romaña's sign). Diagnosis of the chronic phase currently depends largely on suggestive clinical symptoms associated with seropositivity.

7.1.2. Human African Trypanosomiasis (HAT)

The agent of human African trypanosomiasis (HAT) is *Trypanosoma brucei*, of which there are three subspecies, *T. brucei gambiense*, *T. brucei rhodesiense* and *T. brucei brucei*. The majority of HAT is due to *T. b. gambiense*, which has humans as the reservoir of infection (anthroponotic) and occurs principally in West and Central Africa, whereas *T. b. rhodesiense* is endemic in eastern Africa, with cattle as one particularly important zoonotic reservoir. Transmission is from the infected salivary glands of bloodsucking infected tsetse flies (*Glossina* species). The epidemiologies of *T. b. gambiense* and *T. b. rhodesiense* are distinct, with the former transmitted principally by riverine *Glossina palpalis* group flies, whereas *T. b. rhodesiense* is transmitted around savannah thickets by *Glossina morsitans* flies. Progression from the haemolymphatic phase of HAT to the central nervous system (CNS) and the neurological symptoms of sleeping sickness is slower and may take years for *T. b. gambiense*, compared with faster progression for *T. b. rhodesiense*. CNS involvement is fatal unless successfully treated. Currently available drugs are toxic and inadequate and demand distinct chemotherapies for the haemolymphatic and CNS phases, although one promising new candidate drug (fexinidazole) is under trial.[2] Current differential diagnosis between the disease phases depends on lumbar puncture and examination of cerebrospinal fluid. Diagnosis, treatment and vector control have reduced prevalence among the 36 affected sub-Saharan Africa countries.[3] Elimination of HAT as a public health problem remains challenging, in part due to the diagnostic limitations.

7.1.3. Leishmaniasis

Leishmaniasis is one of the most devastating and important human parasitic diseases. *Leishmania* species are transmitted by the bite from the blocked foreguts of infected bloodsucking sand flies. It has diverse clinical presentations, of two main types, cutaneous leishmaniasis (CL), with various forms of skin lesion, and visceral leishmaniasis (VL), in which the bone marrow, spleen and liver are infected. The most severe type of CL is mucocutaneous leishmaniasis (MCL), which occurs in the Americas, and is usually due to metastatic infection with *Leishmania braziliensis* (subgenus *Viannia*) that can destroy the entire nasopharyngeal region. Visceral leishmaniasis, caused by *Leishmania donovani* and *Leishmania infantum* (subgenus *Leishmania*), is widespread, incorporating the Mediterranean region. VL due to *L. donovani* is anthroponotic, whereas dogs are the major and highly significant reservoir host of *L. infantum*. There may be vast epidemics of VL, with 10–15 year cycles, particularly in the Indian subcontinent, eastern Africa and Brazil, the cyclical epidemics being influenced by the fluctuating herd immunity of the human population.[4] Symptomatic VL is fatal unless successfully treated. Like *T. cruzi* infection, the disease is opportunistic, with active VL emerging in asymptomatic carriers who are immunocompromised. Drugs are toxic and not always effective. Post kala-azar dermal leishmaniasis (PKDL) is a possible, often post-chemotherapeutic, sequel of VL, with skin lesions that are infective to sand flies. Diagnosis has been dependent on the indicative but not definitive clinical symptoms, such as fever and hepatosplenomegaly, and associated positive serology. Microscopy of spleen or bone marrow aspirate biopsies may confirm the presence of intracellular *Leishmania* amastigotes.

7.2. The Need for Improved Clinical Diagnostics for Trypanosomiasis and Leishmaniasis

7.2.1. Shared Needs for Chagas Disease, HAT and VL

(1) Diagnostic indicators/biomarkers to confirm cure after chemotherapy, particularly for HAT and VL, for which symptomatic infection is usually fatal.
(2) Molecular markers to identify drug-resistant strains, to guide chemotherapy, avoid treatment with ineffective drugs and reduce fatalities due to drug-resistant infections. This is especially relevant to VL in the Indian subcontinent, where widespread resistance to antimonials renders them largely ineffective as first-line treatment.

(3) Diagnostic tests with improved sensitivities and specificities, for more effective surveillance of prevalence and for detection of residual infections, to facilitate elimination after disease control initiatives. In particular, elimination of anthroponotic HAT and VL due to *T. b. gambiense* and *L. donovani*, respectively, are dependent on specific diagnosis of residual foci of infection.

(4) Point-of-care, rapid diagnostic tests that fulfil ASSURED criteria: affordable; sensitive; specific; user friendly; robust; equipment free; deliverable.[5]

7.2.2. For Chagas Disease

(5) A prognostic indicator to predict carriers of *T. cruzi* infection who are likely to progress to cardiomyopathy and/or megasyndromes, such that those individuals may be closely monitored.

(6) A diagnostic test at birth for congenital Chagas disease that is not impeded by transplacental transmission of maternal antibody.

(7) Diagnostic tests to discriminate between *T. cruzi* genetic lineages that may have distinct epidemiologies and clinical consequences.

(8) An improved serological assay that is equally sensitive in all endemic regions, regardless of differential local prevalence of *T. cruzi* genetic lineages.

7.2.3. For Human African Trypanosomiasis

(9) A non-invasive test, not dependent on lumbar puncture, to distinguish between the haemolymphatic and CNS phases of HAT, which, at present, require such different chemotherapies, until there is validation and deployment of novel non-toxic drugs that are effective against both phases of the disease.

(10) Improved serology for *T. b. rhodesiense* infections, which is less effective than that for *T. b. gambiense*.

(11) Molecular diagnostics, applicable to both humans and mammalian reservoirs, for epidemiological tracking of transmission pathways and outbreaks.

7.2.4. For Visceral Leishmaniasis

(12) A diagnostic test to determine high risk of post-chemotherapy relapse to active fatal VL.
(13) Prediction of high risk of progression from asymptomatic carrier to active VL, which can then provide a new source of infection to sand flies and propagate endemic foci.
(14) Validation of a more effective rapid diagnostic test (RDT) for eastern Africa.
(15) Improved diagnostic tests to discriminate PKDL, which is known to be infective to sand flies, from other confounding dermatological conditions.
(16) A diagnostic test applicable to immunocompromised individuals who may have depleted antibody levels.

7.3. Lineage-Specific Diagnosis of *Trypanosoma cruzi* Infection

The species *T. cruzi* is remarkably diverse and is divided into six intra-species lineages (discrete typing units, DTUs) TcI–TcVI, with a possible seventh, TcBat. TcII, TcV and TcVI are the principal agents of Chagas disease in human populations of the Southern Cone countries of South America. In contrast, TcI is distributed throughout the Americas in zoonotic infections and is the most commonly found lineage of *T. cruzi* among infected humans in northern South America, Mexico and Central America. TcIII is widespread among armadillos of the genus *Dasypus* but as yet rare in humans. TcIV is a secondary agent of Chagas disease in Venezuela and, like TcI, is commonly found in mammals in the USA.[6,7]

Determination of the lineages of *T. cruzi* carried by individual patients with different clinical outcomes has been of interest since description of the disparate geographical distributions of TcII and TcI between patients in Brazil and Venezuela, where prevalence of gastrointestinal syndromes is reported to be quite distinct.[8] If parasite DNA is available, direct genotyping by restriction endonuclease digest profiles of PCR amplicons (PCR–restriction fragment length polymorphism (PCR–RFLP)) can be used to identify *T. cruzi* lineages,[9] as illustrated in Figure 7.1.

However, it is difficult to isolate and genotype *T. cruzi* from patients in the chronic phase, because the parasite multiplies in the tissues and

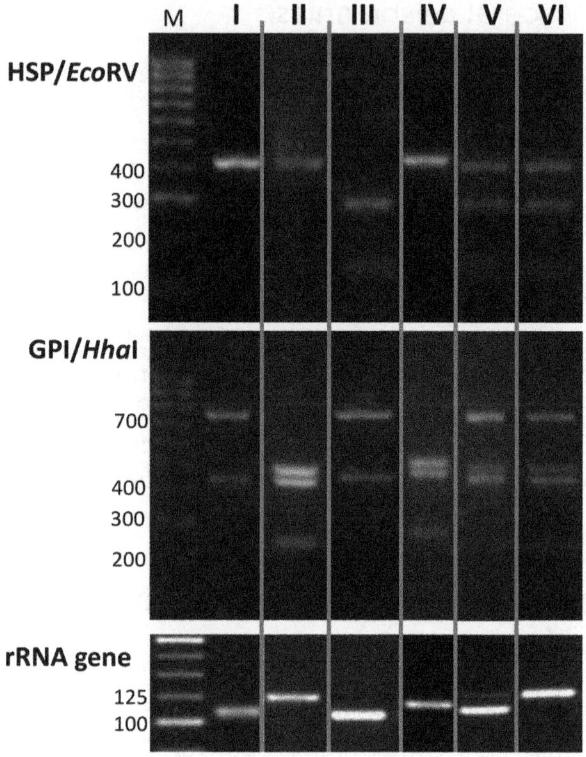

Figure 7.1 Genotyping by PCR–RFLP of *HSP60* and *GPI* genes, and by rRNA gene amplicon size, can distinguish *T. cruzi* genetic lineage without DNA sequencing. M = DNA size marker (bp).

is rare in the blood. Human research studies have revealed mixed infections of *T. cruzi* lineages and the presence in the heart or other tissues of lineages that were undetectable in the blood.[10,11] Similarly, monitoring of transgenic bioluminescent chronic *T. cruzi* infections in experimental mice reveal dynamic foci of replication in the tissues, in animals without detectable blood parasitaemia.[12] Another approach to identification of the lineage of the *T. cruzi* infection is to seek antigens that are lineage-specific and then devise lineage-specific serology to determine which lineages are, or have been, carried by each patient.

Comparative DNA and amino acid sequence analysis revealed that the mucin trypomastigote small surface antigen (TSSA) provided such lineage-specific antigens (Figure 7.2). The TcV/VI locus is heterozygous, with one haplotype shared with TcII and the other specific to TcV/VI.[13,14]

Accordingly, we designed peptides specific to these epitopes for lineage-specific enzyme-linked immunosorbent assays (ELISAs) applicable to humans[15] and several other mammal species, enabling the

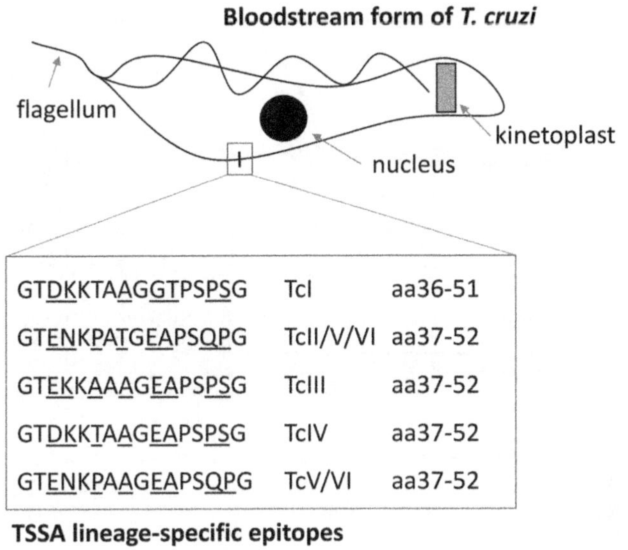

Figure 7.2 Lineage-specific epitopes of *T. cruzi* TSSA are expressed on the surface of bloodstream trypomastigotes; polymorphic residues underlined. aa, amino acid residues.

discovery of corresponding sylvatic and peridomestic reservoirs in Brazil[16] (M. McClean, unpublished work) (Figure 7.3). We adapted the TSSA-II/V/VI specific peptide serology to a novel, low-cost, point-of-care, lateral flow immunochromatographic RDT, called Chagas Sero K-SeT. Furthermore, we demonstrated a significant association between serological response to this RDT and the prevalence of increasing severity of Chagas disease cardiomyopathy.[17] Thus, the Chagas Sero K-Set RDT may provide a clinical diagnostic indicator of presence and risk of chagasic cardiomyopathy. We proposed that repeated *T. cruzi* superinfection of inhabitants of triatomine-infested houses may increase inflammatory responses and risk of cardiac damage, concomitantly boosting the immune response to the TSSA-II/V/VI epitope. We also designed the Chagas Sero K-seT RDT so that it can also be used without modification with the sera of mammals of diverse orders (Figure 7.3).

7.3.1. Detection of *Trypanosoma cruzi* DNA

Trypanosoma cruzi DNA can be detected in the blood of patients by quantitative PCR (qPCR), and this has been used in recent clinical trials to monitor the efficacy of novel anti-trypanosomal drug regimes in chronically-infected adults.[18–22] Prototype loop-mediated isothermal

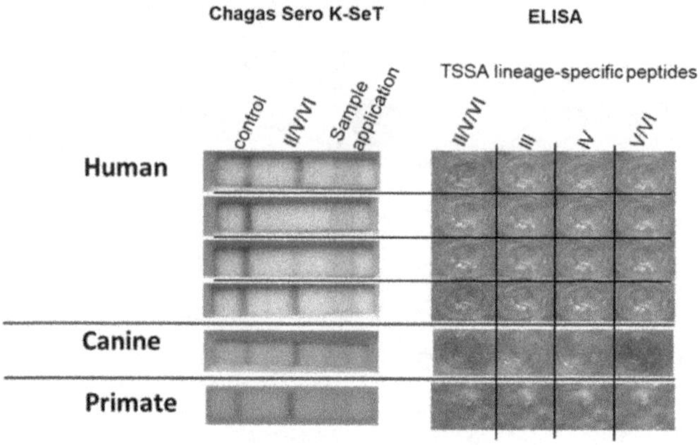

Figure 7.3 Chagas Sero K-SeT is a novel RDT that can identify TSSApep-II/V/VI-specific IgG in humans and other mammals; note the correspondence between results of the RDT and ELISAs. (Colour version available in the ESI.†)

amplification (LAMP) of DNA sequences has also been reported.[23] However, in chronic infections the low circulating parasitaemias and thus stochastic sampling, often from small blood volumes, may be a limitation to these assays. It is therefore advisable to amplify DNA targets that are multicopy, such as the kinetoplast DNA (kDNA) minicircles or repetitive satellite sequences. Schijman has provided an excellent detailed yet concise review of molecular diagnostics for *Trypanosoma cruzi* infection and their clinical applications.[24] An international comparison of PCR targets confirmed that repetitive minicircle kDNA and satellite DNA had superior sensitivities compared with rDNA or spliced-leader sequences; similarly duplex qPCR was most sensitive with kDNA and slightly less so with satellite DNA. kDNA has the disadvantage that some minicircle kDNA sequences are highly conserved, whilst satellite DNA sequences differ in abundance across the TcI–TcVI genetic lineages of *T. cruzi*. The current LAMP prototype kit detects satellite DNA, without cross reactivity to *Trypanosoma rangeli* or *Leishmania*, again with some variable sensitivity between the *T. cruzi* genetic lineages.[23] With both qPCR and LAMP approaches, as with standard PCR, controls must monitor cross contamination and consider the potential presence of inhibitors in the clinical samples. Molecular diagnosis is generally applied to blood samples but may be applicable to other clinical samples in which *T. cruzi* could be sequestered, such as pericardial effusion. Molecular diagnosis has shown that congenital transmission is most likely to occur in PCR-positive mothers and that PCR may provide an early means of detecting infection in immunosuppressed organ recipients or immunocompromised

AIDS patients. Whilst molecular methods have been applied to monitor response to chemotherapy, as with all chronic *T. cruzi* infection the current limitation is still the stochastic and limited volume sampling, which constrains sensitivity, even with repetitive DNA targets.[24] As listed in Section 7.2, a high priority is to devise a method that is applicable to diagnosis of congenital transmission, at birth and with an RDT that fulfils the minimal ASSURED criteria to allow realistic deployment at sites with no laboratory facilities.

7.4. Improved Diagnosis of Human African Trypanosomiasis (HAT)

The three *T. brucei* subspecies infections, and some other species of African trypanosomes, elude the host immune response due to their astounding capacity for antigenic variation. For example, a single infective *T. b. rhodesiense* genome may carry the genetic information and recombination mechanisms to replace the variant surface antigenic coat hundreds of times, evading each new host antibody response, and causing waves of blood parasitaemia. Multiple copies of the variant surface antigen genes (VSGs) can be activated sequentially, for example by duplicative transposition to expression sites, from VSG copies stored in minichromosomes.[25]

However, *T. b. gambiense* is more homogeneous than *T. b. rhodesiense*, with a smaller repertoire of VSGs, consistent with its predominantly clonal population origins and structure. Consequently, a card agglutination test (CATT) with stained whole organisms aids provisional serological surveillance, to guide follow up parasitological diagnosis and clinical evaluation.[26] Furthermore, common surface coat antigens have been isolated *in vitro* and used to produce a lateral-flow RDT. However, this process is cumbersome and costly. Recombinant forms of these proteins, which can be produced in bulk at low cost, match the sensitivity and specificity of the native proteins. The *T. b. gambiense* RDTs, which are in commercial production, have revolutionised serological surveillance and have been validated and deployed, particularly in the Democratic Republic of Congo (DRC).[27] In the latest development, comparative genomic information has been used to produce the corresponding VSGs in the protozoan *Leishmania tarentolae* expression system which, unlike expression in bacteria, has the advantage of glycosylating the recombinant proteins.[28] A parallel approach has used protein-G affinity chromatography to isolate IgG from *T. b. gambiense*-infected patients and then to purify corresponding proteins from *T. b. gambiense* lysates. Subsequent identification of these proteins by mass spectrometry, and trials in diagnostic ELISAs, led to four alternative diagnostic antigens that were incorporated into a lateral-flow diagnostic test.[29,30] The same technological approaches are applicable to diagnosis of animal trypanosomiasis (nagana) such as the disease due to *Trypanosoma congolense* infection[31] and *Trypanosoma vivax*.[32]

T. b. rhodesiense and *T. b. gambiense* escape lysis by human serum due to the presence of the serum resistant-associated (*SRA*) gene and the *TgsGP* gene, respectively, which impede the lytic action of apolipoprotein 1, whereas *T. b. brucei* is lysed by human serum and thus is not a human pathogen. These distinctive genomic features of *T. b. rhodesiense* and *T. b. gambiense* can be used for epidemiological tracking to identify the origin of outbreaks, follow their spread and guide control measures, for example, the surveillance and treatment of cattle in the case of *T. b. rhodesiense*. Furthermore, both *T. b. rhodesiense* and *T. b. gambiense* are present in Uganda, the former in the South and the latter in the North. Such epidemiological tracking can be used to monitor the proximity of the two subspecies and as an early warning system to detect the potential emergence of new hybrid epidemic clones.[33]

There is now considerable interest in developing LAMP for the diagnosis of African trypanosomes, as well as for *Trypanosoma cruzi* (see Section 7.3.1.) and *Leishmania* (see Section 7.5). Mugasa *et al.*[34] compared nucleic-acid-sequence-based amplification (NASBA) combined with oligochromatography (NASBA-OC) against LAMP for diagnosis of *T. b. gambiense* in Uganda and found the sensitivity of NASBA-OC to be significantly higher than that of LAMP. Subsequent studies have focused on LAMP, producing assays that distinguish between the three *T. brucei* subspecies, *T. b. rhodesiense*, *T. b. gambiense* and *T. b. brucei*.[35] LAMP has also been applied to bovine African trypanosomiasis showing sensitivity superior to that of microscopy of blood film, and resolving the species of infection, whether *T. vivax*, *T. congolense* or *T. b. rhodesiense*.[36] The detection of *T. b. rhodesiense* in cattle is of special importance as one measure of the risk of HAT among local communities and visiting tourists. LAMP was also shown to be superior to blood film microscopy in a small group of vervet monkeys followed for 180 days after infection: LAMP showed some potential to detect infection in saliva and urine but not in the later stages of infection.[37] LAMP has been adapted to a digital microfluidic system (DMF) chip, enabling reduction of reagent requirements and lower risk of contamination compared with conventional LAMP.[38] Linking PCR amplicons to matrix-assisted laser desorption ionization-time of flight (MALDI-TOF) has also shown the capacity to distinguish between trypanosomatid species, although in comparison to LAMP, prospects for deployment seem limited.[39]

One mechanism of overcoming the difficulties in discriminating between the many agents of fever syndrome, including *T. brucei* spp. infection, for which a high proportion of cases may lack specific diagnosis due to the diffuse systemic nature of the symptomology, is to devise a single diagnostic device capable of identifying a wide range of pathogens, including those associated with emerging and potentially escalating outbreaks. In pursuit of this aim, Liu *et al.*[40] produced a prototype TaqMan array card capable of simultaneously detecting 26 acute-febrile-illness-associated organisms. Further development of such a low-cost user-friendly system, including an ability to distinguish between susceptible and drug-resistant organisms for which there are molecular markers, would be of great benefit in both response to outbreaks and for surveillance and risk assessment.

7.5. Identifying Disease Agents and Monitoring Clinical Status of Visceral Leishmaniasis (VL)

There are more than 20 named species of *Leishmania*. They are morphologically very similar and comparative morphology does not allow reliable identification to species level. Historically the species were separated based on clinical presentation, geographical distribution, associated hosts or phenotype in experimental models. PCR–RFLP (Figure 7.4),[41] multilocus sequence typing (MLST), multilocus microsatellite typing (MLMT) and comparative genomics have provided analytical methods to discriminate between *Leishmania* species and to resolve taxonomic and phylogenetic relationships.[42] However, genotyping by PCR–RFLP and MLST is currently not-user friendly for routine clinical diagnosis or RDT delivery at point-of-care.

Visceral leishmaniasis was, with rare exceptions, attributed to three species, *L. donovani*, *L. infantum* and *L. chagasi*. However, MLST and MLMT demonstrated that *L. chagasi* in South and Central America was indistinguishable from *L. infantum*. Furthermore, comparative genotyping tracked the Iberian origins of New World *L. infantum* (syn. *L. chagasi*) and dated it to the colonisation of Latin America by the conquistadors, who took their dogs, the principal reservoir hosts of

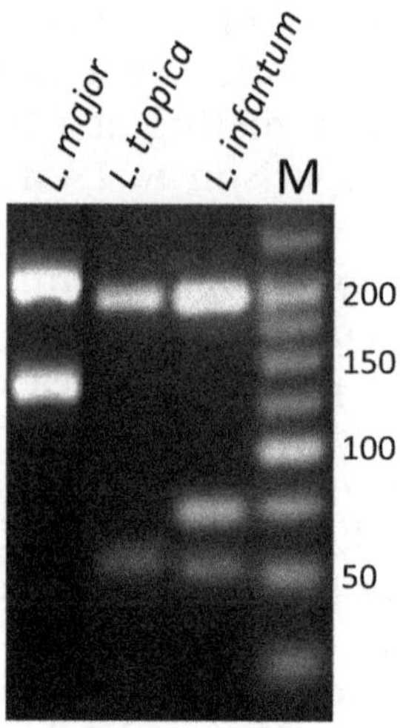

Figure 7.4 *Leishmania* species can be identified by PCR–RFLP using parasite cultures or, with variable sensitivities, clinical samples, *e.g.* skin lesion biopsies, to distinguish *L. major* from *L. tropica*, which may have different clinical presentations and responses to treatment. As an example, PCR–RFLPs for ITS1-*Hae*III of *L. tropica*, *L. infantum* and *L. major* are shown. M = DNA size marker (bp).

L. infantum and which have prolonged VL incubation periods. The permissive sand fly vector species was already present in the New World.[43] Application of these molecular methods has also led to the provisional description of new *Leishmania* taxa, with extraordinary geographical distributions and possibly not transmitted exclusively by sand flies.[44–46]

Monitoring of treatment outcome of VL is currently based on absence of clinical symptoms at a defined time, such as six months after completion of chemotherapy, with parasitological proof dependent on microscopy of high-risk invasive spleen or bone marrow aspirates. Antibodies in VL have been detected by ELISA, indirect fluorescence (IFAT) or direct agglutination (DAT). Screening of a genomic library with patients' serum identified a kinesin-related gene product from a Brazilian '*L. chagasi*' strain (now understood to be European – derived *L. infantum*, see above) as a candidate antigen for serology,[47] and a contiguous fragment consisting of a 46 amino acid (aa) region followed by 6.5 × 39 aa repeats was used to produce the recombinant diagnostic protein rK39. The rK39 protein was adapted to a lateral-flow immunochromatographic point-of-care RDT, which proved to be highly sensitive in the Indian subcontinent but less so in eastern Africa.[48,49] Comparisons of rK39 homologous sequences between

Table 7.1 Polymorphism of rK39 sequences in east African and south Asian *L. donovani*. In these composite data, numbers represent the amino acid residues in the canonical 6.5 × 39 aa repeats of the parent protein (GenBank L07879).

	2	4	5	6	8	11	13	16	17	18	22	24	27	35	38
rK39	E	Q	L	R	S	R	A	A	S	Q	T	A	S M	S	A
Africa	G	L		L H	P			K	A	E		V	P T	N	
Asia		R	F			H	V	M	A R	K	A		T	N	P

Asian and eastern Africa *L. donovani* showed that the lower efficacy of rK39 may in part be explicable by divergent codons, particularly in the 5′ half of the repeats, which in many cases led to non-conservative aa changes (Table 7.1).[50]

It was also shown that significantly higher anti-*Leishmania* antibody levels occur in Indian VL patients compared with Sudanese.[51] A synthetic recombinant protein, rK28, comprising two 39 aa repeats of the homologous kinesin of a Sudanese *L. donovani* and flanking HASPB1 and HASPB2 sequences from Ethiopian *L. donovani*, appears to have improved sensitivity in eastern Africa and is under further validation.[52,53] However, some strains that produced unpredicted size amplicons were shown to have non-canonical HASPB1 and HASPB2 repeat arrangements.[50]

Despite the success of the rK39 RDT in the Indian subcontinent, the limitations apply of failing to detect relapse after chemotherapy or to predict progression of asymptomatic carriers to active VL. Anti-rK39 IgG titres may be sustained for long periods after treatment and seropositivity may be frequent among endemic healthy controls. Accordingly, we have adopted an alternative approach, of comparing antibody subclass profiles between patients with different clinical status, showing that IgG1 levels decline after chemotherapy but may be sustained in relapse, and similarly that high-IgG1 profiles are associated with progression to active VL[54] (T. Marlais, unpublished work). We have produced a corresponding IgG1-specific RDT, called VL Sero K-SeT, currently under trial in India, and are comparing synthetic peptides and recombinant antigens for optimal incorporation into this IgG1 RDT format (G. Mollett and B. Hinckel, unpublished works).

Various molecular methods have been reported that may detect parasite-specific DNA and could provide an empirical test of treatment success, which would be less invasive than aspirate microscopy if the DNA is detectable in the circulating blood of VL patients. Such tests can be applicable to generic diagnosis of a *Leishmania* infection if a genus-specific target is selected. qPCR has enabled highly sensitive detection of parasite DNA in peripheral blood buffy coat in VL and relapsed VL cases.[55] Equally, the ability to quantify parasite DNA from PKDL skin biopsies can enable concurrent monitoring of treatment efficacy in a condition with limited alternative options for follow-up, apart from microscopy.[56]

LAMP of DNA is a technique that can circumvent the need for expensive laboratory PCR equipment. LAMP technology has been applied with very high sensitivity and specificity to peripheral blood using a low-cost DNA extraction method.[57] In monitoring treatment, blood and bone marrow aspirate have had 96.9–100% sensitivity for detecting VL in only 30 minutes amplification time.[58] Equally, LAMP had 97% sensitivity for PKDL when used with tissue biopsy samples[58] and in both VL and PKDL had the potential to indicate residual infection and recurrence of disease. The recombinase polymerase assay (RPA), provides an additional option for molecular diagnosis using a blood sample, without the need for large laboratory equipment or a cold chain and has produced similar results to lab-based qPCR.[59]

Molecular methods have additional potential for diagnosing and confirming VL in HIV co-infected patients, who may have elevated levels of amastigotes compared with immunocompetent individuals and for whom serology is less reliable and aspirate procedures carry more risk than in those who are HIV negative.[60] Of great importance is the prospect of using non-invasive sampling types in sensitive and specific molecular tests with urine or saliva as potential alternatives to blood as sources of parasite DNA. Urine samples have shown promise for diagnosis of VL, including in HIV co-infected individuals.[61,62] HIV-positive individuals may suffer from chronic relapsing VL. Circulating *Leishmania* DNA, detected by qPCR of peripheral blood DNA, has predicted VL relapse within the subsequent 4 months, a useful indication and where a negative result could also support the decision to interrupt VL prophylaxis among certain HIV-VL patients.[63]

7.6. Comparative Genomics, Genetic Crosses and Diagnostics Discovery

A summary of the *T. cruzi*, *T. brucei* and *L. donovani*/*L. infantum* genomes that are in the public domain is shown in Table 7.2.

Genomes of five of the six principal *T. cruzi* genetic lineages, TcI, TcII, TcIII, TcV and TcVI, are publically available. However, assembly of these *T. cruzi* genomes, particularly those of TcV and TcVI, which

Table 7.2 Trypanosomatid whole-genome sequences available for download.

	Strain	Bioproject[e]	Publication	Comments
T. cruzi[a]				
TcI	Silvio X10/1	PRJNA40815	Franzén *et al.*[65,80]	First TcI genome
TcI	JR	PRJNA59941	—	
TcI	Dm28c	PRJNA225928	Grisard *et al.*[81]	
TcII	Esmeraldo	PRJNA50493	—	
TcII	Y	PRJNA393091	—	
TcIII	231	PRJEB9129	Baptista *et al.*[82]	
TcV	Bug2148	PRJNA393091	—	
TcVI	CL Brener	PRJNA11755/ PRJNA15540	El-Sayed *et al.*[64]	First *T. cruzi* genome/RefSeq genome
TcVI	Tula	PRJNA169675	—	
T. cruzi marinkellei	B7	PRJNA77843	Franzén *et al.*[80]	Bat-restricted subspecies
T. brucei[b]				
T. b. gambiense	MHOM/CI/86/ DAL972	PRJEA40697/ PRJNA260635	Jackson *et al.*[83]	RefSeq genome
T. b. rhodesiense[c]	Isolates B17 and Z310	PRJEB3300	Goodhead *et al.*[84]	
T. b. brucei	GPAL/KE/70/ EATRO 1534 (TREU927/4 GUTat10.1)	PRJNA11756	Berriman *et al.*[85]	First *T. brucei* spp. genome
Leishmania[d]				
L. infantum	MCAN/ES/98/ LLM-877 (JPCM5)	PRJEB20254	Peacock *et al.*[86]	First *L. infantum* genome; subsequently re-sequenced *de novo*.
L. donovani	MHOM/ NP/2003/ BPK282	PRJEA61817/ PRJNA171503	Downing *et al.*[87]	First *L. donovani* genome/RefSeq genome
	MHOM/ IN/1983/AG83	PRJNA297706	—	
	1S2D	PRJNA396645	—	

[a]*T. cruzi* genomes (https://www.ncbi.nlm.nih.gov/genome/genomes/25).
[b]*T. brucei* genomes (https://www.ncbi.nlm.nih.gov/genome/genomes/24).
[c]*T. b. rhodesiense* genomes (https://www.ebi.ac.uk/ena/data/view/PRJEB3300).
[d]*Leishmania* genomes (https://www.ncbi.nlm.nih.gov/genome/genomes/3516).
[e]A full list of sequencing projects, including partial sequences and transcriptomes, is available on the NCBI BioProject website by searching for the relevant species name; also *L. donovani*.[68]

are natural hybrids of TcII and TcIII, is generally imperfect, due to the extreme complexity and repetitive nature of the subtelomeric gene families, such as the trans-sialidases,[64,65] although multiple gene families may now be partially resolved by high-coverage PacBio sequencing. Reference genomes of the three subspecies of *T. brucei* are also available, with quite detailed population genetics analyses, particularly of *T. b. gambiense*.[66] The most genomes are available for *Leishmania* species, for example in the context of searching for drug resistance markers in Nepal and India or to understand taxonomic and phylogenetic relationships.[67,68] This comparative genomics research has been pioneered by the Wellcome Trust Sanger Institute. Nevertheless, the genomics research on trypanosomatids lags behind that on viral and bacterial pathogens or on *Plasmodium falciparum*, the agent of the severest form of malaria. Furthermore, the coverage of trypanosomatid genomes is patchy, there are few reliable archives of samples and there are constraints on the international exchange of live strains or DNA extracts. As indicated above, it is difficult to predict reliably the success of diagnostic prototypes, vaccine candidates or new drugs without a wider knowledge of the genomic diversity of these organisms, which warrants a systematic international initiative.

Historically, all trypanosomatid species were considered to be clonal, and even more recently there have been polarized views on the frequency of genetic exchange and whether it has significant evolutionary or epidemiological significance.[69,70] However, improved and wider application of comparative and population genetics have demonstrated abundant genetic exchange among natural trypanosomatid populations. Furthermore, experimental genetic exchange has been achieved for *T. cruzi*, *T. brucei* subspecies and *Leishmania* species.[71–75] This has been greatly facilitated by crossing transgenic fluorescent clones and selection of hybrid populations using double drug-resistance markers.[76] Thus, green and red parents yield hybrid progeny displaying both colours (yellow) (Figure 7.5). In the case of *T. brucei* and *Leishmania* genetic crosses have now at last been achieved with relative facility, in the insect vectors.

Figure 7.5 Confocal microscopy image of a *Leishmania donovani* hybrid clone expressing eGFP and dsRED, and resistant to both Hygromycin B and Geneticin (G418). Clone derived from a genetic cross between two parental clonal populations, one expressing eGFP and resistant to Geneticin, and the other expressing dsRED and resistant to Hygromycin B. From left to right: DAPI DNA stain, eGFP, dsRED and composite image. (Colour version available in the ESI.†)

Genetic exchange and/or hybridisation can lead to radical changes in diversity, re-assorting or disrupting genes, including those encoding diagnostic antigens (Figure 7.6).

Comparative analysis of parents and progeny of genetic crosses can be highly informative, providing a means of understanding the effects of genetic exchange on candidate diagnostic antigens, allowing discarding of those with limited potential and selection of appropriately conserved strain-, species- or genus-specific antigens.

7.7. A Comprehensive 'Omics' Approach

Diagnostics discovery can benefit from multiple comparisons, incorporating genomics, transcriptomics and proteomics. Each of the three parallel analyses can contribute to searches for diagnostic antigens (Figure 7.7). Such a systematic combined 'omics' approach has been developed for identification of vaccine candidates[77,78] and is

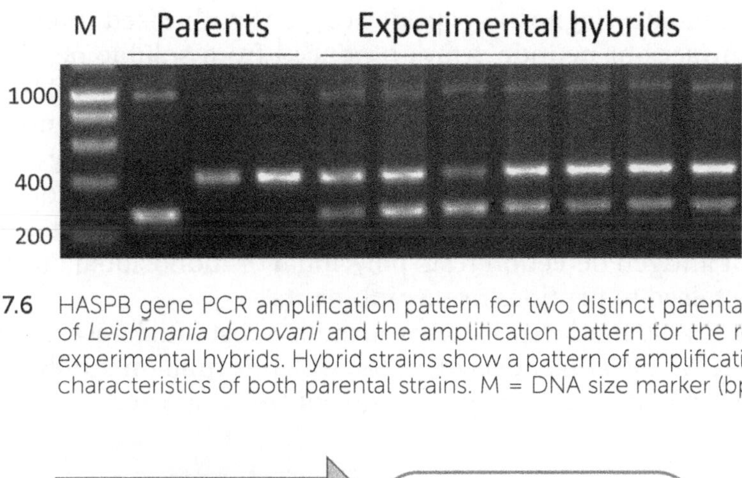

Figure 7.6 HASPB gene PCR amplification pattern for two distinct parental strains of *Leishmania donovani* and the amplification pattern for the resulting experimental hybrids. Hybrid strains show a pattern of amplification with characteristics of both parental strains. M = DNA size marker (bp).

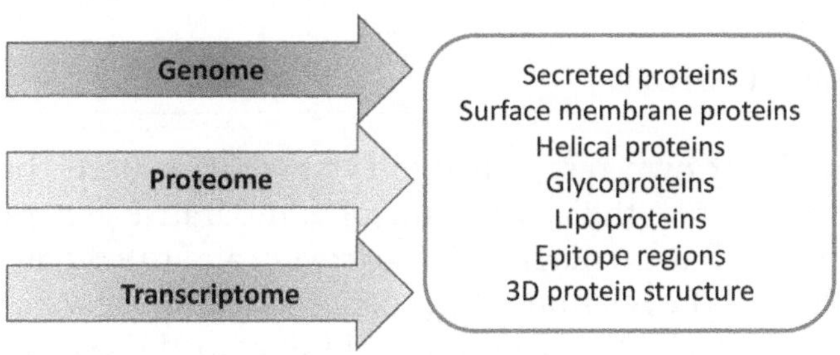

Figure 7.7 A combined 'omics' approach to diagnostic target discovery utilising genomic, proteomic and transcriptomic data and computational tools to predict molecular features.

equally applicable to diagnostics discovery.[79] Comparative genomics enables identification of species-specific or genus-specific genes and/or proteins; species-specific, life-cycle-stage-specific or organ-specific proteins can be defined further by transcriptomics, and excretory–secretory (E/S) proteomes can be identified by mass spectrometry.

Candidate proteins can be roughly resolved by SDS-PAGE or size-exclusion chromatography and the affinity of specific host antibodies to the candidates can be tested by immunoblots or ELISA, respectively. Mass spectrometry allows for the identification of a pool of antigens of interest, while publicly available RNA-seq data can be used to shortlist proteins on the basis of their expression levels. Online bioinformatic tools then allow prediction of protein localisation and other desired properties, such as the presence of tandem repeats (Figure 7.8).

Online epitope prediction servers can be used alone or in combination, to identify consensus epitopes from multiple predictors, to select antigenic peptides from multiple proteins, for which species or wider conservation can be confirmed by gene amplification and sequencing of disparate new or archived samples, if available. Multiple candidate diagnostic peptides can be synthesised directly onto solid surfaces or peptide arrays produced from soluble peptides for pilot serological assays using low-cost reagents (*e.g.* peptides of low purity). Final peptide adjustments, such as higher-purity synthesis of multiple antigenic peptides, may increase the antigenicity of previously selected low-cost antigenic peptides. Alternatively, recombinant proteins may be expressed for pilot serological assays. For development of antigen detection tests polyclonal or monoclonal antibodies may be raised to peptides or recombinants.

Such a systematic approach warrants substantial investment, to deliver ASSURED diagnosis of NTDs, which is fundamentally important to both clinical management and control strategies.

7.8. Conclusions

Here, around our own research interests, we have used the trypanosomatids to illustrate the crucial role of comparative genetic diversity and genomics to inform diagnostics discovery and to validate the scope and specificity of the resultant diagnostic assays. We also highlight the importance of providing ASSURED rapid diagnostic tests, particularly for NTDs. We have shown that RDTs based on partial but focused understanding of genomic diversity can provide life-saving tests, which can also inform disease-control strategies. We also highlight the high-priority gaps in the current range of trypanosomatid diagnostics.

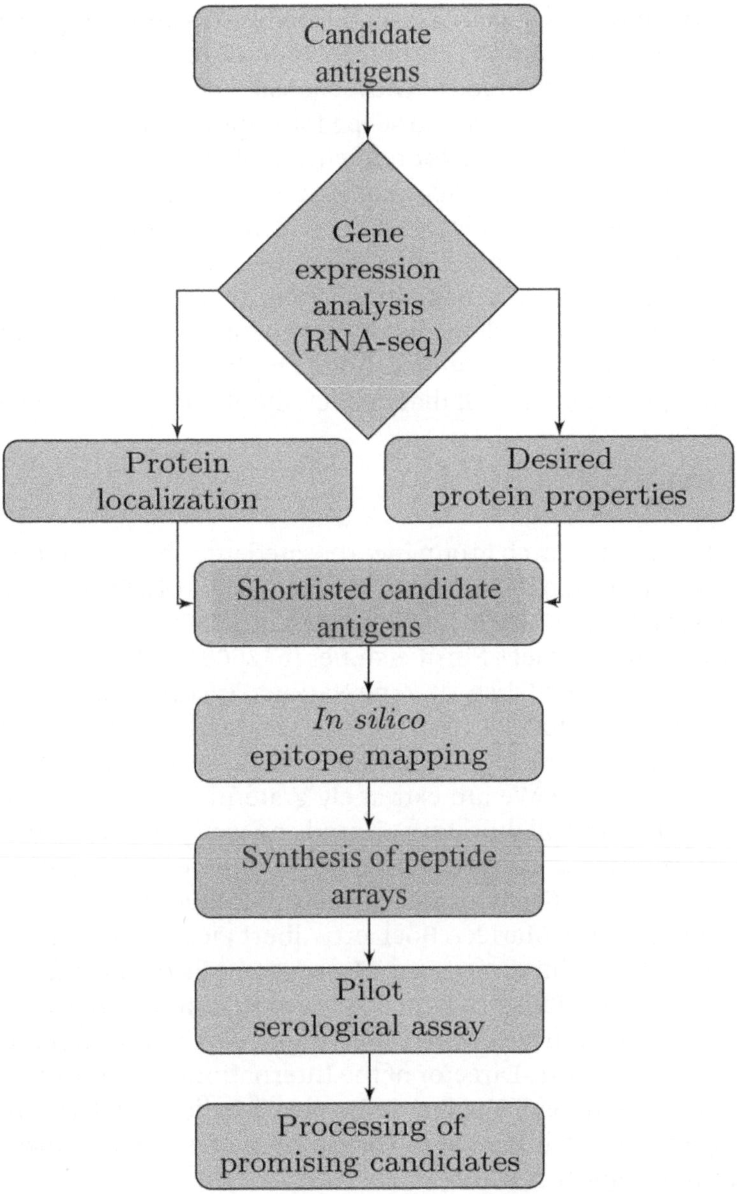

Figure 7.8 Flowchart of the initial processes for the selection of candidate diagnostic antigens.

The systematic *in silico* comparative 'omics' pathway to diagnostics is underexplored and lacks investment, yet is an irresistible route to selecting candidate diagnostics for prototype assessment. There are some limitations to the deployment of DNA-based assays.

Detection of DNA is dependent on the ability to select and process adequate samples. For example, chronic infections, in which disease agents are rarely present in the blood or with sequestered foci deep in the tissues, may not yield DNA. There is also scope for improved sample processing, for example of faeces, a need for reference biobanks and for better interaction between diagnostics laboratories and field research centres.

Highly sensitive diagnostic technological innovations are under development. However, for some discriminative diagnostic tests we have seen here that absolute sensitivity may not be required. Given appropriate antigens, ideally identified by comparative genomics, antibody profiling or antigen detection may fulfil a high-priority need, therefore, the further potential of such diagnostics should not be overlooked.

Acknowledgements

The work by our research group incorporated into this chapter has been principally supported by European Union Framework Programme Projects ChagasEpiNet (223034) and NIDIAG (260260), and the Horizon 2020 Programme Project EuroLeish.net (642609). T. M. was funded by the Sir Halley Stewart Trust, UK. The National Institutes of Health, USA, grant number R01AI07028 contributed to research on the genetic diversity of *T. cruzi*. Historically, funding for M. A. M. was provided by the Wellcome Trust, UK. We are extremely grateful to our many MSc and PhD students and collaborators around the world, without whom our research programme would not have been possible, particularly: Shyam Sundar, O. P. Singh, Sayda El Safi, Alejandro Luquetti, Ana Jansen, Petr Volf, Jovana Sadlova, Marleen Boelaert, Albert Picado, Andrew Falconar, Caryn Bern, Bob Gilman, Louisa Messenger, Martin Llewellyn, Björn Andersson, Carlos Talavera-López, Suman Rijal and other co-authors of our previous publications. We also thank Noah Fongwen and especially Rosanna Peeling, Director of the International Diagnostics Centre at the London School of Hygiene & Tropical Medicine (LSHTM), for her encouragement and initiative in championing research on diagnostics of neglected tropical diseases.

References

1. Weekly Epidemiological Record, 2015, 90, 33–43.
2. M. P. Pollastri, *Trends Parasitol.*, 2018, **34**, 178–179.
3. J. R. Franco, G. Cecchi, G. Priotto, M. Paone, A. Diarra, L. Grout, R. C. Mattioli and D. Argaw, *PLoS Neglected Trop. Dis.*, 2017, **11**, e0005585.
4. O. P. Singh, E. Hasker, M. Boelaert and S. Sundar, *Lancet Infect. Dis.*, 2016, **16**, e304–e309.

5. R. W. Peeling, K. K. Holmes, D. Mabey and A. Ronald, *Sexually Transmitted Infections*, 2006, **82**, v1.
6. L. A. Messenger, M. A. Miles and C. Bern, *Expert Rev. Anti-Infect. Ther.*, 2015, **13**, 995–1029.
7. B. Zingales, M. A. Miles, D. A. Campbell, M. Tibayrenc, A. M. Macedo, M. M. Teixeira, A. G. Schijman, M. S. Llewellyn, E. Lages-Silva, C. R. Machado, S. G. Andrade and N. R. Sturm, *Infect., Genet. Evol.*, 2012, **12**, 240–253.
8. M. A. Miles, R. A. Cedillos, M. M. Povoa, A. A. de Souza, A. Prata and V. Macedo, *Lancet*, 1981, **1**, 1338–1340.
9. M. D. Lewis, J. Ma, M. Yeo, H. J. Carrasco, M. S. Llewellyn and M. A. Miles, *Am. J. Trop. Med. Hyg.*, 2009, **81**, 1041–1049.
10. A. R. Vago, L. O. Andrade, A. A. Leite, D. d'Avila Reis, A. M. Macedo, S. J. Adad, S. Tostes Jr, M. C. Moreira, G. B. Filho and S. D. Pena, *Am. J. Pathol.*, 2000, **156**, 1805–1809.
11. J. M. Burgos, M. Diez, C. Vigliano, M. Bisio, M. Risso, T. Duffy, C. Cura, B. Brusses, L. Favaloro, M. S. Leguizamon, R. H. Lucero, R. Laguens, M. J. Levin, R. Favaloro and A. G. Schijman, *Clin. Infect. Dis.*, 2010, **51**, 485–495.
12. M. D. Lewis and J. M. Kelly, *Trends Parasitol.*, 2016, **32**, 899–911.
13. J. M. Di Noia, C. A. Buscaglia, C. R. De Marchi, I. C. Almeida and A. C. Frasch, *J. Exp. Med.*, 2002, **195**, 401–413.
14. T. Bhattacharyya, J. Brooks, M. Yeo, H. J. Carrasco, M. D. Lewis, M. S. Llewellyn and M. A. Miles, *Int. J. Parasitol.*, 2010, **40**, 921–928.
15. T. Bhattacharyya, A. K. Falconar, A. O. Luquetti, J. A. Costales, M. J. Grijalva, M. D. Lewis, L. A. Messenger, T. T. Tran, J. D. Ramirez, F. Guhl, H. J. Carrasco, P. Diosque, L. Garcia, S. V. Litvinov and M. A. Miles, *PLoS Neglected Trop. Dis.*, 2014, **8**, e2892.
16. C. L. Kerr, T. Bhattacharyya, S. C. Xavier, J. H. Barros, V. S. Lima, A. M. Jansen and M. A. Miles, *Parasites Vectors*, 2016, **9**, 584.
17. T. Bhattacharyya, L. A. Messenger, C. Bern, P. Mertens, Q. Gilleman, N. Zeippen, B. C. Bremer Hinckel, N. Murphy, R. H. Gilman and M. A. Miles, *Clin. Infect. Dis.*, 2018, **67**, 519–524.
18. I. Molina, J. Gomez i Prat, F. Salvador, B. Trevino, E. Sulleiro, N. Serre, D. Pou, S. Roure, J. Cabezos, L. Valerio, A. Blanco-Grau, A. Sanchez-Montalva, X. Vidal and A. Pahissa, *N. Engl. J. Med.*, 2014, **370**, 1899–1908.
19. C. A. Morillo, J. A. Marin-Neto, A. Avezum, S. Sosa-Estani, A. Rassi Jr, F. Rosas, E. Villena, R. Quiroz, R. Bonilla, C. Britto, F. Guhl, E. Velazquez, L. Bonilla, B. Meeks, P. Rao-Melacini, J. Pogue, A. Mattos, J. Lazdins, A. Rassi, S. J. Connolly and S. Yusuf, *N. Engl. J. Med.*, 2015, **373**, 1295–1306.
20. C. A. Morillo, H. Waskin, S. Sosa-Estani, M. Del Carmen Bangher, C. Cuneo, R. Milesi, M. Mallagray, W. Apt, J. Beloscar, J. Gascon, I. Molina, L. E. Echeverria, H. Colombo, J. A. Perez-Molina, F. Wyss, B. Meeks, L. R. Bonilla, P. Gao, B. Wei, M. McCarthy and S. Yusuf, *J. Am. Coll. Cardiol.*, 2017, **69**, 939–947.
21. F. Torrico, J. Gascon, L. Ortiz, C. Alonso-Vega, M. J. Pinazo, A. Schijman, I. C. Almeida, F. Alves, N. Strub-Wourgaft and I. Ribeiro, *Lancet Infect. Dis.*, 2018, **18**, 419–430.
22. J. C. Ramirez, R. Parrado, E. Sulleiro, A. de la Barra, M. Rodriguez, S. Villarroel, L. Irazu, C. Alonso-Vega, F. Alves, M. A. Curto, L. Garcia, L. Ortiz, F. Torrico, J. Gascon, L. Flevaud, I. Molina, I. Ribeiro and A. G. Schijman, *PLoS One*, 2017, **12**, e0188550.
23. S. A. Besuschio, M. Llano Murcia, A. F. Benatar, S. Monnerat, I. Cruz, A. Picado, M. L. A. Curto, Y. Kubota, D. P. Wehrendt, P. Pavia, Y. Mori, C. Puerta, J. M. Ndung'u and A. G. Schijman, *PLoS Neglected Trop. Dis.*, 2017, **11**, e0005779.
24. A. G. Schijman, *Acta Trop.*, 2018, **184**, 59–66.
25. R. Devlin, C. A. Marques and R. McCulloch, *Curr. Genet.*, 2017, **63**, 441–449.
26. V. Lejon, D. M. Ngoyi, M. Boelaert and P. Buscher, *PLoS Neglected Trop. Dis.*, 2010, **4**, e590.

27. P. Buscher, P. Mertens, T. Leclipteux, Q. Gilleman, D. Jacquet, D. Mumba-Ngoyi, P. P. Pyana, M. Boelaert and V. Lejon, *Lancet Glob. Health*, 2014, **2**, e359–363.
28. B. Rooney, T. Piening, P. Büscher, S. Rogé and C. M. Smales, *PLoS Neglected Trop. Dis.*, 2015, **9**, e0004271.
29. L. Sullivan, S. J. Wall, M. Carrington and M. A. J. Ferguson, *PLoS Neglected Trop. Dis.*, 2013, 7, e2087.
30. L. Sullivan, J. Fleming, L. Sastry, A. Mehlert, S. J. Wall and M. A. Ferguson, *PLoS Neglected Trop. Dis.*, 2014, **8**, e2976.
31. J. R. Fleming, L. Sastry, T. W. M. Crozier, G. B. Napier, L. Sullivan and M. A. J. Ferguson, *PLoS Neglected Trop. Dis.*, 2014, **8**, e2936.
32. J. R. Fleming, L. Sastry, S. J. Wall, L. Sullivan and M. A. Ferguson, *PLoS Neglected Trop. Dis.*, 2016, **10**, e0004977.
33. C. D. Kato, V. P. Alibu, A. Nanteza, C. M. Mugasa and E. Matovu, *Parasites Vectors*, 2016, **9**, 259.
34. C. M. Mugasa, D. Katiti, A. Boobo, G. W. Lubega, H. D. Schallig and E. Matovu, *Diagn. Microbiol. Infect. Dis.*, 2014, **78**, 144–148.
35. O. V. Nikolskaia, O. M. Thekisoe, J. S. Dumler and D. J. Grab, *Am. J. Trop. Med. Hyg.*, 2017, **96**, 275–279.
36. I. J. Haji, C. Sugimoto, K. Kajino, I. Malele, H. Simukoko, H. Chitambo and B. Namangala, *Trop. Anim. Health Prod.*, 2015, **47**, 1139–1143.
37. M. Ngotho, J. M. Kagira, B. M. Gachie, S. M. Karanja, M. W. Waema, D. N. Maranga and N. W. Maina, *BioMed Res. Int.*, 2015, **2015**, 867846.
38. L. Wan, T. Chen, J. Gao, C. Dong, A. H. Wong, Y. Jia, P. I. Mak, C. X. Deng and R. P. Martins, *Sci. Rep.*, 2017, **7**, 14586.
39. M. Angelica Luque-Gonzalez, M. Tabraue-Chavez, B. Lopez-Longarela, R. Maria Sanchez-Martin, M. Ortiz-Gonzalez, M. Soriano-Rodriguez, J. Antonio Garcia-Salcedo, S. Pernagallo and J. Jose Diaz-Mochon, *Talanta*, 2018, **176**, 299–307.
40. J. Liu, C. Ochieng, S. Wiersma, U. Stroher, J. S. Towner, S. Whitmer, S. T. Nichol, C. C. Moore, G. J. Kersh, C. Kato, C. Sexton, J. Petersen, R. Massung, C. Hercik, J. A. Crump, G. Kibiki, A. Maro, B. Mujaga, J. Gratz, S. T. Jacob, P. Banura, W. M. Scheld, B. Juma, C. O. Onyango, J. M. Montgomery, E. Houpt and B. Fields, *J. Clin. Microbiol.*, 2016, **54**, 49–58.
41. G. Schonian, A. Nasereddin, N. Dinse, C. Schweynoch, H. D. Schallig, W. Presber and C. L. Jaffe, *Diagn. Microbiol. Infect. Dis.*, 2003, **47**, 349–358.
42. J. Lukes, I. L. Mauricio, G. Schonian, J. C. Dujardin, K. Soteriadou, J. P. Dedet, K. Kuhls, K. W. Tintaya, M. Jirku, E. Chocholova, C. Haralambous, F. Pratlong, M. Obornik, A. Horak, F. J. Ayala and M. A. Miles, *Proc. Natl. Acad. Sci. U. S. A.*, 2007, **104**, 9375–9380.
43. K. Kuhls, M. Z. Alam, E. Cupolillo, G. E. Ferreira, I. L. Mauricio, R. Oddone, M. D. Feliciangeli, T. Wirth, M. A. Miles and G. Schonian, *PLoS Neglected Trop. Dis.*, 2011, **5**, e1155.
44. O. A. Espinosa, M. G. Serrano, E. P. Camargo, M. M. Teixeira and J. J. Shaw, *Parasitology*, 2016, 1–13.
45. A. M. Dougall, B. Alexander, D. C. Holt, T. Harris, A. H. Sultan, P. A. Bates, K. Rose and S. F. Walton, *Int. J. Parasitol.*, 2011, **41**, 571–579.
46. V. Seblova, J. Sadlova, B. Vojtkova, J. Votypka, S. Carpenter, P. A. Bates and P. Volf, *PLoS Neglected Trop. Dis.*, 2015, **9**, e0004060.
47. J. M. Burns, W. G. Shreffler, D. R. Benson, H. W. Ghalib, R. Badaro and S. G. Reed, *Proc. Natl. Acad. Sci. U. S. A.*, 1993, **90**, 775–779.
48. M. Boelaert, S. El-Safi, A. Hailu, M. Mukhtar, S. Rijal, S. Sundar, M. Wasunna, A. Aseffa, J. Mbui, J. Menten, P. Desjeux and R. W. Peeling, *Trans. R. Soc. Trop. Med. Hyg.*, 2008, **102**, 32–40.
49. J. Cunningham, E. Hasker, P. Das, S. El Safi, H. Goto, D. Mondal, M. Mbuchi, M. Mukhtar, A. Rabello, S. Rijal, S. Sundar, M. Wasunna, E. Adams, J. Menten, R. Peeling, M. Boelaert and WHO/TDR Visceral Leishmaniasis Laboratory Network, *Clin. Infect. Dis.*, 2012, **55**, 1312–1319.

50. T. Bhattacharyya, M. Boelaert and M. A. Miles, *PLoS Neglected Trop. Dis.*, 2013, 7, e2057.
51. T. Bhattacharyya, D. E. Bowes, S. El-Safi, S. Sundar, A. K. Falconar, O. P. Singh, R. Kumar, O. Ahmed, M. Boelaert and M. A. Miles, *PLoS Neglected Trop. Dis.*, 2014, **8**, e2675.
52. S. Pattabhi, J. Whittle, R. Mohamath, S. El-Safi, G. G. Moulton, J. A. Guderian, D. Colombara, A. O. Abdoon, M. M. Mukhtar, D. Mondal, J. Esfandiari, S. Kumar, P. Chun, S. G. Reed and A. Bhatia, *PLoS Neglected Trop. Dis.*, 2010, **4**, e822.
53. M. Mukhtar, A. Abdoun, A. E. Ahmed, H. Ghalib, S. G. Reed, M. Boelaert, J. Menten, M. M. Khair and R. F. Howard, *Trans. R. Soc. Trop. Med. Hyg.*, 2015, **109**, 594–600.
54. T. Bhattacharyya, A. Ayandeh, A. K. Falconar, S. Sundar, S. El-Safi, M. A. Gripenberg, D. E. Bowes, C. Thunissen, O. P. Singh, R. Kumar, O. Ahmed, O. Eisa, A. Saad, S. Silva Pereira, M. Boelaert, P. Mertens and M. A. Miles, *PLoS Neglected Trop. Dis.*, 2014, **8**, e3273.
55. F. Hossain, P. Ghosh, M. A. A. Khan, M. S. Duthie, A. C. Vallur, A. Picone, R. F. Howard, S. G. Reed and D. Mondal, *PLoS One*, 2017, **12**, e0185606.
56. S. Moulik, S. J. Chaudhuri, B. Sardar, M. Ghosh, B. Saha, N. K. Das and M. Chatterjee, *Clin. Infect. Dis.*, 2018, **66**, 404–410.
57. M. Mukhtar, S. S. Ali, S. A. Boshara, A. Albertini, S. Monnerat, P. Bessell, Y. Mori, Y. Kubota, J. M. Ndung'u and I. Cruz, *PLoS Neglected Trop. Dis.*, 2018, **12**, e0006264.
58. S. Verma, R. Singh, V. Sharma, R. A. Bumb, N. S. Negi, V. Ramesh and P. Salotra, *BMC Infect. Dis.*, 2017, **17**, 223.
59. D. Mondal, P. Ghosh, M. A. Khan, F. Hossain, S. Bohlken-Fascher, G. Matlashewski, A. Kroeger, P. Olliaro and A. Abd El Wahed, *Parasites Vectors*, 2016, **9**, 281.
60. G. F. Cota, M. R. de Sousa, F. N. Demarqui and A. Rabello, *PLoS Neglected Trop. Dis.*, 2012, **6**, e1665.
61. E. S. R. Pessoa, L. A. Mendonca Trajano-Silva, M. A. Lopes da Silva, S. da Cunha Goncalves-de-Albuquerque, T. C. de Goes, R. C. Silva de Morais, F. Lopes de Melo and M. de Paiva-Cavalcanti, *J. Microbiol. Methods*, 2016, **131**, 34–41.
62. M. S. J. da Costa Lima, A. C. L. Hartkopf, R. A. de Souza Tsujisaki, E. T. Oshiro, J. T. Shapiro, M. de Fatima Cepa Matos and M. E. Cavalheiros Dorval, *Acta Trop.*, 2018, **178**, 248–251.
63. G. F. Cota, M. R. de Sousa, T. S. M. de Assis, B. F. Pinto and A. Rabello, *Acta Trop.*, 2017, **172**, 186–191.
64. N. M. El-Sayed, P. J. Myler, D. C. Bartholomeu, D. Nilsson, G. Aggarwal, A. N. Tran, E. Ghedin, E. A. Worthey, A. L. Delcher, G. Blandin, S. J. Westenberger, E. Caler, G. C. Cerqueira, C. Branche, B. Haas, A. Anupama, E. Arner, L. Aslund, P. Attipoe, E. Bontempi, F. Bringaud, P. Burton, E. Cadag, D. A. Campbell, M. Carrington, J. Crabtree, H. Darban, J. F. da Silveira, P. de Jong, K. Edwards, P. T. Englund, G. Fazelina, T. Feldblyum, M. Ferella, A. C. Frasch, K. Gull, D. Horn, L. Hou, Y. Huang, E. Kindlund, M. Klingbeil, S. Kluge, H. Koo, D. Lacerda, M. J. Levin, H. Lorenzi, T. Louie, C. R. Machado, R. McCulloch, A. McKenna, Y. Mizuno, J. C. Mottram, S. Nelson, S. Ochaya, K. Osoegawa, G. Pai, M. Parsons, M. Pentony, U. Pettersson, M. Pop, J. L. Ramirez, J. Rinta, L. Robertson, S. L. Salzberg, D. O. Sanchez, A. Seyler, R. Sharma, J. Shetty, A. J. Simpson, E. Sisk, M. T. Tammi, R. Tarleton, S. Teixeira, S. Van Aken, C. Vogt, P. N. Ward, B. Wickstead, J. Wortman, O. White, C. M. Fraser, K. D. Stuart and B. Andersson, *Science*, 2005, **309**, 409–415.
65. O. Franzén, S. Ochaya, E. Sherwood, M. D. Lewis, M. S. Llewellyn, M. A. Miles and B. Andersson, *PLoS Neglected Trop. Dis.*, 2011, **5**, e984.
66. J. B. Richardson, K. Y. Lee, P. Mireji, J. Enyaru, M. Sistrom, S. Aksoy, H. Zhao and A. Caccone, *PLoS Neglected Trop. Dis.*, 2017, **11**, e0005949.
67. T. Downing, O. Stark, M. Vanaerschot, H. Imamura, M. Sanders, S. Decuypere, S. de Doncker, I. Maes, S. Rijal, S. Sundar, J. C. Dujardin, M. Berriman and G. Schonian, *Infect., Genet. Evol.*, 2012, **12**, 149–159.

68. H. Imamura, T. Downing, F. Van den Broeck, M. J. Sanders, S. Rijal, S. Sundar, A. Mannaert, M. Vanaerschot, M. Berg, G. De Muylder, F. Dumetz, B. Cuypers, I. Maes, M. Domagalska, S. Decuypere, K. Rai, S. Uranw, N. R. Bhattarai, B. Khanal, V. K. Prajapati, S. Sharma, O. Stark, G. Schonian, H. P. De Koning, L. Settimo, B. Vanhollebeke, S. Roy, B. Ostyn, M. Boelaert, L. Maes, M. Berriman, J. C. Dujardin and J. A. Cotton, *eLife*, 2016, **5**, e12613.
69. M. Tibayrenc and F. J. Ayala, *Mol. Ecol.*, 2015, **24**, 5778–5781.
70. J. D. Ramirez and M. S. Llewellyn, *Mol. Ecol.*, 2015, **24**, 5782–5784.
71. M. W. Gaunt, M. Yeo, I. A. Frame, J. R. Stothard, H. J. Carrasco, M. C. Taylor, S. S. Mena, P. Veazey, G. A. Miles, N. Acosta, A. R. de Arias and M. A. Miles, *Nature*, 2003, **421**, 936–939.
72. L. Peacock, M. Bailey and W. Gibson, *Parasites Vectors*, 2016, **9**, 404.
73. E. Inbar, N. S. Akopyants, M. Charmoy, A. Romano, P. Lawyer, D. E. Elnaiem, F. Kauffmann, M. Barhoumi, M. Grigg, K. Owens, M. Fay, D. E. Dobson, J. Shaik, S. M. Beverley and D. Sacks, *PLoS Genet.*, 2013, **9**, e1003672.
74. W. Gibson, L. Peacock, V. Ferris, K. Williams and M. Bailey, *Parasites Vectors*, 2008, **1**, 4.
75. N. S. Akopyants, N. Kimblin, N. Secundino, R. Patrick, N. Peters, P. Lawyer, D. E. Dobson, S. M. Beverley and D. L. Sacks, *Science*, 2009, **324**, 265–268.
76. J. Sadlova, M. Yeo, V. Seblova, M. D. Lewis, I. Mauricio, P. Volf and M. A. Miles, *PLoS One*, 2011, **6**, e19851.
77. S. J. Goodswen, P. J. Kennedy and J. T. Ellis, *Int. J. Parasitol.*, 2017, **47**, 779–790.
78. A. A. Grassmann, F. S. Kremer, J. C. Dos Santos, J. D. Souza, L. D. S. Pinto and A. J. A. McBride, *Front. Immunol.*, 2017, **8**, 463.
79. L. Gourlay, C. Peri, M. Bolognesi and G. Colombo, *Trends Biotechnol.*, 2017, **35**, 1208–1220.
80. O. Franzén, C. Talavera-Lopez, S. Ochaya, C. E. Butler, L. A. Messenger, M. D. Lewis, M. S. Llewellyn, C. J. Marinkelle, K. M. Tyler, M. A. Miles and B. Andersson, *BMC Genomics*, 2012, **13**, 531.
81. E. C. Grisard, S. M. Teixeira, L. G. de Almeida, P. H. Stoco, A. L. Gerber, C. Talavera-Lopez, O. C. Lima, B. Andersson and A. T. de Vasconcelos, *Genome Announc.*, 2014, 2, e01114–13.
82. R. P. Baptista, J. L. Reis-Cunha, J. D. DeBarry, E. Chiari, J. C. Kissinger, D. C. Bartholomeu and A. M. Macedo, *Microb. Genomes*, 2018, DOI: 10.1099/mgen.0.000156.
83. A. P. Jackson, M. Sanders, A. Berry, J. McQuillan, M. A. Aslett, M. A. Quail, B. Chukualim, P. Capewell, A. MacLeod, S. E. Melville, W. Gibson, J. D. Barry, M. Berriman and C. Hertz-Fowler, *PLoS Neglected Trop. Dis.*, 2010, **4**, e658.
84. I. Goodhead, P. Capewell, J. W. Bailey, T. Beament, M. Chance, S. Kay, S. Forrester, A. MacLeod, M. Taylor, H. Noyes and N. Hall, *mBio*, 2013, **4**, e00197–13.
85. M. Berriman, E. Ghedin, C. Hertz-Fowler, G. Blandin, H. Renauld, D. C. Bartholomeu, N. J. Lennard, E. Caler, N. E. Hamlin, B. Haas, U. Bohme, L. Hannick, M. A. Aslett, J. Shallom, L. Marcello, L. Hou, B. Wickstead, U. C. Alsmark, C. Arrowsmith, R. J. Atkin, A. J. Barron, F. Bringaud, K. Brooks, M. Carrington, I. Cherevach, T. J. Chillingworth, C. Churcher, L. N. Clark, C. H. Corton, A. Cronin, R. M. Davies, J. Doggett, A. Djikeng, T. Feldblyum, M. C. Field, A. Fraser, I. Goodhead, Z. Hance, D. Harper, B. R. Harris, H. Hauser, J. Hostetler, A. Ivens, K. Jagels, D. Johnson, J. Johnson, K. Jones, A. X. Kerhornou, H. Koo, N. Larke, S. Landfear, C. Larkin, V. Leech, A. Line, A. Lord, A. Macleod, P. J. Mooney, S. Moule, D. M. Martin, G. W. Morgan, K. Mungall, H. Norbertczak, D. Ormond, G. Pai, C. S. Peacock, J. Peterson, M. A. Quail, E. Rabbinowitsch, M. A. Rajandream, C. Reitter, S. L. Salzberg, M. Sanders, S. Schobel, S. Sharp, M. Simmonds, A. J. Simpson, L. Tallon, C. M. Turner, A. Tait, A. R. Tivey, S. Van Aken, D. Walker, D. Wanless, S. Wang, B. White, O. White, S. Whitehead, J. Woodward, J. Wortman, M. D. Adams, T. M. Embley, K. Gull, E. Ullu, J. D. Barry, A. H. Fairlamb, F. Opperdoes, B. G. Barrell, J. E. Donelson, N. Hall, C. M. Fraser, S. E. Melville and N. M. El-Sayed, *Science*, 2005, **309**, 416–422.

86. C. S. Peacock, K. Seeger, D. Harris, L. Murphy, J. C. Ruiz, M. A. Quail, N. Peters, E. Adlem, A. Tivey, M. Aslett, A. Kerhornou, A. Ivens, A. Fraser, M. A. Rajandream, T. Carver, H. Norbertczak, T. Chillingworth, Z. Hance, K. Jagels, S. Moule, D. Ormond, S. Rutter, R. Squares, S. Whitehead, E. Rabbinowitsch, C. Arrowsmith, B. White, S. Thurston, F. Bringaud, S. L. Baldauf, A. Faulconbridge, D. Jeffares, D. P. Depledge, S. O. Oyola, J. D. Hilley, L. O. Brito, L. R. Tosi, B. Barrell, A. K. Cruz, J. C. Mottram, D. F. Smith and M. Berriman, *Nat. Genet.*, 2007, **39**, 839–847.
87. T. Downing, H. Imamura, S. Decuypere, T. G. Clark, G. H. Coombs, J. A. Cotton, J. D. Hilley, S. de Doncker, I. Maes, J. C. Mottram, M. A. Quail, S. Rijal, M. Sanders, G. Schonian, O. Stark, S. Sundar, M. Vanaerschot, C. Hertz-Fowler, J. C. Dujardin and M. Berriman, *Genome Res.*, 2011, **21**, 2143–2156.

8 Rare Disease Genomics and Clinical Diagnostics

Arvin M. Gouw*, Lipika R. Pal, Debora Varon, Danielle C. Fumagalli, and Anita Chacko

Rare Genomics Institute, Downey, CA, USA
*E-mail: arvin.gouw@raregenomics.org

8.1. Sequencing Overview

There are about 7000–8000 rare diseases, of which a substantial amount, 80%, seem to have underlying genetic causes.[1,2] About 40% of these diseases are undiagnosed, partly due to the differential diagnoses issue, but also due to the lack of genetic testing in clinical practice.[3–5] Genetic testing has now become very commonly incorporated into clinical use, largely due to Moore's law, with a reduction in price down to less than $1000 today.[1,6,7]

The first DNA sequencing method, Sanger sequencing, was developed in the late 1970s, but was highly tedious and unable to sequence the whole genome.[8] However, with the advent of shotgun sequencing in the 1990s and the completion of the Human Genome Project by Celera and the National Institutes of Health (NIH), led by Francis Collins in the 2000s, the cost of sequencing has been rapidly decreasing. Instead of chemically sequencing each nucleotide, shotgun sequencing shears genomic DNA to create a library of fragments with sticky,

Genomics and Clinical Diagnostics
Edited by David Whitehouse and Ralph Rapley

Published by the Royal Society of Chemistry, www.rsc.org

complementary DNA ends that hybridize. This allows the identification of nucleotides, and thus a DNA sequence.[9,10] However, whole-genome sequencing (WGS) is still not a trivial process. To bypass this problem, most revert to whole-exome sequencing (WES) to look at only exons, which are only about 1% of the whole genome. This may seem irrelevant for pathogenesis, yet studies have shown that exon anomalies are found in 85% of various diseases.[11–15]

8.1.1. Sequencing Technologies

There are several sequencing technologies in the field, and startups are trying new approaches to sequencing, however, this chapter will discuss the current major players only: Illumina and Life Technologies.[16] Though both are similar in their specifications, Illumina is more broadly used in academia for research purposes. The field is still very novel such that there is always room for changes in the landscape.

Illumina has developed the TruSeq technology, which can have maximum read lengths ranging from 2 × 150 bp for HiSeqX to 2 × 300 bp for MiSeq. In general, the platform detects fluorescence emitted from tagged dNTPs that are incorporated in growing DNA strands. In contrast, Life Technologies' Ion Personal and Ion Proton systems detect pH changes and sequence by using synthesis. In terms of maximum read lengths, they range from 200–400 bp for the Ion Personal Genome Machine to up to 200 bp for the Ion Proton system. Thus they are quite similar to Illumina's systems.[16–18]

Both Illumina and Life Technologies' systems are known to have comparatively good sequence read outs. Each technique, fluorescence and pH detection, has its own strengths and weaknesses. Amongst other things, Illumina systems, in general, take much longer, up to 65 hours for MiSeq, while Life Technologies systems take 2–8 hours regardless of the maximum read length.[19]

8.1.2. Capture Technologies

To overcome the limitations of WES, while avoiding the costs of WGS, various enrichment methods have been developed that significantly improve WES, called “capture technologies.” In general, there are two capture technologies pertaining to WES: DNA hybridization and PCR amplification.

Hybridization technologies basically involve an additional step following the shearing of DNA during WES. The additional step includes the addition of DNA or RNA probes or adapters that will then hybridize with the sheared DNA. The probes are customizable to the need of the experiment to increase the chances of identification and binding for sequencing. As a consequence, unprobed segments are washed away, causing a reduction in breadth of sequencing in general. This method is adopted by Agilent, Illumina and NimbleGen.[20]

On the other hand, PCR amplification methodologies rely heavily on developing multiplex PCR with multiple primers in parallel to identify multiple targets simultaneously. This amplification method increases abundance and specificity for the desired targets. However, as with any multiplex PCR technology, amplicon primer efficiency is not trivial. Different primers will always have different affinities to different amplicons, making primer design a challenging task. This technology is currently being developed by Life Technologies.[11]

In both cases, validation of the technologies is tested for in the context of identifying particular targets for specificity and single-nucleotide polymorphism (SNP) arrays for breadth. Though they have their limitations, given some preliminary *a priori* knowledge of the rare disease of interest, these technologies are promising for application to rare disease diagnosis. Having discussed the limitations of the 'hardware' physical experimental challenges with WES, we will discuss the 'software' analysis challenges with WES in general and also in the case of rare diseases specifically.

8.2. Genetic Variant Identification to Clinical Interpretation

Having discussed the basics of genetic sequencing technologies, there are specific challenges pertaining to the use of genetic analyses in clinical diagnosis. With the advances in sequencing technology together with the decreasing costs, next-generation sequencing technologies are rapidly becoming routine tools for diagnosis of rare and undiagnosed diseases. As more and more genomic data related to different diseases becomes available, the burden of accurate identification of causative variants is also on the rise. In this chapter, we will discuss guidelines and different computational methods available for predicting causative variants from genomic data.

8.2.1. Current Practices for Clinical Diagnosis

Traditionally, targeted enrichment of a set of genes relevant to the disease (gene panel) is the optimal choice for rare disease diagnosis. For children with several types of rare genetic diseases, chromosomal microarray (CMA) is the recommended first tier of clinical diagnostics.[21] If no suitable variant is found from a gene panel or CMA, then whole-exome sequencing (WES) or whole-genome sequencing (WGS) can be considered as the second tier of clinical diagnostics.

8.2.2. Which One to Choose: Gene Panel, WES or WGS?

Table 8.1 provides different parameters related to clinical diagnostics from three different NGS technologies – gene panel, WES and WGS. A gene panel, which targets predetermined gene sets related to a specific disease, is often favored over other techniques because of its low sequencing costs, short turnaround time and lower volume of data to be analyzed. With the compromise of read depth, WES has the capability of identifying variants from the genes that are not yet associated with the disease and so has a better chance of identifying diagnostic variants. With WGS, one can analyze an even greater volume of data, including coding, non-coding, copy number variants (CNV) and sequence variants (SV), which are not detectable *via* gene panel or WES. Dillon *et al.* showed that WES has a higher diagnostic yield over gene panels, with a similar financial cost[22] and Lionel *et al.* suggested WGS as the first-tier genetic test compared with a gene panel or WES.[23] The power of WGS really lies in the fact that once generated, this data can be reanalyzed later, if required, for a new gene structure.

8.2.3. Prediction of Causative Variants

While WGS has the capability to provide better clinical care because of the maximum coverage of possible causative variants, the difficulty of the identification of exact causative variants, especially novel ones, also increases with the increase in the coverage. Even for a limited gene panel of WES, the reporting of accurate clinical variants is not a

Table 8.1 Different parameters of gene panels, WES and WGS for clinical diagnosis.

Sequencing techniques	Read depth	Genome coverage	Type of variants
Targeted gene panel	~500 to 1000× or higher	Selected disease-specific genes	Coding
Whole-exome sequencing	~100 to 150×	~2% of genome	Coding
Whole-genome sequencing	~30 to 40×	~95–98% of genome	Coding, non-coding, CNV, SV

trivial matter. The analysis of next-generation sequencing data generally involves three challenges and different computational methods are available to address these.

8.2.4. ACMG Guidelines for Causative Variant Identification

The causative variant identification process is essentially a data reduction and successive variant prioritization exercise, while only one or two variants need to be reported clinically out of approximately 30 000 or more variants from a single exome and approximately three million or more variants from a single genome. The American College of Medical Genetics and Genomics (ACMG) provided standards and guidelines for the interpretation of variants based on a variety of weighted lines of evidence (population data, functional data, segregation studies, *de novo* observations, allelic data and computational predictions) to classify them into pathogenic, likely pathogenic, benign, likely benign and variants of uncertain significance (VUS) categories.[24] Although these guidelines promote consistent variant interpretation, evaluation of this common framework showed that inter-laboratory concordance was 34% – lacking specific details of variant interpretation and failing to capture relevant aspects of clinical genetics.[25] Nykamp *et al.* refined these classification criteria and provided a more consistent and transparent classification approach.[26]

8.2.5. ACMG Guidelines for Incidental Findings

Incidental findings or secondary findings are additional findings from sequencing data that are unrelated to the original phenotypic profile of the patient. An ACMG working group recommended reporting pathogenic or expected pathogenic variants from a set of 59 genes associated with an aggregate total of 27 disorders.[27] These represent medically actionable findings.

8.2.6. Types of Variants Prioritized for Rare Diseases

Exonic variants usually dominate final variant prioritization steps over non-coding variants. Within exonic variants, the majority are missense, followed by nonsense and frameshift insertions or deletions.[23,28,29] There is very little reporting of intronic variants, non-coding RNA variants, small CNVs and structural variants.[23,29] These findings essentially reflect our lack of proper functional interpretation of non-coding variants.

8.3. Challenges of Rare Disease Diagnosis

Research and development, and corresponding clinical diagnosis for rare diseases present a particular set of challenges in addition to the aforementioned general challenges of utilizing genetic sequencing in clinical diagnoses.

8.3.1. Number of Patients by Definition

The patient population for rare diseases is by definition very low. As previously mentioned, there are approximately 7000 rare diseases, which are defined as those diseases for which there are less than 200 000 patients in the USA or that affect no more than one in 2000 in the EU. Because of these unique characteristics of the patient population, rare diseases are a challenging subject for research. There are usually very few cases present at a time upon which researchers may draw any conclusions.

8.3.2. Clinical Trial Accretion Rate

The major challenge in conducting clinical trials in rare diseases is the small number of patients available for clinical studies and as a result obtaining statistically significant results in small cohorts. The patient population is also likely to be geographically diverse, and enrollment of patients at each site is usually very low. These challenges pose a problem for the patient group – more inconvenience, and concerns about safety, the financial burden and the duration of the trial.

8.3.3. Lack of Genetic Definition

Rare diseases are often a combination of different types of phenotypes, most of which have genetic causes as well as non-genetic factors. Mostly, a single variant in a gene is responsible for all phenotypes – sometimes variants from more than one gene are also responsible for all phenotypes, as when two different diseases are present in a single individual. Sometimes non-genetic factors, like infection and nutritional deficiencies during the pre- or post-natal period, also affect rare diseases. The identification of proper disease definition is challenging in rare disease research.

8.3.4. Data Quality Issues

Data quality, quality control (QC) and variant calling is a major challenge in rare disease diagnosis. Usually causative mutations are *de novo* or novel variants, which are not found in any public databases – so it is of utmost importance to check the quality of the data to avoid any false-positive or false-negative findings. Often sequencing is done using different sequencing platforms other than the ones used for public databases, like 1000 genome[30] or ExAC databases,[31] giving rise to a very high number of novel variants for each patient.[28] These false-positive variants can affect incidental findings as well, which is very critical for patients' knowledge and financial burden. Although Sanger sequencing validation is recommended for any high-throughput data generation, careful inspection or appropriate filtering measures at the beginning of data processing can help to reduce the analysis and cost at the reporting stage.

8.3.5. Causative Variants Identification Challenges

Rare diseases are often comprised of a series of symptoms or phenotypes, some of which are very common to many diseases and with only one or two unique features for that particular disease. Often it is not a trivial task to match the causative variant with all the disease symptoms mentioned in the patient's phenotypic profile. While physicians who are generating the phenotypic profile can easily pinpoint which symptom is the most important, that piece of information is usually not available for the researchers who are actually handling the causative variant identification without seeing the patient.

8.3.6. Genomic Annotation

Annotating the variant call format (VCF) file for further downstream analysis is very crucial for rare disease diagnosis. There can be differences in the annotation of the same VCF file with different annotating programs.[32,33] For rare diseases specifically, the correct annotation often helps to diagnose the disease.[34]

8.3.7. Errors in Clinical Databases

We often assign pathogenicity to a variant if it is found in clinical databases, like the human gene mutation database (HGMD).[35] There is evidence that sometimes the relationship of the variant and the disease is not properly curated – thereby leading to erroneous assignment of pathogenicity.[36]

8.3.8. Limitations of Using Model Organism Data

We often use model organism data to increase the performance of predicting causative variants from sequencing data. However, there are differences in model organism genomes and human genomes in terms of disease phenotype. Prediction of causative variant based on model organism phenotypes alone can lead to misdiagnosis and clinically that is not acceptable.[37,38]

8.4. Rare Diseases Are Not So Rare

Despite the seemingly insurmountable challenges of genetic sequencing use for clinical diagnosis and rare disease research, an overview of the various kinds of rare diseases indicate that there are potential solutions to these problems. First, because there are many kinds of rare diseases, actually there are many people who suffer from rare diseases. Second, it follows then that collaborative efforts to pool these rare disease patients would be the first step in overcoming the aforementioned challenges.

8.4.1. Rare Cancers

8.4.1.1. What Are Rare Cancers?

In the USA, a rare cancer is defined as one that affects fewer than 200 000 people; in Europe, it is defined as one that affects less than 1 in 2000 people. The NIH lists on its website for Genetic and Rare Diseases Information Center (GARD) 510 different rare cancers.[39,40] Cancers are also considered rare if they start in an unusual part of the body or if they are of an unusual type and need a special type of treatment.[41,42]

8.4.1.2. Orphan Drugs That Treat Cancers

Orphan products, also known as orphan drugs, are FDA-approved treatments for rare diseases. In 1983, the Orphan Drug Act was passed to give incentives to drug companies to develop new treatments for rare diseases. The Office of Orphan Products Development at the FDA determines whether a drug qualifies as an orphan drug or product. From the onset of the Orphan Drug Act to this day a total of 4462 drugs have been designated as orphan drugs.[43] As of 2016, 36% of all approved orphan drugs are used to treat rare cancers.[44] With the advent of precision medicine, many common diseases can be termed orphan just by molecularly characterizing subsets that can be targeted with a specific drug. As a result, by 2017, seven of the top ten selling drugs qualify as orphan. Even though the number of orphan drugs has increased dramatically since the year 2000, 95% of the more than 7000 rare diseases still have no therapy.[45]

8.4.2. Rare Neurological Diseases

8.4.2.1. Ataxias

Ataxia is a neurodegenerative disease whose symptoms include slurred speech, stumbling, falling, trouble eating and swallowing, and lack of coordination. All ataxias are caused by degeneration of the cerebellum, which is responsible for the coordination of movements. The National Ataxia Foundation lists six types of ataxias: ataxia telangiectasia, episodic ataxia, Friedreich's ataxia, multiple system atrophy, spinocerebellar ataxia, and sporadic ataxia. Symptoms vary for each individual, and the progression may be slow, over decades, or quick, over a few months. Genes SCA1 through SCA36 have been identified for spinocerebellar ataxia. They are autosomal and are dominantly inherited. Friedreich's ataxia is caused by an autosomal recessive gene, *FXN*.[46,47] *FXN* is a nuclear gene which encodes a mitochondrial protein involved in mitochondrial iron transport and respiration.

8.4.2.2. Epilepsies

Rare epilepsy is defined as a disorder that is rare and that is also associated with seizures in the majority of patients. The Rare Epilepsy Network (REN), a patient advocacy group, lists 40 different types of rare epilepsy. Some epilepsies are associated with a gene mutation and are inherited. The spectrum and severity of the symptoms range from mild to severe. Tuberous Sclerosis (TSC) is the most common of the rare epilepsies, affecting 50 000 people in the USA. It is a dominantly inherited disorder with mutations in one of two genes that regulate cell growth, *TSC1* or *TSC2*. TSC can compromise any or all systems in the body and seizures are present in 85% of affected individuals.[48,49]

8.4.2.3. ALS

Amyotropic Lateral Sclerosis (ALS) or Lou Gehrig's disease is a neurodegenerative disease that causes progressive degeneration and the ultimate death of motor neurons. The brain is unable to control muscle movement and patients become paralyzed. More than 20 000 people in the USA have the disease. 90% of ALS cases are sporadic, meaning that only one member in a family is affected. Familial ALS, (FALS), which affects more than one member in a family, also involves frontotemporal dementia (FTD) and symptoms start early in life. FALS is associated with an autosomal dominant mutation.[50,51] Mutations in RNA-binding proteins, such as TDP-43, FUS/TLS, hnRNP A1 and A2B1 are found to be associated with ALS and FTD.[52,53]

8.4.3. Rare Metabolic Disorders

8.4.3.1. Monogenic Diabetes

Monogenic diabetes is a rare form of diabetes. Monogenic diabetes differs from the common diabetes type 1, which is caused by multiple mutations, and from type 2, which is caused by multiple gene mutations in combination with lifestyle factors, such as obesity. Monogenic diabetes is caused by a mutation in a single gene. Most cases of monogenic diabetes are inherited, while a minority of them occur due to a spontaneous mutation. The onset of the disease may be early on as in neonatal diabetes mellitus (NDM), or maturity-onset diabetes of the young (MODY), which first occurs in adolescence or early adulthood. The disease is characterized by the inability of the pancreas to produce insulin. The ensuing high glucose levels in the body cause damage in eyes, kidneys, nerves and blood vessels.[54,55]

8.4.3.2. Mitochondrial Diseases

Mitochondria have been called the energy factories of the body as they produce over 90% of the energy needed to sustain life and support organ function. But mitochondria are much more than that: It takes about 3000 genes to make a mitochondrion. Only 37 of those genes are encoded within the mitochondrion; the rest are encoded in the nucleus and their products are transported to the mitochondrion where they will participate in vital metabolic pathways, from building, breaking down and recycling building blocks to containing the rate-limiting enzymes in RNA, DNA and hemoglobin biosynthesis, among other critical pathways.[56,57] Mitochondrial diseases may be caused by spontaneous or inherited mutations. Because mitochondria perform so many different functions, different mutations may cause the same symptoms (phenocopies of mitochondrial disease), or the same mutations can cause many different symptoms (genocopies of mitochondrial disease). Because of the complexity of the diseases and the difficulty in diagnosing them, multiple tests are needed to diagnose mitochondrial-related diseases. Changes in urinary organic acids are usually found in mitochondrial diseases, as are high levels of fumarate and malate.[58,59]

8.4.3.3. Lipid Diseases

Among the several diseases caused by lipid metabolism malfunction, the most common one is Gaucher disease. Gaucher disease is a lysosomal storage disorder caused by a lack of enzymes that break down lipids and carbohydrates in the cells.[60] Gaucher disease is an autosomal recessive disorder that causes spleen and liver enlargement, low platelet count, bleeding and clotting problems, anemia, fatigue and, in some cases, glucocerebroside accumulation in the lungs, causing difficulty breathing. Bone symptoms are common in Gaucher disease: spontaneous fractures, osteoporosis, joint pain, arthritis and joint damage are some of them.[61–63] More than 400 genetic mutations have been associated with Gaucher disease. Up to 1 in 10 Ashkenazi Jews may be carriers, making Gaucher disease the most common genetic disease in Jews.[64]

8.5. The Necessity for Collaboration

The advances in genomics are crucial in the identification of various kinds of diseases, including rare diseases. As discussed above, rare diseases pose unique challenges, one of which is the rarity of the patients, to generate enough data for genomic analyses. Yet, given the promising prospects of applying genomic analyses to rare disease applications and therapeutic monitoring, various institutions have risen to push the envelope on rare disease genomics, ranging from patient recruitment to data analyses. We will discuss academic and NGO organizations that have collaborative structures.

First, academic centers that focus on rare diseases include the NIH, Hopkins–Baylor, and Duke-Cold Spring Harbor Laboratory (CSHL). The NIH center for rare diseases is called the Undiagnosed Disease Program (UDP), whose mission is to apply genomic analyses for clinical diagnosis. UDP has ventured into SNP, WGS and WES technologies over the years, and they have identified seven new diseases, eight ultra-rare diseases and 15 rare diseases out of their selected 160 patients. These efforts are led by William Gahl and Cornelius Boerkoel.[65,66]

In addition to the UDP, NIH also launched the Center for Mendelian Genetics, where academic centers act as nodes, such as Baylor–Hopkins Center for Mendelian Genetics. Hopkins' previous successful platform with the Online Mendelian in Man database allowed them to focus on improving WES to identify variants involved in rare diseases. Combined with Baylor's exome databases, they were able to analyze 250 rare disease patients. Together, they claim to be able to use WES to identify genetic anomalies in rare disease patients 25% of the time.[67,68]

The Center for Human Genome Variation and the Discovery Genomics program at Duke University, with the help of the private sector, strive to understand the role of genetic variants in predisposition to human disease. This effort is so promising that Duke received $25 million of federal grant funding to investigate the genetics behind epilepsy. Overall, the Duke studies have reported that genetic testing helps in the diagnosis of 25–33% of cases.[69] The proportion of rare diseases that are attributable to genetic anomalies is consistent between Duke and Hopkins – around 25%.

The academic collaborative efforts have demonstrated not only that they are able to gather enough rare disease patients to perform genomic analyses, but that they are able to identify key variants that are causal in the pathogenesis of a substantial number of rare disease cases. As previously discussed, the problems of sample size and variant identification are not mutually exclusive. As more genomic data are compiled through these institutional collaborations, more experts will be able to identify key pathogenic mutational variants.

Nonprofit organizations (NGOs) have been instrumental in addressing the unique problems associated with rare diseases. EURORDIS has collected about 700 rare disease patients from 60 countries around the globe. Similarly, the Rare Genomics Institute in the USA has been actively helping rare disease patients seek diagnoses and effective treatments. On the other hand, there are nonprofits that are more specific disease-wise, *i.e.* Michael J. Fox's center for Parkinson's. The INADCure foundation was built by key family members and caretakers of infantile neuroaxonal dystrophy (INAD) patients that are eager to fundraise for research. INADCure has worked with the Rare Genomics Institute to host an innovative grant funding mechanism for proposals on INAD research (BeHEARD).[14,70,71]

References

1. N. S. Rawson, *Orphanet. J. Rare Dis.*, 2017, **12**(1), 59.
2. Y. R. Rubinstein, S. C. Groft and R. Bartek, *et al.*, *Contemp. Clin. Trials*, 2010, **31**(5), 394–404.
3. M. M. Ward, *Epidemiology*, 2005, **16**(2), 270–271.
4. R. M. Kliegman, B. J. Bordini and D. Basel, *et al.*, *Pediatr Clin North Am*, 2017, **64**(1), 1–15.
5. J. Rigg, H. Lodhi and P. Nasuti, *Value in Health*, 2015, **18**(7), A705.
6. P. Marino, R. Touzani and L. Perrier, *et al.*, *Eur. J. Hum. Genet.*, 2018, **26**(3), 314–323.
7. A. J. Berberich, R. Ho and R. A. Hegele, *Can. Med. Assoc. J.*, 2018, **190**(5), E124–E125.
8. D. Estrada-Rivadeneyra, *FEBS J.*, 2017, **284**(24), 4174.
9. E. Dias Neto, R. G. Correa and S. Verjovski-Almeida, *et al.*, *Proc. Natl. Acad. Sci. U. S. A.*, 2000, **97**(7), 3491–3496.
10. J. C. Venter, M. D. Adams and G. G. Sutton, *et al.*, *Science*, 1998, **280**(5369), 1540–1542.
11. H. Rennert, K. Eng and T. Zhang, *et al.*, *npj Genomic Med.*, 2016, **1**, 16019.
12. C. W. Lindsley, *ACS Chem. Neurosci.*, 2018, **9**(4), 626–627.
13. M. Chopra and T. Duan, Rare genetic disease in China: a call to improve clinical services, *Orphanet. J. Rare Dis.*, 2015, **10**, 140.
14. R. A. Ortiz, S. Witte and A. Gouw, *et al.*, *Interact. J. Med. Res.*, 2018, **7**(1), e3.
15. A. R. Tee, *Semin. Cell Dev. Biol.*, 2016, **52**, 1–2.
16. G. Bertier, A. Cambon-Thomsen and Y. Joly, *Eur. J. Med. Genet.*, 2018, **61**, 634–641.
17. S. Schwartz and Y. Motorin, *RNA Biol.*, 2017, **14**(9), 1124–1137.
18. S. Goodwin, J. D. McPherson and W. R. McCombie, *Nat. Rev. Genet.*, 2016, **17**(6), 333–351.
19. R. K. Ravi, K. Walton and M. Khosroheidari, *Methods Mol. Biol.*, 2018, **1706**, 223–232.
20. K. Bodi, A. G. Perera and P. S. Adams, *et al.*, *J. Biomol. Tech.*, 2013, **24**(2), 73–86.
21. D. T. Miller, M. P. Adam and S. Aradhya, *et al.*, *Am. J. Hum. Genet.*, 2010, **86**(5), 749–764.
22. O. J. Dillon, S. Lunke and Z. Stark, *et al.*, *Eur. J. Hum. Genet.*, 2018, **26**, 644–651.
23. A. C. Lionel, G. Costain and N. Monfared, *et al.*, *Genet. Med.*, 2018, **20**(4), 435–443.
24. S. Richards, N. Aziz and S. Bale, *et al.*, *Genet. Med.*, 2015, **17**(5), 405–424.
25. L. M. Amendola, G. P. Jarvik and M. C. Leo, *et al.*, *Am. J. Hum. Genet.*, 2016, **98**(6), 1067–1076.
26. K. Nykamp, M. Anderson and M. Powers, *et al.*, *Genet. Med.*, 2017, **19**(10), 1105–1117.
27. S. S. Kalia, K. Adelman and S. J. Bale, *et al.*, *Genet. Med.*, 2017, **19**(2), 249–255.
28. L. R. Pal, K. Kundu and Y. Yin, *et al.*, *Hum. Mutat.*, 2017, **38**(9), 1169–1181.
29. D. J. Stavropoulos, D. Merico and R. Jobling, *et al.*, *npj Genomic Med.*, 2016, **1**, 15012.
30. A. Auton, L. D. Brooks and R. M. Durbin, *et al.*, *Nature*, 2015, **526**(7571), 68–74.
31. M. Lek, K. J. Karczewski and E. V. Minikel, *et al.*, *Nature*, 2016, **536**(7616), 285–291.
32. J. L. Yen, S. Garcia and A. Montana, *et al.*, *Genome Med.*, 2017, **9**(1), 7.
33. F. J. Sedlazeck, A. Dhroso and D. L. Bodian, *et al.*, *F1000Research*, 2017, **6**, 1795.
34. C. A. Steward, A. P. J. Parker and B. A. Minassian, *et al.*, *Genome Med.*, 2017, **9**(1), 49.
35. P. D. Stenson, M. Mort and E. V. Ball, *et al.*, *Hum. Genet.*, 2014, **133**(1), 1–9.
36. K. Kundu, L. R. Pal and Y. Yin, *et al.*, *Hum. Mutat.*, 2017, **38**(9), 1201–1216.
37. P. McGonigle and B. Ruggeri, *Biochem. Pharmacol.*, 2014, **87**(1), 162–171.
38. I. W. Mak, N. Evaniew and M. Ghert, *Am. J. Transl. Res.*, 2014, **6**(2), 114–118.
39. R. Groisberg, D. S. Hong and J. Roszik, *et al.*, *Mol. Cancer Ther.*, 2018, **17**, 1595–1601.
40. A. M. Frezza, A. Trama and J. Y. Blay, *et al.*, *Eur. J. Surg. Oncol.*, 2018, DOI: 10.1016/j.ejso.2018.03.030.

41. R. K. Pillai and K. Jayasree, *Indian J. Med. Res.*, 2017, **145**(1), 17–27.
42. C. E. DeSantis, J. L. Kramer and A. Jemal, *Ca-Cancer J. Clin.*, 2017, **67**(4), 261–272.
43. M. Herder, *PLoS Med.*, 2017, **14**(1), e1002191.
44. E. M. Adkins, L. Nicholson and D. Floyd, *et al.*, *ClinicoEconomics and Outcomes Research*, 2017, **9**, 327–342.
45. D. A. Hughes and J. Poletti-Hughes, *PLoS One*, 2016, **11**(10), e0164681.
46. E. Britti, F. Delaspre and A. Feldman, *et al.*, *J. Cell Mol. Med.*, 2018, **22**(2), 834–848.
47. L. Li, L. Voullaire and C. Sandi, *et al.*, *PLoS One*, 2013, **8**(2), e55940.
48. A. Vignoli, F. La Briola and K. Turner, *et al.*, *Epilepsia*, 2013, **54**(12), 2134–2142.
49. C. Shepherd, M. Koepp and M. Myland, *et al.*, *BMJ Open*, 2017, **7**(10), e015236.
50. M. S. C. Mis, S. Brajkovic and F. Tafuri, *et al.*, *Mol. Neurobiol.*, 2017, **54**(6), 4466–4476.
51. V. Hirsch-Reinshagen, C. Pottier and A. M. Nicholson, *et al.*, *Acta Neuropathol. Commun.*, 2017, **5**(1), 96.
52. L. Bargsted, D. B. Medinas and F. Martinez Traub, *et al.*, *Sci. Rep.*, 2017, **7**(1), 14266.
53. D. Ito, M. Hatano and N. Suzuki, *Sci. Transl. Med.*, 2017, **9**, 415.
54. V. Bansal, J. Gassenhuber and T. Phillips, *et al.*, *BMC Med.*, 2017, **15**(1), 213.
55. S. Kherra, J. L. Blouin and F. Santoni, *et al.*, *Swiss Med. Wkly.*, 2017, **147**, w14535.
56. J. Cao, H. Wu and Z. Li, *Exp. Ther. Med.*, 2018, **15**(1), 13–18.
57. R. L. Davis, C. Liang and C. M. Sue, *Handb. Clin. Neurol.*, 2018, **147**, 125–141.
58. M. Kawai and S. Hosaki, *Clin. Biochem.*, 1990, **23**(4), 327–334.
59. A. M. Hall, A. Vilasi and I. Garcia-Perez, *et al.*, *Kidney Int.*, 2015, **87**(3), 610–622.
60. W. L. Stone and S. R. Master, *Gaucher Disease*, StatPearls, Treasure Island (FL), 2018.
61. M. Baldini, G. Casirati and F. M. Ulivieri, *et al.*, *Blood Cells, Mol., Dis.*, 2018, **68**, 148–152.
62. G. Batta, L. Soltesz and T. Kovacs, *et al.*, *Sci. Rep.*, 2018, **8**(1), 157.
63. F. Nascimbeni, A. Dalla Salda and F. Carubbi, *Blood Cells, Mol., Dis.*, 2018, **68**, 74–80.
64. S. Bronstein, M. Karpati and L. Peleg, *Isr. Med. Assoc. J.*, 2014, **16**(11), 683–685.
65. T. Gall, E. Valkanas and C. Bello, *et al.*, *Front. Med.*, 2017, **4**, 62.
66. J. Guzman, E. Lee and D. Draper, *et al.*, *Children*, 2015, **2**(3), 342–357.
67. J. Jurgens, H. Ling and K. Hetrick, *et al.*, *Genet. Med.*, 2015, **17**(10), 782–788.
68. J. X. Chong, K. J. Buckingham and S. N. Jhangiani, *et al.*, *Am. J. Hum. Genet.*, 2015, **97**(2), 199–215.
69. E. Panagiotakaki, E. De Grandis and M. Stagnaro, *et al.*, *Orphanet. J. Rare Dis.*, 2015, **10**, 123.
70. D. C. Fumagalli and A. M. Gouw, *Yale J. Biol. Med.*, 2015, **88**(4), 413–414.
71. P. Wehling, *Poiesis Prax*, 2011, **8**(2–3), 67–82.

9 Molecular Diagnostics†

M. Rabie Al-Turkmani[a,b] and Gregory J. Tsongalis*[a,b]

[a]Department of Pathology, Geisel School of Medicine at Dartmouth, Hanover, NH, USA; [b]Dartmouth Hitchcock Medical Center and Norris Cotton Cancer Center, Lebanon, NH, USA
*E-mail: Gregory.J.Tsongalis@hitchcock.org

9.1. Introduction

Recently, the use of diagnostic testing to understand the molecular mechanisms of numerous diseases and for identification of biological biomarkers involved in multiple signaling pathways has been essential for improving general clinical practice. Applications of modern genomic technologies have allowed clinical laboratories the ability to support clinical management decisions for early detection, diagnosis, prognosis, detection of recurrence after therapy, risk assessment, identification of therapeutic targets, prediction of response to therapy, monitoring clinical outcomes and imaging disease processes.[1]

Currently, molecular technologies are available to identify DNA variation and changes in gene expression associated with infectious diseases, cancer and inherited diseases. The detection of benign and pathogenic alterations in DNA and/or RNA samples facilitates diagnosis, prognosis and monitoring of responses to therapy. The

†This chapter has been reproduced from *Molecular Biology and Biotechnology*, ed. R. Rapley and D. Whitehouse, 2014, 6th edition, copyright The Royal Society of Chemistry.

Genomics and Clinical Diagnostics
Edited by David Whitehouse and Ralph Rapley

Published by the Royal Society of Chemistry, www.rsc.org

discovery of genes and gene-based markers associated with diseases has allowed physicians to assess disease predisposition and to design and implement improved diagnostic algorithms. Advances in molecular diagnostic testing have been used to improve the sensitivity, specificity and turn-around time for actionable test results in clinical laboratories.

For this reason, an exceptional demand has been placed on clinical laboratories to provide increased diagnostic testing with rapid and accurate identification and interrogation of genomic targets. It is evident that molecular diagnostic methods play an increasing and critical role in patient management.

9.2. Technologies

Molecular technologies first entered clinical laboratories in the early 1980's as manual, labor-intensive procedures that required a working knowledge of chemistry and molecular biology as well as an exceptional skill set. However, recently, these technologies have moved very quickly away from manual and labor-intensive procedures to cheaper, faster, more semi-automated and increasingly accurate procedures.[2]

Since the mid-1980s, Southern blotting was the method of choice for a variety of clinical applications, such as detection of gene deletions in Duchenne and Becker muscular dystrophy, repeat expansions in fragile-X syndrome, linkage analysis for cystic fibrosis and other genetic conditions (Figure 9.1). In the early 1990s, the transition of the polymerase chain reaction (PCR) from research to clinical settings began a new era for molecular diagnostics (Figure 9.2). Because PCR is a target-based amplification technology that allows for the detection of specific gene or pathogen sequences with higher sensitivity and specificity it rapidly became the method of choice for diagnostic molecular laboratories. Since then, many modifications to PCR have been introduced, but none as significant as the real-time capability.[3–6] The elimination of post-PCR detection systems and the ability to perform the entire assay in a closed vessel had significant advantages for clinical laboratories.

Real-time PCR allows for the simultaneous amplification and detection of amplified nucleic acid targets. It also increases the specificity of the assay since it can be used with a fluorescent probe that hybridizes to the specific nucleic acid sequence. In routine clinical practice, the main advantages of real-time PCR are (1) high specificity, (2) the

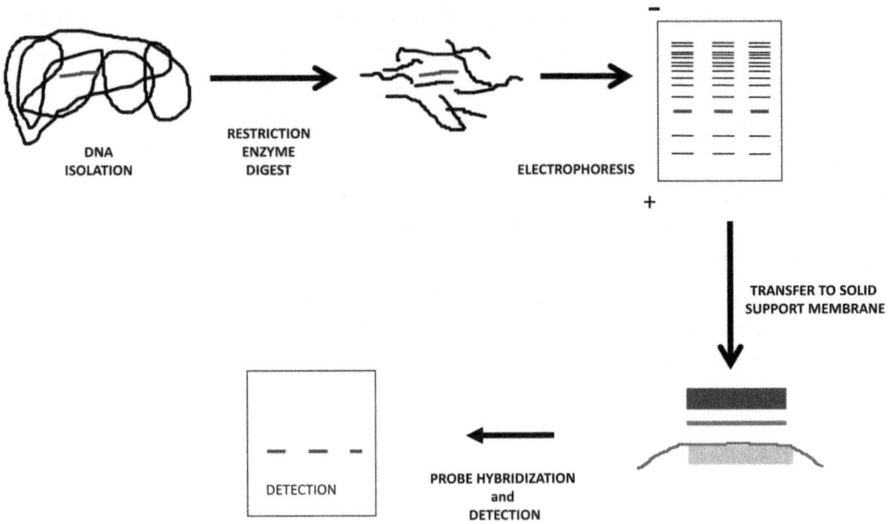

Figure 9.1 Schematic diagram of the Southern blot transfer procedure.

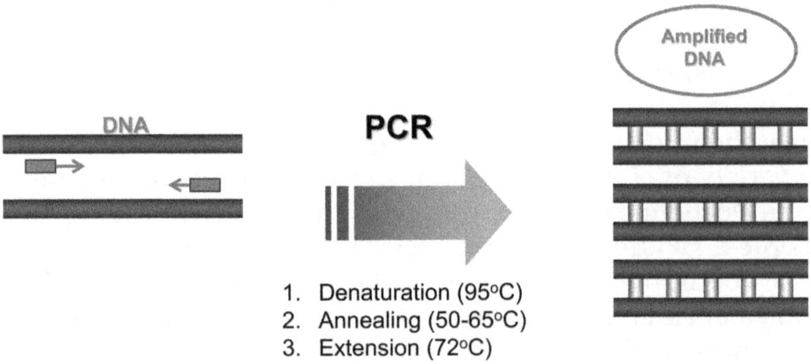

Figure 9.2 Schematic diagram of the polymerase chain reaction (PCR).

speed with which samples can be analyzed, as there are no post-PCR processing steps required, and (3) the "closed-tube" nature of the technology to eliminate potential sources of contamination. The analysis of results *via* amplification curve and melt-curve analysis is very simple and more amenable to routine clinical laboratory quality control procedures.

The rapid development of more complex molecular technologies, such as microarrays, next generation sequencing (NGS) and nanotechnology, have allowed testing of numerous biomarkers and a more detailed classification of diseases, contributing to a personalized prognostic and predictive approach to management.[7] Microarrays can be divided into three main applications: (1) to detect structural alterations (*i.e.* array comparative genomic hybridization (CGH)), (2) to identify phenotype-specific single-nucleotide polymorphisms (SNPs) that are known to drive specific phenotypes, such as panels for retinal degeneration, and (3) to detect genome-wide SNPs that assess risk of multiple common genetic disorders, such as certain cancers and ophthalmologic, cardiac, renal and neurological disorders.[8] NGS uses massively parallel sequencing chemistries to sequence many genes of interest, the whole exome or the whole genome for variants in a broad range of complex disorders. Nanobiotechnology, using a nanometer length scale, can extend the limits of molecular diagnostics to unprecedented levels on the nanoscale as assays and technologies with better performance characteristics are developed.

Clinical applications of molecular technologies continue to grow and expand into non-traditional diagnostic arenas. Recently, at clinical laboratories molecular technologies have been applied to newborn screening for highly penetrant disorders, diagnostic and carrier testing for inherited disorders and predictive and pre-symptomatic testing for adult-onset and complex disorders. In addition, detection of infectious disease pathogens and pharmacogenetic testing, to guide individual drug dosage, selection and response are being performed routinely.[8] Today, molecular diagnostic laboratories are armed with numerous instruments and technologies to address the increasing demand for clinical test results (Table 9.1).

Table 9.1 Current technologies being used in the Molecular Pathology Laboratory at the Dartmouth Hitchcock Medical Center.

Technology	Platform/Instrument
DNA/RNA extraction	1. Manual 2. Spin column 3. Qiagen EZ1 robot 4. Roche ampliprep
Fluorescence *in situ* hybridization (FISH)	1. Thermobrite elite 2. Molecular imaging system
Real time PCR	1. Cepheid GeneXpert 2. AB 7500 real-time PCR system 3. Cepheid smartcycler 4. Roche Taq48
Capillary electrophoresis	1. AB 3500 genetic analyzer
Traditional PCR	1. MJ research 2. DNA engines
Microarray	1. Affymetrix
Next-generation sequencing	1. Ion torrent personal genome machine (PGM) 2. Illumina MiSeq 3. Illumina NexSeq 4. Nanopore MinIon

9.3. Infectious Diseases

For many years, clinical laboratories used traditional methods of microbial identification and differentiation that rely on phenotypic characteristics (*i.e.* protein, bacteriophage and chromatographic profiles), biotyping and susceptibility testing. Recently, with the development and application of molecular diagnostic techniques, a new era in microbial identification and characterization has begun.

The diagnosis of an infectious disease has typically depended upon isolation of the pathogen by a culture technique. Even though this approach is adequate for the identification of the majority of common infections, the approach is less than desirable for the detection of organisms that are difficult to grow *in vitro*, including those with long incubation times. These issues, along with performance characteristics of sensitivity and specificity, brought about a need for alternative techniques that would allow for direct detection of infectious agents in clinical samples while maintaining a rapid turn-around time and high performance.

Molecular diagnostic technologies, such as non-amplified nucleic acid probe and nucleic acid amplification, were rapidly introduced into clinical microbiology laboratories. Non-amplified nucleic acid probe could be bound to the target from direct specimens, blood culture bottles and colony isolates allowing the identification of infectious agents.[9] Nucleic acid amplification methods (*i.e.* traditional PCR, reverse transcriptase PCR, nested PCR, multiplex PCR and real-time PCR) have the ability to selectively amplify specific targets present at low concentrations that could then be identified by various post-PCR methods or probes incorporated into a real time PCR assay (Figure 9.3).

Clinical applications of molecular methods for infectious diseases include (1) virology, where it has been applied to resistance testing, genotyping and viral quantification and viral detection, (2) bacteriology, where it has been applied to resistance testing, detection of infection due to fastidious bacteria and detection of bacterial infection after antibiotics have been administered, and (3) parasitology and mycology, where it has been applied to difficult fungal infections.[10] Many molecular infectious disease tests that use PCR-based techniques are approved or cleared by the US Food and Drug Administration (FDA) and are being performed routinely in clinical laboratories (Table 9.2).

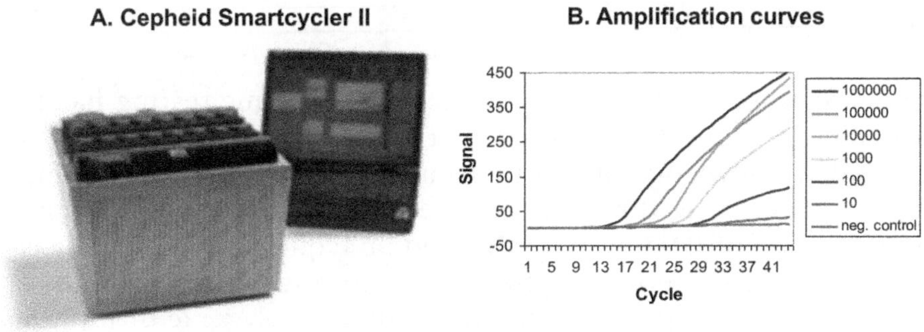

Figure 9.3 (A) The Cepheid Smartcycler II real-time PCR instrument. (B) Real-time PCR amplification curves showing cycle threshold differences for various quantities of target sequence.

Table 9.2 Molecular infectious disease tests that are currently FDA-cleared and being performed in clinical laboratories.

Bacteria	Viruses
Bacillus anthracis	Avian flu
Candida albicans and *Candida* spp.	Cytomegalovirus
Candida albicans and *Candida glabrata*	Enterovirus
Chlamydia trachomatis	HBV (quantitative)
Clostridium difficile	Hepatitis C virus (qualitative and quantitative)
Enterociccous faecalis	HIV drug resistance
Francisella tularensis	HIV (quantitative)
Gardnerella, *Trichomonas* and *Candida* spp.	HBV/HCV/HIV blood screening assay
Group a streptococci	Human papillomavirus
Group B streptococci	Respiratory viral panel
Legionella pneumophila	West nile virus
Methicillin-resistant *Staphylococcus aureus*	
Mycobacterium tuberculosis	
Mycobacterium spp.	
Staphylococcus aureus	

The next generation of these technologies, such as nucleic acid arrays and whole-genome sequencing, will provide a wealth of information for personalized medicine in the area of infectious diseases. Nucleic acid arrays with thousands of sequence-specific oligonucleotide probes have been developed to identify and genotype influenza viruses, respiratory syncytial virus, herpes viruses and HPV. Bacterial, parasitic and fungal organisms can also be detected, genotyped and evaluated for resistance by this method.[9] Whole-genome sequencing can provide megabases of sequence information from the whole organism that can help with identification and resistance testing for specific infectious diseases.

9.4. Genetics

The landscape of molecular genetics and genomic testing has rapidly changed due to advances in DNA analysis technologies and expanded knowledge of the molecular basis of rare and common disorders. Genetic disorders can be defined as conditions due to an alteration of DNA, either inherited or acquired. They can be categorized into three major groups: (1) chromosomal disorders, (2) monogenic or single-gene disorders and (3) polygenic or multifactorial disorders.

Chromosomal disorders are due to the loss, gain or abnormal arrangement of one or more chromosomes, which results in the presence of excessive or deficient amounts of genetic material. These types of alterations usually involve large segments of DNA containing numerous genes and can be organized into two groups: numerical or structural chromosome alterations. Numerical alteration is characterized by an abnormal number of chromosomes, also known as aneuploidy, which occurs when there is a loss or excess of one or more chromosomes (monosomy, disomy, trisomy, tetrasomy, *etc.*). Structural alterations are characterized by modification of the chromosome structure and can be classified into six groups:

(1) Deletion: breakage and/or loss of a portion of a chromosome.

(2) Duplication: a portion of a chromosome is duplicated, resulting in extra genetic material.

(3) Translocation: breakage of two chromosomes with transfer of broken parts to the opposite chromosome.

(4) Inversion: a portion of the chromosome has broken off, turned upside down and reattached to the chromosome.

(5) Insertion: breakage and loss of a portion of one chromosome and insertion of this part into another chromosome.

(6) Isochromosome: splitting at the centromere during mitosis so that one arm is lost and the other duplicated to form one chromosome with identical arms.

Monogenic disorders are the result of a single mutant gene and display traditional Mendelian inheritance patterns including autosomal dominant or recessive and X-linked types. The overall population frequency of monogenic disorders is thought to be approximately 10 per 1000 live births. Polygenic or multifactorial disorders, however, consist of chronic diseases of adulthood, congenital malformations and dysmorphic syndromes. These disorders result from multiple genetic and/or epigenetic factors, which may not conform to traditional Mendelian inheritance patterns.

Recently, the ability to screen individuals for many types of genetic alterations in a clinical setting has been expanding at an enormous rate. Many clinical laboratories offer routine testing for common genetic diseases such as the monogenic disorders cystic fibrosis and fragile-X syndrome and polygenic disorders such as human cancers (Table 9.3).

Cystic fibrosis (CF) is a monogenic autosomal recessive disease caused by multiple mutations in the *CFTR* (cystic fibrosis transmembrane regulator) gene. According to the CFTR mutation database (http://www.genet.sickkids.on.ca/cftr/StatisticsPage.html), 1951 heterogeneous mutations have been documented in different ethnic groups. Over the past few years, the American College of Medical Genetics (ACMG) and the American College of Obstetrics and Gynecology (ACOG) approved guidelines for a national CF screening panel as a standard of care for CF carrier testing in the general population. This panel consists of 23 mutations in the *CFTR* gene with an allele frequency of at least 0.1% in the US general population and a detection rate ranging from 94% to 49% in Ashkenazi

Table 9.3 Genetic and oncological diseases commonly tested for using molecular diagnostic methods.

Alpha1-antitrypsin deficiency
Angelman syndrome
ApoE
Colon cancer: KRAS, EGFR,
Cystic fibrosis
Drug metabolizing enzymes
Duchenne and Becker muscular dystrophy
Factor II (prothrombin)
Factor V Leiden
Fanconi's anemia
Fragile-X syndrome
Gaucher disease
Glioma and Glioblastoma: IDH1, IDH2, SMO
Hemochromatosis
Huntington's disease
HER-2 status
Melanoma: BRAF, KIT
Methylene tetrahydrofolate reductase (MTHFR)
Non-small cell lung cancer (NSCLC): EGFR, KRAS,
Non-small cell lung cancer (NSCLC): ALK fusion
Sex mismatched bone marrow transplantation
Prader–Willi syndrome
Tay–Sachs disease

Table 9.4 Panel of 23 recommended mutations used in screening for cystic fibrosis.

ΔF508	N1303K	A455E	621+1 G > T	2789+5 G > A
ΔI507	R553X	R560T	711+1 G > T	3120+1 G > A
G542X	R347P	R1162X	1898+1 G > A	2184delA
G551D	R117H	G85E	1717-1G > A	3659delC
W1282X	R334W		849 + 10kbC > T	

Jews and Asian Americans, respectively (Table 9.4).[11,12] Recently, a CF carrier screening panel using next-generation sequencing was FDA approved for use with the Illumina MiSeq platform. This panel consists of 156 genomic positions (153 single-nucleotide variants and small insertion/deletion sites, two large deletions and the polyTG/polyT region), which corresponds to 162 unique mutations that provides a more comprehensive detection rate among different ethnic groups.

Fragile-X Syndrome (FraX) is a monogenic X-linked neuro-developmental disorder with a non-Mendelian inheritance pattern. It is caused by an expansion of a trinucleotide (CGG) repeat in the *FMR-1* (Fragile-X mental retardation 1) gene. The repeat is up to 55 trinucleotides long in the normal population, but it can exceed 200 in patients with FraX, resulting in hypermethylation of the promoter region and silencing of the *FMR-1* gene. This disease is a result of lack of expression of the FMR protein, resulting in intellectual disability, hyperactivity and autistic-like behavior.[13] Since this mutation is X-linked, males are more severely affected than females. Affected females have less severe mental retardation and can pass an expanded allele on to their offspring. The original molecular diagnostic assay for FraX was based on Southern blot transfer analysis and now, more commonly, PCR with capillary electrophoresis is used to more accurately size individual alleles (Figure 9.4). Screening for CGG repeats is used as a diagnostic tool and as a carrier screening assay whose results are used in genetic counseling of family members.

9.5. Hematology

Precise diagnosis and classification of hematological malignancies are important since treatment options and prognosis vary considerably. Historically, the diagnosis of myeloid and lymphoid neoplasms was based solely on the histological evaluation of hematoxylin and

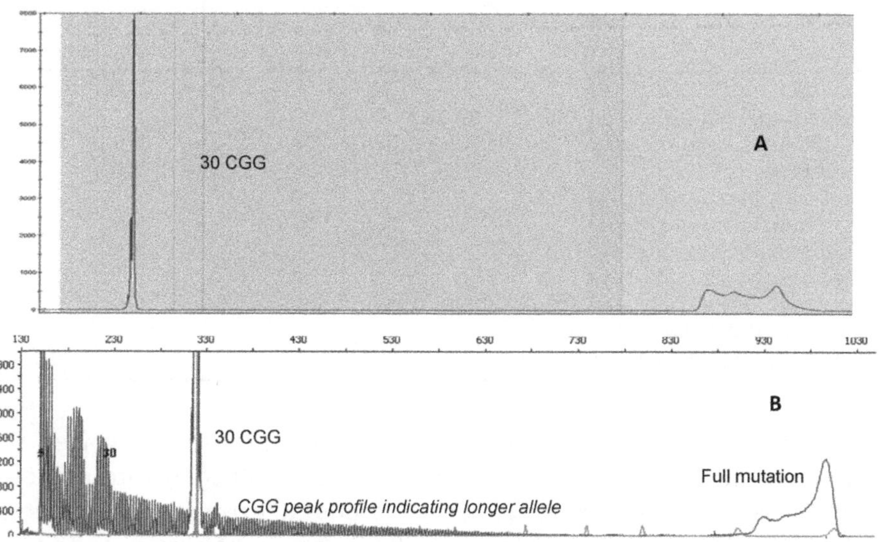

Figure 9.4 Diagnosing fragile X syndrome using capillary electrophoresis. (A) Gene specific FMR1 PCR and (B) CGG repeat primed FMR1 PCR.

eosin (H&E)-stained slides. With the development of immunophenotyping methods, diagnose and classification of leukemia or lymphoma according to the cell lineage (myeloid, B, T, NK) and stage of maturation became possible. Subsequent availability of cytogenetic and molecular genetic methods to identify unique and specific biomarkers increased the ability to diagnose and classify leukemia and lymphoma, providing a newly characterized spectrum of hematological malignancies according to gene mutations, gene deletions or amplifications, gene or microRNA (miRNA) expression profiles and chromosomal translocations.[14–17]

Today, the clinical workup of leukemia and lymphoma requires the correlation of traditional and advanced diagnostic techniques (clinical features, morphology, immunophenotyping with cytogenetic and molecular genetic data) to accurately diagnose and classify the disease. Since the current diagnosis and classification method for hematological malignancies was established by the WHO (World Health Organization's Classification of Tumors of Hematopoietic and Lymphoid Tissues), several molecular-based tests have become routinely available in clinical laboratories (Table 9.5).

Hematological malignancies are divided into myeloid neoplasms (leukemia), lymphoid neoplasms (leukemia and lymphoma), lymphoproliferative disorders and histiocytic or dendritic cell neoplasms.

Table 9.5 Common molecular diagnostic tests applied to hematological malignancies.

Leukemia
B-cell chronic lymphocytic leukemia (B-CLL)
Chromosome 8 enumeration (CML, AML, MPD, MDS)
Lymphoma
BCL-1 translocation (11; 14)
BCL-2 translocation (18; 14)
BCL-6 translocation (3; 14)
c-Myc translocation (8; 14)
Immunoglobulin heavy and light chain rearrangements
NPM–ALK translocation (2; 5)
T-cell receptor gene rearrangements

Leukemia is a myeloid or lymphoid malignancy that involves the peripheral blood and bone marrow and can be divided into acute or chronic conditions. Acute leukemia is characterized by a defect in maturation, leading to an imbalance between proliferation and maturation, and chronic leukemia is characterized by an increase of proliferation of cells that maintain their capacity to differentiate to end cells. Lymphoma is a lymphoid malignancy that involves lymph nodes and/or other extramedullary sites and is classified into the two main categories of Hodgkin or non-Hodgkin.

The myeloid neoplasms are characterized by acute myeloid leukemia (AML), myeloproliferative neoplasms (MPN) (*i.e.* chronic myeloid leukemia or CML) and myelodysplastic syndrome (MDS). AML is an aggressive clonal neoplasm that lacks maturation of myeloid cells which then rapidly proliferate, causing accumulation of myeloblasts in the bone marrow and/or blood. It is a heterogeneous disease with regard to clinical, morphological, immunophenotypic, karyotypic and genetic features (*i.e.* point mutations and chromosomal translocations). The genetic abnormalities involve multiple pathways and have been grouped into the "Class I" oncogenes that confer proliferative advantage to the leukemic cells and "Class II" oncogenes that contribute to myeloid maturation arrest. The association between point mutations and/or translocations of both Class I and Class II oncogenes appears to be important to the development of AML.[18,19] Class I mutations are characterized by point mutations in *FLT3*, *KIT*, *RAS* and *JAK2* genes. Class II are characterized by point mutations in *NPM1* and *CEBPA* genes, as well as chromosomal translocations (*i.e.* *RUNX1–RNNX1T1*, *CBFB–MYH11*, *PML–RARA*) and *MLL* gene rearrangement from t(8; 21) (q22; q22), inv(16) (p13.1q22), t(15; 17) (q22; q12) and 11q23 translocations.[20]

CML is defined as a clonal hematopoietic stem cell disorder defined by the presence of monocytosis in the blood and the presence of myelodysplastic and myeloproliferative features in the bone marrow.[21] CML is characterized by the Philadelphia chromosome or the *BCR–ABL1* fusion gene, which is a result of a translocation between chromosomes 9 and 22 [t(9; 22) (q34; q11)]. However, CML also shows karyotypic and genetic abnormalities (*i.e.* point mutations). Some of the karyotype abnormalities include isochromosome 17 [i(17q)], trisomy 8 and monosomy 9, and point mutations include ones involving genes associated with signaling pathways and proliferation, such as *RAS*, *JAK2* and *CBL*.[22]

The lymphoid neoplasms are characterized by acute lymphoblastic leukemia (ALL), chronic lymphocytic leukemia (CLL) and lymphoma (*i.e.* Hodgkin and non-Hodgkin). ALL is characterized by a group of structural rearrangements, submicroscopic DNA copy number alterations and sequence mutations. Approximately 20% of B-ALL cases have chromosomal rearrangements of *ETV6–RUNX1*, *TCF3–PBX1*, *BCR–ABL1*, rearrangements of MLL and mutations of *PAX5*, *JAK1*, *JAK2*, *IKZF1*, *CRLF2*, *IL7R*, *CREBBP* and *TP53*.[23,24] T-lineage ALL cases are characterized by activating mutations of *NOTCH1* and rearrangements of transcription factors TLX1, TLX3, LYL1, TAL1 and MLL.[23,24] Approximately 75% of the chromosomal alterations are detectable by karyotyping, FISH or other molecular techniques.

The understanding of the biology of CLL B cells provided the identification and characterization of several genetic abnormalities (*i.e.* point mutations and chromosomal translocations) that are prognostic indicators of disease progression and survival. The chromosomal abnormalities commonly identified are deletion at 13q14, trisomy 12, deletion at 11q22–23 and deletion at 17q. CLL patients with the 17q deletion have aggressive clinical disease, most probably because of loss of the *TP53* gene.[25] Recently, next-generation sequencing expanded the knowledge of the genomic alterations in CLL, providing the first comprehensive view of somatic mutation in this disease. *NOTCH1* and *SF3B1* mutations were identified in CLL patients and can be considered new drivers of aggressive forms of the disease.

Non-Hodgkin Lymphomas (NHL) are a heterogeneous group of solid tumors of lymphoid cell origin. They are divided into two subtypes, lymphomas derived from B-cells, which correspond to 85% of the NHL cases, or lymphomas derived from T-cells.

For B-cell neoplasms, clonality can often be determined immunophenotypically by demonstrating the presence of monoclonal surface immunoglobulins or by detecting gene rearrangements in the immunoglobulin gene family. In contrast, for T-cell malignancies there is no immunophenotpyic equivalent to monoclonal surface immunoglobulin. Thus, molecular genetic approaches for the determination of clonality in T-cell lymphoma by detecting gene rearrangements in the T-cell receptor family are especially important.[26] Other molecular genetic applications to the assessment of lymphoid malignancies include detection of chromosomal translocations in the subclassification of non-Hodgkin lymphoma. For example, in a lymph node suspect for follicular lymphoma, the detection of a translocation, t(14; 18), involving the *BCL*-2 proto-oncogene, would confirm this diagnosis. Similarly, the detection of a translocation, t(11; 14), involving the *BCL*-1 proto-oncogene, would confirm a diagnosis of mantle cell lymphoma.[26,27]

9.6. Oncology

Very rapid advances in molecular technologies for identifying somatic and germline mutations in oncology have revolutionized our knowledge base of tumor cell biology. The identification of genetic alterations, such as point mutations, copy number alterations, translocations and/or chromosomal rearrangements, contributed to the characterization of novel biomarkers, leading to a new paradigm for molecular oncology testing. The rapidly growing field of clinical biomarkers, with either a prognostic or predictive importance has contributed to a personalized or precision medicine approach to managing oncology patients.

Although numerous genes and chromosomal alterations have been identified in many human cancer types, few have had an impact and made it to clinical laboratory use. However, this increased medical and scientific knowledge base associated with the need for early detection and selection of the most appropriate treatment for cancer patients has resulted in the rapid uptake of somatic mutation testing for all cancer types. The following examples highlight the utility of molecular diagnostic tools in the assessment of breast, bladder, melanoma, colorectal and lung cancers.

In breast cancer patients, higher resolution imaging, targeted therapies and improved surgical techniques have all contributed to improved patient outcomes by detecting, diagnosing and treating this disease more effectively. Currently, there are two molecular diagnostic assays to identify the presence of metastasis or recurrence in breast cancer patients, MamaPrint® (Agendia, Amsterdam, The Netherlands) and Oncotype DX (Genomic Health, Inc. Redwood City, CA). MamaPrint®, a 70-gene signature, is a microarray gene expression profiling test that identifies the risk of distant recurrence following surgery. It analyzes 70 critical genes that contain a definitive gene expression signature and stratifies patients into two distinct groups, with low risk or high risk of distant recurrence. MamaPrint® was developed by a group of researchers who identified a prognostic signature using a microarray platform in node-negative breast cancer patients under the age of 55.[28] This signature consisted of genes involved in the cell cycle, invasion, metastasis, angiogenesis and signal transduction. The 70-gene prognostic signature was validated in node-negative and node-positive tumors, as well as treated and untreated patients, and proved to be a robust predictor for distant metastatic-free survival, independent of adjuvant treatment, tumor size, histological grade and age.[29] A second validation was performed in node-negative T1–2 breast tumors not treated with chemotherapy and compared with traditional clinical factors included in the Adjuvant Online software.

Oncotype DX® is a 21-gene expression assay to provide a recurrence score (RS) as a prognostic indicator. The RS predicts the probability of distant recurrence in node-negative patients treated with tamoxifen and in ER^+ breast cancer.[30] The assay identifies expression of 21 genes, five reference genes and 16 genes associated with breast cancer, selected from a set of 250 genes previously studied by the NSABP (National Surgical Adjuvant Breast and Bowel Project) clinical studies. Quantitative real-time, reverse transcriptase PCR (qRT-PCR) in formalin-fixed paraffin-embedded (FFPE) samples is performed to quantify the expression of the 21 genes and calculate the RS (recurrence score), which classifies patients into three groups: high-, intermediate- and low-risk. The RS was shown to correlate with distant recurrence, relapse-free interval and overall survival, independent of age and tumor size.

The Genomic Grade Index (GGI) signature was developed to reclassify patients with histological grade 2 tumors, which is informative for clinical decision making. Sotiriou *et al.*[31] analyzed microarray data from 189 invasive breast cancers and identified 97 genes associated with histological grade, most of them involved in cell cycle regulation and proliferation. These genes were differently expressed between high-grade and low-grade breast tumors. The intermediate-grade tumors showed expression patterns similar to either high-grade or low-grade cases. The GGI may increase the accuracy of tumor grading and improve treatment decisions.

Another example of a molecular biomarker for breast cancer is the HER2 (human epidermal growth factor receptor 2) gene. It regulates cell proliferation, survival and other processes important for carcinogenesis. The activation of HER2 occurs through gene amplification, which was initially used as an indicator of poor prognosis. The occurrence of this abnormality in up to 35% of all breast cancers made it a robust target for new therapies. Currently, HER2 gene amplification is used as a targeted pharmacogenomics assay to identify those breast cancer patients who may benefit from treatment with trastuzumab (Herceptin) (Figure 9.5).[32]

Her2 FISH pharmDx™ Kit (Dako Denmark A/S Glostrup, Denmark), PathVysion® HER-2 DNA Probe kit (Abbott Molecular, Inc., Des Palines, IL), SPOT-Light HER2 CISH kit (Invitrogen, Carlsbad, CA) and Inform Dual ISH (Ventana Medical Systems, Roche Diagnostics, Tucson, AZ) are FDA-cleared tests. These tests are performed on FFPE tissue affixed to a slide. The kits have a DNA probe that hybridizes to the *HER2* gene and another probe that hybridizes to the centromere of chromosome 17 (except for the SPOT-Light HER2 CISH kit). The analysis is done using a fluorescence or light microscope to count the number of signals from each probe that will represent the number of HER2 gene copies present in the tumor sample.[33] Since the development of trastuzumab, other small molecular drugs have been designed to target the HER2 receptor when it is overexpressed.[34,35]

Unlike the HER2 example in breast cancer, bladder cancer progression is characterized by increased chromosomal instability and aneuploidy of several chromosomes, including chromosome 3, 7 and 17, and loss of locus 9p21. Since bladder cancer is a chronic disease, the diagnosis of primary and recurrent disease is a significant problem and patients need to be constantly monitored for recurrence of their

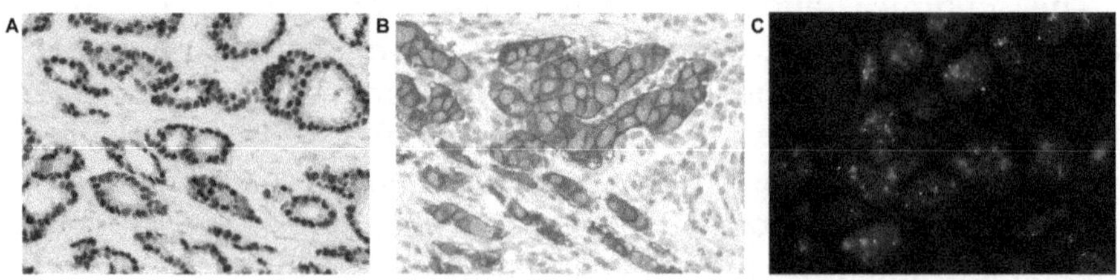

Figure 9.5 Breast cancer biomarkers. (A and B) Immunohistochemistry (A. ER^+ positive. B. HER-2 overexpression). (C) FISH (HER-2 gene amplification).

disease. The UroVysion Bladder Cancer Kit (Abbott Molecular Inc., Des Plaines, IL) is a FDA-cleared assay that uses multicolor fluorescence *in situ* hybridization (FISH) to detect aneuploidy for chromosomes 3, 7 and 17 and loss of the 9q21 locus (P16/CDKN2A gene) in urine specimens. This assay is also available for use, in conjunction with urine cytology and cystoscopy evaluation, as an aid for diagnosing and monitoring recurrence of urothelial carcinoma. Traditional cytological analysis may not be sensitive enough for this application and it has been shown that the combination of FISH with routine cytology evaluation is more sensitive than cytology alone at detecting urothelial carcinoma.[36–40]

Clearly a better understanding of tumor cell biology and the pathways involved in carcinogenesis associated with the advantages of molecular diagnostics has led to the identification of novel biomarkers and therapeutics for human cancers. In the context of patient management, not only will molecular biomarkers be responsible for the reclassification of many of these tumor types, they will also be responsible for a more personalized medicine.[41]

9.7. Pharmacogenomics

Pharmacogenomics (PGX) has been an important molecular application in clinical laboratories that address drug metabolism (PGX_m) and targeted therapies (PGX_t). PGX_m refers to the characterization of whole-genome variation associated with different pharmacokinetics (PK) or pharmacodynamics (PD) of administered drugs. Pharmacokinetics corresponds to drug absorption, distribution, metabolism and elimination, and pharmacodynamics represents the effects on drug receptors and other drug targets. Genetic variants in genes that encode proteins involved in PK and PD of a drug will significantly contribute to differences in drug response. PGX_t refers to the identification of drug targets or genetic variants that confer resistance to therapeutics in a targeted pathway. The overall aim of PGX testing is to decrease adverse responses to therapy and increase efficacy by ensuring the appropriate selection and dose of therapy.[42,43]

Many variations, such as SNPs, deletion, insertion, copy number variations (CNVs) and tandem repeats, in genes may lead to change in protein structure or stability and therefore to protein activity. Thus, genetic alterations in genes responsible for drug-metabolizing enzymes, drug transporters and drug targets, will alter response to therapeutics. These proteins are essential in the absorption, distribution and elimination of various endogenous and exogenous substances, including pharmaceutical agents. Several groups of drug transporters that may be significant in the field of pharmacogenomics exist, including multidrug resistance proteins (MDRs), multidrug resistance-related proteins (MRPs), organic anion transporters (OATs), organic anion transporting polypeptides (OATPs), organic cation transporters (OCTs) and peptide transporters (PepTs).[44–46]

With the knowledge of genetic factors and the rapid advance in molecular technologies, many pharmacogenomics biomarkers (drug-metabolizing enzymes, drug transporters and drug targets) have been characterized, FDA approved and applied in clinical diagnostic laboratories.

With respect to $PGX_{m,}$ most of the enzymes involved in drug metabolism are members of the cytochrome P450 (CYP) superfamily, which encodes 57 CYP enzymes and is responsible for 75–80% of all phase I-independent metabolism and for 65–70% of the clearance of clinically used drugs.[47–49] These cytochrome P450 enzymes are mainly located in the liver and gastrointestinal tract and include more than 30 isoforms. Classification of these enzymes includes nomenclature, for example CYP2D6*1, the name of the enzyme (CYP) is followed by the family (CYP2), subfamily (CYP2D) and gene (CYP2D6) associated with the biotransformation. Allelic variants are indicated by a *, followed by a number (CYP2D6*1). Genetic variants or polymorphisms in these genes can lead to the following phenotypes: poor, intermediate, extensive and ultra-rapid metabolizers. Poor metabolizers (PM) have no detectable enzymatic activity; intermediate metabolizers (IM) have decreased enzymatic activity; extensive metabolizers (EM) are considered normal and have at least one copy of an active gene; ultra-rapid metabolizers (UM) contain duplicated or amplified gene copies that result in increased drug metabolism.

CYP2D6 and *CYP2C19* genes are drug-metabolizing enzymes involved in the metabolism of cholesterol, steroids and other important lipids, such as arachidonic acid.[50] The *CYP2D6* gene, localized on chromosome 22q13.1, encodes a highly polymorphic protein, which contains 497 amino acids. The *CYP2C19* gene, localized on chromosome 10q24, encodes a protein that contains 490 amino acids. More than 60 and 24 allele variants have been described for CYP2D6 and CYP2C19, respectively. These variants are the result of SNPs, deletions, insertions, gene rearrangements or deletion or duplication of the entire gene and result in decreased or increased enzymatic activity.[51] The AmpliChip CYP450 test is a FDA-cleared diagnostic test that can be used to predict how an individual will metabolize drugs that are substrates for CYP2D6 and CYP2C19. It uses the Affymetrix microarray technology that analyses a patient's genotypic profile of the CYP2D6 (27 allelic variations) and CYP2C19 (three allelic variations) genes. This test provides information about the patients' drug metabolizing status (*i.e.* poor, intermediate, extensive or ultra-rapid metabolizer) based on their genetic information, allowing physicians to select relevant medication and doses of medications for some diseases, such as cardiac diseases, pain and cancer.

Another example of PGX_m is the uridine diphosphate (UDP)-glucuronosyltransferase (*UGT1A1*) gene, which plays an important role in the metabolism of drugs and endogenous substances through glucuronidation, allowing these compounds to be more easily excreted. A genetic TA repeat length polymorphism in the promoter of the *UGT1A1* gene has been shown to affect its expression, leading to serious clinical manifestations. The "wild-type" allele, *UGT1A1*1*, has six tandem TA repeats in the regulatory TATA box of the *UGT1A1* promoter. The most common polymorphism associated with low activity of *UGT1A1* is the *28 variant, which has seven TA repeats (Table 9.6). UGT1A1*28 reduces activity of the *UGT1A1* enzyme which can then decrease the conversion of the active metabolite of irinotecan to its glucuronidated form for excretion from the body. Decreased UGT1A1 activity, therefore, results in more of the active irinotecan metabolite being present in the body, with subsequent adverse reactions. Thus, UGT1A1*28 is associated with an increased risk of severe irinotecan toxicity. Irinotecan is a first-line therapy approved for patients with metastatic colorectal cancers and is also used in other forms of cancer as indicated. A single FDA-approved assay for this polymorphism has recently been removed from the market due to the low volumes of testing. We have developed a PCR-based fragment-sizing assay that identifies the different alleles of this gene.[52] This laboratory-developed test (LDT) can identify the 5, 6, 7 and 8 TA repeats (Figure 9.6). Although the clinical significance of the 5 and 8 TA repeat alleles is not well-established, there is clear evidence that TA repeat size inversely affects UGT1A1

Table 9.6 Frequency of UGT1A1 polymorphisms in the general Caucasian population.

TA repeat number	Allele designation	Frequency
6/6	*1/*1	46%
6/7	*1/*28	39%
6/8	*1/*37	3%
7/7	*28/*28	9%
7/8	*28/*37	1.5%

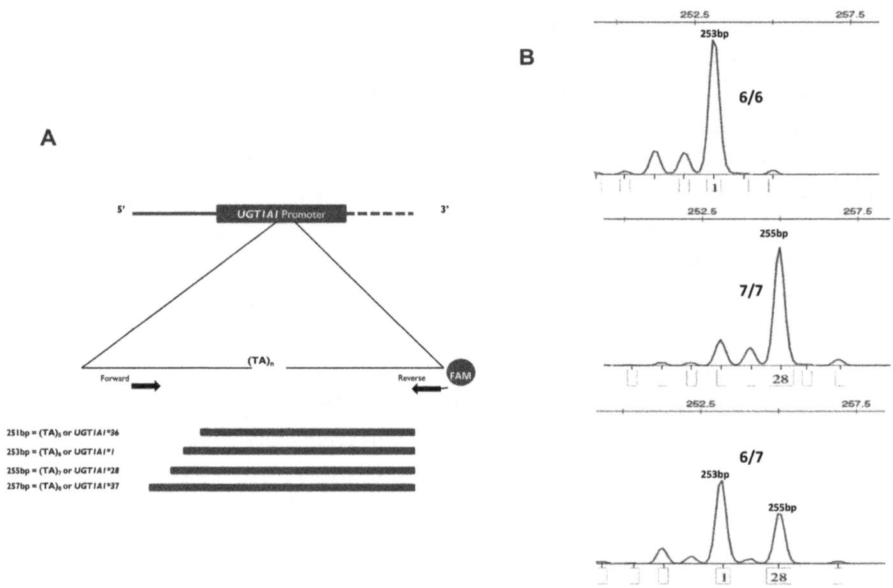

Figure 9.6 Schematic of TA dinucleotide repeat in the UGT1A1 promoter (A). Electropherograms showing different patterns for detection of TA-repeat sizes (B).

expression.[53] Thus, patients with eight TA repeats might have the lowest UGT1A1 expression level and, therefore, might require special care if they show the clinical manifestations. Screening for UGT1A1 polymorphisms has been limited to confirmation of a diagnosis of Gilbert's syndrome.

PGX studies in clinical oncology identified many genetic variants as having potential roles in individualizing therapy.[54] This is the so called PGX_t application. Currently, there are 24 FDA-approved drug labels for anticancer agents that include gene variants, functional deficiencies, expressions changes and chromosomal abnormalities.[54] For example, in breast cancer, HER2 amplification and overexpression is a positive target for eligibility for treatment with trastuzumab therapy and estrogen receptor (ER) expression is considered an important biomarker as it is a positive target for anti-estrogen therapy (tamoxifen) (Figure 9.5). ER-positive breast cancer patients benefit from endocrine therapy, such as tamoxifen (TAM). TAM is a selective modulator of ER that blocks steroid mechanisms, preventing cellular replication and proliferation. CYP2D6 enzyme is responsible for the conversion of TAM to its active metabolites (*i.e.* endoxifen), which is important because they have a greater affinity for the estrogen receptor than

TAM itself.[55] It has been postulated that patients with genetic polymorphism in *CYP2D6* have lower or absent CYP2D6 activity, resulting in lower levels of endoxifen. Thus, these patients would have less benefit from tamoxifen therapy than patients with functional copies of *CYP2D6*.[56,57]

Cetuximab and panitumumab are two monoclonal antibodies that inhibit the growth and survival of tumor cells that have an activated *EGFR* in colon cancers. However, tumors with *KRAS* codon 12 and 13 mutations may not experience any benefit from these therapies. KRAS is a member of the membrane GTPase family that activates many proteins, including *EGFR* signaling pathways, such as c-Raf and PIK3. Only colon cancer patients whose tumors express *EGFR* and whose tumors are negative for *KRAS* mutation will benefit from these drugs.[58–60] A significant number of tumors harbor an activated EGFR pathway and thus therapies targeting the EGFR are being used in numerous tumor types including non-small-cell lung cancers (NSCLC). NSCLCs contain activating mutations, most commonly found in exons 19 and 21 of the EGFR gene, which confer sensitivity to small-molecule anti-EGFR therapies to the tumor cells.[61]

Many laboratories have reverted somatic mutation testing to next-generation sequencing panels that allow for the simultaneous detection of multiple mutations in many genes (Figure 9.7).

Crizotinib is another example of a targeted therapy. It is an *ALK* inhibitor which is FDA-approved to treat NSCLC that has a distinct rearrangement within this gene. Approximately 2–5% of lung adenocarcinomas harbor this novel EML4–ALK gene rearrangement that is

ABL1	*EGFR*	*GNAS*	*KRAS*	*PTPN11*
AKT1	*ERBB2*	*GNAQ*	*MET*	*RB1*
ALK	*ERBB4*	*HNF1A*	*MLH1*	*RET*
APC	*EZH2*	*HRAS*	*MPL*	*SMAD4*
ATM	*FBXW7*	*IDH1*	*NOTCH1*	*SMARCB1*
BRAF	*FGFR1*	*IDH2*	*NPM1*	*SMO*
CDH1	*FGFR2*	*JAK2*	*NRAS*	*SRC*
CDKN2A	*FGFR3*	*JAK3*	*PDGFRA*	*STK11*
CSF1R	*FLT3*	*KDR*	*PIK3CA*	*TP53*
CTNNB1	*GNA11*	*KIT*	*PTEN*	*VHL*

Figure 9.7 The Ion Torrent AmliSeq Cancer Hotspot Panel.

mutually exclusive from EGFR and KRAS mutations. Tumors with ALK rearrangement are eligible for treatment with crizotinib. Therefore, it is important to identify these patients. The EML4–ALK rearrangement may occur at various breakpoints, resulting in multiple possible fusion transcripts. Currently, most laboratories detect ALK rearrangements with an FDA-approved break-apart FISH assay.

9.8. Conclusion

Advances in technologies have led to a better understanding the molecular basis of infectious, genetic and oncological diseases. Highly complex technologies, such as microarrays and next generation sequencing, have resulted in biomarker discoveries that have led to the development of novel therapies and companion diagnostics. Qualitative and quantitative molecular testing continues to be critical in the management of patients with infectious disease, genetic disease and cancer. Drug development strategies that incorporate genomics have required clinical laboratories to be prepared for and routinely test for pharmacogenomic markers that address metabolic and targeted aspects of drug delivery. Clearly the transition to high-complexity and high-data-throughput technologies will once again revolutionize how we approach diagnostic medicine with respect to patient care.

References

1. W. E. Grizzle, S. Srivastava and U. Manne, *Cancer Biomarkers*, 2010, **9**, 7.
2. G. J. Tsongalis and W. B. Coleman, *Clin. Chim. Acta*, 2006, **363**, 127.
3. C. T. Wittwer, M. G. Herrmann, A. A. Moss and R. P. Rasmussen, *BioTechniques*, 1997, **22**, 130.
4. S. B. Parks, B. W. Popovich and R. D. Press, *Am. J. Clin. Pathol.*, 2001, **115**, 439.
5. C. T. Wittwer, G. H. Reed, C. N. Gundry, J. G. Vandersteen and R. J. Pryor, *Clin. Chem.*, 2003, **49**, 853.
6. J. Wilhelm and A. Pingoud, *ChemBioChem*, 2003, **4**, 1120.
7. K. H. Allison, *Am. J. Clin. Pathol.*, 2012, **138**, 770.
8. S. H. Katsanis and N. Katsanis, *Nat. Rev. Genet.*, 2013, **14**, 415.
9. K. L. Muldrew, *Curr. Opin. Pediatr.*, 2009, **21**, 102.
10. D. J. Speers, *Clin. Biochem. Rev.*, 2006, **27**, 39.
11. M. S. Watson, G. R. Cutting, R. J. Desnick, D. A. Driscoll, K. Klinger, M. Mennuti, G. E. Palomaki, B. W. Popovich, V. M. Pratt, E. M. Rohlfs, C. M. Strom, C. S. Richards, D. R. Witt and W. W. Grody, *Genet. Med.*, 2004, **6**, 387.
12. The American College of Obstetricians and Gynecologists Committee Opinion, *Update on Carrier Screening for Cystic Fibrosis*, 2011, vol. 486, p. 1.
13. C. Bagni and B. A. Oostra, *Am. J. Med. Genet., Part A*, 2013, **161**, 2809.
14. H. Kiyoi, M. Yanada and K. Ozekia, *Int. J. Hematol.*, 2005, **82**, 85.
15. J. M. Scandura, *Curr. Oncol. Rep.*, 2005, **7**, 323.

16. M. C. Stubbs and S. A. Armstrong, *Curr. Drug Targets*, 2007, **8**, 703.
17. C. H. Lawrie, *Br. J. Haematol.*, 2007, **137**, 503.
18. Y. Ishikawa, H. Kiyoi, A. Tsujimura, S. Miyawaki, Y. Miyazaki, K. Kuriyama, M. Tomonaga and T. Naoe, *Eur. J. Haematol.*, 2009, **83**, 90.
19. R. P. Hasserjian, *Int. J. Lab. Hematol.*, 2013, **35**, 358.
20. T. Naoe and H. Kiyoi, *Int. J. Lab. Hematol.*, 2013, **97**, 165.
21. J. W. Vardiman, J. Thiele, D. A. Arber, R. D. Brunning, M. J. Borowitz, A. Porwit, N. L. Harris, M. M. Le Beau, E. Hellstrom-Lindberg, A. Tefferi and C. D. Bloomfield, *Blood*, 2009, **114**, 937.
22. S. A. Parikh and A. Tefferi, *Am. J. Hematol.*, 2012, **87**, 610.
23. C. G. Mullighan, *Hematology Am. Soc. Hematol. Educ. Program*, 2012, 389.
24. H. Inaba, M. Greaves and C. G. Mullighan, *Lancet*, 2013, **381**, 1943.
25. N. Chiorazzi, *Hematology Am. Soc. Hematol. Educ. Program*, 2012, 76.
26. A. J. Bench, W. N. Erber, G. A. Follows and M. A. Scott, *Int. J. Lab. Hematol.*, 2007, **29**, 229.
27. Y.-L. Kwong, *Br. J. Haematol.*, 2007, **137**, 273.
28. L. J. van't Veer, H. Dai, M. J. van de Vijver, Y. D. He, A. A. Hart, M. Mao, H. L. Peterse, K. van der Kooy, M. J. Marton, A. T. Witteveen, G. J. Schreiber, R. M. Kerkhoven, C. Roberts, P. S. Linsley, R. Bernards and S. H. Friend, *Nature*, 2002, **415**, 530.
29. N. Snoj, P. Dinh, P. Bedard and C. Sotiriou, *Molecular Biology of Breast Cancer*, ed. W.B. Coleman and G. J. Tsongalis, Elsevier Press, San Diego, 2012, vol. 26, p. 341.
30. S. Paik, S. Shak, G. Tang, C. Kim, J. Baker, M. Cronin, F. L. Baehner, M. G. Walker, D. Watson, T. Park, W. Hiller, E. R. Fisher, D. L. Wickerham, J. Bryant and N. Wolmark, *N. Engl. J. Med.*, 2004, **351**, 2817.
31. C. Sotiriou, P. Wirapati, S. Loi, A. Harris, S. Fox, J. Smeds, H. Nordgren, P. Farmer, V. Praz, B. Haibe-Kains, C. Desmedt, D. Larsimont, F. Cardoso, H. Peterse, D. Nuyten, M. Buyse, M. J. van de Vijver, J. Bergh, M. Piccart and M. Delorenzi, *J. Natl. Cancer Inst.*, 2006, **98**, 262.
32. L. J. Tafe and G. J. Tsongalis, *Clin. Chem. Lab. Med.*, 2011, **50**, 23.
33. A. C. Wolff, M. E. Hammond, D. G. Hicks, M. Dowsett, L. M. McShane, K. H. Allison, D. C. Allred, J. M. Martlett, M. Bilous, P. Fitzgibbons, W. Hanna, R. B. Jenkins, P. B. Mangu, S. Paik, E. A. Perez, M. F. Press, P. A. Spears, G. H. Vance, G. Viale and D. F. Hayes, *et al.*, *J. Clin. Oncol.*, 2013, **31**, 3997.
34. A. Ocana and A. Pandiella, *Curr. Pharm. Des.*, 2013, **19**, 808.
35. G. Tinoco, S. Warsch, S. Gluck, K. Avancha and A. J. Montero, *J. Cancer*, 2013, **4**, 117.
36. K. C. Halling, W. Kine and I. A. Sokolova, *et al.*, *J. Urol.*, 2000, **164**, 1768.
37. M. Skacel, J. D. Pettay and E. K. Tsiftsakis, *et al.*, *Anal. Quant. Cytol. Histol.*, 2001, **23**, 381.
38. M. Skacel, M. Fahmy and J. A. Brainard, *et al.*, *J. Urol.*, 2003, **169**, 2101.
39. V. Vit, D. Pacik, A. Cermak, I. Falkova, J. Smardova, R. Hrabalkova, M. Svitakova, Z. Pavlovsky and M. Votava, *Cesk. Patol.*, 2009, **45**, 46.
40. I. Karaoglu, A. G. van der Heijden and J. A. Witjes, *World. J. Urol.*, 2014, **32**, 651.
41. J. M. Pipas, L. D. Lewis, M. S. Ernstoff and G. J. Tsongalis, *Oncol. Hematol. Rev.*, 2013, **9**, 75.
42. J. Lazarou, B. H. Pomeranz and P. N. Corey, *JAMA*, 1998, **279**, 1200.
43. B. S. Shastry, *Pharmacogenomics J.*, 2006, **6**, 16.
44. A. C. Lockhart, R. G. Tirona and R. B. Kim, *Mol. Cancer Ther.*, 2003, **2**, 685.
45. C. Marzolini, R. G. Tirona and R. B. Kim, *Pharmacogenomics*, 2004, **5**, 273.
46. B. L. Urquhart, R. G. Tirona and R. B. Kim, *J. Clin. Pharmacol.*, 2007, **47**, 566.
47. M. Ingelman-Sundberg, *Trends Pharmacol. Sci.*, 2004, **25**, 193.
48. M. Ingelman-Sundberg, *Pharmacogenomics J.*, 2005, **5**, 6.
49. S. J. Gardiner and E. J. Begg, *Pharmacol. Rev.*, 2006, **58**, 521.

50. D. W. Nebert and D. W. Russel, *Lancet*, 2002, **360**, 1155.
51. S. Bernard, K. A. Neville, A. T. Nguyen and D. A. Flockhart, *Oncologist*, 2006, **11**, 126.
52. A. N. Abou Tayoun, F. B. de Abreu, J. A. Lefferts and G. J. Tsongalis, *Clin. Chim. Acta*, 2013, **422**, 1.
53. E. Beutler, T. Gelbart and A. Demina, *Proc. Natl. Acad. Sci. U. S. A.*, 1998, **95**, 8170.
54. L. Weng, L. Zhang, Y. Peng and R. S. Huang, *Pharmacogenomics*, 2013, **14**, 315.
55. D. W. Lum, P. Perel, A. D. Hingorani and M. V. Holmes, *PLoS One*, 2013, **8**, 76648.
56. Y. Jin, Z. Desta, V. Stearns, B. Ward, H. Ho, K. H. Lee, T. Skaar, A. M. Storniolo, L. Li, A. Araba, R. Blachard, A. Nguyen, L. Ullmer, J. Hayden, S. Lemler, R. M. Weinshilboum, J. M. Rae, D. F. Hayes and D. A. Flockhart, *J. Natl. Cancer Inst.*, 2005, **97**, 30.
57. K. Sideras, J. N. Ingle, M. M. Ames, C. L. Loprinzi, D. P. Mrazek, J. L. Black, R. M. Weinshilboum, J. R. Hawse, T. C. Spelsberg and M. P. Goetz, *J. Clin. Oncol.*, 2010, **28**, 2768.
58. C. J. Allegra, J. M. Jessup, M. R. Somerfield, S. R. Hamilton, E. H. Hammond, D. F. Hayes, P. K. McAllister, R. F. Morton and R. L. Schilsky, *J. Clin. Oncol.*, 2009, **27**, 2091.
59. B. Chibaudel, C. Tournigand, T. Andre and A. de Gramont, *Ther. Adv. Med. Oncol.*, 2012, **4**, 75.
60. J. M. Caretheres, *Clin. Gastroenterol. Hepatol.*, 2014, **12**, 377.
61. M. Roengvoraphoj, G. J. Tsongalis, K. H. Dragnev and J. R. Rigas, *Cancer. Treat. Rev.*, 2013, **39**, 839.

10 Nanomaterials in Human Disease Diagnosis

Ragini Singh and Sanjay Singh*

Division of Biological and Life Sciences, School of Arts and Sciences, Ahmedabad University Central Campus, Navrangpura, Ahmedabad – 380009, Gujarat, India
*E-mail: sanjay.singh@ahduni.edu.in

10.1. Introduction

Nanotechnology can be defined as the manipulation of matter on a molecular, atomic or supramolecular scale to control its size at the nanometer scale (1–100 nm) which leads to the origin of unique properties.[1,2] Recent developments in this field have explored its unique physical, chemical and biological properties and created a new way to use nanomaterials (NMs) in various biomedical applications, particularly in treatment and diagnosis of diseases, implants, prostheses and drug delivery. Metals, metal oxides, carbon-based materials, quantum dots, upconversion materials and even composite nanomaterials have potential to be integrated into various biomedical products and the treatment of complex diseases.[1,3] NMs are well suited to being used in biomedical devices, due to the nanosize of the biological system. Biospecific molecules can be conjugated with NMs *via* different biological reactions *i.e.* receptor–ligand interaction, antigen–antibody interaction and DNA–DNA hybridization

Genomics and Clinical Diagnostics
Edited by David Whitehouse and Ralph Rapley

Published by the Royal Society of Chemistry, www.rsc.org

or by using physical or chemical methods.[3] Various factors, like surface physics (topography and roughness),[4,5] surface chemistry (composition)[6,7] and surface thermodynamics (free energy and wettability),[8] govern the role of NMs in biomedical applications. Due to the drawbacks of conventional diagnosis techniques, such as lack of sensitivity, time and cost and the need for trained manpower, there is an urgent need to realize the medical potential of nanotechnology in prevention and cure of complex diseases, including cancer and neurodegenerative disorders *etc.*[1,9,10] The major hurdles in diagnosis and treatment of human disease are primarily due to the inability to detect changes occurring at the molecular level during the genesis of particular disease.[11] From the past decade, nanoparticles (NPs) have been employed as alternatives to traditional diagnostic and therapeutic methods due to the following important characteristics: (1)Targeted drug delivery: passive targeting has certain drawbacks, like the chance of drug release into the extracellular matrix, which diffuses into tissue,[12,13] limited permeability and low retention effect in tumors.[14] These limitations can be overcome by conjugating specific affinity ligands (antibodies, peptides, aptamers or other small molecules) to the nanocarrier surface. These nanoconjugates can recognize and bind to the receptors present on the cell surface *via* ligand–receptor interaction[1,15] (2) High payload: a systemic approach for drug delivery leads to the low accumulation of drugs in affected areas. Whereas drugs conjugated or encapsulated with NMs possess "enhanced permeability and retention effect (EPR)" due to which they show the promising effect of being passively extravasated through leaky vascularization to accumulate in the tumor region,[11] resulting in high payload delivery. Liposomes of small size (50–100 nm) encapsulating pharmaceutical ingredients are extensively used in drug delivery systems. Thus, nanocarriers improve drug accumulation in affected areas and reduce renal excretion and hepatic degradation, which leads to increased retention time and also reduces the volume distribution, which avoids drug accumulation in non-targeted and healthy tissues.[1] (3) Continuous monitoring and disease progression: certain imaging techniques, like magnetic resonance imaging (MRI), computed tomography (CT), ultrasound *etc.*, play a critical role in monitoring disease and its progression. However, several reports conclude that these techniques have low accuracy and are unable to distinguish between cancerous and healthy cells.[16,17] Optical probes can be easily conjugated on the surfaces of NMs for imaging techniques, such as positron emission tomography (PET), MRI and CT imaging, and used to monitor disease progression and track NPs. (4) Combined diagnosis and therapy: with the use of NPs it is possible to create formulations containing both drug and imaging agent in a single system, known as theranostic nanosystems, which allow both the visual monitoring of biodistribution and target-site accumulation of drugs.[18,19] This aspect is very useful in developing personalized nanomedicine depending on the accumulation of drugs in patients. Non-invasive imaging is used to estimate the therapeutic efficacy and to identify patients responding well to the treatment.[1]

Various biomarkers have been discovered for early detection and prevention of chronic diseases like diabetes, cancer, cardiovascular diseases *etc.*[20–22] Currently, NMs are being used for detection of biomarkers with more sensitivity and specificity[23] which can be enhanced by multiple targeting moieties.[24] A major application of nanomedicine is to develop novel diagnostic tools for "point of care (POC)" use and to develop a new strategy for targeted drug delivery.[25,26] In comparison with cells, nano-sized tools are of small size and thus can be used at both cellular and organelle-level with great potential. In this chapter, recent advances in nanotechnology-based nano-tools for disease diagnosis and their potential application in early disease detection with high sensitivity and specificity are discussed. Applications of NMs in disease diagnosis are described for some common diseases like cancer, diabetes and infectious diseases.

10.2. Nanoparticles for Diagnosis and Treatment of Cancer

Diagnosis of cancer at an early stage is considered to be an urgent need in today's world due to the high death rate when detected at later stages. By utilizing the unique properties of NMs, it is possible to develop nanobiosensors for detection of biological events with high sensitivity.[27] For cancer diagnosis, NPs can be utilized in two ways, by *in vivo* detection, in which particles can accumulate inside tumor cells *via* ligand–biomarker interaction and produce signals according to the changes that occur at molecular levels; or by *in vitro* detection, which involves detection from biological samples, such as urine, saliva and blood.[27,28] Some of the most common biomarkers used for detection of cancer are epidermal growth factor receptor (EGFR) for epithelia cancer,[29] human epidermal growth factor receptor 2 (HER-2) for breast cancer,[30] folate receptors in several tumors, particularly in ovarian and endometrial cancer,[31] and phosphatidylserine (PS) found on tumor vascular endothelium.[32]

Using NPs as probes for cancer cell detection offers several advantages over traditional methods. First, due to their small size, NPs can escape renal clearance and can reach the desired targets.[33] Also, their small size enables the particles to interact with biological systems at the molecular level by penetrating the several biological barriers, like cell and nuclear membrane and endothelial barriers.[34] Second, due to their unique properties (optical, magnetic, acoustic) NPs can be paired with different imaging functionalities and can be used as probes having multimodal abilities with high sensitivity.[27] Third, NPs possesses multivalent properties, which enable the attachment of different targeting ligands to the surface of one NP. This imparts the small dissociation rate and high rate of receptor binding which leads to longer retention time and high uptake in tumor cells.[35,36] Fourth, these particles can be used for theranostic purposes (for diagnostics as well as for therapy) by loading different therapeutics and imaging components in one nanoplatform which enables therapy process to be monitored in real time.[37]

According to reports, oxidative stress (due to the imbalance in free radical production) is one of the major causes of the transformation of normal healthy cells into cancerous cells.[38] In order to protect organisms from the harmful effect of oxidative stress, certain enzymatic natural antioxidants like glutathione peroxidase (GPx), superoxide dismutase (SOD), catalase, glutathione reductase and non-enzymatic antioxidants like glutathione (GSH), vitamin C and E act in a system.[39] Apart from these naturally occurring antioxidants certain NPs, like cerium oxide NPs (CeNPs), iron oxide NPs (IONPs) and gold NPs (AuNPs), exhibit biological enzyme-like activities.[40–44] Among them, CeNPs possesses both SOD[43,45] and catalase mimetic activity[46] and have also been found to retain their activity in biologically relevant liquids,[40,47] due to which they have found major application in free-radical-mediated diseases.[48]

10.2.1. Iron Oxide Nanoparticles

IONPs have the potential to be used for biomedical purposes due to properties like biocompatibility and biodegradability, facile synthesis[49] and high retention times in blood. Spherical IONPs with diameters smaller than 20 nm have exhibited enhanced contrast in MRI due to their superparamagnetic behavior. Due to their well-studied surface properties, IONPs can be efficiently conjugated with tumor-targeting agents, such as peptides, monoclonal antibodies and small molecules, for diagnosis as well as therapeutic agent delivery. IONPs exhibits strong T2 and T*2 contrast, as well as T1 effects at very low concentrations in MRI, which makes them efficient probes for clinical oncology imaging.[50]

Recently, BSA-coated IONPs (BSA–IONPs) were shown to accumulate in tumor cells (C6 glioblastoma, 4T1 mammary adenocarcinoma and RS-1 hepatic mucous carcinoma) and their large vessels, which enabled them to be diagnosed with MRI.[51] These BSA–IONPs were also found to be gradually eliminated from normal tissues whereas they remained accumulated in 4T1 mammary adenocarcinoma 2 hours after injection, which makes the image clear and bright[51] probably due to the EPR effect.[52]

Chen *et al.*[53] synthesized a dual modality probe for precise detection of liver tumor by synthesizing a formulation of indocyanine green (ICG) and liposome-coated IONPs. These NPs were further conjugated with the tumor-targeting agent, Arg–Gly–Asp peptides (RGD) which serve as both near-infrared fluorescence (NIRF) and MRI probes. During *in vitro* and *in vivo* experiments uptake of probe was analyzed by following the fluorescence intensity for ICG and concentration of IONPs by inductively coupled plasma mass spectrometry (ICP-MS). Probes were particularly targeted towards the $\alpha v\beta 3$ receptor of HepG2 cells and internalized into the cell *via* receptor-mediated endocytosis. RGD peptides were combined with neuropilin-1 peptide ligand and targeted towards tumor vasculature, which facilitated the deep penetration of the imaging agent into the tumor.[54,55]

Zhang *et al.*[56] developed an IONP-based colorimetric technique for detection of a cancer biomarker, carcinoembryonic antigen (CEA). This assay is based on red color formation due to coordination between, bathophenanthrolinedisulfonic acid disodium salt (BPT) and Fe^{2+} released from IONPs.[57] The large amount of Fe^{2+} released from IONPs itself acts as an efficient signal amplifier without the use of any signaling cascade, therefore, the signal is directly proportional to targeted ligands for accurate quantifications. This technique has great potential for early cancer diagnosis with a limit of detection (LOD) at 3.6 pg ml^{-1} for CEA under ideal condition.[56]

In another study, PEG-coated IONPs conjugated with chlorotoxin (CTX) were used for targeted delivery to glioma cells and *in vivo* xenograft tumor mouse model.[58] CTX are known to be internalized into cells *via* endocytosis by preferential binding with matrix metalloproteinase 2 (MMP-2).[59] These nanoprobes were found to be accumulated in gliomas and to demonstrate a significant increase in negative contrast (darkening) and also higher R2 relaxivity. Nanoprobes were shown to have long retention times before getting cleared out from the body and resulted in R2 changes of ~16 ms^{-1} for up to 12 hours. No apparent toxicity was observed for nanoprobes in clearance organs (kidney, liver and spleen) of mice, as investigated by histological analysis, the nanoprobes were thus found to be suitable for tumor staging, early cancer detection and therapy assessment.

An epidermal growth factor receptor (EGFR) variant (EGFRvIII) is expressed only in malignant glioma cells and not in normal cells. Therefore, several studies have used these and other similar receptors for targeted delivery. Hadjipanayis *et al.*[60] demonstrated the synthesis of antibody-conjugated IONPs (10 nm) which are targeted and bind selectively to the EGFRvIII present on human glioblastoma multiforme (GBM) cells after convection-enhanced delivery (CED) in both *in vitro* and *in vivo* experimental conditions. The resultant EGFRvIIIAb–IONPs have proficient antitumor effects which caused decreases in glioblastoma cell survival, and concomitant enhanced MRI contrast in U87ΔEGFRvIII cells with T2-weighted MRI. CED allowed the initial dispersion of IONPs into and adjacent to intracranial xenograft tumors.

Lee *et al.*[61] developed Mn-doped iron oxide NPs (MnMEIO) conjugated with the cancer targeting antibody Herceptin and evaluated the cancer detection ability in various cell lines, such as Bx-PC-3, MDA-MB-231, MCF-7 and NIH3T6.7. The MnMEIO–Herceptin conjugate can detect the Bx-PC-3 (lowest expression of HER2/neu marker) with enhanced MR contrast. The increases in R2 (ΔR2/R2Control) were found to be enhanced (~10%) for Bx-PC-3 cells, which increased with high HER2/neu marker expression *i.e.* ~40%, ~70% and ~130% for MDA-MB-231, MCF-7, NIH3T6.7 cell lines respectively. This effect is in agreement with the expression level of Herceptin in these cell lines. They have also tested the efficiency of a MnMEIO–Herceptin conjugate for the detection of small cancers (~50 mg) in a mouse model, which showed enhanced R2 signal (ΔR2/R2Control = ~31%).

Thus, on the basis of the results discussed above, it can be concluded that IONPs can be prepared with the desired targeting moiety at the surface. Furthermore, such IONPs have shown tremendous results in cancer imaging and treatment. It is also expected that in the near future some of these formulations would easily pass through clinical trials and be ready for human use.

10.2.2. Gold Nanoparticles

AuNPs shows surface plasmon resonance (SPR) absorbance depending on the size and shape of particles.[62,63] AuNPs are used as fluorescence quenchers and can be utilized in several "on–off" fluorescence probes. In comparison to organic quenchers, AuNPs are more efficient due to surface energy transfer processes.[63,64] In spite of being used as a fluorescence quencher, small AuNPs or Au nanoclusters composed of gold atoms (less than 3 nm in size) are used as a fluorophore label in imaging applications.[65] Kang and Nguyen[66] developed an AuNP-based contrast agent for detection of cancer cells expressing nucleolin and MMP-14 on the cell membrane (Figure 10.1). It consist of, (1) AuNPs, (2) AS1411 aptamer (target for nucleolin), (3) short spacer (SS, composed of MMP-14 substrate) and long spacer (LS), (4) NIR-fluorophore cypate (at the end of spacer) and (5) biopolymer (11-mercapto undecyl-tri-ethylene glycol) coating. Nucleolin is generally present in the cell nucleus but overexpressed on the cancerous cell surface membrane and binds with the AS1411 aptamer.[67] The SS places the cypate close to AuNPs, whereas the LS keeps it at a distant place where it shows enhanced fluorescence. When it binds to cancer cells *via* AS1411, MMP-14 cleaves the SS, leaving cypate distant from AuNPs attached by the LS, which leads to the generation of enhanced fluorescence.

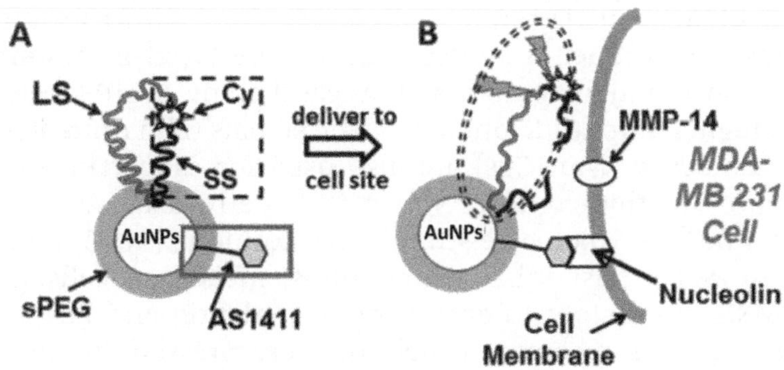

Figure 10.1 Schematic diagram of AuNPs-based optical/CT contrast agent for MDA-MB 231: AS1411 is a cancer cell target (*solid rectangle*) and cypate is conjugated *via* SS/LS. (A) The agent normally emits little fluorescence because SS holds cypate close to the AuNPs (*dotted rectangle*). (B) When the agent is delivered to the cancer cell site, AS1411 binds to the cell. MMP-14 on the cell membrane then cleaves SS, placing cypate from the AuNPs at a distance of the fully stretched LS length (oval area) and cypate fluoresces at an increased level. Reproduced from ref. 66 with permission from Springer Nature, Copyright 2017.

In another attempt, Chen *et al.*[68] demonstrated the use of anti-mucin-antibody-conjugated AuNPs in bioimaging as well as photothermal therapy (PTT) of urothelial cancer. They used green laser light (532 nm) to excite AuNPs, which produced thermal energy to destroy cancerous cells.

To target the tumor-specific antigen selectively and with high sensitivity, Popovtzer *et al.*[69] explored the role of gold nanorods (AuNRs) in X-ray-based CT, which is one the most convenient imaging technique in cancer diagnosis. They conjugated UM-A9 antibody with AuNRs and targeted it towards squamous cell carcinoma (SCC); oral cancer (UM-SCC-1) and larynx cancer (UM-SCC-5) cell lines, known to overexpress the A9 antigen.[70] The results revealed that A9-AuNRs binds specifically to cancerous cells, as indicated by the high change in attenuation coefficient (five times more when compared with non-targeted cancerous cells) and higher CT attenuation number (Hounsfield units (HU)) in comparison with soft tissues. This technique can detect a small tumor (1 mm^3) with sufficient differential signal, which has an approximately similar limit of resolution to clinical CT scanning. AuNPs also serve as a most promising contrast agent for photoacoustic (PA) imaging due to their tunable optical properties and SPR.[71]

Manivasagan *et al.*[72] demonstrated the use of chitosan to synthesize AuNPs loaded with paclitaxel (PTX) and evaluated their use in therapy and PA imaging of MDA-MB-231 cells. They employed 532 nm laser light for photoacoustic tomography (PAT) along with PTX-AuNPs as a contrast agent. Such a strategy can be employed in non-invasive early diagnosis and imaging-based cancer therapy. AuNP-labeled cells showed higher amplitude photoacoustic signals than untreated cells.

Gold nanocages (AuNCgs) are also used for cancer imaging and treatment.[73] In this context, Kim *et al.*[73] derivatized AuNCgs with [Nle4, D-Phe7]-α-melanocyte-stimulating hormone (MSH) and evaluated their PA contrast enhancement on B6 melanoma cells. [Nle4,D-Phe7]–MSH is resistant to enzymatic degradation and has a strong affinity for MSH receptor, which is overexpressed in melanoma cells.[74] PA signal was approximately 300% higher for cells treated with AuNCg-[Nle4, D-Phe7]-MSH than PEG-coated AuNCgs. PA detection limit for AuNCgs was reported to be 4.5 pM, whereas injection dose was found to be as low as 10 nM, employed during an *in vivo* molecular PAT experiment. Bioconjugated AuNCgs combined with PAT can potentially be employed in early-stage melanoma and metastatic lymph nodes detection.[75]

Surface-enhanced Raman scattering (SERS) is designated to the Raman-active vibrations of molecules when they are in close proximity to the metal surface. AuNPs also proved a promising agent for SERS probes, as it has been shown that when a molecule is adsorbed on the AuNPs surface its Raman signal gets enhanced by a factor of 10^{14}–10^{15}. Size, shape and inter-particle distance affect the degree of enhancement.[27,76] Using the SERS property of AuNPs, Zong *et al.*[77] demonstrated the synthesis and targeting of cancer cells by CdTe quantum dots (QDs) conjugated and silica-coated Au@Ag core–shell nanorods (Au@Ag NRs) with the dual property of SERS and fluorescence. Folic acid (FA) was conjugated with a probe and targeted towards HeLa (which express folate receptor) and MRC-5 cells (which display no folate receptor expression). Fluorescence and SERS results confirmed that the probe could serve as a specific and efficient targeting agent for cancerous cell detection.

In another approach, Li *et al.*[78] developed a bifunctional nanocomposite probe consisting of a gold shell and a nickel–iron core (NiFe@Au) with the dual detection property of SERS monitoring *via* magnetic focusing. The detection limit of NiFe@Au towards CEA was reported to be ~0.1 pM and could further be reduced by optimizing particle size.

Circulating tumor cells (CTC) are cells that detach from a solid tumor and circulate in blood and lymph systems to create a secondary tumor. Therefore, it is expected that their detection would help in early diagnosis, monitoring of response to therapy *etc.* In order to detect CTC, Bhattacharyya *et al.*[79] reported a system consisting of AuNPs conjugated with antibody and rhodamine dye. They targeted epithelial cell adhesion molecule (EpCAM) receptor of a breast cancer cell line (T47D) in suspension with 50 nm AuNPs and detected it through a photoacoustic flowmeter. Results revealed the attachment of AuNPs to cells, which was further confirmed by optical contrast and photoacoustic signal generation. They recorded photoacoustic waveforms from 9, 19, 38, 150 and 300 tagged cells and found an increasing trend in total energy specified by an area under the curve of each pressure wave.

Many sensors have been developed where AuNPs act as a fluorescence quencher due to their stronger absorbance and lower background signal than those of traditionally used organic fluorophore quenchers. In one such report, You *et al.*[80] demonstrated the development of an AuNP probe for cancer biomarker detection depending on nonspecific electrostatic adsorption of protein biomarker to a library of fluorophores, tagged with AuNPs. It can simultaneously detect two overexpressed cancer biomarkers *i.e.* acid phosphatase and alkaline phosphatase, which are isoenzymes, upregulated in cancerous cells. This can differentiate between various protein biomarkers, however, for any arbitrary target detection it requires more AuNPs libraries to be screened, which are specific to the target.[63] In a similar study, Huang *et al.* reported on aptamer conjugated AuNPs (Apt–AuNPs) for detection of platelet-derived growth factor (PDGF) through a 'turn-on' fluorescence mechanism. They conjugated Apt–AuNPs with fluorophore *N*,*N*-dimethyl-2,7-diazapyrenium dication (DMDAP), which has a strong binding affinity for DNA. When DMPAD intercalates with aptamer of Apt–AuNPs its fluorescence gets quenched through a fluorescence resonance energy transfer (FRET), mechanism. However, in the presence of PDGF, DMPAD restores its fluorescence due to distortion in aptamer conformation or blocked intercalation of DMPAD. The limit of detection for this assay was reported to be 8 pM.[81]

In spite of using AuNPs as a fluorescence quencher, Cho *et al.*[82] (Figure 10.2) explored the surface enhanced fluorescence (SEF) between AuNPs and Cyanine 3B (Cy3B) to synthesize an 'on–off' probe for detection of vascular endothelial growth factor (VEGF), which is major biomarker for cancer angiogenesis. Here, Cy3-labelled VEGF aptamer was electrostatically attached to AuNPs by positively charged poly-L-lysine and it possesses SEF due to metal interaction and local SPR. However, in the presence of target molecule, VEGF aptamer binds to its target, which causes its conformation to change and hence decreases the electrostatic binding ability. This results in irreversible detachment of aptamer from AuNPs, and leads to decreased Cy3 fluorescence intensity. This strategy can detect the target molecules from a wide range of concentrations from 25 pg ml^{-1} to 25 μg ml^{-1}.

The Janus kinase–signal transducer and activator of transcription (JAK–STAT) pathway is a known oncogenic pathway involving STAT protein. Detection of STAT protein is a useful strategy to investigate tumor status because STAT is an important transcription factor for solid tumor proliferation and survival.[83] In this context, Xue *et al.*[84] developed a beacon for detecting human STAT 5B mRNA expression in MCF-7 cells by using hairpin DNA for AuNPs functionalization (hDAuNPs) which keeps dye and AuNPs in close proximity for quenching of dye fluorescence. In the presence of target mRNA, the DNA sequence get stretched and 5′-attached fluorophores are pulled away from the 3′-attached AuNPs to sufficient distance to regain the dye fluorescence. hDAuNPs were found to be efficiently internalized into MCF-7 cells and their fluorescence correlated with

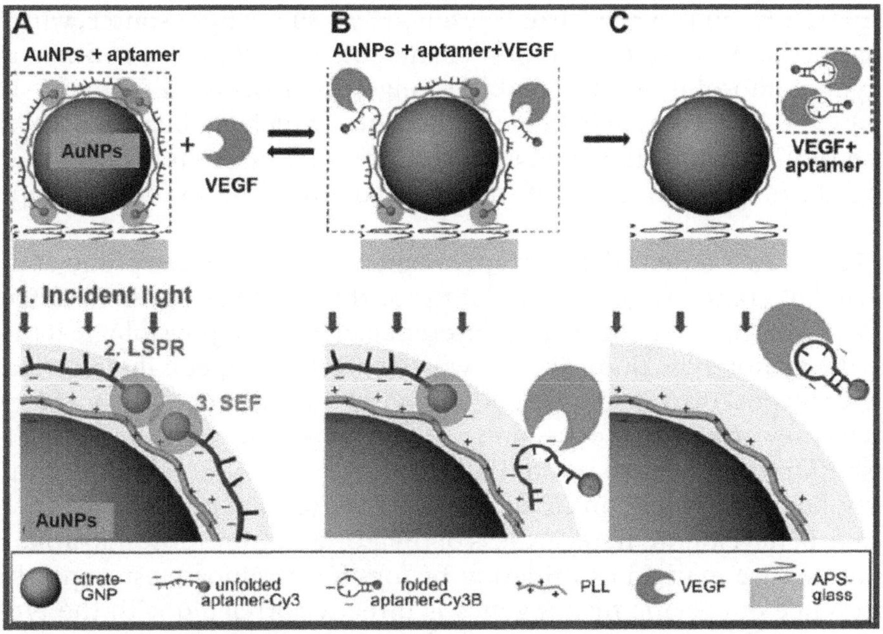

Figure 10.2 Schematic representations of detection mechanism of the aptasensor for VEGF165. (A) In the absence of target molecules, unfolded VEGF165 aptamer is electrostatically bound to a positively charged PLL-coated AuNPs surface and surface-enhanced fluorescence (SEF) of Cy3B conjugated with the VEGF165 aptamer is created by both a metal interaction increasing the radiative fluorescent decay rate of Cy3B and the LSPR enhancing the intensity of an incident light. (B) The interaction of the VEGF165 aptamer with its target induces a reversible conformation change of the aptamer and, consequently, decreased electrostatic binding force. (C) As a result, the target-binding interaction of the aptamer causes the irreversible detachment of the aptamer from the AuNP's surface and avoids the SEF effect of Cy3B. Reproduced from ref. 82, with permission from American Chemical Society, Copyright 2012.

the STAT 5B mRNA expression, similar to PCR results. Thus, the role of AuNPs in cancer diagnosis and treatment has been comprehensively presented above. The multifunctional role of AuNPs has been established in cancer research and they have realized their potential in treatment and imaging, which needs to be applied during clinical trials.

10.2.3. Quantum Dots

Due to their high molar extinction coefficient, quantum yields, tunable emission properties and resistance against photobleaching[85] quantum dots (QDs) are used as excellent agents for detection purposes.[63,86]

Differently emitting QDs can be excited by a single light source, which makes them suitable for multiplex imaging purposes.[27] In contrast to organic fluorophores, QDs exhibit longer fluorescence lifetimes (>10 ns)[87] and better signal to noise (S : N) ratios, which enables them to be used for temporal imaging. QDs are efficiently used in cancer imaging, such as illumination of tumor vasculature[88] and sentinel lymph nodes.[89]

Fan *et al.*[90] reported a novel pH-responsive fluorescent probe composed of graphene QDs (GQDs) for robust and universal tumor detection from tumor xenografts of different cell lines. At physiological pH, GQDs exhibit blue fluorescence, which switches to green fluorescence when exposed to acidic pH. Along with the fluorescence property, GQDs also exhibit upconversion photoluminescence (UCPL), which is useful in non-invasive tumor detection. This dual fluorescence property has its application in sensitive detection of tumors at an early stage of development, regardless of their origin. Chromosomal abnormalities are marked as initiation factors for carcinogenesis, and the presence of specific fusion genes is directly correlated with the type of cancer.[91,92] Determination of different types of fusion genes in the sample is important for the understanding of abnormal gene functionality in cancerous cells. In this context, Lee *et al.*[93] developed oligonucleotide-functionalized NPs for prostate cancer marker detection in various biological samples including buffer, cell lysate and urine. This nanoprobe was composed of oligonucleotide-functionalized QDs and magnetic microparticles for optical detection of rearranged TMPRSS2–ERG (transmembrane protease serine 2–v-ets erythroblastosis virus E26 oncogene homolog) fusion gene in the urine sample. Overexpression of TMPRSS2–ERG, and also of truncated ERG protein product, is the specific and prominent marker of prostate cancer.[94] This system was able to detect different TMPRSS2–ERG fusion genes in a single assay with different color-coded oligonucleotides with a detection limit of 1 fmol.

DNA methylation is one of the distinct epigenetic change in cancer cells,[95] thus, its identification helps in detection of cancerous cells as well as early diagnosis. Using this property, Ma *et al.*[96] also developed a nanosensor based on FRET of QDs for detection of DNA methylation of promoters in the early stages of non-small-cell lung cancer (NSCLC) tissue and non-invasive bronchial brushing specimens. Detection depends on the FRET between positively charged amino-QDs and negatively charged A647-labelled DNA which are in close proximity. Non-methylated DNA was cleaved by a methylation-sensitive

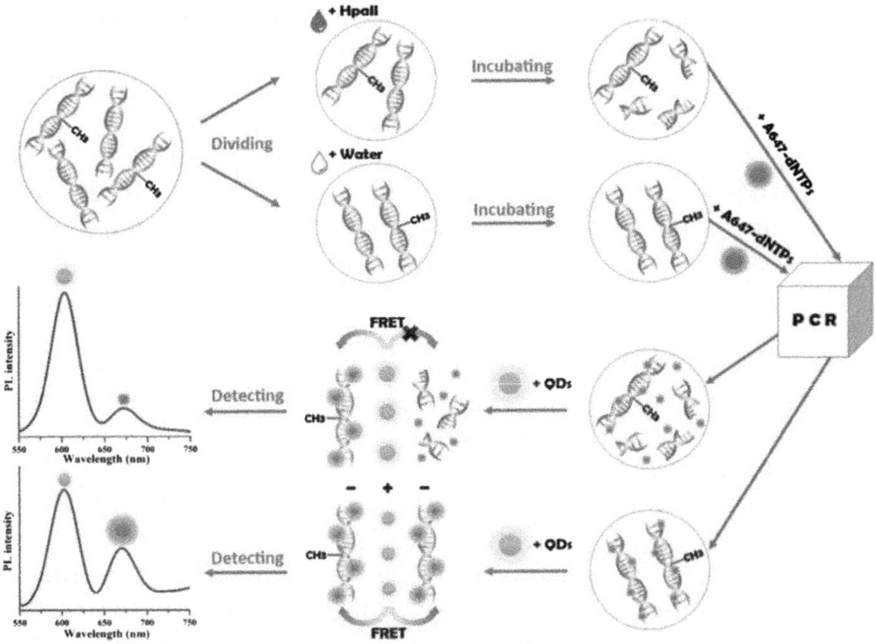

Figure 10.3 Schematic illustration of the detection of DNA methylation based on QDs FRET. Reproduced from ref. 96 with permission from Elsevier, Copyright 2016.

restriction enzyme, *Hpa*II leaving only the A-647 labeled methylated part for PCR amplification, thus producing a stronger FRET signal with more DNA methylation (Figure 10.3).

In another study, Wei *et al.*[97] reported the detection of estrogen receptor-β (ER-β), a tumor suppressor downregulated in advanced cancer stages, *via* sandwich fluoroimmunoassay in solution by utilizing FRET. In this assay, QDs were attached with anti ER-β monoclonal antibody, whereas, AlexaFluor (AF) was labeled with anti ER-β polyclonal antibody. In the presence of ER-β antigen, QDs and AF bind with it at a distance of 80–90 Å and undergo FRET, which can be measured by confocal microscopy, and have very high sensitivity, with a detection limit of 0.05 mM (2.65 ng ml^{-1}).

Similarly, Wegner *et al.*[98] developed immunoassays for detection of the cancer biomarker, prostate-specific antigen (PSA) by using terbium–QDs FRET pairs. Six different primary antibodies were used to bind with PSA, enabling the detection limit to be low as 1.6 ng ml^{-1}.

In another approach, Kim *et al.*[99] reported a sandwich immunoassay technique for CEA detection based on QDs in combination with a vertical ZnO nanowire array on a glass substrate which provides a larger surface area for biomolecule immobilization than conventional glass-based immunoassays. To preserve the original fluorescence of QDs, FRET phenomena between QDs and ZnO nanowire were suppressed by introducing an anti-FRET layer composed of the biotin–streptavidin complex. CEA analyte was sandwiched between monoclonal anti-CEA antibody conjugated to ZnO nanowire and secondary antibody-conjugated QDs to give fluorescence (Monoclonal Ab – CEA – polyclonal Ab – QDs labeled with secondary Ab).

A multimodal nanosensor, exhibiting both FRET and chemiluminescence resonance energy transfer (CRET) properties for more sensitive detection of micro-RNA-141 (miR-141), a prominent biomarker of prostate cancer, was developed by Jou *et al.*[100] In this assay, FRET-quencher-functionalized nucleic acid, which recognizes miR-141, was attached to CdSe–ZnS QDs. Upon binding of miR-141,the quencher is released, which leads to increased fluorescence of QDs and enables the detection of miR-141 with a limit of detection of 1 pM. Furthermore, to increase the detection sensitivity, cleaved nucleic acid acts as the primer for telomerase, overexpressed in cancerous cells.[101] It leads to the formation of G-quadruplexes with the telomeric repeating unit in which hemin gets intercalated and yields a telomeric–hemin G-quadruplex HRP (horseradish peroxidase)-mimicking DNAzyme unit. It catalyzes the oxidation of luminol in the presence of H_2O_2 and produces chemiluminescence, which enhances the sensitivity of the assay and reduces the detection limit to 0.28 pM. The first step is non-selective and serves only as an indicator signal for the presence of mi-R141, whereas the second step provides a specific signal for the existence of miR-141. The surface plasmon-coupled emission (SPCE) property of QDs was used for PSA detection by Jin *et al.*[102] In this study, PEGylated CdSe/ZnS QDs were conjugated with PSA antibody and used as a fluorescent probe with a low detection limit of 10 fg ml^{-1}. PSA were immobilized on the SiO_2 gold substrate and Ab-QDs were used to analyze specific antigen–antibody interaction.

The presence of alpha-fetoprotein (AFP) in the blood is indication of some of the specific cancer types, like stomach, liver, pancreas, testicular or ovarian.[103] Considering this, Chen *et al.*[103] reported the development of a QD-based homogenous time-resolved fluorescence assay for detection of AFP, a prominent marker for many type of cancers. In this assay, QDs doped with micro particles (QPs) were conjugated with anti-AFP antibody and acted as energy donors, whereas, luminescent terbium chelates (LTCs) conjugated with different anti-AFP antibody served as energy acceptors. AFP was detected in a double antibody sandwich structure in which time-resolved Förster resonance transfer (TR-FRET) was activated and used to detect AFP. QDs co-polymerized with a cancerous cell binding aptamer *via* DNA-programmed hybridization chain reaction form linear QD–aptamer polymers (QAPs) and exhibit stronger photoluminescence (PL) than QD–aptamer monomers (QAMs) and show strong cell imaging signals at low concentrations of QDs (5 nM). In this assay, QAPs were targeted towards the protein kinase 7 receptor (PTK7) of human acute lymphoblastic leukemia (CCRF-CEM).[104]

10.2.4. Upconversion Nanoparticles

Upconversion nanoparticles (UCNPs) exhibit unique optical properties, therefore being useful in multicolor imaging applications. UCNPs have anti-Stokes shift and emit photons of higher energy than the absorbed photon. Anti-stoke shift occurs due to a nonlinear optical process by the multiphoton excitation process (where two or more photons get simultaneously absorbed and produce a sufficient amount of energy for upconversion), and UCNPs can be excited with near-infrared (NIR) light (approximately 980 nm).[27,63] This property of UCNPs has several biological applications, which lead to decreases in autofluorescence of cells or tissues. Several reports have shown the use of UCNPs in deeper penetration in tissue and minimum interference of autofluorescence in the tissue-transparent NIR window.[105]

The tunable emission property of UCNPs provide cancer detection by multicolour imaging. In this context, Wang *et al.*[106] (Figure 10.4) reported fluorescent UCNPs composed of $NaYbF_4$:RE (RE = Er^{3+}, Tm^{3+}, Ho^{3+}, or a combination of these ions) which emit light in visible range upon excitation with NIR irradiation (980 nm). They targeted HeLa cells by conjugating anti-CEA8 antibody with nanoprobe. The emission range of these probes can be tuned simply either by varying codopant concentration or dopant types. The results indicated that these UCNPs were specifically attached to HeLa cells and could also be used as labels for different cancer cells depending on the presence of different types of cell surface markers.

A mRNA-based (PCA-3) sensor for prostate cancer was developed by Vilela *et al.*,[107] which was composed of graphene oxide (GO) and UCNPs where, GO and UCNPs acts as energy transfer acceptor and donor, respectively. UCNPs were conjugated with nucleotide sequences complementary to the PCA3 mRNA sequence. In this two-step assay,

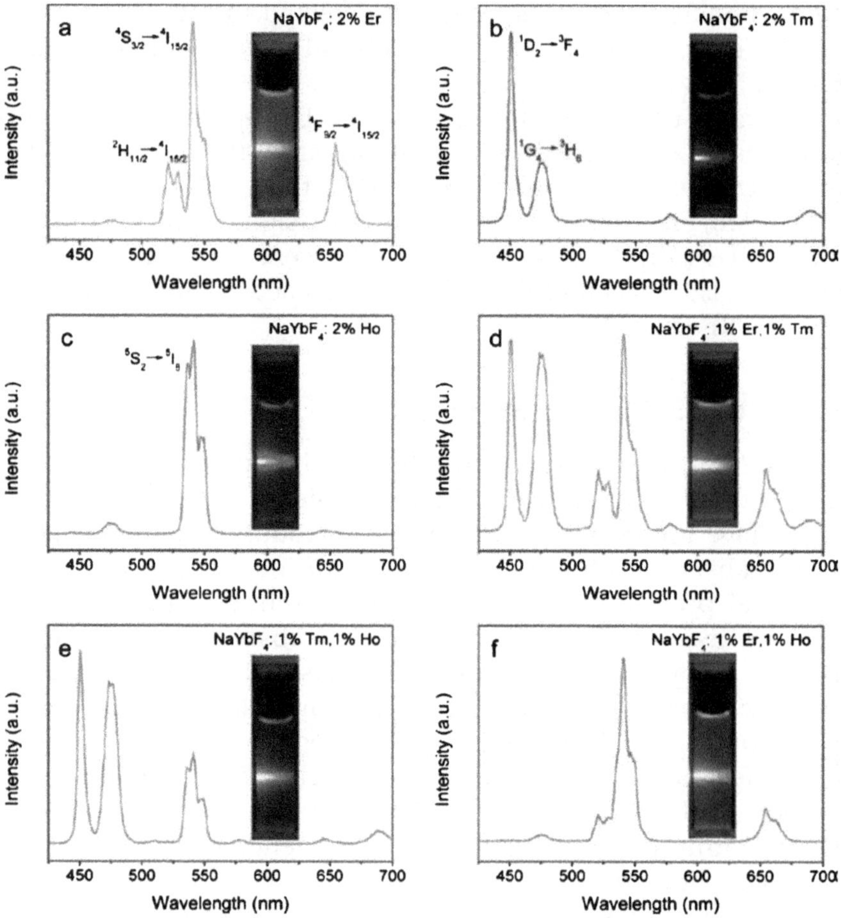

Figure 10.4 UC fluorescence spectra of 1% (by weight) colloidal solutions of NaYb0.98F$_4$:RE0.02@SiO$_2$ NPs in water excited with a 980 nm laser: (a) NaYb0.98F$_4$:Er0.02@SiO$_2$, (b) NaYb0.98F4:Tm0.02@SiO$_2$, (c) NaYb0.98F4:Ho0.02@SiO$_2$, (d) NaYb0.98F4:Er0.01,Tm0.01@SiO$_2$, (e) NaYb0.98F4:Tm0.01,Ho0.01@SiO$_2$, and (f) NaYb0.98F4:Er0.01,Ho0.01@SiO$_2$. Insets are the photographs of 1 wt% colloidal solutions of corresponding NPs in water, excited with a 980 nm laser. Reproduced from ref. 106 with permission from American Chemical Society, Copyright 2009.

UCNPs were mixed with cell lysate in the presence of target oligonucleotide, followed by hybridization between nucleotide chains, which results in lower adsorption of UCNPs to the GO surface. This event was recorded by detecting fluorescent signals of UCNPs mediated through excitation with a 980 nm laser. This strategy overcomes several drawbacks, such as biomolecules' absorbance interference or photo blinking, and has high sensitivity with a limit of detection in the femtomolar range.

To avoid protein adsorption over UCNPs, Rao *et al.*[108] coated them with erythrocyte membrane. They found that even after exposure to 100% human plasma red blood cell (RBC)-membrane-capped UCNPs (RBC–UCNPs) did not adsorb any protein on their surface. Furthermore, RBC-UCNPs were conjugated with FA for its targeted delivery towards cancerous cells (MCF-7). In comparison to FA–UCNPs, FA–RBC–UCNPs showed significantly better targeting efficiency, probably due to the absence of protein corona on the NPs surfaces.

Recently, Shen *et al.*[109] have prepared sub-20 nm (~16 nm) multifunctional nanoprobes with a sandwich structure for *in vivo* bioimaging under the dual-mode of imaging upconversion luminescence (UCL)/CT. A $NaLuF_4$:Yb/Tm (~3 nm) shell layer was formed over ultrasmall $NaGdF_4$:Yb/Tm (~5 nm).

An *in vivo* multicolor imaging application of UCNPs [sodium yttrium fluoride ($NaYF_4$) nanocrystals] was demonstrated by Cheng *et al.*[110] which is based on the luminescence resonance energy transfer (LRET) property. NaYF4 UCNPs doped with Er^{3+}/Yb^{3+} and Er^{3+}/Tm^{3+} were synthesized, which emit in green (UCNP1) and NIR (UCNP2) light respectively, upon excitation at 980 nm. UCNPs modified by conjugating two fluorescence dyes, Rhodamine B (RhB), Rhodamine 6G (Rh6G), and one quencher, (tide quencher 1) served as acceptor in the UCNP–LRET system. Five-color, UCL *in vivo* imaging was achieved by using UCNPs and dye-conjugated UCNPs by spectral deconvolution.

Similarly, Liu *et al.*[111] designed a LRET-based nanosensor composed of UCNP ($NaYF_4$:Yb/Tm@$NaYF_4$) as a core and mesoporous silica shell containing tris (4, 7-diphenyl-1,10-phenanthroline) ruthenium(II) dichloride ($[Ru(dpp)_3]^{2+}Cl_2$) as an oxygen indicator. The assay is based on the fact that UCNPs acts as an energy donor in the LRET process, whereas the red luminescence of $[Ru(dpp)_3]^{2+}Cl_2$ was quenched strongly in the presence of oxygen due to a reversible photochemical oxidation process. However, in the absence of oxygen, (hypoxic conditions) $[Ru(dpp)_3]^{2+}Cl_2$ accepts energy transfer from UCNPs and shows red fluorescence. The results can be directly correlated with the presence or absence of oxygen. This assay was positively tested in zebrafish embryo model of hypoxia as well as in U87MG cells.

In another study, Xiong *et al.*[112] developed RGD-labelled upconversion nanophosphors (UCNPhs) (with high affinity for the α_v β_3 integrin receptor) for targeted imaging of cancerous cells U87MG (expressing high α_v β_3 integrin receptor) and MCF-7 cells (expressing low levels of α_v β_3 integrin receptor) at the surface. Multicolor UCNPs were composed of $NaYF_4$: 20%Yb, 1.8%Er, 0.2% Tm, and possessed green, red and NIR UCL with different penetration depths, which can be used for combined application *in vivo*, in cells and in tissues. The results revealed that UCNPhs can penetrate up to ~600 μm deep in the tissue and do not suffer from an autofluorescence signal, while UCL imaging can efficiently be used for targeted imaging. Signal-to-noise ratios were recorded to be ~24 between tumor and background as analyzed by UCL signal in the region of interest. Upon excitation with a 980 nm laser, an intense light signal was recorded in mice having U87MG cells, whereas, no significant signal was observed in the MCF-7 cells.

Application of magnetic and fluorescent lanthanide-doped gadolinium oxide (Gd_2O_3) UCNPs in dual modal imaging with MRI/UCL and photodynamic therapy (PDT) was demonstrated by Liu *et al.*[113] As a MRI contrast agent, it exhibited a high longitudinal relaxivity (R1) value and T1-weighted contrast enhancement in mice with tumor xenografts. As synthesized Gd_2O_3:Yb^{3+}/Tm^{3+}, Gd_2O_3:Yb^{3+}/Ho^{3+} and Gd_2O_3:Yb^{3+}/Er^{3+} UCNPs exhibited strong emission band in the blue, green and red regions, respectively. When tested in a murine macrophage cell line (RAW 264.7 cells), results indicated that after incubation with synthesized UCNPs cells showed their respective fluorescences after excitation at 980 nm, with no autofluorescence. The study also revealed that, at the same Gd concentration, UCNPs serves as a better MRI contrast agent than gadolinium–diethylenetriamine penta-acetic acid (Gd–DTPA; a clinical contrast agent). Thus, this product can be used as fluorescent nanoprobes for live cell imaging as well as a better MRI contrast agent. In addition to multicolor imaging, UCNPs also found an application in probe design for cancer diagnosis by hypoxia detection, as tumor tissues are known to suffer with low oxygen conditions due to their abnormal vasculature (Table 10.1).

10.3. Nanoparticles for Diagnosis and Treatment of Diabetes

Diabetes is one of the most prominent diseases in the world, affecting about 347 million people worldwide. It is caused by insulin disorder in the body and classified into type 1 and type 2 according to its underlying mechanism.[127,128] Type 1 diabetes (insulin-dependent diabetes) is caused by the destruction of pancreatic beta cells by autoimmune attack, whereas type 2 diabetes (non-insulin-dependent diabetes) arises after a metabolic disorder detected by high blood glucose. Conventional monitoring methods are invasive, painful (resulting in non-compliance) and intermittent, which leads to the need

Table 10.1 Nanoparticles in disease imaging.

Imaging modality	Nanoparticle used	Targeting ligand	Result	Reference
MRI	IONPs	RGD	RGD–IONPs efficiently label $\alpha_v \beta_3$ integrins expressed on endothelial cells and have potential for monitoring angiogenesis in tumor therapy	114,115
	Gadolinium loaded apoferritin probe	C3d peptide	Target to neuronal cell adhesion molecule and visualize tumor endothelial cells *in vitro* and *in vivo* [Severe combined immunodeficiency (SCID) mouse]	
CT	AuNPs	RGD	Tumor site targeted by probes in 10 min after injection and can circulate in whole body, including renal filtration	116–118
	Bismuth sulphide NPs	—	High X-ray absorption, long circulation time, enhanced *in vivo* imaging of vasculature	
	Barium- and ytterbium-based NPs	—	Longer circulation time *in vivo* and contrast enhancement observed within 2 hours	
Ultrasound	Nanobubbles	—	Intravenous injection reveals selective, long lasting and strong contrast in tumor volume which indicate extravasation through the defective tumor microvasculature	119,120
	Nanobubbles from gas-generating polymeric NPs (GGPNPs)	—	GGPNPs penetrate tumor interstitium and merge to form nanobubbles, thus, causing vibrating reactivity seen under cadence mode	
Optical imaging	Chitosan based NPs(CNPs)		CNPs efficiently accumulated to tumor site, showed high therapeutic efficacy due to more cellular uptake	121,122
	ZnS capped CdSe QDs	Prostate-specific membrane antigen monoclonal antibody	QDs probe can targeted to tumor specific site more efficiently *via* active targeting	
Photoacoustic imaging	Gold nanocages	—	Due to large optical absorption cross-section 50 nm AuNCgs serve as better contrasting agents for sentinel lymph node (SLN) imaging than 30 nm AuNCgs	123,124
	Single-wall carbon nanotubes	RGD	Intravenous injection shows eight times stronger PA signal	
PET	Radiolabelled gold nanoshell	—	Internalized in tumor cells 20 hours after injection and provide good resolution in PET imaging	125,126
	Dextran NPs (DNPs)	—	13 nm DNPs enable noninvasive assessment of inflammation in experimental atherosclerotic plaques	

for development of a method that offers non-invasive monitoring of diabetes. Treatment of diabetes suffers from many other problems, which are expected to be resolved with the help of nanomedicine *i.e.* biocompatible nanofilms can be utilized for continuous glucose monitoring, islet nanoencapsulation helps in islet cell transplantation, NPs carrying insulin may help in sustained insulin delivery.[129] Some of the frequently used NPs in diabetes diagnosis and treatment are discussed below.

10.3.1. Iron Oxide Nanoparticles

Due to their low toxicity, strong superparamagnetic property and intrinsic peroxidase-like activity, IONPs represent one of the best materials to be used in biosensing. Several biosensors have been developed for hydrogen peroxide (H_2O_2) detection by using IONPs modified with carbon electrodes[130] and with chitosan–Fe_3O_4 nanocomposites.[131]

The presence of glucose in urine is an important indication of diabetes, therefore, Su *et al.*[132] reported the intrinsic peroxidase-like activity of $ZnFe_2O_4$ magnetic NPs and its application in glucose detection in urine samples. The sensing mechanism is based on the fact that $ZnFe_2O_4$ catalyzes the oxidation of peroxidase substrate 3,3′,5,5′-tetramethylbenzidine (TMB), in the presence of H_2O_2. In this assay, H_2O_2 was produced from catalytic activity of glucose oxidase (GOx) over glucose in a concentration-dependent manner and can be directly utilized for a peroxidase-like-activity-based glucose sensor with a detection limit of 3.0×10^{-7} mol l^{-1}.

Similarly, Shi *et al.*[133] also utilized the peroxidase-like activity of Fe_3O_4–Pβ-CD (Fe_3O_4 magnetic microsphere + β-cyclodextrin polymer) nanocomposite for rapid, selective and sensitive glucose detection from a human serum sample. The sensing mechanism was dependent on the presence of H_2O_2 and Fe_3O_4–Pβ-CD which catalyzes the oxidation of benzoic acid (BA) to form a strong fluorescent product (OHBA). In this assay, GOx catalyzes the oxidation of glucose to produce gluconic acid and H_2O_2, which was utilized in above mentioned reaction. The detection limits for this assay were calculated as 0.015 μM for H_2O_2 and 0.03 μM for glucose.

In another glucose detection approach, Zhang *et al.*[134] demonstrated the Fe_3O_4 NPs modified chitosan/graphene nanocomposites (Fe_3O_4–CG hybrid). GOx enzyme was immobilized on nanocomposite *via* covalent linkage and the presence of glucose was detected with a cyclic voltameter (CV). There was no obvious redox peak in the voltammogram in the absence of glucose, whereas in the presence of glucose a prominent redox peak was seen, which could be due to the reaction between glucose and GOx. This sensor had detection limit of 16 μM with linear range of detection up to 26 mM glucose.

In another study, Wu *et al.*[135] demonstrated a free-radical-polymerization-based fabrication of iron oxide core–shell microgels, generated due to the cascade reaction of GOx and HRP. Thermal stability and reusability of enzymes was better in the presence of IONPs. The detection principle was based on the enzyme cascade reaction, where H_2O_2 is generated by reaction of glucose and oxygen *via* GOx catalysis. Subsequently, the H_2O_2 produced reacts with *o*-phenylenediamine (OPD) in the presence of HRP to form a colored product, 2,3-diaminophenazine (DAP), (absorbance at 450 nm). The limit of detection was 0.8 μM for glucose, and the method is highly selective, as the presence of other saccharides *i.e.* fructose, lactose or maltose, in the reaction showed negligible absorbance.

A novel potentiometric glucose biosensor was developed by Yang *et al.*,[136] which was based on the excellent magnetism and conductivity of core–shell Fe_3O_4–enzyme (GOx)–polypyrrole (Ppy) NPs. This complex was attached to the surface of magnetic glassy carbon electrode (MGCE) for glucose detection. The assay is based on the fact that the change in the concentration of ions surrounding Fe_3O_4–GOD–Ppy/MGCE will change the electrode potential. This sensor showed a wide and linear range of glucose detection (0.5 μM to 34 mM) with a detection limit of 0.3 μM, and has very short response time of 6 s, with high selectivity and stability. In the above discussed context, it can be concluded that IONPs hold great promise for application in the diagnosis and treatment of diabetes.

10.3.2. Gold Nanoparticles

Metallic NPs, such as, AuNPs, silver NPs (AgNPs), copper NPs (CuNPs) and platinum NPs (PtNPs), are extensively being explored for developing strategies of non-invasive detection of low amount of glucose.[137–140] Among these metallic NPs, AuNPs have been investigated for construction of several biosensors, predominantly, due to their plasmon absorption in the visible range, which depends on their shape, size and dielectric constant.[137,141,142] Aggregation of AuNPs shifts the absorption maxima to longer wavelengths and changes the color from red to blue due to mutually induced dipoles that depend on interparticle distance and aggregate size.[142] Enzyme functionalized AuNPs undergo color change upon interaction with glucose which can help patients to perform self-monitoring of disease. In this context,

Radhakumary and Sreenivasan.[142] devised a simple method for visual glucose detection in urine by the assay, in which GOx was conjugated to the thiol capped AuNPs. The presence of glucose induces aggregation of particles, which leads to a visible change in color of the GOx–AuNPs suspension from red to blue. The redshift in the plasmon peak can be directly correlated with the change in concentration of glucose and therefore, the method can be employed for quantitative estimation. It was observed that the AuNPs suspension changes its color in the presence of 100 μg ml^{-1} of glucose.

Glycated hemoglobin (HbAI1c) reflects the average plasma glucose level over the last 8–12 weeks.[143] A colorimetric immunoassay using AuNPs for detection of the diabetes-specific biomarker, HbA1c, was developed by Wangoo *et al.*[144] In this assay, antibody-functionalized AuNPs were influentially aggregated by increasing the concentration of salt and showed a blue color, whereas in the presence of HbA1c AuNPs did not aggregate and thus retained their red color. Aggregation kinetics of the AuNPs-Ab probe were calculated by following their zeta potential values. The decrease in zeta potential of particles was directly correlated with increased concentration of HbA1c. This assay exhibited excellent sensitivity, with a detection limit of 0.0015 mg ml^{-1} and required less than 15 min to perform, which makes it a promising strategy for incorporation in a "POC" device.

AuNPs conjugated with single-walled carbon nanotubes modified edge plane pyrolytic graphite electrode (AuNPs–SWCNT/EPPGE) were developed by Bishoni *et al.*[145] Using this material they developed an assay to study oxidation of 8-hydroxyguanine (8-OH-Gua), found in urine of diabetic patients as a biomarker for oxidative DNA damage. AuNPs-SWCNT/EPPGE showed an anodic peak for oxidation of 8-OH-Gua at ~221 mV potential, measured in square wave voltammograms, at pH 7.2 with a detection limit and sensitivity of 5.0 (±0.1) pM and 4.9 (±0.1) μA nM^{-1}, respectively.

In order to avoid the finger-prick test, Manju and Sreenivasan[141] demonstrated the use of functionalized AuNPs in the fluorescent detection of glucose in aqueous fluid (tears). In this study, dextran-coated AuNPs were conjugated with aminophenyl boronic acid followed by addition of fluorescent rhodamine B isothiocyanate (RBITC) on the surface, which leads to a reduced RBITC fluorescence signal. In the presence of glucose, the complex recovered its fluorescence signal, which increases with increasing concentrations of glucose. Glucose has strong affinity to boronic acid and due to this it exhibited competitive binding and formed glucose boronate ester, which further leads to the detachment of fluorophore from AuNPs. Mechanistically, with the addition of glucose, the distance between AuNPs and fluorophore increases because glucose binding induces restructuring of the surface molecule assembly which results in an overall increase in the size of the complex. The utility of this complex was assessed in healthy human tear samples, and it was observed that the recorded glucose concentration was in agreement with the same range as reported earlier,[146] with a detection limit of 0.005 ± 0.002 μM.

Glucose detection from clinical samples has also been performed by several investigators, among them, Gao *et al.*[137] demonstrated colorimetric detection of glucose in clinical human serum samples by a plasmonic sensing platform composed of small-sized AuNPs (~4 nm) coupled with AgNPs. The sensing mechanism was based on the fact that in the presence of oxygen, smaller AuNPs exhibit GOx-like activity and catalyze glucose oxidation, to produce gluconic acid and H_2O_2. The H_2O_2 produced dissolves AgNPs (due to an etching effect) which leads to color change of AgNPs (from yellow to red) due to a change in local surface plasmon resonance (LSPR) band strength. This assay can detect glucose in a concentration range of 5–70 μM, with detection limit of 3 μM. Furthermore, L-cysteine addition was shown to improve the selectivity of the assay towards glucose detection.

10.3.3. Quantum Dots

QDs are well known to exhibit unique optical and electronic properties, therefore, being extensively used for imaging applications in biological and medical sciences. Certain properties, such as high photostability and quantum yield, size-dependent tunable PL with broad excitation and narrow emission bandwidth, mean that QDs can be used for evaluation of biomolecular signatures both *in vitro* and *in vivo*.[147–149] QDs are used in the construction of biosensors for detection of biomolecules, metal ions, multiple biomarkers in cell biology and organic compounds.[150,151]

For diabetes detection, Chopra *et al.*[152] developed a CdTe QDs-based nanoprobe to detect HbA1c, in a sandwich type immunoassay model. In this assay, HbA1c antibody acts as a specific receptor and QDs-labeled secondary antibody acts as a dual tracer (fluorescence as well as electrochemical) to quantify the analyte concentration (sandwiched between two antibodies). Fluorescence signal and cadmium ions produced after the dissolution of Cd [quantified using the ASV (anodic stripping voltammetry) technique] were found to be directly proportional to the concentration of HbA1c and require much lower amounts of sample for quantification.

In another study, Maaoui *et al.*[153] demonstrated the use of copper oxide (Cu_2O)-decorated carbon quantum dots (CQDs) for glucose sensing applications in aqueous solution and in human serum by improving the electrocatalytic activity of Cu_2O NPs through QDs hybridization. CQDs–Cu_2O-modified GC electrodes sense glucose with excellent electrocatalytic activity in basic environments and in a non-enzymatic manner in the complex human serum sample. The so obtained electrocatalytic oxidation peak was suggested to be due to the conversation of Cu(III) to Cu(II) in basic medium and the detection mechanism was explained by eqn (10.1)–(10.3). This assay has a detection limit of 6 μM at a signal-to-noise ratio (S : N) = 3.

$$\text{CQDs–Cu}_2\text{O} + \text{OH}^- \rightarrow \text{CQDs–CuOOH} + \text{e}^- \quad (10.1)$$

$$\text{CQDs–Cu}_2\text{O} + \text{H}_2\text{O} + 2\text{OH}^- \rightarrow \text{CQDs–Cu(OH)}_4^- + \text{e}^- \quad (10.2)$$

$$\text{Cu}^{3+} + \text{glucose} + \text{e}^- \rightarrow \text{Cu}^{2+} + \text{gluconolactone} \quad (10.3)$$

In another study, CdTe–CdS QDs-based enzymatic detection of glucose in blood and urine sample was developed by Hu *et al.*[154] Here, the mechanism is based on the PL quenching of QDs in the presence of H_2O_2, which is produced due to oxidation of glucose by GOx. The QDs surface was blocked by 10% BSA to avoid non-specific interactions with protein when analyzed in the blood sample. The results indicated that the decrease in PL intensity of QDs was directly proportional to glucose concentration. This assay has a detection limit of 1.8 μM and can be performed in 96-well plates, making it more convenient and portable.

Since H_2O_2 plays an important role in regulating physiological processes, including diabetes, a sensitive probe for H_2O_2 detection is imperative. In this context, Wang *et al.*[155] reported the application of hybrid nanocomposite comprised of AgNPs and GQDs mediated by DNA (AgNPsDNA@GQDs), for fluorescent sensing of H_2O_2. The detection mechanism is based on the etching effect of H_2O_2 over AgNPs and DNA cleavage by hydroxyl radical ($^{\bullet}OH$), produced as a reaction product between H_2O_2 and AgNPs. In this nanocomposite, AgNPs quench the fluorescence of QDs by resonance energy transfer, whereas in the presence of H_2O_2 the energy transfer among AgNPs and GQDs, mediated by DNA, is reduced due to the cleavage of DNA by $^{\bullet}OH$, resulting in the recovery of the fluorescence signal. Under optimal conditions, this assay exhibit a detection limit of 0.10 μM

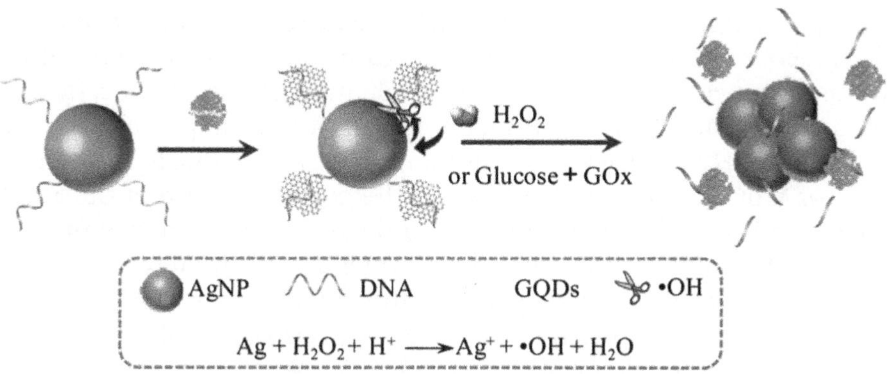

Figure 10.5 Schematic description of H_2O_2 and glucose detection Based on AgNP-DNA@GQDs. Reproduced from ref. 155 with permission from American Chemical Society, Copyright 2014.

H_2O_2. At the next level, this method was extended to glucose detection in a human urine samples in the presence of GOx enzyme, which oxidizes glucose and forms H_2O_2. It can detect glucose with satisfactory recoveries of 94.6–98.8%, and has significant potential for ultrasensitive quantitative analysis of glucose and also in clinical diagnosis of diabetes (Figure 10.5).

10.3.4. Upconversion Nanoparticles

UCNPs are an attractive agent to be used in the construction of luminescent biosensors due to the advantages of UCNPs over traditional methods *i.e.* no photobleaching, no autofluorescence, low damage to sample and high sensitivity. Although there are several types of UCNPs reported in the literature, we will discuss here some limited examples, which hold immense potential in diabetes research. The properties, such as high sensitivity and excellent optical contrast, further make UCNPs an attractive material for diagnosis and treatment of diabetes. Thus, researchers have utilized this property of UCNPs in the detection of H_2O_2 and glucose in biological samples.[156] In this context, Hu *et al.*[157] developed $NaGdF_4$:48% Yb^{3+}, 2% Er^{3+}-based UCNPs array for highly sensitive HbA1c detection, based on the LRET property. In this assay, UCNPs acts as a fluorescence donor, whereas HbA1c acts as an acceptor, therefore, the presence of HbA1c strongly quenches the green luminescence of UCNPs. In comparison with direct detection, the UCNPs array shows high sensitivity (approximately four times more than in solution), better reproducibility and reusability.

In another approach, Yuan *et al.*[158] reported a detection method for glucose monitoring in human serum and even in whole blood by using upconversion fluorescence of manganese dioxide (MnO_2) nanosheets-modified UCNPs, where MnO_2 acts as a quencher. In the presence of H_2O_2, UCNPs recover their fluorescence due to the conversion of MnO_2 to Mn^{2+} which leads to the decomposition of the MnO_2 sheet, and thus the glucose level can be detected due to the production of H_2O_2 in an enzymatic reaction catalyzed by GOx to glucose. The results revealed that upconversion fluorescence of UCNPs was gradually recovered by increasing the concentration of glucose and the detection limit was calculated to be 3.7 μM.

Detection of H_2O_2 and glucose in human serum was carried out by Liu *et al.*,[156] based on the quenching of UCNPs PL by oxidized TMB. In this study, the prepared $NaYF_4:Yb^{3+}/Er^{3+}$ UCNP exhibit green PL, which in the presence of HRP and H_2O_2, oxidizes the colorless substrate TMB, into blue oxTMB. Here, H_2O_2 was generated by the enzymatic action of GOx on glucose, which further reacts with TMB for detection purposes with a limit of detection 64 nM and detection sensitivity in the range of 0.1–5.0 μM.

Wang *et al.*[159] reported upconversion carbon dots (CDs) embedded in a boronic acid-functionalized polymer microgel network to form a hybrid microgel system. The system comprised of UCDs and functional co-monomers of 4-vinylphenyl boronic acid (VPBA) and acrylamide (AAm). Glucose reversibly binds to the VPBA through the boronic acid moiety. This binding induces a reversible volume phase transition of the poly (VPBA–AAm)–CDs hybrid microgels, due to changes in the physicochemical environment of the CDs in response to variations in glucose concentration. The generated biochemical signal was translated into the fluorescent signal with NIR light excitation. Binding of glucose with the *cis*-diol of boronic acid produces negatively charged boronates, which leads to microgel swelling due to the Donnan potential produced. Hydrodynamic size of poly (VPBA-AAm)–CDs hybrid microgels increases gradually with increasing glucose concentration and shows less significant swelling at >20 mM glucose concentration. Glucose-responsive reversible swelling and deswelling transition of the poly(VPBA-AAm) network chains also induces an alteration in the fluorescence properties of the hybrid microgel due to changed surface states of the hydrogen-bonded CDs embedded in the networks. NIR upconversion fluorescence gradually decreases with increasing glucose concentration (up to 25 nM) (Table 10.2).

Table 10.2 Different nanoparticles used in diabetes detection.[a]

Substrate	Nanoparticle type	Limit of detection (LOD)	Detection method	Reference
I27L gene	AuNPs functionalized ITO electrode	8.1×10^{-12} M	Electroluminescent biosensor	160
Glucose	Nitrogen, boron co-doped fluorescent carbon NPs	1.8 μM	Fluorescence quenching by glucose	161
HbA1c	PtNPs/rGO-MWCNT nanocomposite	0.1 μM	Voltametric sensing	162
Acetone gas	Na doped p-Type Na:ZnO nanoflowers	0.2 ppm	Photoactivated sensing	163
HbA1c	AuNPs embedded N-doped graphene nanosheets	0.2 μM	Electrochemical impedance spectroscopy (EIS) study	164
Insulin level	IONPs	5 pM	SPR microarray	165
Salivary glucose	Self-assembled CNTs and AuNPs	0.41 mg dl^{-1}	Electrochemical method	166
Type 2 diabetes mellitus-related SNP detection	Growth of AgNPs on SWCNTs	5 pM	Electrochemical sensing	167
Glucose	AuNPs	3 mM	Colorimetric assay	168
Glucose	IONPs	50 μM	Colorimetric assay	42
Glucose	Au–CeO_2 core shell (CSNPs)	100 μM to 1 mM	Colorimetric assay	169

[a]PtNPs/rGO-MWCNT, reduced graphene oxide, multiwalled carbon nanotubes and platinum NPs.

10.4. Use of Nanoparticles for Diagnosis and Treatment of Infectious Diseases and Pathogens

Pathogenic microorganisms, such as bacteria, viruses, fungi and parasites are the main cause of infectious diseases, including dengue fever, hepatitis, cholera and avian influenza. These diseases cause a severe threat to human life and health due to some of their unique characteristics, such as rapid multiplication, an evolutionary advantage and unpredictability.[170,171] These diseases are generally located in certain geographical regions but they can spread worldwide due to global travel facilities. Conventional detection techniques, like immunology, culture and microscopy and PCR, have contributed significantly to the diagnosis of infectious diseases, but they suffer from certain drawbacks, such as expense, low accuracy, slow speed and requiring a skilled technician to perform them. In this context, it is imperative to develop novel diagnostic techniques that possess qualities like high sensitivity, early detection, low cost and user-friendliness and have the potential to be developed as "POC" devices.[170] Several efforts have been made in this direction to detect infectious disease by the use of different NMs. Researchers have developed a smartphone dongle as a "POC" system for infectious disease detection, by integrating the traditional immunoassay into a smartphone acessory.[172] Special emphasis is given to fluorescent, metallic and magnetic NPs. Nanodiagnostics have shown that simple and portable devices platforms can be developed, which can analyze blood or urine sample from patients.[170]

10.4.1. Iron Oxide Nanoparticles

IONPs have been reported to be used in the preparation of novel colorimetric protease biosensors with high capture efficiency and more sites for binding.[173] Biorecognition elements can be eventually conjugated with IONPs to form a sensing probe which can be readily utilized in the preparation of "POC" diagnostic biosensors, such as development of colorimetric biosensors for detection of PSA.[174]

In this context, Wignarajah *et al.*[175] also reported the development of a colorimetric biosensor to detect periodontitis by using two typical inflammatory salivary biomarkers, human neutrophil elastase (HNE) and cathepsin-G, which has high proteolytic activity. The biosensor was comprised of IONPs conjugated with HNE (or cathepsin-G) substrate which was covalently attached to a gold sensing platform, hiding its golden color. In the presence of HNE (or cathepsin-G) the peptide linkage between IONPs and the gold sensor was cleaved and magnetic beads were collected by an external magnet, which subsequently results in the visual appearance of the golden color of the gold-based sensor. This assay has a detection limit of 1 pg ml^{-1} and 100 fg ml^{-1} for HNE and cathepsin G respectively.

In another study, an IONPs-based assay for detection of enteric fever was developed by Park *et al.*[176]. Magneto-DNA complex was used to detect the amplified *Salmonella* mRNA and analyzed by a miniaturized micro-NMR (μNMR) system. Targeted DNA was amplified and allowed to attach to beads conjugated with captured probes and hybridized with a biotinylated detection probe. This system was finally attached to IONPs coated with streptavidin to form a magnetic sandwich complex and analyzed by a μNMR device, which showed a detection limit of 0.01–1.0 colony forming units (CFU) ml^{-1}.

Syphilis is one of the most common of several infectious diseases caused by *Treponema pallidum*. Yang *et al.*[177] reported a novel lateral flow immunoassay (LFIA) strip test system for detection of syphilis by using gold iron oxide NPs (AuIONPs). In this assay, *T. pallidum* recombinant antigens (r-TP) were conjugated on AuIONPs surfaces to construct probes for detection of anti-Tp antibodies in sera of patients. Here, AuIONPs–rTP and AuIONPs–IgG conjugates were utilized. In the presence of an analyte (TP antibodies), AuIONPs–rTP probe was able to recognize and bind to Tp antibodies. This complex moved towards the 'test line' (T-line) and 'control line' (C-line) (rTP antigen and IgG immobilized to nitrocellulose membrane, respectively). These conjugates agglutinate as a triplet construct of AuIONPs–rTp antigens and anti-Tp antibodies in a sandwich format and develop red color on the T-line. The intensity of color development is directly proportional to the concentration of analyte present in the sample.

A chemiluminescence-based method by using IONPs for detection of hepatitis B virus by a DNA hybridization technique was developed by Wang *et al.*[178] where the complementary DNA was immobilized onto carboxylated IONPs and mixed with biotinylated target DNA for hybridization. Avidin–alkaline phosphatase (ALP) complex was attached to the hybrid conjugated IONPs *via* the biotin–streptavidin linkage. Finally, chemiluminescence signal was obtained by adding 3-[2-spiroadamatane]-4-methoxy-4-[3-phosphoryloxy]-phenyl-1,2-dioxetane (AMPPD) which can be catalyzed by ALP.

10.4.2. Gold Nanoparticles

AuNPs have been used to develop sensitive and selective detection methods for many infectious diseases by rational modification of their surfaces by aptamers, antibodies and certain ligands. Aggregation-dependent alteration of the color of AuNP suspension has had great success for identification of antigens and nucleic acid with naked eyes. In certain resource-limited countries, it has great importance because it does not require a power source and is an instrument-free method.[179] In this context, Ahmed *et al.*[180] reported the synthesis of AuNPs and carbon nanotubes (CNTs) hybrids for detection of influenza virus A (H3N2) from human serum. Au–CNTs nanohybrids were further conjugated with anti-influenza A (H3N2) antibody and exhibit peroxidase-mimetic activity to oxidize TMB in the presence of H_2O_2 and develop a blue color. In the presence of influenza virus A (H3N2), Ab conjugated Au–CNTs hybrid were allowed to selectively bind with it which gives a deeper blue color after addition of TMB and H_2O_2. The

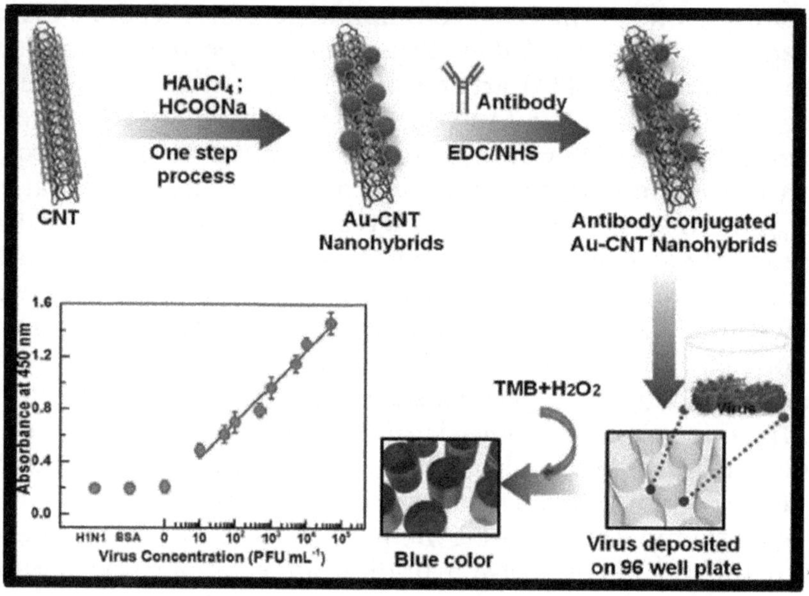

Figure 10.6 Schematic representation of detection of virus by peroxidase-like activity. Reproduced from ref. 180 with permission from Elsevier, Copyright 2016.

optical density of the blue color depends on the virus concentration (Figure 10.6). Limit of detection for this assay was calculated as 3.4 plaque-forming units (PFU) ml^{-1}, which is 385 times lower than the detection limit of conventional enzyme-linked immunosorbent assay (ELISA; 1312 PFU ml^{-1}).

Another highly infectious pathogen, cyprinid herpesvirus-3 (CyHV-3), was detected by Saleh and El-Matbouli[181] using an AuNPs-based hybridization assay. This is a colorimetric, label-free nano-diagnostic detection method for unamplified CyHV-3 DNA. In the presence of appropriate CyHV-3 DNA target, DNA probes bind through the complementary sequence, which leads to the aggregation of AuNPs with concomitant change in color from red to blue. However, in the presence of non-complementary sequence it remains red. This is due to the adsorption of ssDNA on the surfaces of AuNPs to stabilize them against aggregation. This assay exhibits high sensitivity and can detect 50 fg of CyHV-3 DNA, which is equivalent to 150 virions, and also has high specificity as AuNPs remain red in the presence of CyHV-1 and CyHV-2 DNA. In contrast to PCR, this assay is more rapid to perform (20 min) after DNA extraction, and also does not require sophisticated instruments and hence can be applied as a "POC" device.

In another study, Driskell *et al.*[182] developed an AuNPs-based assay for detection of influenza-A virus by using dynamic light scattering (DLS). AuNPs were conjugated with an influenza-specific antibody, and the presence of target virus in solution induced aggregation, which could be further measured by DLS, in which the extent of AuNPs aggregation can be directly correlated with the virus concentration. Under optimum conditions, the assay has a detection limit of <100 $TCID_{50}$ ml^{-1} (tissue culture infectious dose), which is better than commercially available kit by 1–2 orders of magnitude.

An AuNPs-based FRET assay for sensitive detection of *Mycobacterium tuberculosis* was developed by Mansour *et al.*[183] by targeting its 16S rDNA gene. In this assay, extracted DNA was exposed to Cy-3-labelled probe and further AuNPs were added into it. In the positive sample, probes were unavailable for AuNPs and hybridized with targeted DNA, which led to the generation of intense fluorescence, however, in negative samples probes were forced to get adsorbed on the AuNPs surface thus, quenching the fluorescence. This can be explained by the fact that ssDNA can uncoil, exposing its bases, and can efficiently bind to citrate-capped AuNPs. Moreover, native surface charge of AuNPs and bacteria can also be utilized for their colorimetric detection.

Electrochemical and optical properties of gold nanostars are governed by the degree of branching and their size.[184] Verma *et al.*[185] reported a CTAB-capped gold-nanostar-based colorimetric assay for rapid detection of *Staphylococcus aureus*. This assay is based on the fact that positively charged CTAB-coated AuNPs can electrostatically interact with the negatively charged cell wall of *S. aureus* (due to the presence of teichoic acid) and induce aggregation of AuNPs, which leads to rapid change of solution color from red to blue.

In another study, de la Rica and Stevens[186] have reported an assay based on plasmonic ELISA for detection of HIV-1 capsid antigen p24 in human serum. In the presence of analyte, catalase enzyme utilizes H_2O_2 present in the system, leading to slow crystal growth of AuNPs and abnormal morphology, with aggregated particles of blue colored suspension. Whereas, in the absence of analyte, H_2O_2 rapidly reduces gold ions and quasi-spherical, non-aggregated AuNPs are formed. This assay can detect p24 antigen at a very low concentration (1×10^{-18} g ml^{-1}). Clinical samples from HIV-positive donors showed a blue color, whereas those from HIV-negative donors exhibited the red color of dispersed NPs.

In another approach, Kim *et al.*[187] developed a plastic-chip-based magnetophoretic immunoassay (pcMPI) for detection of *M. tuberculosis* by using magnetic microparticles and AuNPs.

10.4.3. Quantum Dots

QDs can serve as an alternative to organic fluorophores and have been reported to be used in membrane labeling, antibody tagging, fluorescent microscopy *etc.* In their native conformation, QDs are considered to be biologically non-reactive, they can be conjugated with biological intermediates to act in biological assay systems *i.e.* disease diagnosis.[188]

Yu-Hong *et al.*[189] reported the use of a hybridized material consisting of QDs and IONPs for detection of human papillomavirus (HPV 16) from cervical swab samples. In this assay, IONPs were attached with streptavidin for detection purpose, and two types of probes were used, one was labeled with biotin and other was QD-labeled. In the first step, both the probes were mixed with sample DNA and denatured at high temperature for attachment of probes, further IONPs were added into the hybridization mixture. IONPs get attached to the biotin-labeled sample DNA due to streptavidin–biotin conjugation. This hybridization complex was collected with the help of a neodymium–iron–boron (NdFeB) magnet and its relative fluorescence intensity was measured. The main advantage of this method over the conventional PCR method is that it requires less time for detection (~1 hour) and is a simpler process.

Another sensitive immunoassay method was reported by Zhang *et al.*,[190] using quantum dots nanobeads (QDNBs) as a reporter for detection of hepatitis B surface antigen (HBsAg). Here, the antigen was immobilized on the PVDF membrane by electrostatic and hydrophobic interaction and detected with a probe comprised of anti-HBsAg antibody-conjugated QDs. Fluorescence of QDNBs attached with protein was followed by UV lamp illumination and recorded with a charge coupled device (CCD) camera. It was found that the fluorescence intensity of captured QDs was dependent on the concentration of antigen present on the membrane. In contrast to single QDs, the QDNBs assay is more sensitive due to increased QDs and the amount of antibody–antigen-binding events, and the detection limit recorded was as low as 78 pg of HBsAg (Figure 10.7).

In another study, Takemura *et al.*[191] described a rapid and ultrasensitive, LSPR-induced immunofluorescence biosensor for detection of influenza virus H1N1 by using AuNPs to induce CdSeTeS QDs fluorescence signal. This assay is based on the principle of LSPR-induced fluorescence changes where AuNPs and QDs are conjugated with anti-neuraminidase (NA) antibody (anti-NA Ab) and anti-hemagglutinin (HA) antibody (anti-HA Ab), respectively. In the presence of target

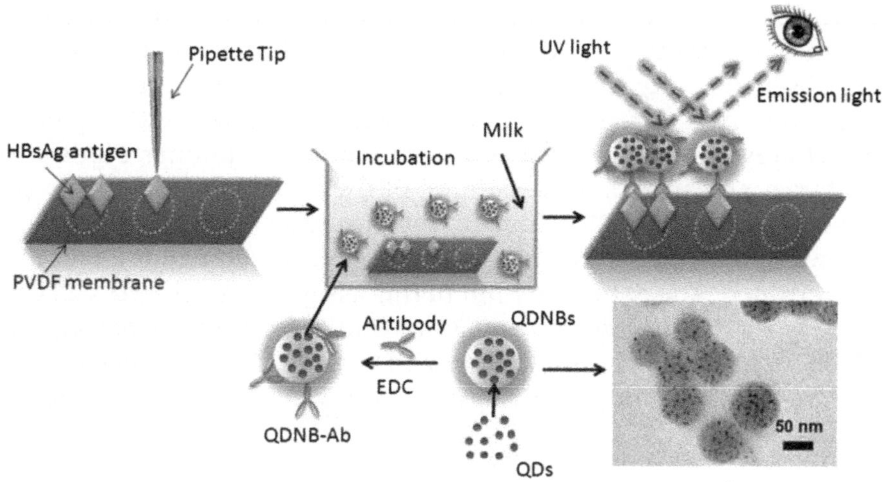

Figure 10.7 Schematic illustration of fluorescent detection of HBsAg proteins by dot-blot immunoassay with QDNBs antibody conjugates. Reproduced from ref. 190 with permission from Ivyspring International Publisher.

influenza virus, these anti-NA Ab-conjugated AuNPs and anti-HA Ab-conjugated QDs bind with NA and HA antigen present on the surface of virus due to their high affinity and specificity. Based on the concentration of influenza virus (antigen) and their interaction, this triggers changes in fluorescence of QDs. The detection limit of this assay was recorded to be 0.03 pg ml^{-1} in deionized water, 0.4 pg ml^{-1} in human serum and 10 PFU ml^{-1} for clinically isolated H3N2.

A nanodevice was generated by Klostranec *et al.*[192] with multiplex detection and high-throughput capability with the help of microfluidics and QDs. This device was used for detection of blood-borne infectious disease [*i.e.* hepatitis B, hepatitis C and human immunodeficiency virus (HIV)] with the help of their corresponding biomarkers HBsAg, HCV non-structural protein 4 (NSP4), and HIV glycoprotein 41 (gp41), respectively. QDs were coated with the above-mentioned antibodies conjugated with a fluorophore (AF-488 organic dye conjugated to goat anti-mouse antibody IgG) to give a fluorescence signal peak at ~520 nm. Serum volume for detection was as low as 100 μl and only 1 hour was required to complete the test.

Pathogens like *Vibrio parahaemolyticus* and *Salmonella typhimurium* can be detected by an approach with dual FRET obtained by using green- and red-emitting QDs (gQDs and rQDs) as a donor, where carbon NPs (CNPs) act as acceptors.[193] gQDs and rQDs were conjugated with aptamers recognizing *V. parahaemolyticus* and *S. typhimurium*, respectively. The principle of this strategy is based on the difference in binding affinity of CNPs towards QDs–aptamer in the presence and absence of target bacteria. In the absence of target bacteria, QDs and CNPs show self-assembly, which leads to efficient FRET quenching. However, in the presence of bacteria, QDs–aptamer binds with the target and results in weakening of the π-stacking effect between aptamers and CNPs. Thus, due to the formation of QDs–aptamer–target complexes, quenching by CNPs was suppressed and fluorescence was regained. The detection limit of this assay was recorded as 25 CFU ml^{-1} for *V. parahaemolyticus* and 35 CFU ml^{-1} for *S. typhimurium*, with a linear increase in fluorescence of QDs in the range of 50–106 CFU ml^{-1}. In real food samples, analysis results were similar to the conventional plate count method.[193]

10.4.4. Upconversion Nanoparticles

Lack of autofluorescence is one of the major advantages, which results in improvement of the signal to noise ratio. UCNPs can have tunable optical properties which can be obtained by lanthanide doping, such as with Er^{3+}, Tm^{3+}, and Ho^{3+}. Multicolor UCNPs can be used to label multiple targets simultaneously. Researchers have developed systems with the conjugation of aptamers and UCNPs.[194]

Kurt *et al.*[195] developed dual-excitation sensing method for detection of multiplexed food pathogens (*Staphylococcus aureus* and *Salmonella typhimurium*). It consists of aptamer-functionalized QDs and UCNPs with stokes and anti-stokes excitation profiles, respectively. The downconversion character of CdTe QDs and upconversion of $NaYF_4$:YbTm UCNPs result in a detection spectrum utilized twice, with the elimination of crosstalk signals. The detection limits for this assay were found to be 16 and 28 CFU ml^{-1} for *S. aureus*, and *S. typhimurium*, respectively, with good specificity.

A luminescence bioassay platform based on multicolor labelling of UCNPs and separation of IONPs for detection of foodborne pathogenic bacteria (*S. aureus*, *V. parahaemolyticus* and *S. typhimurium*) was developed by Wu *et al.*[194] Bacteria-specific aptamers were conjugated to the UCNPs, whereas IONPs were attached to the cDNA. Thus, UCNPsTm–IONPs complex was assembled through aptamer and cDNA hybridization. In the absence of bacteria, the background luminescence was at maximum due to the high number of UCNPsTm–IONPs signal probes collected by magnetic separation from the system. Whereas in the presence of bacteria, aptamers bind with the bacterial cells, which causes dissociation of some cDNA and liberates some

Table 10.3 Different nanoparticles used in detection of various pathogens causing infectious diseases[196]

Pathogen	Nanoparticle type	Method	Limit of detection (LOD)	Reference
Campylobacter jejuni	Bimetallic Au@ Ag core–shell structures	Immuno-magnetic separation–polymerase chain reaction method	10^3 CFU μl^{-1}	197
S. aureus	Au-coated magnetic NPs conjugated with *S. aureus* antibody	SERS	10^{-9} M or 10 pg ml^{-1}	198
E. coli	Cysteine gold NPs	Colorimetric assay	100 cells ml^{-1}	199
Hepatitis B virus	Anti-HBsAg coated IONPs	Cyclic voltammetry	0.9 pg ml^{-1}	200
Dengue virus	Zinc oxide/platinum–palladium (ZnO/Pt–Pd) nanocomposites	Electrochemical genosensor	4.3×10^{-5} M	201
Influenza virus	CdZnSeTeS QDs	Fluorescence enhancement signal transduction	~1 copy ml^{-1}	202
HIV	Fluorescent AgNPs	Fluorescence based sandwich immunoassay	10 and 1000 pg ml^{-1}	203
Tuberculosis antigen	GQDs and magnetoplasmonic (MagPlas) NPs	SERS	0.0511 pg ml^{-1}	204
E. coli	SWCNTs	Changes in electric current (ΔI)	10^2 CFU ml^{-1}	205

S. aureus–Apt1–UCNPsTm, which leads to decreased intensity of the emission peak due to the reduced concentration of UCNPsTm–IONPs signal probes. The limits of detection were found to be 25, 10 and 15 CFU ml^{-1}, for *S. aureus*, *V. parahaemolyticus* and *S. typhimurium*, respectively (Table 10.3).

10.5. Nanopatterned Devices

Due to the vast use of nanotechnology in disease diagnosis, nanoscale devices with a high degree of accuracy are needed to be developed. Either "top-down" or "bottom-up" approaches are used to construct nanopatterned devices. "Top-down" methods are employed to form biosensors with micro and nanoscale features, whereas, "bottom-up" approach uses the thermodynamic accumulation of molecules or atoms self-assembly under specific conditions, which leads to the formation of materials with internal and external nanoscale structures, such as gold nanocrystals.[1,206] These nanodevices are known to have great potential for the identification of biological species and quantitative detection of target samples, without employing any radioactive or fluorescent labeling. Biosensing devices show a great effect in "POC" diagnosis devices due to their high specificity and sensitivity and their capability for high-throughput biological sample screening. The role of nanopatterned devices in disease diagnosis will be discussed under the following three broad topics.

10.5.1. Micro- and Nano-cantilevers-based Sensing Platforms

Biosensing devices are expected to be label-free, sensitive, small in size and portable for their household utilization. Microcantilever and nanocantilever technology enables better biosensing performance by these devices. Micro-fabrication of such devices requires silicon or piezoelectric materials, such as quartz, and the analyte detection can be performed by measuring the deflection of the cantilever caused due to the binding of target molecules on the receptor present at the cantilever surface and also by analyzing the resultant frequency shift.[207] Both nanocantilevers and microcantilevers work under two operation modes which are used for biomarker detection: (1) Dynamic mode, in which variation in the resonance frequency of the cantilever is measured, which is due to the absorbed mass. (2) Static mode, here the target molecules bind with the receptor immobilized on the cantilever surface, which induces stress and causes bending of the cantilever.[208–210]

A cantilever array has been used for detection of BRAF mutation in RNA of melanoma cells, which was based on a BRAF-specific oligonucleotide probe. It can detect mutant BRAF at low concentration of 500 pM in 20 ng μl^{-1} of RNA. It is associated with several advantages, such as cost efficiency, no requirement for labeling or pre-amplification and that multiple mutations can also be interrogated simultaneously.[211]

Yin *et al.*[212] successfully developed a cantilever-based sensor for immunoassay of troponin I, which was dependent on the surface stress changes induced by analytes leading to its static bending. This sensor showed label-free and real-time detection of troponin I in a few minutes with a detection limit of 1 pg ml^{-1}.

In a similar way, a microcantilever was also developed by Wu and co-workers[213] for detection of PSA from blood protein mixtures using anti-PSA functionalized cantilever surface. It can detect PSA at a concentration range of 0.2 to 60 $\mu g\ ml^{-1}$.

With the help of a cantilever, lung cancer detection was also possible from blood droplets using nanomechanical bioassay. The frequency shift of the biosensor was measured, which takes place due to MMP-2-induced proteolysis of peptide chains (attached to the biosensor's surface) in order to monitor the secretion level of MMP-2 and thus cancer metastasis. This assay is highly sensitive and showed a detection limit of 0.05 nM.[214]

10.5.2. Use of Nano Chips as Patterned Biosensors

The programmable Bio-Nano-Chip (p-BNC) system combines unique chemical and biological sensing capabilities with a proficient machine-based algorithm to offer novel and intuitive single-value indices across several major diseases.[215,216] BNC has been used in several biomedical measurements and contains both biological specified recognition elements as well as pattern recognition features for identification of analytes and is designed as an "electronic taste chip".[216,217] The BNC consist of two units, dedicated for tailored applications, the first is the "cellular processing unit" with a micro-membrane filter integrated into a fluidics structure, and performs cell counting applications,[218] the second employs a microbead array where silicon or stainless steel wafers have indentations with chemically sensitized bead reactors. These chemical sensors are used to isolate and quantitate macromolecules from complex samples.[219]

Salivary biomarkers related to an acute myocardial infarction (AMI) were measured by Floriano and co-workers[220] by novel nano-biochips, devised for patients suffering from chest pain. Nanodevices exhibited an excellent diagnostic capability to distinguish between the patients with or without AMI on the basis of cardiac biomarker concentrations. They used Luminox and lab-on-chip methods to detect 21 proteins in serum and unstimulated whole saliva obtained from AMI patients and healthy controls. A saliva-based biomarker panel of myoglobin, C-reactive protein and myeloperoxidase possesses great detection activity [Area under curve (AUC) 0.96], in comparison with the panel of creatine kinase-MB, troponin-I and brain natriuretic peptide; (AUC 0.98), and its screening capacity exceeded that of the electrocardiogram (ECG; AUC ~0.6).

Rodriguez *et al.*[218] developed a microchip for counting CD4 lymphocytes for monitoring of HIV. In this assay, a small amount of blood in microlitre volumes was stained with fluorescently tagged antibodies, further captured on a membrane in a miniaturized flow cell and images were recorded by using microscopic optics with CCD. The

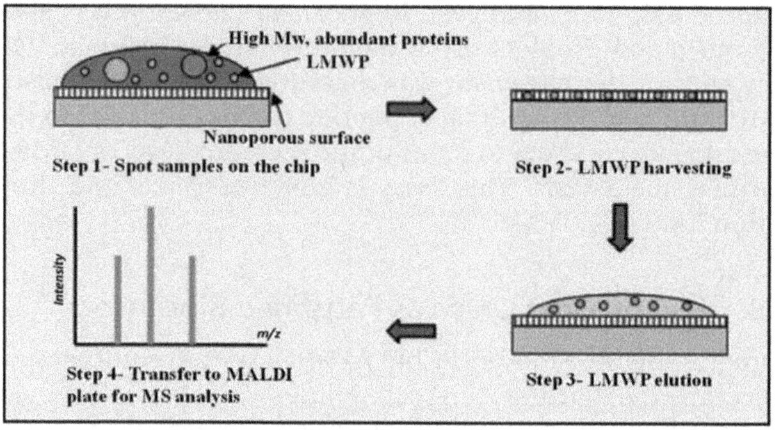

Figure 10.8 Schematic representation of the harvesting protocol consisting of (step-1) deposition of plasma directly onto the chip surface; (step-2) washing of unbound substances; (step-3) extraction of bound molecules; and (step-4) MS analysis. Reproduced from ref. 1 with permission from Bentham Science Publishers, Copyright 2017 Eureka Science Ltd.

accuracy of this technology was calculated to be similar to that of standard flow cytometry ($r = 0.95$), over absolute CD4 count range and to have high specificity and sensitivity to clinically discriminate CD4 count threshold.

Low molecular weight protein (LMWP), found in blood serum, can be used to recognize circulating biomarkers associated with disease and also in real-time monitoring of therapeutic efficacy of a treatment. LMWP are small proteins or peptide fragments derived by proteolytic action on larger proteins or shed from tissues.[1] Separation of LMWP was shown by Fan and co-workers[221] using nanoporous silica-chips. They exhibit periodic nanostructure and have a uniform size distribution and can be used to isolate LMW proteins and peptides from the serum of nude mice having lung metastasis induced by human breast cancer cell line MDA-MB-231.

Hu *et al.*[222] developed mesoporous silica chips in which tri-block copolymers (Pluoronic F127) were used as a synthetic template. The microfluidic channel was fabricated with polydimethylsiloxane (PDMS) by a rapid prototyping technique and formed as rectangular cross sections (75 μm wide × 20 μm deep). LMWP from biological samples can be collected by utilizing size exclusion properties of silica nanopores in a pressure-driven fractionation step (Figure 10.8).

Microfluidic components exhibit various advantages, like precise quantification, the need for small volumes of sample and high reproducibility. Such devices can be accepted as clinical diagnostic and

"POC" tools in the near future and could rationalize high-throughput proteomics research on many fronts.

10.5.3. Carbon-nanotube-based Biosensors

Carbon nanotubes (CNTs) possess some excellent properties, like exceptional fluorescence in the infra-red region and great photostability, which make them suitable for use as biosensors.[1,223] CNTs fluorescence occurs in the infra-red region, where biological tissues absorb minimally and thus have the least interference in the 800–1600 nm region.[224] All the above-mentioned properties are the key element for the design of an efficient biosensor. The current scenario of biosensor development must overcome several key issues, like the potential for real-time monitoring, minimum cost and non-invasive "POC" devices. Electrochemical approaches have great specificity and sensitivity and, thus, are considered as a promising candidate in this field.[1]

Drouvalakis *et al.*[225] developed a peptide-coated-CNTs-based immunosensor for detection of rheumatoid arthritis (RA). Here, CNTs were functionalized with RA-specific (cyclic citrulline-containing) peptide, deposited on a quartz crystal microbalance (QCM) sensing crystal. Antibody binding from patient serum to the nanotube probe was detected by a QCM-sensing-based approach. This technique is superior to the traditional ELISA and microarray assay in terms of sensitivity and can detect 34.4 and 37.5% more RA patients with anti-citrullinated peptide antibodies. In this context, Yang *et al.*[226] developed a new class of fluorescent biosensor which consists of SWNTs and fluorescent dye labeled ssDNA oligonucleotide. In the absence of target molecule, SWNTs and oligonucleotide lie in close proximity, resulting in fluorescence quenching. Whereas in the presence of target there is a competition between SWNTs and target molecule to bind with DNA oligonucleotide, which leads to increased fluorescence relative to the situation without target. In this study, the target was detected by fluorescence anisotropy and fluorescent intensity measurement for DNA-hybridization and aptamer–protein interaction studies (Figure 10.9).

Intrinsic properties of CNTs, such as high conductivity and fast electron transfer rate, render them a suitable material for an electrochemical-based detection system. Such electrodes possess intermediate properties between glassy carbon, gold or platinum electrodes and can be used for bio-recognition.[227] CNTs-based electrodes are of low cost in comparison with platinum or gold and also possess high

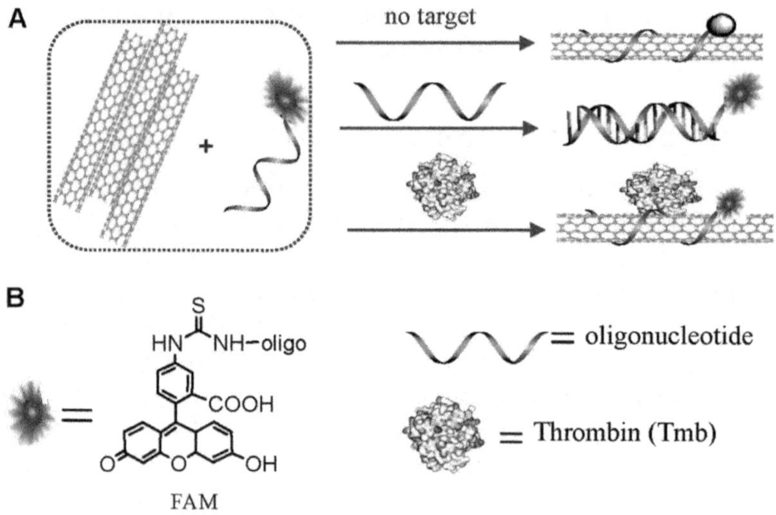

P_1: 5'-CAGTTACATTCTC**C**CAGTTGATT-FAM-3'

P_2: 5'-TCTCTCAGTCCGTGGTAGGGCAGGTTG GGGTGACT-FAM-3'

T_1: 5'-AATCAACTG**G**GAGAATGTAACTG-3'

T_2: 5'-AATCAACTG**T**GAGAATGTAACTG-3'

Figure 10.9 (A) Scheme for signaling biomolecular interactions by the assembly of SWNTs and dye-labeled ssDNA. The illustration of the dye-DNA functionalized SWNT is only a graphic presentation and does not represent the precise way that DNA binds on SWNTs. (B) Structures of FAM, the FAM-labeled oligonucleotides, P1 and P2, the perfect cDNA (T1) and one mismatched DNA (T2) of P1, and thrombin. Oligo represents the position of attachment of tethered oligonucleotides. Reproduced from ref. 226 with permission from American Chemical Society, Copyright 2008.

sensitivity and specificity due to easy bio-functionalization. It is possible to tune the enzyme loading efficiency within a very thin layer of CNTs due to their well-studied surface chemistry. Thus, CNTs can be used for the efficient detection of several biomolecules.[228]

10.6. Conclusion

NPs-based formulations have the potential to be utilized for various disease diagnosis and therapeutic purposes. Metal-, metal oxides-, carbon- and polymers-based NMs play important role in the development of a better lifestyle for society. Theranostic NMs are incorporated to visualize biodistribution, drug release, target site accumulation and treatment response. They are applied in the "POC" devices for the development of biosensors, which are used to detect clinically important molecules, like cancer and other disease-specific biomarkers, pathogens and biomolecules. NMs incorporating electrochemical biosensors have the ability to detect disease-specific biomarkers at very low concentrations with excellent sensitivity and specificity. Multiple NPs can be utilized in order to enhance the sensitivity of the device. Extraordinary properties result from their nanoscale dimensions and ability to transfer electrons between an enzyme and signal transducer. Theranostic NMs also enable the pre-selection of patients for optimal therapeutic formulations and facilitate the concept of personalized medicine. Thus, nanotechnology proves to be a great technology, which can expand the limits of current diagnosis, therapeutics, "POC" devices and theranostics.

Abbreviations

DAP	2,3-diaminophenazine
TMB	3,3′,5,5′-tetramethylbenzidine
VPBA	4-vinylphenyl boronic acid
AMI	Acute myocardial infraction
AF	Alexa Fluor
ALP	Alkaline phosphatase
AFP	Alpha-fetoprotein
AUC	Area under curve
RGD	Arg-Gly-Asp peptides
BPT	Bathophenanthrolinedisulfonic acid disodium salt
BA	Benzoic acid
CEA	Carcinoembryonic antigen
CeNPs	Cerium oxide nanoparticles
CCD	Charged couple device
CRET	Chemiluminescence resonance energy transfer
CTC	Circulating tumor cells
CT	Computed tomography
CLSM	Confocal laser scanning microscopy
CV	Cyclic voltameter
CyHV-3	Cyprinid herpes virus-3
DLS	Dynamic light scattering
EIS	Electrochemical impedance spectroscopy
EPR	Enhanced permeability and retention effect
EGFR	Epidermal growth factor receptor

FRET	Fluorescence resonance energy transfer
GOx	Glucose oxidase
GSH	Glutathione
GPx	Glutathione peroxidase
HbA1c	Glycated hemoglobin
AuNCgs	Gold nanocage
AuNRs	Gold nanorods
AuNPs	Gold nanoparticles
GO	Graphene oxide
HA	Hemagglutinin
HBsAg	Hepatitis B surface antigen
HRP	Horseradish peroxidase
HIV	Human immunodeficiency virus
HPV 16	Human papilloma virus
H_2O_2	Hydrogen peroxide
$^{\bullet}OH$	Hydroxyl radical
IMS-PCR	Immuno-magnetic separation-polymerase chain reaction
ICG	Indocyanine green
IONPs	Iron oxide nanoparticles
JAK–STAT	Janus kinase–signal transducer and activator of transcription
LFIA	Lateral flow immunoassay
LOD	Limit of detection
LSPR	Local surface plasmon resonance
LMWP	Low molecular weight protein
LRET	Luminescence resonance energy transfer
MRI	Magnetic resonance imaging
MMP-2	Matrix metalloproteinase-2
MTB	*Mycobacterium tuberculosis*
DMDAP	*N*,*N*-dimethyl-2,7-diazapyrenium dication
NMs	Nanomaterials
NPs	Nanoparticles
NIRF	Near-infrared fluorescence
NA	Neuraminidase
NCAM	Neuronal cell adhesion molecule
NSCLC	Non-small-cell lung cancer
PAT	Photoacoustic tomography
PS	Phosphatidylserine
PL	Photoluminescence
PDGF	Platelet-derived growth factor
PDMS	Polydimethylsiloxane
PET	Positron emission tomography
p-BNC	Programmable Bio-Nano-Chip
PSA	Prostate specific antigen
QDs	Quantum dots
QCM	Quartz crystal microbalance

RA	Rheumatoid arthritis
Rh6G	Rhodamine 6G
RhB	Rhodamine B
SLN	Sentinel lymph node
AgNPs	Silver nanoparticles
SWNTs	Single-wall carbon nanotubes
SOD	Superoxide dismutase
SPIONS	Superparamagnetic iron oxide nanoparticles
SEF	Surface enhanced fluorescence
SPR	Surface plasmon resonance
SPCE	Surface plasmon-coupled emission
SERS	Surface-enhanced Raman scattering
TR-FRET	Time-resolved Förster resonance energy transfer
TEC	Tumor endothelial cells
UCNPs	Upconversion nanoparticles
UCNPhs	Upconversion nanophosphors
UCPL	Upconversion photoluminescence
VEGF	Vascular endothelial growth factor

References

1. R. Savaliya, D. Shah, R. Singh, A. Kumar, R. Shanker, A. Dhawan and S. Singh, *Curr. Drug Metab.*, 2015, **16**, 645–661.
2. R. Bayford, T. Rademacher, I. Roitt and S. X. Wang, *Physiol. Meas.*, 2017, **38**, R183–R203.
3. A. P. Ramos, M. A. E. Cruz, C. B. Tovani and P. Ciancaglini, *Biophys. Rev.*, 2017, **9**, 79–89.
4. L. E. McNamara, R. J. McMurray, M. J. Biggs, F. Kantawong, R. O. Oreffo and M. J. Dalby, *J. Tissue Eng.*, 2010, **2010**, 120623.
5. E. K. Yim, E. M. Darling, K. Kulangara, F. Guilak and K. W. Leong, *Biomaterials*, 2010, **31**, 1299–1306.
6. D. G. Castner and B. D. Ratner, *Surf. Sci.*, 2002, **500**, 28–60.
7. D. F. Moyano and V. M. Rotello, *Langmuir: ACS J. Surf. Colloids*, 2011, **27**, 10376–10385.
8. K. L. Menzies and L. Jones, *Optom. Vis. Sci.: Off. Publ. Am. Acad. Optom.*, 2010, **87**, 387–399.
9. D. Frank, C. Tyagi, L. Tomar, Y. E. Choonara, L. C. du Toit, P. Kumar, C. Penny and V. Pillay, *Int. J. Nanomed.*, 2014, **9**, 589–613.
10. A. Babu, A. K. Templeton, A. Munshi and R. Ramesh, *AAPS PharmSciTech*, 2014, **15**, 709–721.
11. S. Singh, A. Sharma and G. P. Robertson, *Cancer Res.*, 2012, **72**, 5663–5668.
12. H. Li, D. E. Marotta, S. Kim, T. M. Busch, E. P. Wileyto and G. Zheng, *J. Biomed. Opt.*, 2005, **10**, 41203.
13. J. K. Kim, H. Yuan, J. Nie, Y. T. Yang, M. Leggas, P. M. Potter, J. Rinehart, M. Jay and X. Lu, *Small*, 2012, **8**, 2895–2903.
14. R. K. Jain, *Sci. Am.*, 1994, **271**, 58–65.
15. J. Sudimack and R. J. Lee, *Adv. Drug Delivery Rev.*, 2000, **41**, 147–162.
16. J. F. Ward 3rd, J. C. Rewcastle, O. Ukimura and I. S. Gill, *Curr. Opin. Urol.*, 2012, **22**, 104–108.

17. M. R. Cooperberg, P. R. Carroll and L. Klotz, *J. Clin. Oncol.: Off. J. Am. Soc. Clin. Oncol.*, 2011, **29**, 3669–3676.
18. A. K. Silva, R. Di Corato, T. Pellegrino, S. Chat, G. Pugliese, N. Luciani, F. Gazeau and C. Wilhelm, *Nanoscale*, 2013, **5**, 11374–11384.
19. J. Gautier, E. Allard-Vannier, E. Munnier, M. Souce and I. Chourpa, *J. Controlled Release: Off. J. Controlled Release Soc.*, 2013, **169**, 48–61.
20. H. J. Issaq, T. J. Waybright and T. D. Veenstra, *Electrophoresis*, 2011, **32**, 967–975.
21. T. J. Lyons and A. Basu, *Transl. Res. J. Lab. Clin. Med.*, 2012, **159**, 303–312.
22. E. P. Rhee and R. E. Gerszten, *Clin. Chem.*, 2012, **58**, 139–147.
23. N. Dasilva, P. Diez, S. Matarraz, M. Gonzalez-Gonzalez, S. Paradinas, A. Orfao and M. Fuentes, *Sensors*, 2012, **12**, 2284–2308.
24. S. S. Dharap, Y. Wang, P. Chandna, J. J. Khandare, B. Qiu, S. Gunaseelan, P. J. Sinko, S. Stein, A. Farmanfarmaian and T. Minko, *Proc. Natl. Acad. Sci. U. S. A.*, 2005, **102**, 12962–12967.
25. C. P. Chan, W. C. Mak, K. Y. Cheung, K. K. Sin, C. M. Yu, T. H. Rainer and R. Renneberg, *Annu. Rev. Anal. Chem.*, 2013, **6**, 191–211.
26. D. J. Crommelin and A. T. Florence, *Int. J. Pharm.*, 2013, **454**, 496–511.
27. H. Chen, Z. Zhen, T. Todd, P. K. Chu and J. Xie, *Mater. Sci. Eng. R, Rep.: Rev. J.*, 2013, **74**, 35–69.
28. M. Ferrari, *Nat. Rev. Cancer*, 2005, **5**, 161–171.
29. S. Sebastian, J. Settleman, S. J. Reshkin, A. Azzariti, A. Bellizzi and A. Paradiso, *Biochim. Biophys. Acta*, 2006, **1766**, 120–139.
30. L. Coussens, T. L. Yang-Feng, Y. C. Liao, E. Chen, A. Gray, J. McGrath, P. H. Seeburg, T. A. Libermann, J. Schlessinger and U. Francke, *et al.*, *Science*, 1985, **230**, 1132–1139.
31. S. Jaracz, J. Chen, L. V. Kuznetsova and I. Ojima, *Bioorg. Med. Chem.*, 2005, **13**, 5043–5054.
32. S. Ran and P. E. Thorpe, *Int. J. Radiat. Oncol., Biol., Phys.*, 2002, **54**, 1479–1484.
33. H. S. Choi, W. Liu, P. Misra, E. Tanaka, J. P. Zimmer, B. Itty Ipe, M. G. Bawendi and J. V. Frangioni, *Nat. Biotechnol.*, 2007, **25**, 1165–1170.
34. M. M. Cheng, G. Cuda, Y. L. Bunimovich, M. Gaspari, J. R. Heath, H. D. Hill, C. A. Mirkin, A. J. Nijdam, R. Terracciano, T. Thundat and M. Ferrari, *Curr. Opin. Chem. Biol.*, 2006, **10**, 11–19.
35. H. Y. Lee, Z. Li, K. Chen, A. R. Hsu, C. Xu, J. Xie, S. Sun and X. Chen, *J. Nucl. Med.: Off. Publ., Soc. Nucl. Med.*, 2008, **49**, 1371–1379.
36. X. Montet, K. Montet-Abou, F. Reynolds, R. Weissleder and L. Josephson, *Neoplasia*, 2006, **8**, 214–222.
37. S. Mura and P. Couvreur, *Adv. Drug Delivery Rev.*, 2012, **64**, 1394–1416.
38. S. Reuter, S. C. Gupta, M. M. Chaturvedi and B. B. Aggarwal, *Free Radical Biol. Med.*, 2010, **49**, 1603–1616.
39. H. Sies, *Am. J. Med.*, 1991, **91**, 31S–38S.
40. R. Singh and S. Singh, *Colloids Surf. B*, 2015, **132**, 78–84.
41. J. Shah, R. Purohit, R. Singh, A. S. Karakoti and S. Singh, *J. Colloid Interface Sci.*, 2015, **456**, 100–107.
42. N. V. Vallabani, A. S. Karakoti and S. Singh, *Colloids Surf. B*, 2017, **153**, 52–60.
43. C. Korsvik, S. Patil, S. Seal and W. T. Self, *Chem. Commun.*, 2007, 1056–1058.
44. A. S. Karakoti, S. Singh, A. Kumar, M. Malinska, S. V. Kuchibhatla, K. Wozniak, W. T. Self and S. Seal, *J. Am. Chem. Soc.*, 2009, **131**, 14144–14145.
45. R. Singh, A. S. Karakoti, W. Self, S. Seal and S. Singh, *Langmuir: ACS J. Surf. Colloids*, 2016, **32**, 12202–12211.
46. T. Pirmohamed, J. M. Dowding, S. Singh, B. Wasserman, E. Heckert, A. S. Karakoti, J. E. King, S. Seal and W. T. Self, *Chem. Commun.*, 2010, **46**, 2736–2738.
47. S. Singh, T. Dosani, A. S. Karakoti, A. Kumar, S. Seal and W. T. Self, *Biomaterials*, 2011, **32**, 6745–6753.

48. Y. Gao, K. Chen, J. L. Ma and F. Gao, *OncoTargets Ther.*, 2014, **7**, 835–840.
49. R. A. Revia and M. Zhang, *Mater. Today*, 2016, **19**, 157–168.
50. X. H. Peng, X. Qian, H. Mao, A. Y. Wang, Z. G. Chen, S. Nie and D. M. Shin, *Int. J. Nanomed.*, 2008, **3**, 311–321.
51. A. S. Semkina, M. A. Abakumov, N. F. Grinenko, A. A. Lipengolts, N. V. Nukolova and V. P. Chekhonin, *Bull. Exp. Biol. Med.*, 2017, **162**, 808–811.
52. H. Maeda, H. Nakamura and J. Fang, *Adv. Drug Delivery Rev.*, 2013, **65**, 71–79.
53. Q. Chen, W. Shang, C. Zeng, K. Wang, X. Liang, C. Chi, X. Liang, J. Yang, C. Fang and J. Tian, *Oncotarget*, 2017, **8**, 32741–32751.
54. J. S. Desgrosellier and D. A. Cheresh, *Nat. Rev. Cancer*, 2010, **10**, 9–22.
55. O. Feron, *Sci. Transl. Med.*, 2010, **2**, 34ps26.
56. Z. Zhang, J. Xie, J. Yu, Z. Lu and Y. Liu, *J. Mater. Chem. B*, 2017, **5**, 1454–1460.
57. Y. Liu, J. Xie, Z. Zhang and Z. Lu, *RSC Adv.*, 2016, **6**, 89484–89491.
58. C. Sun, O. Veiseh, J. Gunn, C. Fang, S. Hansen, D. Lee, R. Sze, R. G. Ellenbogen, J. Olson and M. Zhang, *Small*, 2008, **4**, 372–379.
59. S. A. Lyons, J. O'Neal and H. Sontheimer, *Glia*, 2002, **39**, 162–173.
60. C. G. Hadjipanayis, R. Machaidze, M. Kaluzova, L. Wang, A. J. Schuette, H. Chen, X. Wu and H. Mao, *Cancer Res.*, 2010, **70**, 6303–6312.
61. J. H. Lee, Y. M. Huh, Y. W. Jun, J. W. Seo, J. T. Jang, H. T. Song, S. Kim, E. J. Cho, H. G. Yoon, J. S. Suh and J. Cheon, *Nat. Med.*, 2007, **13**, 95–99.
62. J. N. Anker, W. P. Hall, O. Lyandres, N. C. Shah, J. Zhao and R. P. Van Duyne, *Nat. Mater.*, 2008, **7**, 442–453.
63. A. B. Chinen, C. M. Guan, J. R. Ferrer, S. N. Barnaby, T. J. Merkel and C. A. Mirkin, *Chem. Rev.*, 2015, **115**, 10530–10574.
64. K. A. Willets and R. P. V. Duyne, *Annu. Rev. Phys. Chem.*, 2007, **58**, 267–297.
65. H. Qian, M. Zhu, Z. Wu and R. Jin, *Acc. Chem. Res.*, 2012, **45**, 1470–1479.
66. K. A. Kang and M. D. Nguyen, *Adv. Exp. Med. Biol.*, 2017, **977**, 399–407.
67. M. T. Malik, M. G. O'Toole, L. K. Casson, S. D. Thomas, G. T. Bardi, E. M. Reyes-Reyes, C. K. Ng, K. A. Kang and P. J. Bates, *Oncotarget*, 2015, **6**, 22270–22281.
68. C. H. Chen, Y. J. Wu and J. J. Chen, *BioMed Res. Int.*, 2015, **2015**, 813632.
69. R. Popovtzer, A. Agrawal, N. A. Kotov, A. Popovtzer, J. Balter, T. E. Carey and R. Kopelman, *Nano Lett.*, 2008, **8**, 4593–4596.
70. K. A. Kimmel and T. E. Carey, *Cancer Res.*, 1986, **46**, 3614–3623.
71. W. Li and X. Chen, *Nanomedicine*, 2015, **10**, 299–320.
72. P. Manivasagan, S. Bharathiraja, N. Q. Bui, I. G. Lim and J. Oh, *Int. J. Pharm.*, 2016, **511**, 367–379.
73. C. Kim, E. C. Cho, J. Chen, K. H. Song, L. Au, C. Favazza, Q. Zhang, C. M. Cobley, F. Gao, Y. Xia and L. V. Wang, *ACS Nano*, 2010, **4**, 4559–4564.
74. T. K. Sawyer, P. J. Sanfilippo, V. J. Hruby, M. H. Engel, C. B. Heward, J. B. Burnett and M. E. Hadley, *Proc. Natl. Acad. Sci. U. S. A.*, 1980, **77**, 5754–5758.
75. K. H. Song, C. Kim, C. M. Cobley, Y. Xia and L. V. Wang, *Nano Lett.*, 2009, **9**, 183–188.
76. Y. Lu, G. L. Liu, J. Kim, Y. X. Mejia and L. P. Lee, *Nano Lett.*, 2005, **5**, 119–124.
77. S. Zong, Z. Wang, J. Yang, C. Wang, S. Xu and Y. Cui, *Talanta*, 2012, **97**, 368–375.
78. J. Li, Z. Skeete, S. Shan, S. Yan, K. Kurzatkowska, W. Zhao, Q. M. Ngo, P. Holubovska, J. Luo, M. Hepel and C.-J. Zhong, *Anal. Chem.*, 2015, **87**, 10698–10702.
79. K. Bhattacharyya, B. S. Goldschmidt, M. Hannink, S. Alexander, A. Jurkevic and J. A. Viator, *Clin. Lab. Med.*, 2012, **32**, 89–101.
80. C. C. You, O. R. Miranda, B. Gider, P. S. Ghosh, I. B. Kim, B. Erdogan, S. A. Krovi, U. H. Bunz and V. M. Rotello, *Nat. Nanotechnol.*, 2007, **2**, 318–323.
81. C. C. Huang, S. H. Chiu, Y. F. Huang and H. T. Chang, *Anal. Chem.*, 2007, **79**, 4798–4804.
82. H. Cho, E. C. Yeh, R. Sinha, T. A. Laurence, J. P. Bearinger and L. P. Lee, *ACS Nano*, 2012, **6**, 7607–7614.

83. E. M. Fox, T. M. Bernaciak, J. Wen, A. M. Weaver, M. A. Shupnik and C. M. Silva, *Mol. Endocrinol.*, 2008, **22**, 1781–1796.
84. J. Xue, L. Shan, H. Chen, Y. Li, H. Zhu, D. Deng, Z. Qian, S. Achilefu and Y. Gu, *Biosens. Bioelectron.*, 2013, **41**, 71–77.
85. X. Gao, L. Yang, J. A. Petros, F. F. Marshall, J. W. Simons and S. Nie, *Curr. Opin. Biotechnol.*, 2005, **16**, 63–72.
86. A. P. Alivisatos, *Science*, 1996, **271**, 933.
87. A. L. Efros and M. Rosen, *Annu. Rev. Mater. Sci.*, 2000, **30**, 475–521.
88. W. Cai, D.-W. Shin, K. Chen, O. Gheysens, Q. Cao, S. X. Wang, S. S. Gambhir and X. Chen, *Nano Lett.*, 2006, **6**, 669–676.
89. B. Ballou, L. A. Ernst, S. Andreko, T. Harper, J. A. J. Fitzpatrick, A. S. Waggoner and M. P. Bruchez, *Bioconjugate Chem.*, 2007, **18**, 389–396.
90. Z. Fan, S. Zhou, C. Garcia, L. Fan and J. Zhou, *Nanoscale*, 2017, **9**, 4928–4933.
91. F. Mitelman, B. Johansson and F. Mertens, *Nat. Rev. Cancer*, 2007, **7**, 233–245.
92. F. Mertens, B. Johansson, T. Fioretos and F. Mitelman, *Nat. Rev. Cancer*, 2015, **15**, 371–381.
93. H. Lee, C. Kim, D. Lee, J. H. Park, P. C. Searson and K. H. Lee, *Int. J. Nanomed.*, 2017, **12**, 4397–4407.
94. S. A. Tomlins, A. Bjartell, A. M. Chinnaiyan, G. Jenster, R. K. Nam, M. A. Rubin and J. A. Schalken, *Eur. Urol.*, 2009, **56**, 275–286.
95. R. Lister, M. Pelizzola, R. H. Dowen, R. D. Hawkins, G. Hon, J. Tonti-Filippini, J. R. Nery, L. Lee, Z. Ye, Q. M. Ngo, L. Edsall, J. Antosiewicz-Bourget, R. Stewart, V. Ruotti, A. H. Millar, J. A. Thomson, B. Ren and J. R. Ecker, *Nature*, 2009, **462**, 315–322.
96. Y. Ma, Y. Bai, H. Mao, Q. Hong, D. Yang, H. Zhang, F. Liu, Z. Wu, Q. Jin, H. Zhou, J. Cao, J. Zhao, X. Zhong and H. Mao, *Biosens. Bioelectron.*, 2016, **85**, 641–648.
97. Q. Wei, M. Lee, X. Yu, E. K. Lee, G. H. Seong, J. Choo and Y. W. Cho, *Anal. Biochem.*, 2006, **358**, 31–37.
98. K. D. Wegner, Z. Jin, S. Linden, T. L. Jennings and N. Hildebrandt, *ACS Nano*, 2013, **7**, 7411–7419.
99. J. Kim, S. Kwon, J. K. Park and I. Park, *Biosens. Bioelectron.*, 2014, **55**, 209–215.
100. A. F. Jou, C. H. Lu, Y. C. Ou, S. S. Wang, S. L. Hsu, I. Willner and J. A. Ho, *Chem. Sci.*, 2015, **6**, 659–665.
101. S. B. Cohen, M. E. Graham, G. O. Lovrecz, N. Bache, P. J. Robinson and R. R. Reddel, *Science*, 2007, **315**, 1850–1853.
102. L. H. Jin, S. M. Li and Y. H. Cho, *Biosens. Bioelectron.*, 2012, **33**, 284–287.
103. M. J. Chen, Y. S. Wu, G. F. Lin, J. Y. Hou, M. Li and T. C. Liu, *Anal. Chim. Acta*, 2012, **741**, 100–105.
104. Z. Li, X. He, X. Luo, L. Wang and N. Ma, *Anal. Chem.*, 2016, **88**, 9355–9358.
105. F. Wang, D. Banerjee, Y. Liu, X. Chen and X. Liu, *Analyst*, 2010, **135**, 1839–1854.
106. M. Wang, C. Mi, Y. Zhang, J. Liu, F. Li, C. Mao and S. Xu, *J. Phys. Chem. C*, 2009, **113**, 19021–19027.
107. P. Vilela, A. El-Sagheer, T. M. Millar, T. Brown, O. L. Muskens and A. G. Kanaras, *ACS Sens.*, 2017, **2**, 52–56.
108. L. Rao, Q. F. Meng, L. L. Bu, B. Cai, Q. Huang, Z. J. Sun, W. F. Zhang, A. Li, S. S. Guo, W. Liu, T. H. Wang and X. Z. Zhao, *ACS Appl. Mater. Interfaces*, 2017, **9**, 2159–2168.
109. J.-W. Shen, J. Wang, D. Kong and X.-P. Yan, *RSC Adv.*, 2014, **4**, 5088–5091.
110. L. Cheng, K. Yang, M. Shao, S.-T. Lee and Z. Liu, *J. Phys. Chem. C*, 2011, **115**, 2686–2692.
111. J. Liu, Y. Liu, W. Bu, J. Bu, Y. Sun, J. Du and J. Shi, *J. Am. Chem. Soc.*, 2014, **136**, 9701–9709.
112. L. Xiong, Z. Chen, Q. Tian, T. Cao, C. Xu and F. Li, *Anal. Chem.*, 2009, **81**, 8687–8694.

113. J. Liu, L. Huang, X. Tian, X. Chen, Y. Shao, F. Xie, D. Chen and L. Li, *Int. J. Nanomed.*, 2017, **12**, 1–14.
114. C. Zhang, M. Jugold, E. C. Woenne, T. Lammers, B. Morgenstern, M. M. Mueller, H. Zentgraf, M. Bock, M. Eisenhut, W. Semmler and F. Kiessling, *Cancer Res.*, 2007, **67**, 1555–1562.
115. S. Geninatti Crich, B. Bussolati, L. Tei, C. Grange, G. Esposito, S. Lanzardo, G. Camussi and S. Aime, *Cancer Res.*, 2006, **66**, 9196–9201.
116. Y. H. Kim, J. Jeon, S. H. Hong, W. K. Rhim, Y. S. Lee, H. Youn, J. K. Chung, M. C. Lee, D. S. Lee, K. W. Kang and J. M. Nam, *Small*, 2011, **7**, 2052–2060.
117. O. Rabin, J. Manuel Perez, J. Grimm, G. Wojtkiewicz and R. Weissleder, *Nat. Mater.*, 2006, **5**, 118–122.
118. Y. Liu, K. Ai, J. Liu, Q. Yuan, Y. He and L. Lu, *Adv. Healthcare Mater.*, 2012, **1**, 461–466.
119. Z. Gao, A. M. Kennedy, D. A. Christensen and N. Y. Rapoport, *Ultrasonics*, 2008, **48**, 260–270.
120. E. Kang, H. S. Min, J. Lee, M. H. Han, H. J. Ahn, I. C. Yoon, K. Choi, K. Kim, K. Park and I. C. Kwon, *Angew. Chem.*, 2010, **49**, 524–528.
121. K. Kim, J. H. Kim, H. Park, Y. S. Kim, K. Park, H. Nam, S. Lee, J. H. Park, R. W. Park, I. S. Kim, K. Choi, S. Y. Kim, K. Park and I. C. Kwon, *J. Controlled Release: Off. J. Controlled Release Soc.*, 2010, **146**, 219–227.
122. X. Gao, Y. Cui, R. M. Levenson, L. W. Chung and S. Nie, *Nat. Biotechnol.*, 2004, **22**, 969–976.
123. X. Cai, W. Li, C. H. Kim, Y. Yuan, L. V. Wang and Y. Xia, *ACS Nano*, 2011, **5**, 9658–9667.
124. A. De la Zerda, C. Zavaleta, S. Keren, S. Vaithilingam, S. Bodapati, Z. Liu, J. Levi, B. R. Smith, T. J. Ma, O. Oralkan, Z. Cheng, X. Chen, H. Dai, B. T. Khuri-Yakub and S. S. Gambhir, *Nat. Nanotechnol.*, 2008, **3**, 557–562.
125. H. Xie, Z. J. Wang, A. Bao, B. Goins and W. T. Phillips, *Int. J. Pharm.*, 2010, **395**, 324–330.
126. M. D. Majmudar, J. Yoo, E. J. Keliher, J. J. Truelove, Y. Iwamoto, B. Sena, P. Dutta, A. Borodovsky, K. Fitzgerald, M. F. Di Carli, P. Libby, D. G. Anderson, F. K. Swirski, R. Weissleder and M. Nahrendorf, *Circ. Res.*, 2013, **112**, 755–761.
127. G. Danaei, M. M. Finucane, Y. Lu, G. M. Singh, M. J. Cowan, C. J. Paciorek, J. K. Lin, F. Farzadfar, Y. H. Khang, G. A. Stevens, M. Rao, M. K. Ali, L. M. Riley, C. A. Robinson, M. Ezzati and Global Burden of Metabolic Risk Factors of Chronic Diseases Collaborating, *Lancet*, 2011, **378**, 31–40.
128. P. Makaram, D. Owens and J. Aceros, *Diagnostics*, 2014, **4**, 27–46.
129. J. C. Pickup, Z. L. Zhi, F. Khan, T. Saxl and D. J. Birch, *Diabetes/Metab. Res. Rev.*, 2008, **24**, 604–610.
130. X. Sun, S. Guo, Y. Liu and S. Sun, *Nano Lett.*, 2012, **12**, 4859–4863.
131. A. Kaushik, R. Khan, P. R. Solanki, P. Pandey, J. Alam, S. Ahmad and B. D. Malhotra, *Biosens. Bioelectron.*, 2008, **24**, 676–683.
132. L. Su, J. Feng, X. Zhou, C. Ren, H. Li and X. Chen, *Anal. Chem.*, 2012, **84**, 5753–5758.
133. Y. Shi, J. Huang, J. Wang, P. Su and Y. Yang, *Talanta*, 2015, **143**, 457–463.
134. W. Zhang, X. Li, R. Zou, H. Wu, H. Shi, S. Yu and Y. Liu, *Sci. Rep.*, 2015, **5**, 11129.
135. Q. Wu, X. Wang, C. Liao, Q. Wei and Q. Wang, *Nanoscale*, 2015, **7**, 16578–16582.
136. Z. Yang, C. Zhang, J. Zhang and W. Bai, *Biosens. Bioelectron.*, 2014, **51**, 268–273.
137. Y. Gao, Y. Wu and J. Di, *Spectrochim. Acta Part A*, 2017, **173**, 207–212.
138. M. Ghiaci, M. Tghizadeh, A. A. Ensafi, N. Zandi-Atashbar and B. Rezaei, *J. Taiwan Inst. Chem. Eng.*, 2016, **63**, 39–45.
139. D. Jiang, Q. Liu, K. Wang, J. Qian, X. Dong, Z. Yang, X. Du and B. Qiu, *Biosens. Bioelectron.*, 2014, **54**, 273–278.
140. W. Liu, H. Wu, B. Li, C. Dong, M. M. F. Choi and S. Shuang, *Anal. Methods*, 2013, **5**, 5154–5160.

141. S. Manju and K. Sreenivasan, *Talanta*, 2011, **85**, 2643–2649.
142. C. Radhakumary and K. Sreenivasan, *Anal. Chem.*, 2011, **83**, 2829–2833.
143. D. M. Nathan, H. Turgeon and S. Regan, *Diabetologia*, 2007, **50**, 2239–2244.
144. N. Wangoo, J. Kaushal, K. K. Bhasin, S. K. Mehta and C. R. Suri, *Chem. Commun.*, 2010, **46**, 5755–5757.
145. S. Bishnoi, R. N. Goyal and Y. B. Shim, *Bioelectrochemistry*, 2014, **99**, 24–29.
146. K. M. Daum and R. M. Hill, *Invest. Ophthalmol. Visual Sci.*, 1982, **22**, 509–514.
147. M. Howarth, W. Liu, S. Puthenveetil, Y. Zheng, L. F. Marshall, M. M. Schmidt, K. D. Wittrup, M. G. Bawendi and A. Y. Ting, *Nat. Methods*, 2008, **5**, 397–399.
148. J. Yan, M. Hu, D. Li, Y. He, R. Zhao, X. Jiang, S. Song, L. Wang and C. Fan, *Nano Res.*, 2008, **1**, 490–496.
149. X. Michalet, F. F. Pinaud, L. A. Bentolila, J. M. Tsay, S. Doose, J. J. Li, G. Sundaresan, A. M. Wu, S. S. Gambhir and S. Weiss, *Science*, 2005, **307**, 538–544.
150. E. R. Goldman, I. L. Medintz, J. L. Whitley, A. Hayhurst, A. R. Clapp, H. T. Uyeda, J. R. Deschamps, M. E. Lassman and H. Mattoussi, *J. Am. Chem. Soc.*, 2005, **127**, 6744–6751.
151. A. C. Vinayaka, S. Basheer and M. S. Thakur, *Biosens. Bioelectron.*, 2009, **24**, 1615–1620.
152. A. Chopra, S. Tuteja, N. Sachdeva, K. K. Bhasin, V. Bhalla and C. R. Suri, *Biosens. Bioelectron.*, 2013, **44**, 132–135.
153. H. Maaoui, F. Teodoresu, Q. Wang, G. H. Pan, A. Addad, R. Chtourou, S. Szunerits and R. Boukherroub, *Sensors*, 2016, **16**, 1720.
154. M. Hu, J. Tian, H. T. Lu, L. X. Weng and L. H. Wang, *Talanta*, 2010, **82**, 997–1002.
155. L. Wang, J. Zheng, Y. Li, S. Yang, C. Liu, Y. Xiao, J. Li, Z. Cao and R. Yang, *Anal. Chem.*, 2014, **86**, 12348–12354.
156. J. Liu, L. Lu, A. Li, J. Tang, S. Wang, S. Xu and L. Wang, *Biosens. Bioelectron.*, 2015, **68**, 204–209.
157. P. Hu, X. Wu, S. Hu, Z. Tang, G. Dai and Y. Liu, *RSC Adv.*, 2016, **6**, 102226–102230.
158. J. Yuan, Y. Cen, X. J. Kong, S. Wu, C. L. Liu, R. Q. Yu and X. Chu, *ACS Appl. Mater. Interfaces*, 2015, **7**, 10548–10555.
159. H. Wang, J. Yi, Y. Yu and S. Zhou, *Nanoscale*, 2017, **9**, 509–516.
160. S. Zhai, C. Fang, J. Yan, Q. Zhao and Y. Tu, *Anal. Chim. Acta*, 2017, **982**, 62–71.
161. M. Liang, Y. Ren, H. Zhang, Y. Ma, X. Niu and X. Chen, *Luminescence: J. Biol. Chem. Lumin.*, 2017, **32**, 1031–1038.
162. U. Jain, S. Gupta and N. Chauhan, *Int. J. Biol. Macromol.*, 2017, **101**, 896–903.
163. R. Jaisutti, M. Lee, J. Kim, S. Choi, T. J. Ha, J. Kim, H. Kim, S. K. Park and Y. H. Kim, *ACS Appl. Mater. Interfaces*, 2017, **9**, 8796–8804.
164. U. Jain and N. Chauhan, *Biosens. Bioelectron.*, 2017, **89**, 578–584.
165. V. Singh, C. Rodenbaugh and S. Krishnan, *ACS Sens.*, 2016, **1**, 437–443.
166. Y. Du, W. Zhang and M. L. Wang, *Biosensors*, 2016, **6**, 10.
167. J. Tao, P. Zhao, J. Zheng, C. Wu, M. Shi, J. Li, Y. Li and R. Yang, *Chem. Commun.*, 2015, **51**, 15704–15707.
168. S. Unser, I. Campbell, D. Jana and L. Sagle, *Analyst*, 2015, **140**, 590–599.
169. S. Bhagat, N. V. Srikanth Vallabani, V. Shutthanandan, M. Bowden, A. S. Karakoti and S. Singh, *J. Colloid Interface Sci.*, 2017, **513**, 831–842.
170. Y. Wang, L. Yu, X. Kong and L. Sun, *Int. J. Nanomed.*, 2017, **12**, 4789–4803.
171. B. Ray, E. Ghedin and R. Chunara, *J. Biomed. Inf.*, 2016, **64**, 44–54.
172. T. Laksanasopin, T. W. Guo, S. Nayak, A. A. Sridhara, S. Xie, O. O. Olowookere, P. Cadinu, F. Meng, N. H. Chee, J. Kim, C. D. Chin, E. Munyazesa, P. Mugwaneza, A. J. Rai, V. Mugisha, A. R. Castro, D. Steinmiller, V. Linder, J. E. Justman, S. Nsanzimana and S. K. Sia, *Sci. Transl. Med.*, 2015, **7**, 273re271.
173. S. P. Ravindranath, L. J. Mauer, C. Deb-Roy and J. Irudayaraj, *Anal. Chem.*, 2009, **81**, 2840–2846.
174. G. A. Suaifan, C. Esseghaier, A. Ng and M. Zourob, *Analyst*, 2013, **138**, 3735–3739.

175. S. Wignarajah, G. A. Suaifan, S. Bizzarro, F. J. Bikker, W. E. Kaman and M. Zourob, *Anal. Chem.*, 2015, **87**, 12161–12168.
176. K. S. Park, H. J. Chung, F. Khanam, H. Lee, R. Rashu, M. T. Bhuiyan, A. Berger, J. B. Harris, S. B. Calderwood, E. T. Ryan, F. Qadri, R. Weissleder and R. C. Charles, *Sci. Rep.*, 2016, **6**, 32878.
177. D. Yang, J. Ma, Q. Zhang, N. Li, J. Yang, P. A. Raju, M. Peng, Y. Luo, W. Hui, C. Chen and Y. Cui, *Anal. Chem.*, 2013, **85**, 6688–6695.
178. F. Wang, C. Ma, X. Zeng, C. Li, Y. Deng and N. He, *J. Biomed. Nanotechnol.*, 2012, **8**, 786–790.
179. M. Lin, H. Pei, F. Yang, C. Fan and X. Zuo, *Adv. Mater.*, 2013, **25**, 3490–3496.
180. S. R. Ahmed, J. Kim, T. Suzuki, J. Lee and E. Y. Park, *Biosens. Bioelectron.*, 2016, **85**, 503–508.
181. M. Saleh and M. El-Matbouli, *J. Virol. Methods*, 2015, **217**, 50–54.
182. J. D. Driskell, C. A. Jones, S. M. Tompkins and R. A. Tripp, *Analyst*, 2011, **136**, 3083–3090.
183. A. Mansour, S. Tammam, A. Althani and H. M. E. Azzazy, *J. Microbiol. Methods*, 2017, **139**, 165–167.
184. L. Shao, A. S. Susha, L. S. Cheung, T. K. Sau, A. L. Rogach and J. Wang, *Langmuir: ACS J. Surf. Colloids*, 2012, **28**, 8979–8984.
185. M. S. Verma, P. Z. Chen, L. Jones and F. X. Gu, *RSC Adv.*, 2014, **4**, 10660–10668.
186. R. de la Rica and M. M. Stevens, *Nat. Nanotechnol.*, 2012, **7**, 821–824.
187. J. Kim, M. Jang, K. G. Lee, K. S. Lee, S. J. Lee, K. W. Ro, I. S. Kang, B. D. Jeong, T. J. Park, H. J. Kim and J. Lee, *ACS Appl. Mater. Interfaces*, 2016, **8**, 23489–23497.
188. K. Kampani, K. Quann, J. Ahuja, B. Wigdahl, Z. K. Khan and P. Jain, *J. Virol. Methods*, 2007, **141**, 125–132.
189. W. Yu-Hong, C. Rui and L. Ding, *Nanoscale Res. Lett.*, 2011, **6**, 461.
190. P. Zhang, H. Lu, J. Chen, H. Han and W. Ma, *Theranostics*, 2014, **4**, 307–315.
191. K. Takemura, O. Adegoke, N. Takahashi, T. Kato, T. C. Li, N. Kitamoto, T. Tanaka, T. Suzuki and E. Y. Park, *Biosens. Bioelectron.*, 2017, **89**, 998–1005.
192. J. M. Klostranec, Q. Xiang, G. A. Farcas, J. A. Lee, A. Rhee, E. I. Lafferty, S. D. Perrault, K. C. Kain and W. C. Chan, *Nano Lett.*, 2007, **7**, 2812–2818.
193. N. Duan, S. Wu, S. Dai, T. Miao, J. Chen and Z. Wang, *Microchim. Acta*, 2015, **182**, 917–923.
194. S. Wu, N. Duan, Z. Shi, C. Fang and Z. Wang, *Anal. Chem.*, 2014, **86**, 3100–3107.
195. H. Kurt, M. Yuce, B. Hussain and H. Budak, *Biosens. Bioelectron.*, 2016, **81**, 280–286.
196. T. Mocan, C. T. Matea, T. Pop, O. Mosteanu, A. D. Buzoianu, C. Puia, C. Iancu and L. Mocan, *J. Nanobiotechnol.*, 2017, **15**, 25.
197. C. Cao, L. C. Gontard, L. Thuy Tram le, A. Wolff and D. D. Bang, *Small*, 2011, **7**, 1701–1708.
198. J. Wang, X. Wu, C. Wang, Z. Rong, H. Ding, H. Li, S. Li, N. Shao, P. Dong, R. Xiao and S. Wang, *ACS Appl. Mater. Interfaces*, 2016, **8**, 19958–19967.
199. V. Raj, A. N. Vijayan and K. Joseph, *Sens. Biosens. Res.*, 2015, **5**, 33–36.
200. S. Nourani, H. Ghourchian and S. M. Boutorabi, *Anal. Biochem.*, 2013, **441**, 1–7.
201. C. Singhal, C. S. Pundir and J. Narang, *Biosens. Bioelectron.*, 2017, **97**, 75–82.
202. O. Adegoke, T. Kato and E. Y. Park, *Biosens. Bioelectron.*, 2016, **80**, 483–490.
203. A. D. Kurdekar, L. A. A. Chunduri, S. M. Chelli, M. K. Haleyurgirisetty, E. P. Bulagonda, J. Zheng, I. K. Hewlett and V. Kamisetti, *RSC Adv.*, 2017, **7**, 19863–19877.
204. F. Zou, H. Zhou, T. V. Tan, J. Kim, K. Koh and J. Lee, *ACS Appl. Mater. Interfaces*, 2015, **7**, 12168–12175.
205. K. Yamada, C. T. Kim, J. H. Kim, J. H. Chung, H. G. Lee and S. Jun, *PLoS One*, 2014, **9**, e105767.
206. T. S. Wong, B. Brough and C. M. Ho, *Mol. Cell. Biomech.*, 2009, **6**, 1–55.
207. K. M. Hansen and T. Thundat, *Methods*, 2005, **37**, 57–64.

208. J. Fritz, M. K. Baller, H. P. Lang, H. Rothuizen, P. Vettiger, E. Meyer, H. Guntherodt, C. Gerber and J. K. Gimzewski, *Science*, 2000, **288**, 316–318.
209. T. Braun, V. Barwich, M. K. Ghatkesar, A. H. Bredekamp, C. Gerber, M. Hegner and H. P. Lang, *Phys. Rev. E, Stat., Nonlinear, Soft Matter Phys.*, 2005, **72**, 031907.
210. T. Y. Kwon, K. Eom, J. H. Park, D. S. Yoon, T. S. Kim and H. L. Lee, *Appl. Phys. Lett.*, 2007, **90**, 223903.
211. F. Huber, H. P. Lang, N. Backmann, D. Rimoldi and C. Gerber, *Nat. Nanotechnol.*, 2013, **8**, 125–129.
212. T. I. Yin, Y. Zhao, J. Horak, H. Bakirci, H. H. Liao, H. H. Tsai, Y. Z. Juang and G. Urban, *Lab Chip*, 2013, **13**, 834–842.
213. G. Wu, R. H. Datar, K. M. Hansen, T. Thundat, R. J. Cote and A. Majumdar, *Nat. Biotechnol.*, 2001, **19**, 856–860.
214. J. W. Choi, H. Lee, G. Lee, Y. R. Kim, M. J. Ahn, H. J. Park, K. Eom and T. Kwon, *Theranostics*, 2017, **7**, 2878–2887.
215. J. McDevitt, M. McRae, G. Simmons and N. Christodoulides, *J. Biosens. Bioelectron.*, 2015, **6**, e137.
216. J. V. Jokerst, J. W. Jacobson, B. D. Bhagwandin, P. N. Floriano, N. Christodoulides and J. T. McDevitt, *Anal. Chem.*, 2010, **82**, 1571–1579.
217. J. J. Lavigne, S. Savoy, M. B. Clevenger, J. E. Ritchie, B. McDoniel, S.-J. Yoo, E. V. Anslyn, J. T. McDevitt, J. B. Shear and D. Neikirk, *J. Am. Chem. Soc.*, 1998, **120**, 6429–6430.
218. W. R. Rodriguez, N. Christodoulides, P. N. Floriano, S. Graham, S. Mohanty, M. Dixon, M. Hsiang, T. Peter, S. Zavahir, I. Thior, D. Romanovicz, B. Bernard, A. P. Goodey, B. D. Walker and J. T. McDevitt, *PLoS Med.*, 2005, **2**, e182.
219. A. P. Goodey and J. T. McDevitt, *J. Am. Chem. Soc.*, 2003, **125**, 2870–2871.
220. P. N. Floriano, N. Christodoulides, C. S. Miller, J. L. Ebersole, J. Spertus, B. G. Rose, D. F. Kinane, M. J. Novak, S. Steinhubl, S. Acosta, S. Mohanty, P. Dharshan, C. K. Yeh, S. Redding, W. Furmaga and J. T. McDevitt, *Clin. Chem.*, 2009, **55**, 1530–1538.
221. J. Fan, X. Deng, J. W. Gallagher, H. Huang, Y. Huang, J. Wen, M. Ferrari, H. Shen and Y. Hu, *Philos. Trans. Ser. A, Math., Phys., Eng. Sci.*, 2012, **370**, 2433–2447.
222. Y. Hu, A. Gopal, K. Lin, Y. Peng, E. Tasciotti, X. J. Zhang and M. Ferrari, *Biomicrofluidics*, 2011, **5**, 13410.
223. H. He, L. A. Pham-Huy, P. Dramou, D. Xiao, P. Zuo and C. Pham-Huy, *BioMed Res. Int.*, 2013, **2013**, 12.
224. M. F. De Volder, S. H. Tawfick, R. H. Baughman and A. J. Hart, *Science*, 2013, **339**, 535–539.
225. K. A. Drouvalakis, S. Bangsaruntip, W. Hueber, L. G. Kozar, P. J. Utz and H. Dai, *Biosens. Bioelectron.*, 2008, **23**, 1413–1421.
226. R. Yang, Z. Tang, J. Yan, H. Kang, Y. Kim, Z. Zhu and W. Tan, *Anal. Chem.*, 2008, **80**, 7408–7413.
227. J. Wang, A. N. Kawde and M. Musameh, *Analyst*, 2003, **128**, 912–916.
228. D. Fu and L. J. Li, *Nano Rev.*, 2010, **1**, 5354.

11 Molecular Microbial Diagnostics†

Karl-Henning Kalland*[a,b], Øyvind Kommedal[a], and Elling Ulvestad[a,b]

[a]Department of Microbiology, Haukeland University Hospital, Laboratory Building 5. etg. Helse Bergen, No-5021 Bergen, Norway; [b]Department of Clinical Science, University of Bergen, Bergen, Norway
*E-mail: Kalland@gades.uib.no

11.1. Introduction

The suffering and deaths caused by epidemics remain unsurpassed by any other malady afflicting the human species.[1] Several attempts at explaining the dramatic outbreaks were made in the prescientific era, and proposals included mysterious forces associated with witchcraft, earthquakes and miasmas residing in foul and decaying air. These forces were not dismissed until the second half of the 19th century when ingenious work by Louis Pasteur, Robert Koch and other continental researchers made it clear that the infectious diseases were of microbial etiology.[2] This recognition quickly and forcefully paved the way for the establishment of diagnostic laboratories in major hospitals, and thus for the coupling of microbiological science to medical practice.

†This chapter has been reproduced from Molecular Biology and Biotechnology, ed. R. Rapley and D. Whitehouse, 2014, 6th edition, copyright The Royal Society of Chemistry.

Genomics and Clinical Diagnostics
Edited by David Whitehouse and Ralph Rapley

Published by the Royal Society of Chemistry, www.rsc.org

The microbiologist's repertoire of diagnostic tests was initially rather limited, consisting predominantly of specimen examination through the microscope, colony growth on agar plates and serological tests. The laboratories were nevertheless very successful at their task, and so provided scientifically based support for hygienic movements attempting to identify and control carriers of venereal, diarrheal and respiratory diseases. These methods of investigation are still being utilized in diagnostic laboratories. But owing to several shortcomings, including insufficient diagnostic sensitivity and specificity, they are increasingly being supplemented with newer methods of detection.

Foremost amongst the new methods are tests based on analyses of nucleic acids – including DNA and RNA – from the causative microbial agents. Not only are these techniques orders of magnitude faster than the older techniques, they are also more sensitive and specific and enable the detection of non-cultivable microbes as well as microbes that have been rendered non-cultivable by antibiotics. Owing to the ubiquitous presence of nucleic acids in all varieties of life, the diagnostic principles can be utilized to analyze any major pathogen afflicting humans, including bacteria, viruses, parasites and fungi. The progress has made diagnosis of microbial agents ever more relevant, especially in the field of virology, where quick results derived from molecular analyses have virtually revolutionized clinical therapy and disease monitoring.

Microbes have traditionally been associated with disease and malfunctioning but are now increasingly also being recognized as important promoters of health. Lively research activity has provided evidence that absence of symbiotic microbes, or a skewed distribution of microbes in a community of microbes, may be of relevance for a variety of diseases, including asthma, autoimmunity, malnutrition and obesity. The older methods of detection were way too insensitive to quantitate such skewedness, whereas the newer molecular techniques seem perfect for the task.

The molecular methodology associated with detection of infectious agents is continuously improving, and so is the capacity to handle ever increasing numbers of patient specimens. However, owing to an extreme sensitivity, the same methodology is liable to produce erroneous results if left unattended. Diagnosticians thus need to understand the shortcomings just as well as they understand the opportunities provided by the tests. In the following sections we will give an overview of the strengths and weaknesses of the most prevalent and useful nucleic-acid-based diagnostic tests. First we give an overview of sampling techniques, thereafter we review relevant techniques for the detection of microbial agents and, finally, we briefly demonstrate the tests' utility on diagnostically relevant clinical cases.

11.2. Sample Collection and Nucleic Acid Purification

11.2.1. Sample Collection and Transport

Virtually any human secretion, fluid or tissue sample can be collected and examined for nucleic acids of suspected infectious agents, in addition to bacterial colonies and infected cell cultures. Microbiological diagnostic laboratories provide kits for sampling and shipping patient specimens. Some kits contain buffers that inactivate infectious agents, inhibit nucleases and lyse membranes to facilitate subsequent nucleic acid extraction. Whenever patient samples are collected and sent without additives, speedy delivery to the laboratory must be ensured and great care must be taken to label the sample appropriately and to avoid contamination of the outside of the collecting vessel. Several companies supply swab kits and buffer kits for stabilizing samples during transport.[3]

11.2.2. Extraction of Nucleic Acids

Different nucleic acid extraction protocols are required for different samples and different microbiological agents. RNA is more susceptible than DNA to rapid degradation by nucleases, and the best way to counteract the degradation problem is to use transport buffers that stabilize the intact infectious particle, or alternatively, to dissolve the sample in guanidine isothiocyanate-based lysis buffer that inhibits RNase enzymes. Viral nucleic acids usually can be extracted in the same way that host cell nucleic acids are extracted. Sometimes, *e.g.* for faecal (stool) samples, gentle vortexing followed by low-speed centrifugation can pellet debris while viral particles remain in the supernatant. Viral particles (virions) require high speed or ultracentrifugation to be pelleted. The thick cell wall of Gram-positive bacteria is more difficult to disrupt than the relatively thinner cell wall of Gram-negative bacteria and needs special procedures.[4,5]

Haem in blood, bile in stool and components of urine and additives such as heparin may inhibit the enzymes used in nucleic acid amplification, such as Taq polymerase, and must be removed by a purification procedure prior to the amplification step. Purified nucleic acids should be eluted into a small volume of nuclease-free H_2O or 0.1 mM EDTA (ethylenediaminetetraacetate). Ethanol or isopropanol precipitation may be used to achieve smaller volumes, but alcohols must be completely removed by pipetting to avoid subsequent enzymatic inhibition.

11.2.3. Manual and Automated Extraction of Nucleic Acids

Several commercial manual extraction kits are available for use by clinical laboratories. These kits vary regarding method, cost and time required for extraction. Processing by manual methods involves a higher risk of contamination and switching of patient samples. The US Clinical Laboratory Improvement Amendments of 1988 (CLIA) regulations provide guidelines regarding manual extraction of nucleic acids for diagnostic purposes (http://www.cms.hhs.gov/clia/).

Automated extraction instruments are manufactured by many different companies, and, like manual methods, vary in technique, cost and time requirements. Additionally, these instruments vary regarding specimen capacity and sample size. Automated procedures are in several ways superior to manual methods (see ref. 6 and 7), although this is not always so.[8] Recovery of nucleic acids from automated instruments is consistent and reproducible. Automated extraction systems keep sample manipulation to a minimum, thus reducing the risk of cross-contamination of samples. They are most economical when instruments are fully loaded, although smaller and more versatile instruments now have become available.

11.3. Nucleic Acid Amplification Techniques

11.3.1. Polymerase Chain Reaction (PCR)

PCR represents the prototype of nucleic acid amplification techniques. Excellent reviews on its applicability in clinical microbiology have been published.[6,7,9] Due to its extreme sensitivity nucleic acid amplification has revolutionized identification of infectious agents in patient samples. Several alternative nucleic acid amplification methods have been developed for use in clinical microbiological laboratories, such as isothermal strand displacement amplification (SDA), isothermal transcription mediated amplification (TMA/NASBA) and loop-mediated isothermal amplification (LAMP).[10]

11.3.2. The Contamination Problem

Owing to the exquisite sensitivity of nucleic acid amplification techniques great care must be taken to minimize the problem of sample contamination. Sterile filter tips must be employed for all manual pipetting steps. The logistics of the laboratory procedures are of fundamental importance. One separate room should be reserved for work with buffer components, oligonucleotide primers, probes and master mixes. No samples or amplified products should be allowed into this room. Higher air pressure than in the surrounding rooms, restricted access, visible written hygiene guidelines on the door and use of separate clothing and shoes are advantageous. A separate room is necessary for receiving and organizing patient samples and a third room for setting up the PCR reactions. Using conventional PCR, a fourth separate room will be necessary for the detection steps for the amplified products, the amplicons. Commonly used methods for the visualization of amplicons are agarose gel electrophoresis or microplate systems with a detection probe and a streptavidin–enzyme conjugate that binds to biotinylated nucleotides incorporated into the amplicon. Following the washing step the enzyme bound by the biotin–streptavidin interaction will change a colourless substrate into a coloured derivative in a quantitative way. Using real-time PCR (see below) the amplification and detection can be completed in a closed system and for space considerations may be performed in the same room as the reaction setup. If possible, one-way flow of persons and samples from the first to the last room is desirable. Decontamination of critical work space using UV light or 5% bleach represents additional precautions against contamination of samples. If deoxy-uridine-5′-triphosphate (dUTP) has been used in the deoxynucleotide mix of previous amplification reactions, a pre-amplification step using uracil-*N*-glycosylase will efficiently destroy amplicon contamination of the subsequent reaction.

11.3.3. Reverse PCR – cDNA Synthesis

Recovery of microbial DNA is suitable for the diagnosis of bacteria, parasites and many viruses. An additional large proportion of pathogenic viruses carry RNA-genomes instead of DNA-genomes. The genomes of such RNA viruses must be reverse transcribed to complementary DNA (cDNA) before amplification and identification. When PCR DNA amplification is preceded by cDNA synthesis, the combined protocol is named reverse PCR. Reverse PCR may also be useful in order to quantify transcription (mRNA synthesis) of microbial DNA to show the activity of an infectious genome. Using thermostable enzymes, which have both reverse transcriptase and DNA polymerase activity (such as rTth, from *Thermus thermophilus*), reverse transcription can be achieved in one reaction and one tube. Otherwise the reverse transcription is often carried out as a separate first step using a thermolabile reverse transcriptase (typically a cloned retroviral enzyme). If random primers (hexamers or nonamers) or oligo(dT) primers are used in the initial reverse transcription, then the subsequent PCR can be performed with different specific primer pairs and designed for a variety of RNA viruses. In the rTth reverse PCR, the same reverse primer sequence will prime both cDNA synthesis and PCR amplification. This may be a quicker procedure, but the cDNA has reduced versatility compared with cDNA primed by random hexamers or oligo(dT).

11.3.4. Nested PCR

When patient samples are amplified by PCR, the amplicon very often exhibits a smear in contrast to a distinct band in agarose gel electrophoresis. This is due to non-specific priming and amplification of host-cell nucleic acids. Although the first PCR amplification results in a smear in the agarose gel, the specific nucleic acids sought for will usually be highly enriched compared with the starting material. If a small amount of the first PCR is used for a second PCR amplification with a new set of internal primers, then a well-defined PCR-fragment typically results if the infectious agent is present in the initial sample. The use of a second round of PCR with a new set of internal primers is called nested PCR. If only one internal primer is used for the second round of thermocycling, it is called seminested PCR. Nested and seminested PCR increase sensitivity as much as 1000 times compared with conventional PCR[11] and increase specificity, but the techniques require more work and longer assay times and increase the risk of contamination.

11.3.5. Real-time PCR

Compared with conventional PCR, and in particular compared with nested PCR, real-time PCR has the advantage that the result of the amplification process is recorded in real-time (*i.e.* while the amplification reaction is running). The risk of contamination is very much reduced since there is no need to open the reaction tube once the amplification has started, *i.e.* the post-amplification detection step is omitted. Therefore, and also because more of the process can be automated, it takes less time and labour to do real-time PCR than nested PCR. For the same reasons the danger of sample switching is less in real-time PCR. The sensitivity is similar for real-time PCR and nested PCR.

11.3.6. Visualization of Real-time PCR Amplification

Different fluorescence-based principles are employed for real-time detection. SYBR green represents a class of DNA-binding compounds which emit increased fluorescence when intercalated into double-stranded DNA as compared with the unbound chemical in solution (Figure 11.1). As a result of this, there is a proportional increase in SYBR green fluorescence when DNA amplification products increase. The disadvantage of using SYBR green is that non-specifically amplified DNA and amplification artefacts, such as primer dimers, will also generate increased fluorescence, thus reducing the specificity compared with nested PCR. Real-time PCR assays utilizing an amplicon-specific probe increase the specificity very much, because the sequence between the two primers must be complementary to the probe in order to obtain a positive assay. A number of smart assays have been invented based upon altogether three oligonucleotides, *i.e.* two primers and one probe. These assays include TaqMan assays (Figure 11.2), FRET (fluorescence resonance energy transfer) (Figure 11.3) assays, Eclipse probes, Scorpion probes and molecular beacon probes.

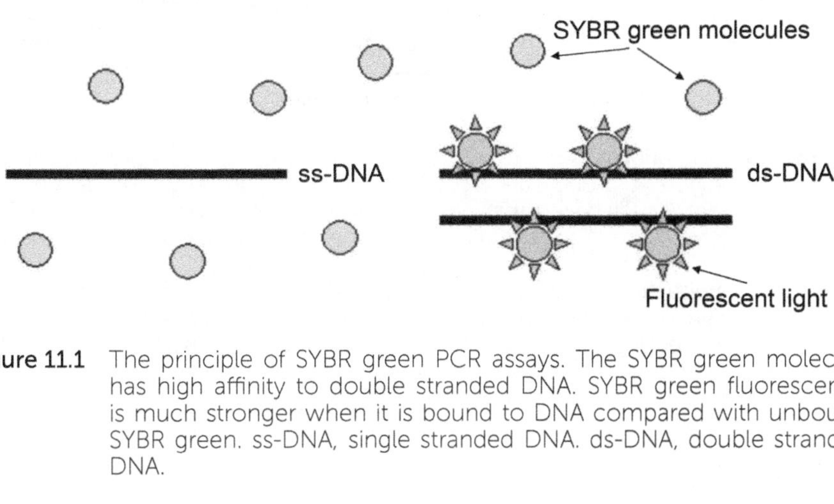

Figure 11.1 The principle of SYBR green PCR assays. The SYBR green molecule has high affinity to double stranded DNA. SYBR green fluorescence is much stronger when it is bound to DNA compared with unbound SYBR green. ss-DNA, single stranded DNA. ds-DNA, double stranded DNA.

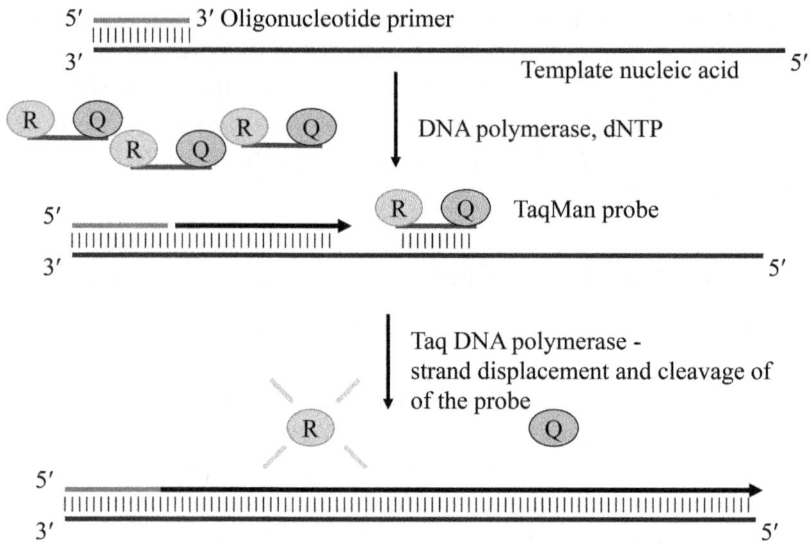

Figure 11.2 The principle of TaqMan real-time PCR assays. The probe is labelled with both a fluorescent chemical group (R) and a quencher (Q). When the Taq polymerase elongates the complementary strand from the primer, its 5′–3′ exonuclease activity cleaves any annealed probe, thus separating R from Q. Freed from its quencher, the R fluoresces when exciting light is sent in. The amount of free R, and thus fluorescence, increases exponentially along with amplicons until the PCR reaction approaches the plateau stage.

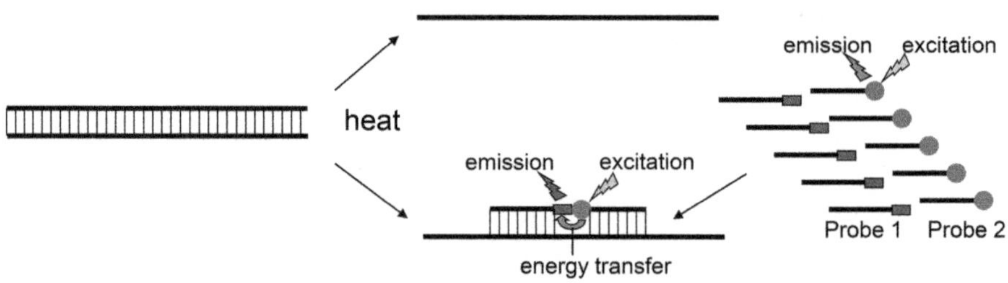

Figure 11.3 The principle of FRET (fluorescence resonance energy transfer). Two different probes are complementary to the target sequence. One probe is labelled with fluorescein and the other with LC Red fluorochrome. The hybridization to the target sequence brings the two fluorochromes physically close enough for the emission energy of fluorescein to excite the LC red so that red fluorescent light is emitted. When PCR generates an increasing concentration of target, an increasing fraction of the probes can be positioned to achieve FRET. This can be monitored as increasing red fluorescence.

11.3.7. Real-time PCR Equipment

Compared with conventional PCR real-time quantitative PCR (qPCR) needs more equipment. Lasers are needed to excite fluorophores and recording devices are required to collect and digitize the fluorescent signals. The choice of thermocycler depends upon a large number of variables, *e.g.* the estimated number of samples per run.

11.3.8. Real-time Quantitative PCR

The increasing fluorescence signal plotted on the *Y*-axis and the increasing thermocycle numbers along the *X*-axis generate a sigmoidal curve. The threshold cycle (C_t) is defined as the cycle number when the fluorescent signal exceeds the detection threshold. During the following 4–8 cycles the fluorescence typically approaches exponential increase before approaching a plateau phase.[12] During the exponential phase the logarithm of the fluorescent signal along the *Y*-axis and the cycle number along the *X*-axis generate a straight line that can be used for quantification. In this phase of the real-time PCR the fractional threshold cycle number is inversely related to the logarithmic number of starting templates.[13] Parallel PCR-amplification of a dilution series of known amounts of the same template therefore allows the construction of a standard curve from which the exact amount of starting template can be calculated.[9] It must be assumed that the amplification efficiency is equal for all samples and standards. One sign of acceptable quantification is that the slope of the standard curve is between minus 3.3 and minus 3.4. More detailed guidelines to quantification and quality control can be found in user bulletins provided by the manufacturers of thermocyclers and software.

11.3.8.1. Absolute and Relative Quantifications

In order to achieve absolute quantification a known amount of reference standard is required for the dilution series and standard curve generation. Synthetic oligonucleotides, purified PCR fragments, plasmids or *in vitro*-transcribed RNA can be quantified using light absorption readings at 260 nM wavelength (OD_{260}). Dilution series of known amounts of target allow absolute quantification. Relative quantification, relating the amount of template to an accepted standard, is often sufficient, and this can be achieved using the standard curve method or the comparative threshold cycle (C_t) method (see ref. 9 and references therein).

11.3.9. Determination of "Viral Load" in Clinical Microbiology

Determination of "viral load", which refers to the number of viral genomes in a defined volume of a patient sample, typically in serum or plasma, is becoming increasingly important with the increasing use of antiviral medication. Real-time quantification of viral load is often required before the start of antiviral measures against human immunodeficiency virus (HIV), hepatitis C- and hepatitis B-viruses, against cytomegalovirus (CMV) in patients with compromised immune systems and against Epstein–Barr virus (EBV) in patients following organ transplantations. Reduction of viral load indicates efficient response to treatment, and quantification is of great value to monitor development of viral resistance to the medication or relapse after cessation of therapy. Commercial kits are widely used for quantification of viral load in clinical microbiology. A known amount of synthetic standard is typically included in each reaction to establish a competitive PCR amplification. The resulting amounts of viral and standard amplicons are compared with a standard dilution curve in order to establish the viral load. It is important to be aware that different manufacturers may have different standardization units so that results obtained using different kits may not be comparable. Progress has recently been made to establish internationally accepted standard units for different viruses, *e.g.* for hepatitis B-virus,[14] hepatitis C-virus[15] and parvovirus B19.[16] The WHO provides guidelines and established standards in order to calibrate different methods to obtain comparable International Units per millilitre (IU ml^{-1}; http://www.who.int/biologicals/reference_preparations/distribution/en/index.html).

11.3.10. Internal Controls in Microbiological Real-time qPCR

Both negative and positive controls are important in nucleic-acid-based diagnosis in microbiological laboratories.[3] The negative control consists of a parallel amplification reaction that does not contain the specific target. There is no rule as to how many negative controls are needed, and there is always a trade-off between the number of negative controls and the cost. Recommendations for a negative control for every fifth tube may come into conflict with available resources.[3] Positive controls can be any target sequence in a separate tube. The best positive control templates are included in the same tube as the diagnostic assay and contain flanking sequences complementary to the primers of the diagnostic assay but with a unique sequence in the region recognized by the probe. Since real-time qPCR assays are typically designed with amplicons shorter than 100 nucleotides it is easy to obtain the desired control template sequence by oligonucleotide synthesis. A more difficult challenge is if the laboratory wants to spike the patient sample with a control sequence to control for the entire process from sample handling, *via* nucleic acid extraction through PCR amplification. For RNA viruses, which represent the greatest challenge, this can be achieved if a known amount of an *in vitro*-synthesized RNA is spiked into the patient sample. Any exogenous sequence may be chosen if it can be detected separately from the sequence of the infectious agent. However, the optimal control RNA will be transcribed from a vector that can generate an RNA that contains more than 1500 nucleotides and includes a target sequence that can be amplified with the same primers as the diagnostic assay but differs in the region recognized by the probe.

In all same-tube controls it is very important to titrate carefully the amount of control target. Otherwise, the control may out-compete the diagnostic assay and lead to false negatives.

A very important example of a positive control RNA is exemplified by human reference (endogenous) genes. Typically, the endogenous control gene should be ubiquitously expressed by human cells. Examples include RNAse P, GAPDH or β-actin. When the control PCR reaction for the endogenous gene is positive, it indicates that representative patient material was sampled and that the sample does not contain substances that inhibit the PCR amplification.

11.3.11. Multiplex Real-time PCR

Multiplex PCR refers either to the use of more than one primer pair in one reaction tube in order to amplify more than one target sequence, or to the use of more than one primer/probe set in the case of real-time PCR. The internal control assay described above is one example of multiplex PCR. In microbiological diagnostics the general aim of multiplex PCR is to simultaneously test for different agents in the same reaction. This requires the use of fluorophores and equipment with the ability to distinguish between the different assays. Non-specific amplification products seem to be the most limiting factor of multiplex PCR as the numbers of oligonucleotide primers and probes increase. Several multiplex assays for the simultaneous detection of two or more different agents are, however, well established. A number of multiplex real-time PCR assays have been published including simultaneous amplification and subtyping of herpes simplex virus 1 and 2 (HSV-1 and HSV-2) and varicella zoster virus (VZV) using FRET probes[17,18] or the combination of influenza A subtypes H1N1 and H3N2, influenza B and respiratory syncytial virus (RSV) subtypes A and B[18,19] and additional respiratory infection agents.[20]

11.3.12. Melting Curve Analysis

Melting curve analysis has made possible an increase in the number of agents detected in multiplex real-time PCR, in particular the ability to distinguish between closely related viruses such as HSV-1 and HSV-2 or influenza types A and B. FRET probes (Figure 11.3) are useful for this type of analysis. When one FRET probe is designed to be complementary to one viral subtype but contains one or very few mismatches to the other viral subtype, melting curve analysis is possible due to different melting points (T_m) dependent upon the exact target sequence. Melting curve analysis is performed following completion of PCR. The probe pair is first allowed to anneal to the target amplicons and thereafter the temperature is increased while the fluorescence is continuously recorded. The viral subtype with mismatch to the FRET probe will exhibit a reduced fluorescence at a lower T_m than the viral subtype with perfect homology to the FRET probe. If both subtypes are present in the same mix, this may be visualized as a curve with two different peaks when the fluorescence is plotted on the *Y*-axis and the temperature is plotted along the *X*-axis.

11.3.13. Genotyping

Many viruses evolve at high rates after having infected their human hosts. In particular RNA viruses that replicate by RNA polymerases which lack 5′ proof reading in contrast to DNA polymerases, are prone to point mutation errors in the order of 10^{-4} per genome replication. However, DNA viruses may also change rapidly upon changes in the selection pressure, owing to their short generation times and enormous number of progeny in each infection. As a consequence, most viral species come in multiple serotypes and genotypes. Some of these types are more efficient at producing disease than others, and it is therefore clinically relevant to distinguish between such genotypes. 12

different main genotypes are presently described for hepatitis C virus. Both prognosis and the recommended duration of PEG (polyethylene glycol)-interferon α–ribavirin combination therapy vary between HCV genotypes. The prevalent genotypes 1 and 4 usually require 48 weeks of combination therapy while less than half that duration of therapy may be sufficient for genotypes 2 and 3. Also, sustained viral response, in which HCV RNA cannot be detected in serum 6 months following cessation of treatment, is achieved in a higher proportion of cases for genotypes 2 and 3. Likewise, more than 130 genotypes of human papillomaviruses (HPV) are currently recorded, and these types are separated into high-risk types and low-risk types concerning their ability to cause cervical cancer (cancer cervicis uteri). Genotyping can be achieved using DNA sequencing. Due to co-infections with multiple genotypes at different concentrations, other methods have been developed for genotyping in clinical microbiology laboratories, such as sandwich hybridization assays (Figure 11.4) or line probe assays (Figure 11.5).

11.4. Other Techniques Used in Clinical Microbiology

11.4.1. Hybridization Techniques

The principle of complementarity, according to which A (adenine) pairs with T (thymine) and C (cytosine) pairs with G (guanine) in a nucleic acid duplex, is the core principle of heredity, all hybridization techniques and PCR alike. The A–T base pairing involves two and the

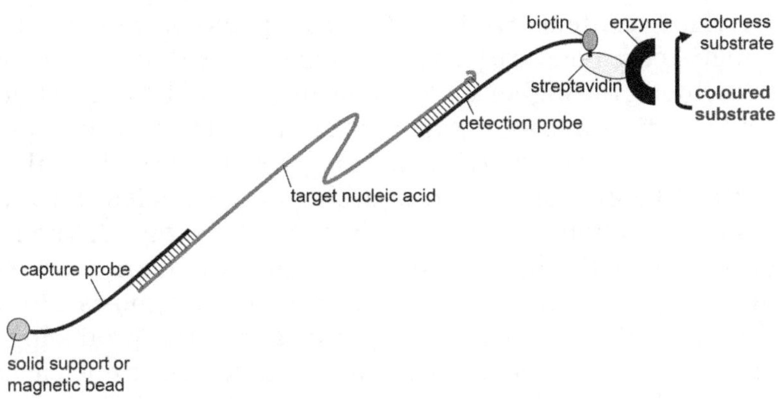

Figure 11.4 The principle of sandwich hybridization.

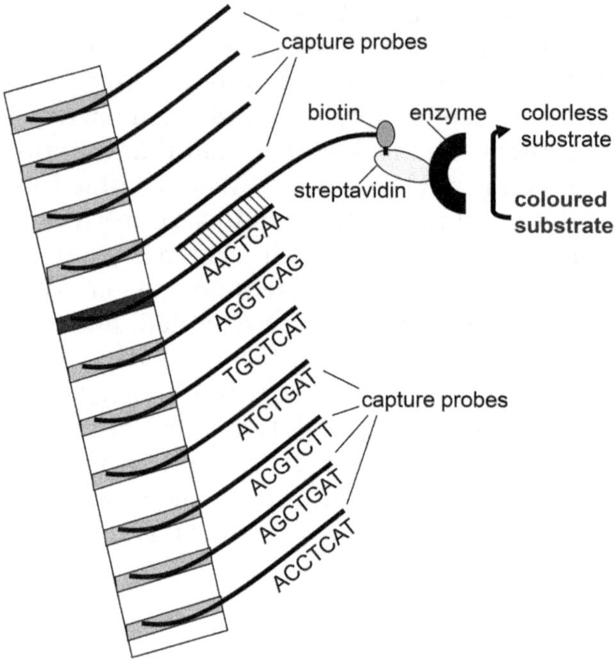

Figure 11.5 The principle of the line probe assay. Different capture probe sequences are fixed in different bands in a membrane strip. The amplification product hybridizes with the capture probe of complementary sequence. Utilizing biotin-labelled PCR primers the position of the complementary capture probe can be visualized as a coloured band following addition of streptavidin-bound enzyme that will change a non-coloured substrate to a coloured product that precipitates in the band.

G–C base pairing involves three hydrogen bonds, respectively. Therefore the relative content of A–T and G–C base pairs contributes strongly to the melting temperature (T_m) of the nucleic acid duplex. The higher the G–C content, the higher the T_m of the nucleic acid duplex. Different types of hybridizations were among the first techniques attempted in laboratory detection of nucleic acids. In general it turned out that the sensitivity of hybridization techniques was not sufficient for routine detection of infectious agents directly in patient samples. The invention of nucleic acid amplification techniques therefore represented a breakthrough in nucleic-acid-based laboratory diagnosis. However, in some cases the sensitivity of hybridization was promising and encouraged further development of methods to detect nucleic acids, in particular for bacteria containing multicopy plasmids or repeated genomic sequences.[21–23]

11.4.1.1. Sandwich Hybridization

In the classical hybridization techniques the nucleic acid target was usually fixed on a solid phase (nitrocellulose, nylon membrane, glass slide) and detected by a labelled probe added in the hybridization buffer to the solid phase. These techniques, such as dot blots, colony blots, Southern blots, northern blots and *in situ* hybridization usually either lacked sufficient sensitivity due to background hybridization or were too laborious for the routine setting. The sandwich hybridization (Figure 11.4) technique turned out, however, to have sufficiently low background and strong enough signal to be employed for some routine laboratory purposes. In this technique the capture probe is fixed to the solid phase and available to the target in the hybridization buffer. Following annealing between the capture probe and the specific target, unbound material is removed by washing and a new hybridization buffer containing a detection probe is added. The detection probe is typically labelled with either biotin or an enzyme and is designed to bind to a different region of the target nucleic acid than the capture probe. The detection probe will therefore stick to the capture probe and the solid phase only if it is bridged by the specific target. Following washing, retained detection probe can be visualised by addition of a colourless substrate that will change to a coloured product by the enzyme conjugated to the detection probe or by a streptavidin–enzyme complex in the case of a biotin-labelled detection probe. The solid phase may be a magnetic particle or the bottom of a 96-well plate to facilitate automation. The sandwich hybridization technique is used in commercial tests for the detection of human papillomaviruses (HPV) in samples from the female cervix (Digene test), and in the visualization step of conventional PCR assays (*e.g.* several Roche assays). The principle is also utilized in line probe assays.

11.4.1.2. Line Probe Assays

In the line probe assays (Figure 11.5) capture probes are printed in parallel bands (lines) on a membrane. Line probe assays have been developed to distinguish strains or genotypes of infectious agents. Each band or lane contains capture probes that differ only in nucleotides that are characteristic for each strain or genotype. The target nucleic acid is typically first amplified by PCR and next added to the line probe membrane in a hybridization buffer. Following washing, a labelled detection probe is added and visualized as described for the sandwich-hybridization above. The line or band corresponding to one specific genotype will then become coloured (Figure 11.5). Commercial line probe assays have been used for the detection of genotypes of HPV and hepatitis C-virus (HCV), strains of atypical mycobacteria and for detection of HTLV-1 (human T-cell lymphotropic virus).

11.4.1.3. *Peptide Nucleic Acid Fluorescent* In Situ *Hybridization (PNA-FISH)*

PNA-FISH is one type of *in situ* hybridization, utilizing hybridization probes in which the sugar–phosphate backbone of DNA is replaced by a polyamide backbone (Figure 11.6).[24,25] Peptide nucleic acids exhibit rapid hybridization kinetics and, when coupled to fluorescent reporter molecules, offer a rapid and sensitive verification of the infectious agent in positive blood cultures.[25] Once a blood culture turns positive, staining of cells on a slide is performed, and, based on the results, the appropriate PNA-FISH test is selected. Identification results are available within just a few hours and can be reported to the attending physician, *e.g.* in the case of serious yeast infections.[26]

11.4.2. TaqMan Low-Density Arrays (TLDA)

The TaqMan Low Density Array is a 384-well microfluidic card manufactured by Applied Biosystems where up to 384 simultaneous real-time PCR reactions can be carried out without the need for

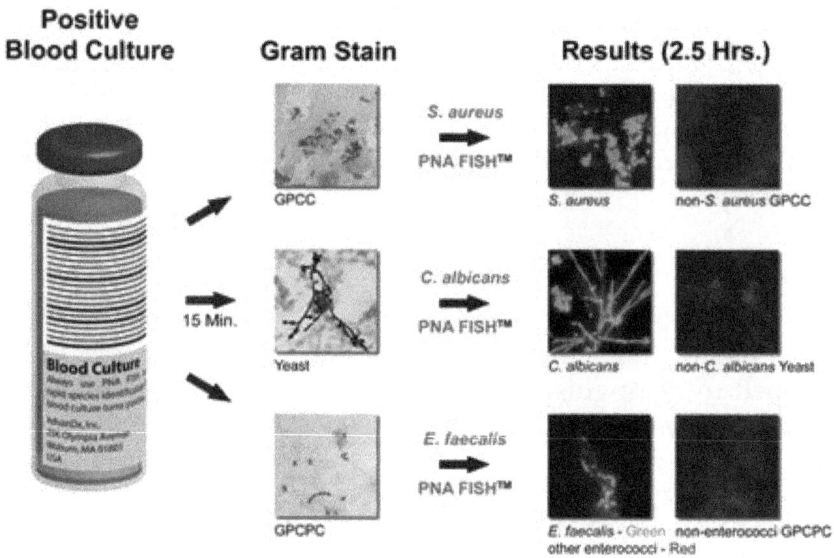

Figure 11.6 Schematics of PNA-FISH.

liquid-handling robots or multi-channel pipettes to load samples. Each card allows one to eight samples to be run in parallel with 12 to 384 different TaqMan Gene Expression Assays that are pre-loaded into separate 1–2 µl wells on the card. The TaqMan Low Density Array is customizable for human, mouse and rat genes and has been evaluated for the simultaneous detection of multiple infectious agents.[27] The TaqMan Array is designed for use on the Applied Biosystems 7900HT Fast Real-Time PCR system. The TLDA format increases by one order of magnitude the number of different real-time PCR assays that are feasible in a given time and is suitable for the validation of DNA microarray gene expression results (Figure 11.7).

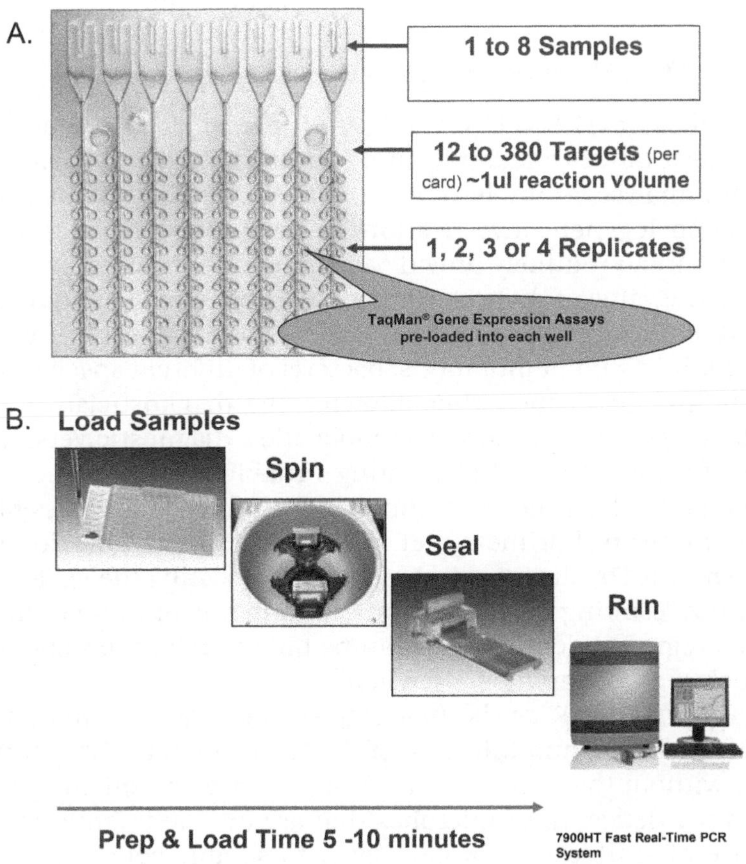

Figure 11.7 (A) The TaqMan Low Density Array (TLDA). (B) About 100 µl master mix and hexamer primed cDNA are loaded into each well and distributed into the reaction wells by low-speed centrifugation and the card is sealed and thermocycled for about 40 cycles.

11.4.3. Nucleic-acid-based Typing of Bacteria

Remarkable progress is currently being achieved in nucleic acid sequencing and associated bioinformatic formatting, storage, analysis and visualization of sequencing data. The progress is variably being referred to as massive parallel sequencing or next-generation sequencing (NGS) and is covered in Chapter 3. It is not possible, at the present time, to foresee exactly how NGS will change microbiological diagnostics, but its potential is formidable. Owing to this potential, several of the nucleic-acid-based techniques covered in the previous edition of this book, will only be listed here. These include random amplified polymorphic DNA (RAPD) and amplified fragment-length polymorphism (AFLP).[28,29] Instead, more emphasis will be put on recent progress in bacterial nucleic acid sequencing.

11.4.3.1. The 16S Ribosomal RNA Gene

The 16S ribosomal RNA (16S rRNA) gene is present in all bacteria, often existing as a multi-gene family and is a nature's gift to bacteriology. Sequence analysis of the bacterial 16S rRNA gene has revolutionized bacterial phylogenetic investigations as well as the clinical microbiologist's ability to routinely provide accurate identification of bacterial isolates. The value of the 16S rRNA gene is linked to its ubiquity, to its patterns of phylogenetic divergence, as well as to its versatility. Owing to accumulating point mutations, bacteria of different species express different 16 rRNA genes, thus allowing the diagnostician to differentiate between the species. The molecule's diagnostic versatility is coupled to the pattern of alternating variable and highly conserved regions along the gene. The conserved regions make possible the design of universal primers that will allow PCR-amplification of the 16S rRNA genes of almost any bacterium. By locating the PCR-primers so that the resulting PCR amplicons contain one or more of the variable 16S regions, identification can be obtained by subsequent DNA sequencing, as exemplified elsewhere.[30]

The universal PCR as the first step of the sequence analysis also allows for selective amplification of the 16S rRNA gene directly from a sample without the need for cultivating the bacteria on an agar. This enables the detection of microbes that are unable to grow as a consequence of antibiotic treatment prior to sample collection, for the detection of atypical bacteria with special growth requirements and for bacteria that are vulnerable to conditions during sample collection procedures and transportation, *e.g.* anaerobic bacteria. Furthermore, the method has allowed the detection of previously unknown causes of disease, including the up to then uncultured causative organisms for cat-scratch disease (*Bartonella henselae*)[31] and Whipple's disease (*Tropheryma whippelii*).[32]

When mixed bacterial populations are being analysed, sequencing based on the Sanger method has been found to be insufficient for identifying individual species. This is partly owing to the generation of mixed sequences and partly due to the phenomenon that signals from DNA belonging to the dominant species completely mask signals from DNA belonging to lower abundance species at ratios higher than 1 : 10.[33] Although these problems could be reduced to some extent by algorithm-based deconvolution of mixed chromatograms[34] or cloning the PCR-product and re-sequencing individual clones,[35] the limitations have seriously hampered effective culture-independent investigations of complex microbial communities like those found in the human colon.

Some recently introduced sequencing techniques have resolved these limitations of Sanger sequencing. In massively parallel sequencing (also known as next-generation sequencing (NGS)) the DNA from a reaction tube is distributed on a chip and then each single DNA is re-amplified and sequenced individually in millions of picoliter reactions. Since each of the picoliter sequencing reactions originates from a single piece of DNA and is being read independently from the neighbouring reactions, the restrictions of the Sanger method are eliminated.

When massive parallel sequencing is used for 16S rRNA analysis of a sample, the gene will be amplified by PCR prior to sequencing like in Sanger sequencing. This pre-amplification will provide the best sequencing coverage since all the sequencing wells will be used for the gene of interest (deep sequencing) assuring a high sensitivity also for the minor components of complex microbial communities like the human gut microbiome.

In a shotgun approach all DNA in a sample is fragmented and sequenced directly without selective amplification of any specific genes. Although this can allow for culture-independent analysis of larger parts of the bacterial genome and even culture-independent whole-genome sequencing in some instances, it is not optimal for culture-independent microbial identification from clinical infectious disease samples or description of complex microbial communities, since a large number of sequencing wells will be "wasted" on non-microbial DNA and bacterial genes not informative for species identification. In addition to compromising sequencing coverage it also significantly increases the complexity of the subsequent data-analysis.

11.4.3.2. Restriction Endonuclease Analysis – "DNA Fingerprinting"

Restriction enzymes are a group of enzymes that each can recognize one specific short double-stranded DNA sequence of four to nine base pairs and cleave the DNA at this site. The enzymes work by cleaving the bonds in the phosphate backbone of the DNA molecule. Depending upon the exact DNA sequence of the bacterial genome different mixtures of restriction fragments will result following digestion with one or more defined restriction enzymes. The restriction fragments can be separated using agarose gel electrophoresis with a fluorescent DNA-intercalating compound followed by visualization of the separated restriction fragments (DNA bands) under UV light. A 'DNA fingerprint' can thus be obtained for each species of bacterium. The method requires relatively high amounts of purified DNA and the resolution is not high.

11.4.3.3. Pulse Field Gel Electrophoresis (PFGE)

PFGE uses rare-cutting restriction enzymes to fragment the entire bacterial genome into large DNA fragments, which are next embedded in agarose plugs and placed in wells of the electrophoresis gel where they become part of the gel. The large DNA fragments are separated in the agarose gel by alternate pulses of perpendicularly oriented electrical fields. Following electrophoresis combined with a DNA-intercalating fluorescent compound in the gel, characteristic DNA fragment patterns can be visualized under UV-light (Figure 11.8). The principle of PFGE is that large DNA fragments require more time to reverse direction in an electric field than do small DNA fragments and thus are more retarded in the agarose gel. By alternating the direction of the electric current during gel electrophoresis, it is possible to resolve DNA fragments of 100–1000 kilobases. PFGE has been used extensively for typing of bacterial species, and is often considered to be the "gold standard" of genomic typing methods. DNA degradation may represent a problem.

11.4.3.4. Multilocus Sequence Typing (MLST)

The MLST technique is well described elsewhere.[36] There are three elements to the design of a new MLST system: the choice of the isolates to be used in the initial evaluation; the choice of the genetic loci to be characterized; and the design of primers for gene amplification and nucleotide sequence determination. It is advisable to

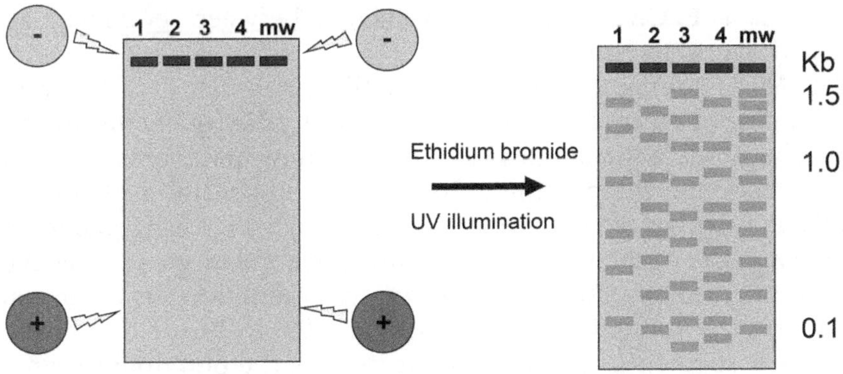

Figure 11.8 Schematic of pulse field gel electrophoresis. Restriction-enzyme-digested DNA is loaded into wells and separated in the agarose gel by electrical fields that alternate between the poles every 90 seconds for 24 hours. UV illumination visualizes DNA with an intercalating fluorescent compound. Different samples were loaded in wells 1–4 with a fragment size marker in the rightmost lane.

assemble a diverse isolate collection on the basis of existing typing information or epidemiological data. This should comprise around 100 isolates (95 is a good number for high-throughput sequencing in 96-well microtitre plates) to ensure that the primers developed will be applicable to as many isolates as possible and to establish the levels of diversity present at each of the loci to be examined. Furthermore, the collection will ideally be representative of the bacterial population, rather than comprising a subset, such as human disease isolates. Housekeeping genes, flanked by genes of similar function, are good targets for MLST and the availability of complete genome sequences has greatly facilitated the identification of candidate loci. Experience with several bacterial species has indicated that PCR fragments of housekeeping genes around 450 base pairs in length are suitable for MLST and a nested strategy is highly recommended for sequencing.[36] The great advantage of MLST is that sequence data are unambiguous and the allelic profiles of isolates can easily be compared with those in a large central database *via* the Internet (in contrast to most typing procedures which involve comparing DNA fragment sizes on gels). Allelic profiles can also be obtained from clinical material by PCR amplification of housekeeping loci directly from patient samples. Thus isolates can be precisely characterised even when they cannot be cultured from clinical material. MLST can effectively distinguish strains that possess high degrees of homology within the compared gene sequences. This technique is not as laborious as PFGE and it provides balance between sequence-based resolution and technical feasibility.

The MLST-principle is highly compatible with new next-generation sequencing platforms that offer the possibility of sequencing an entire bacterial genome in a single run. Whole-genome databases for species of interest open the possibility to explore and continuously improve MLST as a typing tool for a wide number of species, *e.g.* by changing or increasing the number a targets included (extended MLST). Given the development of user-friendly software that automatically detects and compares the defined genome regions and the inclusion of more variable gene segments, MLST-based typing has the potential to replace all other typing-methods for bacteria in the future. These whole-genome databases can also be used to investigate antibiotic susceptibility, vaccine coverage or pathogenicity.

11.5. Selected Examples of Clinical Nucleic-acid-based Diagnosis

11.5.1. Respiratory Infections

Rhinoviruses and coronaviruses are among the most prevalent upper respiratory tract pathogens and exist in too many serotypes to make serological diagnosis possible. The common cold therefore remains a clinical diagnosis.

Lower respiratory tract infections cause a high degree of morbidity and mortality. Traditional culture often fails, is too time consuming or not even possible for many bacterial agents. Serology has limited sensitivity and does not provide information in the acute phase of infection. Real-time PCR and other nucleic amplification techniques targeting core pathogens such as *Chlamydophila pneumoniae*, *Mycoplasma pneumoniae*, *Bordetella pertussis*, *Legionella pneumophila*[37–41] or *Mycobacterium tuberculosis* in respiratory samples have represented a diagnostic revolution for these important infections. In children hospitalized with respiratory difficulties due to bronchiolitis, reverse PCR-tests for human metapneumovirus and respiratory syncytial virus (RSV) have proven to be more sensitive than previous tests based on immunofluorescence or immunochromatography.[42–44] For the diagnosis of influenza A and B, real-time PCR assays have completely replaced viral isolation as the reference standard in the acute phase of the disease.[18,19,45] Real-time PCR assays are available for the specific analyses of avian influenza H5N1,[46,47] the swine flu H1N1 and the SARS virus.[48,49]

A recent development is 'PCR-in-a-bag'-assays: multiplex PCR tests that integrate sample preparation, amplification, detection and analysis all into one system with a turnaround time down to one hour and designed to be performed in the emergency room or on the ward.[50] These easy-to-perform assays have demonstrated sensitivity similar to that obtainable in the lab.[50] The FilmArray (Biofire) respiratory panel offers identification of more than 20 respiratory agents within an hour,[51–53] and GeneXpert (Cepheid) offers *e.g.* simultaneous detection of influenza A, B and the 2009 pandemic H1N1 strain.[54] The cost-effectiveness of these assays still remains a controversy, but they do offer guidance for early treatment as well as isolation of patients. The WHO and Cepheid have initiated a programme making the GeneXpert Mtb/RIF assay available at a lower price for a defined group of low-resource countries. This assay offers the simultaneous detection of *M. tuberculosis* and mutations coding for rifampicin resistance.[55]

11.5.2. Central Nervous System (CNS) Disease

Viruses are the most common causes of encephalitis (inflammation of the brain) in humans and more than 100 encephalitogenic viruses are known[56,57] followed by bacterial, parasitic and fungal agents.[58] The aetiology remains unknown in more than half of all cases.[58] Encephalitis is an uncommon disease, but because early start of specific therapy may prevent serious and lasting damage to the brain, early and correct identification of the infectious agent may represent a dramatic difference to each patient. If for example an enterovirus is present in the CSF, the prognosis is usually good without specific treatment. In contrast, when a member of the herpes virus family infects the brain, early antiviral (acyclovir or ganciclovir) treatment is urgent to avoid lasting damage to the brain or even death. In particular herpes simplex virus type 1 (HSV-1), but also HSV-2 and varicella zoster virus (VZV), account for a large proportion of cases of infectious encephalitis, and may occur in young and otherwise healthy people. As such, amplification of nucleic acids from cerebrospinal fluid (CSF) obtained by lumbar puncture represents a diagnostic breakthrough.

In patients with bacterial meningitis universal amplification of the 16S rRNA gene has been demonstrated to increase diagnostic sensitivity as compared with culture alone, in particular for patients that have received antibiotics prior to lumbar puncture. Since the main proportion of bacterial meningitis is caused by seven to eight bacterial species, multiplex PCR would represent a more sensitive and more rapid diagnostic alternative. There are several new commercial platforms that should have the potential to deliver robust assays at this level of complexity, maybe even including the most common viral agents (*e.g.* FilmArray, Luminex, GeneXpert).[59]

11.5.3. Hepatitis

Serological tests detecting specific antibodies and antigens using enzyme-linked immunosorbent assay (ELISA) and micro-EIA are adequate for the primary diagnosis of hepatitis A, B, C and E viruses (HAV, HBV, HCV and HEV). The detection of HCV RNA is a very useful complement to the standard screening and confirmation test (recombinant immunoblot assay (RIBA)) to establish the diagnosis of hepatitis C. In particular the reverse PCR is useful for the testing whether a newborn child is infected. This diagnosis is difficult to ascertain by serology owing to maternal–foetal transfer of maternal anti-HCV antibodies. Detection of HCV RNA in donor blood or in pools of donor blood may be useful. But the most important role of real-time PCR and other nucleic-acid-based tests for hepatitis C and B viruses relates to therapy. Commercial HCV reverse PCR tests come in both qualitative and quantitative versions. Before PEG-interferon α–ribavirin therapy is attempted against hepatitis C, an anti-HCV-positive patient should in addition test positive in the HCV RNA qualitative test. The next step is then to perform a quantitative reverse PCR-test to establish the HCV RNA IU ml^{-1} of serum. The amplification product is employed in a line probe assay or sequenced in order to determine the genotype (see above). Different genotypes require different kinds of treatment, and the quantitative reverse PCR test is used to monitor the response to treatment and to decide whether or not therapy is successful. When therapy is completed, the HCV PCR test is used to make sure that HCV does not relapse in the next six months and that a sustained viral response has thereby occurred.

Antiviral treatment is also attempted in hepatitis B in order to prevent accelerating damage to the liver, and quantitative PCR to determine viral load in serum prior and during treatment is important. Only occasionally is HBV eradicated as a result of treatment. Instead, drug-resistant mutants commonly develop and are associated with specific point mutations that can be determined by DNA sequencing.

11.5.4. Gastroenteritis

Acute gastroenteritis causes almost three million deaths each year worldwide, mostly among children below 5 years of age in developing countries. The known causes of viral gastroenteritis include the RNA-viruses rotavirus, astrovirus, sapovirus and norovirus and the DNA adenoviruses. Enteroviruses do not usually, despite their name, cause gastroenteritis, but replicate in the intestines and can be isolated from faeces in patients with meningitis, myocarditis and symptoms from many other organs. Electron microscopy has been valuable for the discovery and diagnosis of viruses in diarrhoea and the names of rotavirus ("wheel") and astrovirus ("star") refer to the electron microscopy images. Agglutination tests provide rapid diagnosis of rotavirus and adenovirus gastroenteritis. Reverse PCR tests are increasingly coming into use for the rapid diagnosis of viral gastroenteritis. The incubation time of norovirus gastroenteritis may be less than 1 day, and noroviruses commonly cause outbreaks of gastroenteritis in hospitals and in other settings where groups of people live closely together. By means of reverse PCR and DNA sequencing the hundreds of strains of noroviruses (also known as (caliciviruses/Norwalk agents)) have been identified and classified into genogroups and genetic clusters.[60] Reverse PCR tests using nucleic acids extracted from diarrheal faecal specimens have come into broad use to diagnose norovirus infection.[61,62] The genetic variability of viruses causing gastroenteritis represents a great challenge to molecular diagnosis, and new protocols take advantage of new chemistry and computer analysis to design multiplex PCR protocols directed at different strains of each virus and to cover several different viruses in one reaction.[63–65]

11.5.5. Sexually Transmitted Diseases

11.5.5.1. Human Papillomaviruses (HPVs)

HPVs are among the most commonly sexually transmitted infectious agents. Molecular cloning and DNA sequencing have been crucial for the identification and classification of the more than 130 HPV types presently discovered. The high-risk types, including the most prevalent types 16 and 18, are the cause of cervical cancer. The low-risk types cause benign papillomas and warts and include HPV type 6 and 11 that cause genital warts or condylomas. The cloning and recombinant expression of HPV L1 proteins in yeast and insect cells have provided successful vaccines against HPV 16 and 18, and one vaccine additionally protects against HPV 6 and 11. The cytological smear remains the most important screening test for pre-stages of cervical cancer and relies on microscopic evaluation of abnormalities in cells taken from the surface of the cervix. Microscopy of cytology samples is, however, time-consuming and requires much experience. Atypical cervical cells are usually caused by HPVs. Therefore, whether to use HPV nucleic acid detection in the primary screening of cells from the cervix, followed up by microscopy only of the samples with positive HPV nucleic acid detection, is currently being considered. Many commercial nucleic-acid-based tests are on the market for HPV screening and subtyping. Comparison between these tests is methodologically interesting since different assay principles are used (listed elsewhere[66]). Many issues in HPV nucleic acid testing remain open for the future, including any additional value of focusing the diagnosis on the HPV oncogenes E6 and E7, in particular to monitor their transcriptional activity, the role of next generation sequencing methods and the role of HPV detection in several different human cancer types.[67]

11.5.5.2. Chlamydia trachomatis *and Genital Mycoplasmas and Ureaplasmas*

Nucleic acid detection methods have revolutionized the diagnosis of genital *Chlamydia trachomatis*, one of the most ubiquitously sexually transmitted agents and associated with deep pelvic inflammation and female sterility. Efficient antibiotic treatment is available if the diagnosis is made early. Screening and diagnosis at an early stage is very important. At later stages with established deep pelvic infection the sensitivity of all nucleic acid detection tests become much lower than during early infection.[68] During early stages of infection self-sampling using vaginal swabs or first-pass urine samples provide adequate alternatives to sampling by health professionals.[69,70] Several commercial high-capacity *Chlamydia trachomatis* nucleic acid amplification tests, streamlined with both positive and negative controls are now available.[71,72]

There is now an increasing awareness that in addition to *Chlamydia trachomatis* several other organisms may be involved in non-gonococcal urethritis, including *Mycoplasma genitalium*, *Mycoplasma hominis*, *Ureaplasma urealyticum* and *Ureaplasma parvum*. In particular, *Mycoplasma genitalium* may be prevalent in some sexually active populations and with many of the same symptoms and complications as *Chlamydia trachomatis*, but, importantly, with different therapeutic requirements.[73] In-house PCR assays have been published by many diagnostic laboratories for the sensitive and specific diagnosis of these agents in urine or semen.[74–76]

11.5.6. HIV Infection and AIDS

The screening (ELISA) and confirmation (western blot) tests for HIV-1 and HIV-2 infections provide highly reliable laboratory diagnosis of this serious infection. Nucleic-acids-based amplification tests are therefore not presently important for the establishment of the diagnosis. As opposed to the frustrations encountered in HIV vaccine development, drug treatment of HIV-1 infection and AIDS has been a remarkable success. However, the virus still cannot be eradicated from the body, and life-long combination therapy using three different drugs is necessary to suppress disease progression. By now multiple drugs have been approved for the treatment of HIV infection and inhibit defined molecular stages of the HIV life cycle, including the reverse transcriptase, the protease, the integrase, the coreceptors (CCR5/CXCR4) and the membrane fusion.[77,78] $CD4^+$ cell counts and HIV viral load in plasma must be determined prior to starting and during therapy. When therapy is effective, HIV RNA should be below the detection limit of commercial standardized nucleic acid amplification platforms.[79] If the HIV RNA viral load increases during therapy, it is either a sign that the patient does not take the medication properly or that treatment-resistant HIV strains are evolving. Drug resistance is associated with HIV genome point mutations which are characteristic for each drug. DNA sequencing combined with algorithmic analysis of databases is therefore efficient in the choice of a different combination of drugs that may continue to suppress the HIV infection.

11.5.7. Bacterial Antibiotic Resistance and Virulence Factor Genes

Detection of bacterial genes that confer antibiotic resistance is important for the choice of proper treatment, to help decide which patients need to be isolated and to trace epidemics. At the same time it may be necessary to identify virulence (disease causing) genes. In the following sections only a few key examples have been selected.

11.5.7.1. Methicillin-resistant Staphylococcus aureus

Staphylococcus aureus has long been a notorious problem among hospitalized patients. Whenever bacteraemia (bacteria in the blood) is suspected, the typical procedure is to identify the type of bacteria using automated blood culture systems followed by microscopy of positive blood cultures. When the microscopy shows Gram-positive cocci in clusters it is a main goal to differentiate between *S. aureus* and coagulase-negative staphylococci and to verify methicillin-resistant *S. aureus* (MRSA).[80] MRSA strains are resistant against all beta-lactam antibiotics. A number of different nucleic-acid-based techniques have been used to identify the *mecA* gene (the methicillin resistance gene) from putative MRSA colonies and at the same time to identify *S. aureus* based upon specific genes, such as the *nuc* or *orfX* genes.[25,81,82] Since this verification may take more than 2 days, it is common to start treatment of patients using glycopeptide antibiotics until the results of the *mecA* test are available, and this represents a selection pressure for the increasing vancomycin resistance. Glycopeptide-resistant MRSA strains are an emerging problem.[83,84] One important prophylactic measure is to screen and identify asymptomatic carriers of the *mecA* gene among hospital workers and patients. Colonies grown from nasal swab samples are then analysed for *mecA* and *S. aureus*-specific genes using PCR, real-time PCR or several other nucleic-acid-based techniques.[25,81,82] MRSA-detection was one of the first targets for commercial 'PCR-in-a-bag' developments. Today sensitive screening of patients for MRSA can be performed in the emergency room and provide results in about 1 hour.[85,86]

More recently, serious MRSA infections have appeared outside hospitals; these so-called community-associated MRSA strains harbour a set of phage-carried PVL genes in contrast to the hospital MRSA strains. It is unclear why certain MRSA strains predominate in hospitals and others in the community, but nucleic-acid-based detection of PVL genes represents a very important tool for epidemiology and surveillance.[87–89] It has also been shown that different MRSA strains differ in their virulence-gene profiles and that this contributes to the complexity of defining the disease-causing abilities of *S. aureus*.[87]

11.5.7.2. Antibiotic Resistance and Virulence in Gram-negative Intestinal Bacteria

Extended-spectrum β-lactamase (ESBL)-producing Gram-negative bacteria (such as *Escherichia coli*) are resistant to both penicillins and cephalosporins. The responsible bacterial β-lactamases, *e.g.* TEM, SHV and CTX-M enzymes can be typed by molecular techniques in order to track epidemics and monitor outbreaks.[90] PFGE is the current 'gold standard' for molecular typing, but multilocus sequence typing (MLST) may offer advantages.[91] Real-time PCR combined with deep sequencing may increase speed and capacity for the identification of the antibiotic resistance genes.[92,93] *Klebsiella pneumoniae* and other bacteria containing carbapenemases (KPC) are a further extension of the antibiotic resistance problem.[94]

Although *E. coli* is prevalent in the normal intestinal flora, some strains are highly pathogenic and cause diarrhoea or systemic disease due to acquisition of virulence genes encoded by plasmid, chromosome or bacteriophage DNA. A large number of virulence factor genes separate pathogenic *E. coli* strains into enterotoxigenic (ETEC), enteroinvasive (EIEC), enteropathogenic (EPEC), enterohemorrhagic (EHEC) and enteroaggregative (EAEC).[95] The development of multiplex PCR and other molecular assays has been very useful for the survey of such virulence factors.[96–99]

11.6. Conclusions and Future Development

Nucleic-acid-based methods have revolutionized routine laboratory diagnosis of infectious agents. In particular, the DNA amplification techniques have contributed to this by providing sufficient sensitivity for the direct examination of samples from infected patients. As a result, clinicians can rely much more on microbiology laboratories in the acute differential diagnosis of each patient. Nucleic-acid-based techniques have offered much quicker and more reliable diagnosis of the individual patient and improved the tracing and monitoring of epidemic outbreaks and the discovery of emerging infectious agents. Still, there are many problems, such as contamination during nucleic acid amplification and the continuous biological selection and evolution in the microbiological world that without warning may cause a previously well-designed test to fail. The potential for improvement of nucleic-acid-based laboratory diagnostics is therefore vast. There is a need for back up-tests directed to different regions of the target nucleic acids. We presently see only the beginning of automation and reduction of reaction volumes. Costs already now severely restrict the repertoire of new tests in the clinical microbiological laboratory – even in developed countries. The future roles of nucleic acid amplification, next-generation sequencing and novel technologies incorporating nanotechnology, microfluidics or biosensors remain undetermined. Great challenges still exist in microbiology laboratories regarding logistics and quality control of diagnostic procedures. It is therefore certain that nucleic-acid-based techniques will continue to revolutionize the way we diagnose and treat infections.

References

1. M. Oldstone, *Viruses, Plagues, and History*, Oxford University Press, Oxford New York, 2010.
2. M. Karamanou, G. Panayiotakopoulos, G. Tsoucalas, A. A. Kousoulis and G. Androutsos, *Infez Med*, 2012, **20**, 58–62.
3. M. J. Espy, J. R. Uhl, L. M. Sloan, S. P. Buckwalter, M. F. Jones, E. A. Vetter, J. D. Yao, N. L. Wengenack, J. E. Rosenblatt, F. R. Cockerill III and T. F. Smith, *Clin. Microbiol. Rev.*, 2006, **19**, 165–256.
4. K. Rantakokko-Jalava and J. Jalava, *J. Clin. Microbiol.*, 2002, **40**, 4211–4217.
5. T. Schuurman, R. F. de Boer, A. M. Kooistra-Smid and A. A. van Zwet, *J. Clin. Microbiol.*, 2004, **42**, 734–740.
6. K. B. Barken, J. A. Haagensen and T. Tolker-Nielsen, *Clin. Chim. Acta*, 2007, **384**, 1–11.
7. M. A. Espy, H. Sandin, C. Carr, C. J. Hanson, M. D. Ward and R. H. Kraus Jr, *Cytometry, Part A*, 2006, **69**, 1132–1142.
8. T. Schuurman, A. van Breda, R. de Boer, M. Kooistra-Smid, M. Beld, P. Savelkoul and R. Boom, *J. Clin. Microbiol.*, 2005, **43**, 4616–4622.
9. B. Kaltenboeck and C. Wang, *Adv. Clin. Chem.*, 2005, **40**, 219–259.
10. P. J. Asiello and A. J. Baeumner, *Lab Chip*, 2011, **11**, 1420–1430.
11. J. Ikewaki, E. Ohtsuka, R. Kawano, M. Ogata, H. Kikuchi and M. Nasu, *J. Clin. Microbiol.*, 2003, **41**, 4382–4387.
12. J. L. Vaerman, P. Saussoy and I. Ingargiola, *J. Biol. Regul. Homeostatic Agents*, 2004, **18**, 212–214.
13. R. G. Rutledge and C. Cote, *Nucleic Acids Res.*, 2003, **31**, e93.
14. J. C. Servoss and L. S. Friedman, *Clin. Liver Dis.*, 2004, **8**, 267–281.
15. J. Saldanha, N. Lelie and A. Heath, *Vox Sang.*, 1999, **76**, 149–158.
16. J. Saldanha, N. Lelie, M. W. Yu and A. Heath, *Vox Sang.*, 2002, **82**, 24–31.
17. J. Burrows, A. Nitsche, B. Bayly, E. Walker, G. Higgins and T. Kok, *BMC Microbiol.*, 2002, **2**, 12.
18. R. M. Ratcliff, G. Chang, T. Kok and T. P. Sloots, *Curr. Issues Mol. Biol.*, 2007, **9**, 87–102.
19. B. Stone, J. Burrows, S. Schepetiuk, G. Higgins, A. Hampson, R. Shaw and T. Kok, *J. Virol. Methods*, 2004, **117**, 103–112.
20. S. Leekha, C. L. Irish, S. K. Schneider, E. C. Fernholz, M. J. Espy, S. A. Cunningham, R. Patel, Y. J. Juhn, B. Pritt, T. F. Smith and P. Sampathkumar, *Diagn. Microbiol. Infect. Dis.*, 2013, **75**, 169–173.
21. K. H. Kalland, L. S. Havarstein and H. Sommerfelt, *Tidsskr. Nor. Laegeforen.*, 1987, **107**, 2510–2512, 2543.
22. K. H. Kalland and G. Haukenes, *Tidsskr. Nor. Laegeforen.*, 1999, **119**, 802–809.
23. K. H. Kalland, H. Myrmel and S. A. Nordbo, *Tidsskr. Nor. Laegeforen.*, 2005, **125**, 3110–3114.
24. M. Sogaard, H. Stender and H. C. Schonheyder, *J. Clin. Microbiol.*, 2005, **43**, 1947–1949.
25. F. C. Tenover, *Clin. Infect. Dis.*, 2007, **44**, 418–423.
26. N. R. Stone, R. L. Gorton, K. Barker, P. Ramnarain and C. C. Kibbler, *J. Clin. Microbiol.*, 2013, **51**, 1301–1302.

27. M. Kodani, G. Yang, L. M. Conklin, T. C. Travis, C. G. Whitney, L. J. Anderson, S. J. Schrag, T. H. Taylor Jr, B. W. Beall, R. F. Breiman, D. R. Feikin, M. K. Njenga, L. W. Mayer, M. S. Oberste, M. L. Tondella, J. M. Winchell, S. L. Lindstrom, D. D. Erdman and B. S. Fields, *J. Clin. Microbiol.*, 2011, **49**, 2175–2182.
28. S. Bensch and M. Akesson, *Mol. Ecol.*, 2005, **14**, 2899–2914.
29. F. Saghrouni, J. Ben Abdeljelil, J. Boukadida and M. Ben Said, *J. Appl. Microbiol.*, 2013, **114**, 1559–1574.
30. U. Vogel, R. Szczepanowski, H. Claus, S. Junemann, K. Prior and D. Harmsen, *J. Clin. Microbiol.*, 2012, **50**, 1889–1894.
31. R. A. Tapp, A. F. Roy, R. E. Corstvet and V. L. Wilson, *J. Vet. Diagn. Invest.*, 2001, **13**, 219–229.
32. D. A. Relman, T. M. Schmidt, R. P. MacDermott and S. Falkow, *N. Engl. J. Med.*, 1992, **327**, 293–301.
33. O. Kommedal, B. Karlsen and O. Saebo, *J. Clin. Microbiol.*, 2008, **46**, 3766–3771.
34. O. Kommedal, K. Kvello, R. Skjastad, N. Langeland and H. G. Wiker, *J. Clin. Microbiol.*, 2009, **47**, 3562–3568.
35. M. Al Masalma, F. Armougom, W. M. Scheld, H. Dufour, P. H. Roche, M. Drancourt and D. Raoult, *Clin. Infect. Dis.*, 2009, **48**, 1169–1178.
36. R. Urwin and M. C. Maiden, *Trends Microbiol.*, 2003, **11**, 479–487.
37. S. Kumar and M. R. Hammerschlag, *Clin. Infect. Dis.*, 2007, **44**, 568–576.
38. M. Khanna, J. Fan, K. Pehler-Harrington, C. Waters, P. Douglass, J. Stallock, S. Kehl and K. J. Henrickson, *J. Clin. Microbiol.*, 2005, **43**, 565–571.
39. D. R. Murdoch, *Clin. Infect. Dis.*, 2005, **41**, 1445–1447.
40. Y. R. Chan and A. Morris, *Curr. Opin. Infect. Dis.*, 2007, **20**, 157–164.
41. K. A. Beynon, S. A. Young, R. T. Laing, T. G. Harrison, T. P. Anderson and D. R. Murdoch, *Emerging Infect. Dis.*, 2005, **11**, 639–641.
42. I. Lee and T. D. Barton, *Drugs*, 2007, **67**, 1411–1427.
43. L. C. Jennings, T. P. Anderson, A. M. Werno, K. A. Beynon and D. R. Murdoch, *Pediatr. Infect. Dis. J.*, 2004, **23**, 1003–1007.
44. C. Deffrasnes, M. E. Hamelin and G. Boivin, *Semin. Respir. Crit. Care Med.*, 2007, **28**, 213–221.
45. M. Petric, L. Comanor and C. A. Petti, *J. Infect. Dis.*, 2006, **194**(Suppl 2), S98–S110.
46. M. D. Curran, J. S. Ellis, T. G. Wreghitt and M. C. Zambon, *J. Med. Microbiol.*, 2007, **56**, 1263–1267.
47. D. L. Suarez, A. Das and E. Ellis, *Avian Dis.*, 2007, **51**, 201–208.
48. J. K. Louie, J. K. Hacker, J. Mark, S. S. Gavali, S. Yagi, A. Espinosa, D. P. Schnurr, C. K. Cossen, E. R. Isaacson, C. A. Glaser, M. Fischer, A. L. Reingold and D. J. Vugia, *Emerging Infect. Dis.*, 2004, **10**, 1143–1146.
49. C. Drosten, W. Preiser, S. Gunther, H. Schmitz and H. W. Doerr, *Trends Mol. Med.*, 2003, **9**, 325–327.
50. A. Endimiani, K. M. Hujer, A. M. Hujer, S. Kurz, M. R. Jacobs, D. S. Perlin and R. A. Bonomo, *Clin. Infect. Dis.*, 2011, **52**(Suppl 4), S373–S383.
51. M. R. Couturier, T. Barney, G. Alger, W. C. Hymas, J. B. Stevenson, D. Hillyard and J. A. Daly, *J. Clin. Lab. Anal.*, 2013, **27**, 148–154.
52. S. P. Hammond, L. S. Gagne, S. R. Stock, F. M. Marty, R. S. Gelman, W. A. Marasco, M. A. Poritz and L. R. Baden, *J. Clin. Microbiol.*, 2012, **50**, 3216–3221.
53. M. Xu, X. Qin, M. L. Astion, J. C. Rutledge, J. Simpson, K. R. Jerome, J. A. Englund, D. M. Zerr, R. T. Migita, S. Rich, J. C. Childs, A. Cent and M. A. Del Beccaro, *Am. J. Clin. Pathol.*, 2013, **139**, 118–123.
54. S. Miller, M. Moayeri, C. Wright, L. Castro and M. Pandori, *J. Clin. Microbiol.*, 2010, **48**, 4684–4685.
55. G. Meyer-Rath, K. Schnippel, L. Long, W. MacLeod, I. Sanne, W. Stevens, S. Pillay, Y. Pillay and S. Rosen, *PLoS One*, 2012, **7**, e36966.
56. R. J. Whitley and J. W. Gnann, *Lancet*, 2002, **359**, 507–513.

57. R. L. Debiasi and K. L. Tyler, *Clin. Microbiol. Rev.*, 2004, **17**, 903–925, table of contents.
58. C. A. Glaser, S. Honarmand, L. J. Anderson, D. P. Schnurr, B. Forghani, C. K. Cossen, F. L. Schuster, L. J. Christie and J. H. Tureen, *Clin. Infect. Dis.*, 2006, **43**, 1565–1577.
59. J. K. Moller, *Methods Mol. Biol.*, 2012, **799**, 37–53.
60. R. Goodgame, *Curr. Infect. Dis. Rep.*, 2007, **9**, 102–109.
61. M. Hoehne and E. Schreier, *BMC Infect. Dis.*, 2006, **6**, 69.
62. W. Hymas, A. Atkinson, J. Stevenson and D. Hillyard, *J. Virol. Methods*, 2007, **142**, 10–14.
63. E. Royuela, A. Negredo and A. Sanchez-Fauquier, *J. Virol. Methods*, 2006, **133**, 14–19.
64. A. Stals, E. Mathijs, L. Baert, N. Botteldoorn, S. Denayer, A. Mauroy, A. Scipioni, G. Daube, K. Dierick, L. Herman, E. Van Coillie, E. Thiry and M. Uyttendaele, *Food Environ. Virol.*, 2012, **4**, 153–167.
65. H. Yan, T. A. Nguyen, T. G. Phan, S. Okitsu, Y. Li and H. Ushijima, *Kansenshogaku Zasshi*, 2004, **78**, 699–709.
66. M. Torres, L. Fraile, J. Echevarria, B. Hernandez Novoa and M. Ortiz, *Open Virol. J.*, 2012, **6**, 144–150.
67. H. M. Wood, R. Bolt and K. D. Hunter, *Expert Rev. Mol. Diagn.*, 2012, **12**, 663–665.
68. C. Rodriguez-Cerdeira, E. Sanchez-Blanco, A. Molares-Vila and A. Alba, *ISRN Obstet. Gynecol.*, 2012, **2012**, 581725.
69. I. J. Bakken and S. A. Nordbo, *Tidsskr. Nor. Laegeforen.*, 2007, **127**, 3202–3205.
70. V. Levy, C. S. Blackmore and J. D. Klausner, *Methods Mol. Biol.*, 2012, **903**, 407–418.
71. A. Cheng, Q. Qian and J. E. Kirby, *J. Clin. Microbiol.*, 2011, **49**, 1294–1300.
72. A. L. Harkins and E. Munson, *ISRN Obstet. Gynecol.*, 2011, **2011**, 279149.
73. S. A. Weinstein and B. G. Stiles, *Sex Health*, 2011, **8**, 143–158.
74. S. R. Lee, J. M. Chung and Y. G. Kim, *J. Microbiol.*, 2007, **45**, 453–459.
75. M. Unemo, P. Olcen, I. Agne-Stadling, A. Feldt, M. Jurstrand, B. Herrmann, K. Persson, P. Nilsson, T. Ripa and H. Fredlund, *Eurosurveillance*, 2007, **12**, E5–E6.
76. S. Yokoi, S. Maeda, Y. Kubota, M. Tamaki, K. Mizutani, M. Yasuda, S. Ito, M. Nakano, H. Ehara and T. Deguchi, *Clin. Infect. Dis.*, 2007, **45**, 866–871.
77. E. De Clercq, *Nat. Rev. Drug Discov.*, 2007, **6**, 1001–1018.
78. L. Menendez-Arias, *Antiviral Res.*, 2013, **98**, 93–120.
79. S. Palmer, *Curr. Opin. HIV AIDS*, 2013, **8**, 87–92.
80. A. van der Zee, W. D. Hendriks, L. Roorda, J. M. Ossewaarde and J. Buitenwerf, *Am. J. Infect. Control*, 2013, **41**, 204–209.
81. K. Levi and K. J. Towner, *J. Clin. Microbiol.*, 2003, **41**, 3890–3892.
82. Y. Misawa, A. Yoshida, R. Saito, H. Yoshida, K. Okuzumi, N. Ito, M. Okada, K. Moriya and K. Koike, *J. Infect. Chemother.*, 2007, **13**, 134–140.
83. P. C. Appelbaum, *Int. J. Antimicrob. Agents*, 2007, **30**, 398–408.
84. G. M. Rossolini, E. Mantengoli, F. Montagnani and S. Pollini, *Curr. Opin. Microbiol.*, 2010, **13**, 582–588.
85. J. Davies, C. L. Gordon, S. Y. Tong, R. W. Baird and J. S. Davis, *J. Clin. Microbiol.*, 2012, **50**, 2056–2058.
86. D. H. Spencer, P. Sellenriek and C. A. Burnham, *Am. J. Clin. Pathol.*, 2011, **136**, 690–694.
87. B. A. Diep, H. A. Carleton, R. F. Chang, G. F. Sensabaugh and F. Perdreau-Remington, *J. Infect. Dis.*, 2006, **193**, 1495–1503.
88. J. A. McClure, J. M. Conly, V. Lau, S. Elsayed, T. Louie, W. Hutchins and K. Zhang, *J. Clin. Microbiol.*, 2006, **44**, 1141–1144.
89. R. R. McDonald, N. A. Antonishyn, T. Hansen, L. A. Snook, E. Nagle, M. R. Mulvey, P. N. Levett and G. B. Horsman, *J. Clin. Microbiol.*, 2005, **43**, 6147–6149.

90. S. N. Seiffert, M. Hilty, V. Perreten and A. Endimiani, *Drug Resist. Updates*, 2013, **16**, 22–45.
91. L. L. Nemoy, M. Kotetishvili, J. Tigno, A. Keefer-Norris, A. D. Harris, E. N. Perencevich, J. A. Johnson, D. Torpey, A. Sulakvelidze, J. G. Morris Jr and O. C. Stine, *J. Clin. Microbiol.*, 2005, **43**, 1776–1781.
92. C. I. Birkett, H. A. Ludlam, N. Woodford, D. F. Brown, N. M. Brown, M. T. Roberts, N. Milner and M. D. Curran, *J. Med. Microbiol.*, 2007, **56**, 52–55.
93. T. Naas, C. Oxacelay and P. Nordmann, *Antimicrob. Agents Chemother.*, 2007, **51**, 223–230.
94. P. Nordmann, M. Gniadkowski, C. G. Giske, L. Poirel, N. Woodford and V. Miriagou, *Clin. Microbiol. Infect.*, 2012, **18**, 432–438.
95. M. M. Levine, *J. Infect. Dis.*, 1987, **155**, 377–389.
96. L. T. Brandal, B. A. Lindstedt, L. Aas, T. L. Stavnes, J. Lassen and G. Kapperud, *J. Microbiol. Methods*, 2007, **68**, 331–341.
97. S. L. Foley, A. M. Lynne and R. Nayak, *Infect. Genet. Evol.*, 2009, **9**, 430–440.
98. V. K. Sharma, *Mol. Cell. Probes*, 2006, **20**, 298–306.
99. J. R. Yang, F. T. Wu, J. L. Tsai, J. J. Mu, L. F. Lin, K. L. Chen, S. H. Kuo, C. S. Chiang and H. S. Wu, *J. Clin. Microbiol.*, 2007, **45**, 3620–3625.

12 Parasite Genomes and Diagnostics

Spencer D. Polley

Hospital for Tropical Diseases, Capper Street, Mortimer Market, London, WC1E 6JB, UK
*E-mail: spencer.polley@lshtm.ac.uk

12.1. The Current Diagnostic Landscape for Parasitology

Parasitic diseases, such as malaria, remain an ongoing challenge for clinicians to diagnose by purely clinical symptoms. For many years parasitic infections have often been treated on the basis of clinical diagnosis alone. The significant overlap of many different infections renders this a potentially inaccurate methodology, and indeed the comparison of clinical *versus* biological diagnosis has shown the requirement for the latter in many endemic situations. The use of lateral flow immunochromatographic strips (also known as rapid diagnostic tests or RDTs) has proved that as many as 50% of clinical diagnoses of malaria were incorrect in controlled studies. This figure can be significantly higher dependent upon the timing of the study and the location. It is no surprise therefore that the World Health Organisation (WHO) recommends "prompt parasite-based diagnosis by microscopy or malaria rapid diagnostic test (RDT) in

Genomics and Clinical Diagnostics
Edited by David Whitehouse and Ralph Rapley

Published by the Royal Society of Chemistry, www.rsc.org

all patients suspected of malaria before antimalarial treatment is administered".

Clinical diagnostics has seen a major increase in the use of technologies employing genomic-based approaches for the identification of many pathogenic organisms. The technologies employed have including whole-genome sequencing to diagnose unknown pathogens, or to elucidate the genomic structure of these pathogens. Matrix-assisted laser desorption/ionization-time of flight (MADLI-TOF) has allowed the rapid identification of pathogens based upon their protein make up, or proteome. The single-nucleotide polymorphisms may also be identified for drug resistance analysis by targeted approaches or whole-genome sequencing. Many of these technologies require prior amplification of the pathogen through routine culture. The expansion of these technologies is driven by increased speed of diagnosis, improved specificity of diagnosis and the capacity to identify resistance genes so as to ensure efficacious treatment regimes. The greatest expansion of these technologies has undoubtedly occurred in those disciplines involved with the identification of prokaryotic and viral pathogens.

Parasitology as a discipline has lagged significantly behind the field of microbiology as a whole with regards to the incorporation of new diagnostic technology. To understand why, we need to look at the relative strengths and weaknesses of traditional diagnostic technologies they are supplementing and, in some situations, superseding.

12.2. Light Microscopy, the Old Favourite of the Parasitologist

One of the major tools used in the diagnosis of parasitic infections, the light microscope, has been around ever since Antonie Philips van Leeuwenhoek first described giardia like particles in his own samples of diarrhoea. The inability to culture many of the human parasitic pathogens required the identification of these pathogens by direct observation of the parasite, its ova, cysts or eggs through light microscopy. Thankfully, the large size of many human parasites enables light microscopy to be an excellent tool for this purpose, allowing the diagnosis of gastro-intestinal parasites through the use of stool samples, blood-borne parasites, such as malaria, through slide microscopy and even tissue borne parasites through bone marrow aspirates, punch biopsies or slit skin smears.

Possibly the greatest advantage of light microscopy is that the observation of a parasite is *de facto* evidence that it is present in a patient and as such highly likely to be the causative agent of the patients' symptoms. From a financial viewpoint the capital expenditure for light microscopy is very low, with the operational expenditure being almost entirely the cost of staff wages.

The sensitivity and specificity of light microscopy is entirely dependent upon the skill of the microscopist, as shown by studies on the diagnosis of malaria using light microscopy in different settings. The limit of detection achieved by reference level microscopists (who will often require constant targeted training for up to one month before achieving this level of skill) may be tenfold better than that achieved by microscopists in district-level hospitals. The sensitivity of diagnostics in more rural settings (as is often used in field-based clinics) may be as much as 100 times worse. The sensitivity of microscopy, even in the hands of a highly skilled microscopist, is also hampered by phenomena such as intermittent shedding of parasites (seen with several enteric pathogens such as *Giardia intestinalis*) or sequestration of parasites in tissues (such as occurs in toxoplasmosis).

In addition, different organisms may appear morphologically identical, which make it difficult to differentiate between pathogenic and non pathogenic organisms, such as the cysts of *Entamoeba dispar*, a non-pathogenic amoeba, and *Entamoeba histolytica*, the causative agent of amoebic dysentery, liver abscess and even in rare cases primary amoebic meningitides. However, the ability to analyse a wide range of samples, and an even larger range of parasitic infections using equipment with low capital expenditure, combined with the historical utility of microscopy, means that even more than 100 years after Leeuwenhoek's first self diagnosis of "Pissabeds" in his own stool samples, the light microscope still features heavily in the pathologists tool kit for the diagnosis of parasitic infections.

12.3. Serological Detection of Anti-parasite Antibodies

The second major traditional technology used for the diagnosis of parasitic infections is the detection of host anti-parasite antibodies by the use of serological techniques. Such technologies were traditionally enzyme-linked immunosorbent assay (ELISA)-based, and even today ELISA offers the capacity to screen large numbers of patient sera in parallel to make this a cost-effective and high-throughput diagnostic technology. These tests often utilize whole soluble antigen preparations from parasites harvested from cultures or artificial infections.

The great utility of these techniques is that the diagnosis of infections is not reliant on the sampling of the actual parasite. Where the parasite is sequestered in a tissue or intermittently shed, it is still possible that certain anti-parasite antibodies are generated at a level that is detectable. This can produce an apparent increase in sensitivity in serological techniques compared with microscopy, as seen with the diagnosis of leishmaniasis, schistosomiasis and several other infections. The major drawbacks to traditional ELISA-based techniques are the potential for cross reactivity between different parasitic infections, and the persistence of anti-parasite antibodies even after successful treatment. For instance, a common diagnostic kit, the Trinity Biotek, utilizes two recombinant antigens from *Plasmodium falciparum*, MSP1-19 and MSP2. Antibodies against MSP1-19 have been shown to persist for over a year following successful treatment to clear the blood-stage infections of *P. falciparum*. There is therefore a very real possibility of false positive results with serology-based diagnostics, which may, in part, account for some of the apparent increase in sensitivity of these techniques over microscopy. These traditional techniques for detecting anti-parasite antibodies have been supplemented by the use of western blots to detect antibodies against specific epitopes, and the use of lateral flow immunochromatographic strips to detect anti-parasite antibodies in a highly simplified assay that can be done as a point of care test. The advantage of the western blot technique is that specificity of reaction can be improved if given moieties are shown to be targeted by human antibodies, and often kits will require a strong positive with more than one target. Examples of CE-marked immunoblots for parasitic infections include kits for toxocariasis and hydatid disease (both *Echinococcus granulosus* and *Echinococcus multilocularis*). The incorporation of antigen targets into a RDT format allows the expansion of diagnostics to laboratories where the infrastructure required for more sophisticated diagnostics, including light microscopy, is lacking. The most widely employed example of this is rK39-based lateral flow immunochromatographic test (an RDT) to detect the presence of antibody against visceral leishmaniasis.

12.4. Detection of Parasite Antigens by ELISA and RDTs

A more recent addition to the detection of whole parasites is the analysis of parasite antigens (Table 12.1). These tests can come in the format of high-throughput ELISA tests for screening large numbers of samples, such as the Techlab Crypto and Giardia check. Such tests may be carried out on automated ELISA platforms, such as the DS2 and the Gemini automated ELISA work stations, allowing for high-throughput analysis of samples with automated sample identification, testing, analysis and reporting through laboratory information management system (LIMS). Such automation significantly reduces operator time and error. Alternatively, tests for parasite antigens can be incorporated into a lateral-flow format. Such RTDs for malaria were first produced commercially in the early 1990s. The strips commonly contain three antibodies, two against a malaria antigen, and one against a recombinant protein used as an internal control. The sample is applied to the strip and lysed to release its antigens, one of which is bound by a chromatogenically labelled antibody. The antigens migrate up the strip, where the bound antibody is captured by a secondary immobilised antibody to form a coloured line that may be read by eye. RDTs are currently formatted to detect a range of different parasite antigens for the malaria parasite, each of which has advantages and disadvantages in terms of sensitivity, longevity post infection and, in rare cases, even deletion from the parasite metabolome. Antigen detection for parasite infections has been extended to a number of diseases, including those causing gastro-intestinal infection, blood-borne and even tissue-based infections in the case of cysticercosis.

The advantage of these RDT tests is minimal start-up costs [capital expenditure (capex)] and the capacity to implement them with much reduced operator training compared with light-microscopy-based technologies. Such technologies are therefore easily transferrable to smaller diagnostic units or even non-specialist units in smaller hospitals. In addition, they have a great role to play in developing countries, where there is a distinct lack of diagnostic infrastructure in many locations where, for instance, malaria is endemic. As mentioned previously, such technology can significantly reduce the over prescribing of anti-malarials to those with overlapping symptoms caused by other pathogens, for instance scrub typhus. Not only does this conserve more expensive modern anti-malarials,

Table 12.1 ELISA and RDT tests.

Parasite	Products	Format
Malaria	Numerous – see http://www.who.int/malaria/areas/diagnosis/rapid-diagnostic-tests/rdt-evaluation-programme/en/ for overview	Pan *Plasmodium* and *P. vivax*-specific, pan *Plasmodium* and *P. falciparum*-specific
Giardiasis	Giardia/Cryptosporidium Quik Chek (Techlab Inc., Blacksburg, Virginia, USA),ImmunoCardSTAT!®CGE (Meridian Bioscience Inc., Cincinnati, Ohio, USA), Crypto/Giardia Duo-Strip (Coris Bioconcepts, Gembloux, Belgium), CoproStrip™Cryptosporidium kit (Savyon Diagnostics Ltd, Israel) Xpect™ Giardia/Cryptosporidium Test (Remel)	*G. intestinalis*-specific, combined *Cryptosporidium* spp. and *G. intestinalis* test and combined *Cryptosporidium* spp., *E. histolytica* and *G. intestinalis* test
Cryptosporidiosis	Giardia/Cryptosporidium Quik Chek (Techlab Inc., Blacksburg, Virginia, USA),ImmunoCardSTAT!®CGE (Meridian Bioscience Inc., Cincinnati, Ohio, USA), Crypto/Giardia Duo-Strip (Coris Bioconcepts, Gembloux, Belgium), CoproStrip™Cryptosporidium kit (Savyon Diagnostics Ltd, Israel) Xpect™ Giardia/Cryptosporidium Test (Remel)	*G. intestinalis*-specific, combined *Cryptosporidium* spp. and *G. intestinalis* test and combined *Cryptosporidium* spp., *E. histolytica* and *G. intestinalis* test
Amoebic dysentery	*E. histolytica* QUIK CHEK (Alere), RIDA®QUICK Cryptosporidium/Giardia/Entamoeba Combi (R-BioPharm, Darmstadt, Germany)	*E. histolytica*-specific test and combined *Cryptosporidium* spp., *E. histolytica* and *G. intestinalis* test
Filariasis	Filariasis Test Strip (FTS) (Alere), Brugia Rapid point-of-care cassette test (BRT) (Brezon diagnostics)	*Wuchereria bancrofti*-specifc and *Brugia* spp.-specific
HAT	SD BIOLINE HAT 2.0	*Trypanosoma brucei* spp.
Schistosomiasis	Urine-CCA (Circulating Cathodic Antigen) cassette test	*Schistosoma* spp. (designed for *S. mansoni* but cross reactive with CCA from other species
Cysticercosis	Ap Dia Cysticercosis circulating antigen	*Taenia solium*

currently Artemether combined therapies (ACTs) are used as a front line therapy, for those who would truly benefit from them, but it also may reduce the spread of drug resistance. In such situations there has also been shown to be a significant improvement in prognosis for patients who do not have malaria and are switched early to alternative investigations or treatments following a negative RDT result compared with those who would have empirical treatment for malaria.

12.5. The Advances Afforded by Genome Based Technologies in Pathology

When evaluating the utility of a diagnostic test, a number of different criteria must be evaluated by a diagnostic laboratory. The benchmarks for clinical utility are specificity and sensitivity. Both of these metrics may be further divided into analytical and clinical moieties. Clinical sensitivity may be defined as the proportion of true positives identified as positive by the technique under review (true positives are often determined by the current "gold standard" technique or sometimes a composite standard where a single technique of excellence does not exist). Clinical specificity is the proportion of true negatives identified as negative by the new technique (again determined by the gold standard). Both these should ideally be calculated when deployed in a situation simulating the intended use in routine clinical testing. From the same set of data, the negative and positive predictive values may also be calculated (Table 12.2).

The Standards for Reporting of Diagnostic Accuracy (STARD) guidelines detail a number of criteria that should be satisfied when carrying out such analyses, including sequential samples submitted for routine diagnostic testing and the testing of samples being blinded. Too often sensitivity and specificity are calculated by using the "sickest of the sick" and the "wellest of the well", which

Table 12.2 Values used to calculate clinical performance of a test in comparison with the current gold standard[a].

	True positive	True negative
Test positive	*a*	*b*
Test negative	*c*	*d*

[a]Clinical specificity = $b/b + d$, clinical sensitivity = $a/a + c$, positive predictive value = $a/a + b$, negative predictive value = $c/c + d$

may significantly skew the data in favour of the test under review. Analytical specificity is the performance of the assay with non-specific sample types (*e.g.* other enteric pathogens for a test for Giardiasis) whilst analytical sensitivity is the limit of detection of an assay using serially diluted analyte (the level of analyte where 95% of the test replicates return a positive result. In addition to the clinical and analytical performance the laboratory also has to consider a number of other factors involved in testing before introducing a new assay. These include: turnaround time (how long it takes to produce a final result from receipt of sample), reliability, reproducibility, ease of implementation, initial start-up costs (Capex) and operational costs (Opex).

Finally, a regret-theory approach to decision curve analysis may be implemented to weigh up the advantages and disadvantages induced by false negatives and false positives to evaluate the utility of a new diagnostic test: Where treatment has few side effects and drugs are suitably cost effective it may be tolerable to implement a technique with a lower than optimal specificity. For a condition such as human African trypanosomiasis, where treatment for central nervous system (CNS) infections of *Trypanosoma brucei rhodesiensis* and all stages of *Trypanosoma brucei gambiense* have utilized the highly toxic drug melarsoprol, it is essential that specificity is as near 100% as possible. We will see in the next chapters how genome-based technologies may provide advantages in these different criteria and therefore be attractive to diagnostic laboratories.

12.6. Conventional PCR, the First Parasite-genome-based Diagnostics

Only in the last ten or so years has routine parasite diagnosis begun to actively incorporate techniques involving the parasite genome. The very first techniques involved PCR and were pioneered by research laboratories rather than routine diagnostic facilities, interested in the epidemiology of parasites rather than seeking to diagnose them for patient treatment. The publication in 1982 of George Snounous' nested PCR for the analysis of *Plasmodium* species was published as a tool for population surveillance studies rather than routine clinical work, however, over 30 years later it still remains the gold standard against which most many new diagnostic techniques are measured. The nested PCR involves a common first-round set of primers which targets the 18S region conserved amongst all species of primate malaria parasites. A second round using species-specific primers allowed the definitive diagnosis of four of the different species capable of establishing human infection: *Plasmodium falciparum*, *Plasmodium ovale curtisi*, *Plasmodium malariae* and *Plasmodium vivax*. The alternation of the primers for *Plasmodium ovale* spp. allows the detection of *Plasmodium ovale wallikeri*, whilst a fifth set of primers for *Plasmodium knowlesi* allows the amplification of genomes from this zoonotic parasite from the Sarawak area of Malaysia.

A large number of alternative diagnostic PCRs, both nested and single-round, with conventional detection by gel electrophoresis have been published since Snounou's first molecular paper, but the Snounou assay remains a routine diagnostic technique in many laboratories utilizing parasite genome amplification, most probably due to its historical utility and successful implementation in many different laboratories across the world. With a limit of detection around 0.5 parasites per microliter, when DNA is extracted from 200 microliters of blood using Qiagen DNA mini kits, it is over 10 times more sensitive that the very best microscopist and can definitively identify species in multiple infections that would be difficult to detect using single-round competitive PCRs or using light microscopy. The major drawback of the Snounou PCR is the length of time required to perform it. The assay requires at least 10 hours from first receipt of sample until provision of final result, with DNA extraction taking approximately 1 hour, first round amplification 3–4 hours using the published conditions, and a second-round amplification of 3–4 hours using the published conditions makes it hard to complete this within a single day and allow sufficient time for analysis by gel electrophoresis. Because of this, the second round is commonly run overnight, producing a final result more than 24 hours after receipt of sample. The WHO recommendation that a malaria result is given within 2 hours of the patient supplying the sample makes nested PCR untenable as a front-line diagnostic technology. For those patients, however, where differentiation of *P. vivax* and the *P. ovale* subspecies is difficult on the basis of morphology, or where a very low-level of parasite density is expected on the basis of one or two trophozoites of aberrant appearance in a thick film, the timings of this technique are sufficient as an adjunct to front-line diagnostic technologies.

Conventional PCR assays for a wide range of other parasites have also been published, including for gastro-intestinal parasites (both protozoal and helminth in nature), *Leishmania* spp. *Trypanosoma* spp. (both African and South American), *Toxoplasma* and microsporidia (see Table 12.3). The microsporidial parasites may be identified in stool samples using fluorescent brightener, which can also pick up fungal and bacterial moieties. A modified Trichrome stain is used to evaluate stool samples positive by fluorescent brightener to look for small rose-coloured bodies with a terminal vacuole 7 microns in size. This is a challenging technique, requiring excellent operator training. Even with such training the actual sensitivity and specificity of light microscopy may be quite poor. The utility of light microscopy is further limited by its inability to distinguish between different species of microsporidial pathogens. The following species all are capable of infecting human hosts: *Enterozoon bienusi*, *Enterocytozoon cuniculli*, *Enterocytozoon hellen* and *Enterocytozoon intestinalis*. Of these *E. bienusi* has been reported to show resistance to albendazole, the usual drug treatment for microsporidiosis, requiring second-line drugs, such as fumagillin.

It is almost impossible to differentiate between different species of microsporidia on the basis of light microscopy, it can only be reliably done using electron microscopy to count the number of polar filaments in the cell. The extended cost and turnaround time of electron microscopy prevents its implementation as a front-line diagnostic in this way. PCR assays by comparison are able to definitively identify the infecting microsporidial species with a high level of specificity and a sensitivity that is far greater than that of light microscopy. The SSU-RNA assay listed by the CDC as the gold standard for microsporidiosis utilizes size polymorphism of this locus amplified by a pair of conserved primer sequences. This amplified fragment length polymorphism (AFLP) allows the definitive identification of all four species listed above. In addition, this may be confirmed by additional specific PCR primer sets should greater specificity be sought for the diagnosis.

The intestinal parasite *Giardia intestinalis* is endemic to the UK, being found, for instance, in contaminated well water. Oral ingestion of cysts leads may lead to the establishment of infective trophozoites, which replicate in the upper gastro-intestinal tract. As trophozoites migrate down the tract they develop into cysts for completion of the life cycle when they are shed into the environment in stool samples ready to infect new hosts. Conventional diagnosis is by light-microscopy-based analysis of stool samples for the presence of cysts. The sensitivity of the technique is limited by intermittent

Table 12.3 Organisms for which conventional PCR diagnostic assays have been published and the DNA loci targeted by these assays.

Species	Target	Sample
Ascaris lumbricoides	Internal transcribed spacer (ITS)-1	
Opisthorchis viverrini	Repetitive DNA, ITS2,Cox1, pOV-A6, 28S	Stool
Clonorchis sinensis	28S, CsRn1, ITS	Stool
Haplorchis taichui	28S, ITS	Stool
Strongyloides stercoralis	Cytochrome c oxidase subunit 1 (mitochondrial), ITS and 18S	Stool
Fasciola hepatica	Repetitive element, ITS-2, 18S	Worms, miracidia, eggs, stool
Paragonimus westermani	cox1, ITS, repetitive element	
Fasciola gigantica	ITS-2, 18S	Worms, miracidia, eggs, stool
Fasciolopsis buski		
Taenia solium, *Taenia* spp., *Echinococcus granulosus* and *Echinococcus multilocularis*	Mitochondrial DNA (12s, cox1, rrnS and nadI), HDP2, HDP1 repeats, pTso19 repeat	Stool and parasite segment
Trichinella spiralis and Trichinella spp.	Random amplified polymorphic DNA and 43 kDa excretory–secretory (E–S) protein gene	Muscle biopsy (infected meat)
Necator americanus	ITS1, ITS2 and 28S	Stool
Giaridia intestinalis	18Sc, tpi, gdh, 552-bp intergenic spacer and ITS	Stool, bronchiolar alveolar lavage, and biopsy
Cryptosporidium spp.	18S, oocyst wall protein, 60-kDA glycoprotein, HSP70, Laxer locus and microsatellite loci.	Stool,
Cystoisospora	18S	Stool, duodenal biopsy
Entamoeba histolytica	HLY6, 18S, chitinase gene, adh112 and serine-rich protein	Stool, liver aspirate
Cyclospora catyensis and Cyclospora spp.	18S, ITS, HSP70	Stool
Enterocytozoon bieneusi, *Encephalitozoon intestinalis*, *Encephalitozoon hellem* and *Encephalitozoon cuniculi*		Stool
Schistosoma mansoni, *Schistosoma japonicum* and *Schistosoma haematobium*	121-bp tandem repeat DNA sequence, Mitochondrial DNA, 28S and ITS–rDNA	Stool, plasma, serum and urine (dependent upon species)
Strongyloides stercoralis	18S	Stool

Plasmodium spp. (both genera-wide and species-specific)	18S, var genes, CytB, mitochondrial DNA, telomeric repeats	Blood (finger prick and venipuncture onto filter papers), saliva and plasma
Leishmania spp. (both genera-wide and species-specific)	Mini-circle DNA, ITS1, CytB, HSP70, gp63 intergenic regions	Blood (whole and buffy coat), punch biopsy and liver trephine and aspirate
Trypanosoma cruzi	Kinetoplast minicircle, 13 satellite DNA, 18S and 24S	Whole blood
Trypanosoma brucei	ESAG6, ESAG7, ITS, 177 bp repetitive DNA and 18S	Whole blood
Toxoplasma gondii	B1 gene, 18S, ITS and REP-529	Blood, placenta and cerebrospinal fluid
Babesia microti	18S, β-tubulin gene	Whole blood
Trichomonas vaginalis	Genomic repeat sequences,	Vaginal swabs and urine

shedding of cysts in stool samples, such that stool samples over three sequential days are recommended by many health care systems in order to definitively eliminate giardiasis as a cause of a patients' symptoms. There is even some evidence that some infections never seem to encyst, producing only trophozoites, which disintegrate as they pass down the intestinal tract. Because of this empirical treatment for giardiasis may be undertaken in many health care facilities based only on clinical symptoms such as travel history, diarrhoea, abdominal distention and gaseous foul-smelling stools. The limit of detection (LOD) of conventional PCR has been shown in some studies to be several orders of magnitude superior to light microscopy both in terms of analytical and clinical sensitivity. As such, the improved LOD may therefore allow the identification of many more patients who have giardiasis without the presence of visible cysts in their stools.[1]

In addition, the negative predictive value can remove the need for testing across multiple days and, in some settings, may result in a decrease in empirical testing in the absence of a positive result. Such an outcome is advantageous for the patient in that it may prevent unnecessary treatment (and any resultant side effects), speed up the implementation of second-line investigations and thereby a final correct diagnosis, and, where possible, treatment and abolition of symptoms. For the health provider the removal of unnecessary empirical treatment reduces the wastage of resources of unnecessary drug treatment, whilst increased speed of the correct diagnosis reduces the cost associated with repeat visits from the patient. Finally the reduction of unnecessary drug treatment reduces the possibility of the fixation of drug resistance genes in the pathogen population. A recent study of returning travellers showed that up to 30% of patients have infections that are refractory to Tinidazole, the current front line drug used to treat giardiasis in the UK.

Human leishmaniasis may be caused by a number of morphologically identical parasite species. The established diagnostic techniques are light microscopy for examination of skin biopsies, bone marrow and splenic aspirates and bone marrow trephines. In addition, culture may also be used to amplify parasite numbers in patient samples to detectable numbers. The diagnosis of this condition by light microscopy requires the Giemsa staining of material to enable identification of both the kinetoplast and chromosomal DNA in the amastigote forms of the parasites, unless a sample is culture-positive. It is not surprise that serological techniques have a far higher sensitivity but are only useful for visceral forms of the disease, and, in addition, suffer from all the drawbacks mentioned previously for the serological diagnosis of parasitic infections. The need to differentiate between morphologically identical parasites is due to the capacity of *Leishmania viannia* sub species to cause mucocutaneous infection, destroying the soft tissues of the pallet and nose many years after the original cutaneous infection has been treated.

Cutaneous infections by other species may be treated by localized injection of a pentavalent antimonial, such as sodium IV stibogluconate, into the lesion(s). In order to fully eliminate *L. viannia* sub species, and therefore safeguard against the occurrence of mucocutaneous infection, patients require 21 days of hospitalization for intravenous treatment with the sodium IV stibogluconate, which is highly phlebotoxic under these conditions. It is therefore exceptionally important for patient welfare to differentiate between *Leishmania donovani*, *Leishmania mexicana* and *L. viannia* sub species, all of which can occur in the new world. The first published assay to do this was conventional-PCR-based amplifying the mini-circle DNA for the kinetoplast. Available at up to 1000 copies per cell this is a true multi-copy locus. AFLPs allow the discrimination of the old world *Leishmania tropica*, *Leishmania major* and the new world *Leishmania amazonensis* from other species of the genus *Leishmania*. The differentiation of *L. mexicana* and *L. donovani* from *L. viannia* sub species may be achieved by the use of second-round PCRs amplifying sequences conserved only in the kinetoplast of *L. viannia* sub species, or by restriction fragment length polymorphism analysis of the IT1–IT2 locus. Many of the conventional PCR assays that were developed towards clinical usage were initially designed as research assays. Because of this the assay may have significant draw backs that limit its utility in a routine clinical service. This is true for the IT1–IT2 assay for leishmaniasis, which generates a long amplicon (1000+ bp) from a locus with a much lower copy number than that of the kinetoplast DNA. The result of this is an assay that performs particularly badly with formalin-fixed tissue (formalin cross links DNA to proteins and fragments it, thereby significantly lowering DNA yield during its purification from the clinical sample). Since it common practice to fix biopsy material as formalin-fixed paraffin-embedded (FFPE) prior to histological analysis, many samples supplied for the diagnosis of leishmaniasis may be sub optimal for this assay, creating a problem for the pathologist charged with identifying the species present.

12.7. Whole-genome Sequencing Benefits Assay Development

The majority of conventional PCR-based diagnostic assays have been targeted to well-sampled genes in the genome. For example, many assays utilize the 18S gene, which is ubiquitous amongst eukaryotes. The choice of this locus in most likely to be due to its presence in a great many species in the readily accessible EMBL database. These sequences allow *in silico* analyses to look for conserved loci with suitable kinetics for probe binding sites that are sufficiently divergent from homologs in related species. Through these analyses, it is possible with well-sampled loci to maximize the likelihood of high analytical sensitivity and specificity. The number of loci which can be investigated is significantly increased when a whole-genome sequence is available. The capacity to sequence genomes has dramatically increased with the development of next-generation and even third-generation sequencing.

The first parasite genome to undergo complete sequencing was that of *Plasmodium falciparum*, although the sequence of the *Encephalitozoon cuniculi* was the first to be released as a completed genome in 2001. With the implementation of massively parallel sequencing, such as that achieved with the MySeq and Illumina platforms, the volumes of read numbers generated are orders of magnitude greater than could be achieved with the previous generation of automated capillary sequencing. Because of this increase in output the costs per base were reduced significantly, further increasing the capacity to sequence parasites' genomes. The development of bench-top affordable sequencing platforms has also allowed the migration of whole-genome sequencing away from specialist units into research facilities, allowing the depth of sequence information to be increased significantly for a given species. Therefore, anyone interested in the *in silico* analysis of targets for new genome-amplification-based tests has the capacity to analyse many different isolate genomes to ensure that single nucleotide polymorphisms or even sequence insertions/deletions (indels) will not adversely affect the clinical sensitivity of an assay. An example of where an assay is affected by such polymorphism is seen with the original Snounou malaria PCR. As previously stated a large number of different *P. ovale* isolates failed to amplify with this assay. It was known that sequence diversity existed within this locus for *P. ovale*, but only later was it proved that sequence diversity is in complete linkage disequilibrium across the entire genome, to an extent that the so called "Classical" and "Variant" types were reclassified as two different species. The ability to search for multiple loci when designing primers also allows the development of assays with increased sensitivity and specificity.

The 18S locus in *Plasmodium falciparum* is listed as having five different copies, resulting in it being referred to as a multicopy gene in many publications and therefore ideal for sensitive diagnostic assays. What these papers fail to state is that the gene has significant polymorphism, such that many assays target only the asexually expressed or sexually expressed forms of the gene. As a result, there may only be as few as two copies of a given 18S locus targeted by primer sets. A number of other of assays such as loop-mediated amplification of DNA (LAMP; see later) are even more susceptible to the quality of the genome sequence upon which they are based. As more and more sequences are deposited, so knowledge on the diversity of sequences is increased, as is the density of high-quality sequence that can be mined *in silico* for assay design.[5]

12.8. Real-time PCR, the Next Breakthrough

The simultaneous amplification and detection of genomes through real-time PCR assays has been shown to significantly improve the turnaround time of genomics-based diagnostics. Techniques for the detection of amplicons during the PCR reaction include the use of intercalating fluorescent dyes, such as SYBR green, the digestion of probes by the 5′–3′ exonuclease activity of *Taq* Polymerase used in TaqMan based technologies (including probe modification by linked nucleic acids [LNAs] and the minor groove binder protein [MGB]), Scorpion probes, molecular beacons and fluorescence resonance energy transfer (FRET) probes.

Intercalating-dye-based assays, the most common of which utilize SYBR green, exploit the capacity of molecules to fluoresce when located within the major grove of the DNA duplex. As more double stranded molecules are created so the amount of intercalated dye increases as does the fluorescent signal. This technology is very simple to translate from conventional-PCR-based assays, given the binding of only two primer sites (often those previously used for conventional PCR). The ability to disassociate double-stranded amplicon in a controlled manner allows for the high resolution of the amplicon melt temperature. This can be used to determine the correct amplicon sequence, differentiate between amplicons from related species amplified by a common primer set and even to look for single-nucleotide polymorphisms. SYBR green assays have been developed and tested against a range of parasitic infections. For microsporidia infections, the modification of a conventional single-round PCR into a SYBR green format has allowed the simultaneous amplification and detection of genomes from different microsporidia species. By the incorporation of a melt curve analysis amplicons from the four major species of microsporidia parasites capable of establishing human infection may be detected. The removal of gel-based analysis of products not only decreases the turn-around time of the assay, but also provides a semi quantitative signal allowing the response to drug treatment to be monitored, removes the need for toxic intercalating dyes used in gel electrophoresis, such as ethidium bromide, and, most importantly, limits the possibility of amplicon cross contamination due to opening of tubes to load the products into gels. This last point has significant operational effects, since a single room is needed for the amplification and analysis, whereas conventional PCR requires the segregation of these two activities by physical and procedural boundaries.

For those techniques that use probe-based analysis, where additional binding targets are required for the probes, then greater depth of analysis is important to ensure that clinically effective diagnostics are developed. This is especially true where reactions are to be multiplexed to detect a number of different targets simultaneously in order to stop cross reactivity of primers and probes and normalize the melt temperature kinetics for the different primer and probe combinations. Probe-based assays offer significantly more options for the simultaneous detection of multiple targets (multiplexing) than SYBR-green-based assays. The latter has been shown to be capable of detecting multiple targets through the use of differential melt curves, but is more commonly used for monoplex reactions.

12.8.1. Taqman Probe Technology

The Taqman technology utilizes the 5′ to 3′ endonuclease activity of *Taq* polymerase to digest probes annealed to a sequence on the amplicon between the primer sites (see Chapter 1). The probe contains a fluorescent moiety in close proximity to a quencher molecule. When the probe is intact the fluorophore absorbs laser light energy emitted by the PCR machine and re-emits it a different spectral frequency. The quencher is sufficiently close to the fluorophore to absorb this light by FRET, significantly reducing the signal emitted by the fluorophore. Upon digestion of annealed probe the fluorophore is released from close proximity to the quencher and the energy emitted by fluorophore is no longer absorbed by the quencher and may therefore be registered by the PCR machine using a charge coupled device (CCD) camera. Each fluorophore has a spectrum at which is can absorb energy from the excitation laser and a different spectrum of light at which that energy is emitted. The capacity of different fluorophores to absorb and emit light at different frequencies (the excitation and emission frequencies respectively) and of different quencher molecules to absorb energy at these excitation frequency means a series of probes can be created which can monitor the production of different amplicons simultaneously within a single tube. This means that with appropriately equipped real-time-PCR machines up to six different targets may be multiplexed using the Taqman Technology.

Multiplexed real-time reactions using the Taqman technology have been developed for a large number of parasite targets, including enteric pathogens such as *Giardia lamblia* (see Table 12.4) in order to simultaneously detect multiple pathogens within a single sample with a high sensitivity, such as *Giardia lamblia*, *Cryptosporidium* spp., and *Entamoeba histolytica*. Although the limit of detection may not be improved compared with conventional PCR (in some cases it may actually be slightly poorer), the increased speed with which multiple targets may be tested means a faster result for the patient, increased efficiency for the laboratory and potentially lower cost per test for users of the service due to the reduction in man hours required.[2]

Apart from the increased number of pathogens that may be simultaneously detected, multiplexing allows for the inclusion of in-tube internal control(s). These controls may be a housekeeping human gene (such as B-tubulin) to detect the extraction of human DNA in the sample, a locus from a spiked analogous organism, such as phocine herpes virus, or a locus synthetically integrated into plasmid spiked into the sample, such as green fluorescent protein. All these targets can all monitor the efficiency of DNA extraction and amplification. Where the control is a plasmid spiked directly into the amplification reaction, it may only be used to detect aberrant functioning of the reaction, through lack of mastermix or the presence of inhibitors for instance. The inclusion of these controls may identify samples which have produced false positives to allow for their retesting. Such a process improves confidence in negative results due to the inclusion of an internal control, thereby increasing confidence in

Table 12.4 Parasites and related commensal organisms for which Taqman real-time PCR assays have been published and the DNA loci targeted.

Organism	Target	Sample type analysed	Best sensitivity reported
Ancylostoma ceylanicum	Repetitive DNA	Stool	<1 genome per reaction
Ascaris lumbricoides	ITS, nad5	Treated water and stool	100 copies per reaction (plasmid)
Necator americanus	Repetitive DNA, ITS	Stool	100 copies per reaction (plasmid)
Opisthorchis viverrini	Repetitive DNA	Tumour tissue	Not given
Clonorchis sinensis	ITS, cox1	Stool	5 eggs per g of faeces
Schistosoma genus and species	Tandem repeat, ITS, 28S, nad1, cox1	Stool, urine, serum and semen	10 fg genomic DNA per reaction
Strongyloides stercoralis	18S	Stool	0.004 larvae μl^{-1}
Ancylostoma duodenale	ITS	Stool	Not given
Taenia solium Taenia sagninata	ITS, HDP2 and Ptsol9	Stool, proglottid	10 fg genomic DNA per reaction
Trichuris trichuria	ITS	Stool	Not given
Cryptosporidium spp.	18S, DNA J-like protein, nft2 and COWP, HSP70	Stool, duodenal biopsy, broncho-alveolar lavage (BAL), purified oocysts	1 cell per reaction (purified oocyst)
Giardia intestinalis	18S, β-giardin, HSP70	Stool, duodenal biopsy, BAL	50 trophozoites per g stool
Cyclospora catyensis	HSP70, 18S	Stool	50 oocysts per ml faeces
Entamoeba histolytica Entamoeba dispar	18S, episomal repeats	Stool, liver aspirate, duodenal biopsy, BAL	0.5 cells per ml stool
Microsporidia species	16S	Stool	20 spores per ml faeces
Echinococcus granulosis and *Echinococcus multilocularis*	Mitochondrial 12S	Cyst fluid and stool	1 copy (plasmid)
Plasmodium genus- and species-specific	18S, CytB, var, stevor, msp1, msp2, telomere-associated repetitive element 2 Taqman array card for multiple resistance loci	Blood, plasma, saliva	0.002 parasites per µl blood

Babesia microti	18S, sa1	Blood	2.5 parasites per µl blood (peer reviewed), 20 parasites per ml blood (commercial website)
Leishmania genus- and species-specific	HSP70, REPL repeats, AAP3	Buffy coat, tissue samples	0.1 parasites per reaction
Trypanosoma cruzi	Mini-satellite TCZ region DNA, kinetoplast DNA, 18S	Blood, buffy coat	0.70 parasite equivalents per ml
Wuchereria bancrofti	Long DNA repeat (LDR), Wool-bachia 16S	Dried blood spots	Not given in terms of parasites per ml
Brugia malayi	HhaI repeat element	Dried blood spots	Not given in terms of parasites per ml
Onchocerca vovulus	5S, O-150 tandem repeat	Skin snips	Six copies per assay (plasmid)
Toxocara spp.	18S	Soil, animal faeces and BAL (infected animals only)	Two eggs per g of soil

the negative predictive value of the test. Current UK Standards for Microbiological Investigation (UK SMI) regarding molecular pathology testing stipulate that such internal controls should be used in pathology testing in order to reduce the incidence of false negatives in laboratory testing.

Since TaqMan technology has a high historical utility, it is no surprise that this is the technique used in the highest number of published assays for the diagnosis and analysis of parasites (both human and veterinary). For instance, the poor performance of IT1–IT2 sequence amplification for the differentiation of *L. viannia* sub species from *L. mexicana* has lead to the development of an HSP70 real-time amplification followed by amplicon sequencing to detect and differentiate infections with different species of *Leishmania*.

12.8.2. Molecular Beacon

Molecular beacons use the same type of chemistry for fluorophore and quencher moieties that the Taqman probes use. The difference is in how the two are disassociated in order to allow the fluorophore to absorb and re-emit light energy. Both moieties are on opposite ends of a probe which has a stem and loop structure (or pan handle). This structure keeps them in close association in unbound copies of the probe. Upon hybridization of the probe, the increased binding kinetics of the full probe sequence is greater than that of the stem, resulting in a change of conformation of the probe. These change forces the fluorophore and quencher apart by a sufficient amount that the fluorophore is free to absorb and re-emit light energy at a unique frequency that may be registered by the real-time PCR machine. Parasite diagnostic assays employing molecular beacon probes have been published for detection of malaria parasites (with a surprising claim of an LOD of 0.004 parasites per microliter of blood), for a few different parasites including *Toxoplasma gondii*, *Entamoeba histolytica* and *Babesia microti*.

12.8.3. Scorpion Probes

Scorpion probes also use a fluorophore and quencher held in tight association by a loop–stem structure as per molecular beacon probes, with the exception that the quencher is upstream of a sequence complimentary to the target locus. When in close association the quencher is able to absorb the energy emitted by the fluorophore. Upon binding to the primer to the target sequence and complementary sequence is synthesised, within which is a sequence complimentary to the loop structure. Following denaturation of the double stranded DNA during the PCR reaction the loop is able to bind to this complimentary sequence, dissociating the fluorophore and quencher and allowing the fluorophore energy emittance to be registered. A blocker molecule attached to the quencher prevents the synthesis of a complimentary strand to the primer which would otherwise be allowed. Here too, the quencher absorbs light at a spectral frequency that is the same as the fluorophore emits upon laser excitation. A blocker separates the sequence from the 3′ sequence that is complementary to the target. The loop is complementary to a sequence 3′ to the primer site in the genome, and therefore is capable of annealing to this sequence upon its synthesis by primer extension. The blocker prevents the synthesis of a complementary primer sequence to the PCR primer. Scorpion-probe-based assays have so far only been reported in assays for *Giardia intestinalis* sub type analysis as well as the detection and species identification of *Cryptosporidium* spp.

12.8.4. FRET Probes

Fluorescence resonance energy transfer (FRET) probes utilize fluorophores in a way that is very different from the previously listed systems. In FRET the donor molecule absorbs energy from the excitation laser, and emits light energy at a specific wavelength. When the donor is in close proximity to the fluorophore the light energy emitted by the donor excites the second fluorophore, which will emit the light at a different wavelength to the donor and can be differentiated from it by a CCD. If the energy can be passed between the two, then fluorescent light energy will be emitted. The donor and fluorophore are on different oligonucleotides, which are only adjacent to each other (less than 5 bp apart) when they are bound to the complimentary amplicon sequence. The ability to denature the resultant structures means that melt curves analyses may be performed to determine the specificity of the amplicon. FRET probes have been used in reported assays to detect various protozoa, including *Plasmodium* spp., *Entamoeba histolytica*, south American *Leishmaniasis*, old world *Leishmaniasis* and *T. vaginalis* along with the helminths *Paragonimus* spp., *Toxocara* spp., *O. viverrini* and *C. sinensis*.

12.8.5. Methods for Primer and Probe Modification

The above methodologies can be further optimized by the modification of the primer and probe sequences. LNA probes are modified RNA structures produced by an extra bridge connecting the 2′ oxygen and 4′ carbon. The resultant nucleotides may be used to replace either RNA or DNA bases and significantly increase the hybridization temperature of the oligonucleotide structure. This allows for a shortened oligonucleotide structure with a higher melt temperature compared with non-LNA probes.

Minor groove binder probes use a dihydrocyclopyrroloindole tripeptide (DPI_3) at the 3′ of the oligonucleotide to stabilize the structure of the primer-template double stranded DNA. They do so by folding into the minor groove formed by the terminal 5–6 base pairs of the primer–target duplex and stabilizing this structure with Van der Waals forces. A 12 base pair MGB probe may have the same melting temperature as a 27 base pair probe without this moiety. In addition, the increased stabilization of A–T duplexes simplifies the requirements for probe design.

The increased binding affinity of these LNA and MGB probes allows for the normalisation of probe and primer melting temperatures whilst using shorter allele-specific sequences. This is of great use where multiplex reactions are required, whereby similar melting temperatures are required for multiple targets without primer–probe cross reactivity. It may also be used to enhance allele-specific hybridisation should drug-sensitivity mutations be of interest to the pathologist.

Modified probes have been described in Taqman assays for a number of different parasites including: *Brugia malayi* and *Brugia timori* (causative agents of human filariasis), *Leishmania* spp. (and *Leishmania infantum*-specific probes), *Plasmodium* spp., *Cryptosporidium* spp. (and *C. parvum*- and *C. hominis*-specific probes, *E. histolytica* (as well as the non pathogenic *E. dispar*), *Anisakis* spp. and *Pseudoterranova* spp., whilst this technology has been developed into CE-marked kits in the case of *Toxoplasma gondii*.

12.9. DNA Extraction

As shown previously with nested PCR analysis of malaria parasites in blood, the requirement for highly purified DNA in most assays involves a significant time component of the assay. Some samples are less complex and challenging for DNA extraction than others. Serum and plasma are relatively simple matrices that do not provide much of a challenge for DNA extraction. Unfortunately for the parasitologist, few parasites exist as extracellular bodies within this matrix, with the exception of microfilarial helminths. Sedimentation of erythrocytes, as performed by virologists and microbiologists, will result in the loss of the majority of parasite cells and therefore genomic DNA. Many *Taq* polymerases are sensitive to haemoglobin, and it is therefore essential that good quality DNA clean-up is achieved. Other samples may be even more challenging, such as stools and sputum. Stool samples contain a large amount of human and gut flora DNA, together with bilirubin, bile salt and other potent inhibitors of *Taq* polymerase. The majority of assays published utilise Qiagen-developed DNA extraction, either silica-based spin columns or bead-based processes. Such extractions can be done manually with kits such as DNA blood mini kits, semi automated through the use of robotic platforms, such as the Qiacube or EZ1, or fully automated through the use of Qiasymphony platforms, Magna Pure or Promega Maxwell systems. Alternatively, laboratories can design and build their own bespoke platforms for sample and fluid handling with the help of commercial companies. Such platforms allow for the batching of DNA extraction, allowing for economy of scale, but bespoke platforms may be highly dependent upon companies without depth of resources for replacement parts should the system go wrong. Less bespoke platforms may be sourced by a reagent rental system in order to lower the start-up CapEx costs normally encountered when initiating molecular diagnostic testing. The parasitologist is faced with a whole range of potential samples, including blood, urine, semen, aspirates, stools, solid tissue biopsies, scrapes, snips or even whole or incomplete parasites. Hence, a range of extraction techniques may need to be developed.

Some of the parasites may be incredibly tough, such as microsporidia with their chitin endospores, since they are designed to survive in challenging environments outside the host if they are spread by the faecal–oral route. Therefore, published assays may need to be adapted in order to ensure a maximized yield of amplifiable DNA or RNA. The adaptation of extraction techniques may also increase the size of samples that are processed to produce the final DNA extract. So long as inhibitors are not proportionately increased, this will have a dramatic effect on the LOD of an assay. For malaria, most assays utilize a 200 microlitre sample size, but for artificial infection studies, where increased sensitivity is required, up to 2 ml of blood may be processed, resulting in a tenfold improvement in LOD.

To further complicate matters, samples arriving at parasitology laboratories in the past were often pre-treated with formalin to deactivate any pathogens in the sample and preserve the morphology of any parasites present. Formalin acts to cross link DNA to proteins and methylate DNA, thereby significantly reducing the amount of DNA that can be purified by standard processes. In some studies, the fixation of stools in formal-saline reduced DNA yield by 1000-fold. The simplest way to counteract this is to request that sending laboratories or clinics do not formalin fix their samples in any way, but some samples are fixed by their very nature, such as formalin-fixed paraffin-embedded tissues. In such cases the pathologist must be willing to use a target which may still be positive, even with such reduced DNA yields, or employ specialized extraction techniques to maximize the yield of DNA from these samples, with extended pre-incubation and washing to remove the formalin and reverse cross linking.

12.10. Commercial Assays

The commercialization of genomic-based assays for parasite detection is in its infancy compared with other diagnostic disciplines, such as virology. The market for such diagnostics is far more limited in developed countries (which have the budget to purchase such assays) due to the lack of autochthonous infections. However, the last few years have seen a proliferation of commercial assays to detect parasite pathogens, both in humans and in animals. ISO-accredited pathology laboratories are bound by new European regulations to buy in only CE-marked diagnostics from external sources. In-house assays developed by reference centres may continue to be used so long as a suitable program of validation and re-verification is adhered to.

What the CE marking process involves is dependent upon which annex the organisms fall into under *In Vitro* Diagnostic Directive (98/79/EC) regulations. List A organisms in Annex II, require verification of new lots of reagents by notifiable bodies before their release. As such it requires a great economy of scale to make these products viable, limiting the number of manufacturers who can provide such products. Thankfully all parasitic infections are so far within Annex III, removing the need for such measures and bringing down the cost of product development and release. Taqman-based technologies are the most common commercial products to be offered for the detection of parasites, probably due to the long-established utility of this methodology and the well-developed parameters for primer and probe design, many of which are automated by *in silico* analysis programs.

A number of commercial CE-marked real time assays have been released in the last five or so years targeting parasitic infection. Fast Track Diagnostics have a portfolio of tests for enteric parasites (*E. histolytica*, *G. intestinalis* and *Cryptosporidium* spp.), malaria (including a screening assay for all species of *Plasmodium* and a malaria differentiation test to definitively identify *P. malariae*, *P. vivax*, *P. falciparum* and both subspecies of *P. ovale*). Qiagen have released the Artus Malaria PCR kit, whilst Vacunek, Altona Diagnostics and BioSB have also released kits for malaria diagnosis. Clonit have released the AMS94/F assay for *T. gondii*. Amplidiag have released a real-time PCR assay for *Cryptosporidium* spp. *G. intestinalis*, *E. histolytica* and *D. fragilis*. Diagnode diagnostics have released assays for *Trichomonas vaginalis* and an enteric assay for *E. histolytica*, *G. intestinalis* and *Cryptosporidium parvum*.

The Stat Nat Leishmania diagnostic is a lyophilized real-time PCR reaction aimed at all species of *Leishmania* (marketed by Alpha biosciences), whilst Progenie Molecular offer a catalogue for the molecular detection of *Leishmania donovani* complex, *T. gondii*, *Plasmodium* spp. and *Trypanosoma cruzi*. In addition to tests for *Cryptosporidium* spp. *D. fragilis*, *E. histolytica*, *G. intestinalis* (split between two different linkages), *Leishmania* spp., *Leishmania infatum–donovani* sub complex and malaria, bioactiva also offer a test for ascariasis. RIDAGENE also offer an enteric test for *Cryptosporidium* spp. *D. fragilis*, *E. histolytica* and *G. intestinalis*. As the vast majority of these tests have appeared on the market within the last five years the number is only likely to grow. As occurred during the large-scale increase in diagnostic manufacturers offering RDTs for malaria, there is the potential for the release of products with less than optimal performance and reliability, making it essential that laboratories all verify these tests before attempting to implement them locally for the diagnosis of human infections.

12.11. Syndromic Assays

Because many parasites may be grouped with other pathogens (viral, bacterial and fungal) by the symptoms they cause and may be detected by the same molecular methodology, a growth in so called syndromic assays has occurred in recent years. Of these syndromic panels, only those involving gastroenteritis have so far incorporated parasite detection, however, blood-borne causes of fever may also be a growth market in the future. The following section is by no means an exhaustive list of these but lists the platforms most commonly deployed in European diagnostic laboratories.

12.11.1. Enteric Bio System

The Enteric Bio system is a syndrome-based assay that allows for the simplified detection of pathogens capable of causing gastro-enteritis. Stool samples are tested by lightly coating a swab in fresh stool material, and then resuspending this material in a large volume of extraction buffer. The samples are then placed onto a heating array and heated to release nucleic acid, before the DNA extract is added to lyophilized multiplexed PCR reactions. A number of different panels are available for bacterial and viral pathogens: The Gastro Panel 2 detects *Cryptosporidium* spp. and *G. intestinalis* in addition to the bacterial targets of *Salmonella*, *Shigella* and *Campylobacater*. Separate parasite-specific panels also exist, with one parasite panel detecting the presence of *Cryptosporidium*, *G intestinalis* and *Entamoeba histolytica* and a second panel detecting all three pathogens together with *Dientamoeba fragilis*, *Blastocystis hominis* and *Cyclospora cayetanensis*. The choice of *Cryptosporidiu*m and *Giardia* in the Gastro Panel 2 is a sensible one, since the IIB1 and IIB2 studies both identified these parasites as causing over 95% of parasitic infections found in visitors and residents tested for gastrointestinal infections in the UK.

The significant reduction in the complexity of the DNA extraction process and the full automation of the assay set up following stool lysis significantly reduces operator time and error. The reconstituted reactions with stool DNA extracts are then capped and run on a modified Roche light cycler. The software automatically analyses samples to identify positive, negative or equivocal results. Once validated by a human operator the results may be automatically released *via* the Laboratory Information Management System (LIMS), again reducing operator time and potential error. The sensitivities for the Enteric Bio Panel 2 parasite targets are between 95% and 100% for *Giardia* and *Cryptosporidium* in various studies, but for *Giardia* an increased number of positives has been seen compared with microscopy, again most probably due to the limited LOD of conventional microscopy.

12.11.2. Biofire Film Array

The film array is marketed as a syndromic multiplex PCR system. It combines automated sample extraction, rapid amplification and analysis and automated results release *via* LIMS. The system uses bead beating and magnetic particle DNA purification. Through a two-stage

PCR reaction and target-specific second-round PCR primers in contained reaction chambers a highly sensitive and specific reaction is said to be achievable. Amplicon production is monitored by the use of an intercalating fluorescent dye with the capacity therefore to perform melt curve analysis to determine the correct identity of the amplicons.

The gastrointestinal panel covers bacterial, viral and parasite targets, most notably *Cryptosporidium* spp., *Cyclospora*, *E. histolytica* and *Giardia intestinalis*. Because of the integrated sample processing, DNA extraction, amplification and analysis the system is very fast, producing a result within 1 hour of the start of sample processing. A multicentre analysis of the Biofire Film Array system resulted in a sensitivity for of 100% for *Cyclospora*, *Giardia* and *Cryptosporidium* when tested against real-time PCR with DNA extracted using a MagNA Pure LC 2.0 automated platform. *Entamoeba histolytica* was not present in any of the samples tested in this analysis. Specificity was between 99.5% and 100% for these parasites. The Biofire has low hands-on time for the operator and the time to result is approximately 60 minutes, however, the biofire system does not allow samples to be batch processed, limiting the throughput achievable with a single machine to 7–8 samples per 8 hour shift.

12.11.3. Luminex xTAG System

Luminex xTAG Gastrointestinal Pathogen Panel (GPP) is another syndromic panel test using the proprietary XTAG detection system. The GPP assay simultaneously diagnoses 15 different causative agents of gastrointestinal disease. After routine DNA extraction all 15 targets are simultaneously amplified in a PCR reaction which randomly incorporates fluorescent cytosine's into the amplicons. The resultant amplicons are denatured and hybridized to complementary sequences on microspheres, which are then subject to flow cytometry. Each set of microspheres is a different size and so they migrate independently from each other. The fluorescent signals on each bead cluster are read to detect the presence of complementary amplicon by laser excitation. Alternatively the reactions may be analysed with a fluorescent image analyser or even a fluorimeter. In an independent analysis of the Xtag system, sensitivity for *Giardia* diagnosis was 100%, compared with only 67% using microscopy. Sensitivity for *Cryptosporidium* diagnosis was 93%, identical to microscopy. The XTAG system has approximately 60 minutes hands-on time, with a time to result of under 6 hours. The capacity to batch process samples on the XTAG system means that 96 samples could be processed within an 8 hour shift. However, several publications have suggested an increased false positivity rate with the XTAG system compared with other nucleic amplification technologies.

12.11.4. BD Max

Becton Dickinson's BD MAX system is a third syndromic panel tied to proprietary technology. Again, the system has on-board DNA extraction, amplification and analysis, like the aforementioned systems. Viral, bacterial and parasite panels are available for the diagnosis of gastroenteritis, with the parasite panel containing *G. intestinalis*, *Cryptosporidium* spp. (*C. parvum* and *C. hominis*) and *Entamoeba histolytica*. Results are ready within 4–5 hours of receipt of sample. In an independent analysis sensitivity was calculated at 100% for both *Cryptosporidium* and *G. intestinalis* with relative specificities of 99.7 and 98.9% when compared against stool microscopy. Similar results have been seen in other assays using microscopy as a comparator, although when in-house PCRs are used as the gold standard the performance is lower.

12.11.5. EasyScreen Protozoan Detection Kit

The EasyScreen detection kits incorporate a novel nucleic acid clean up and amplification technique utilising the 3base technology. All cysteines are first converted to uracil and then to thymine. This has the effect of normalizing primer melt temperatures in order to allow for easier multiplexing of reactions. Cell lysis and base conversion is complete within 15 minutes and is integrated into an automated workflow using DNA purification platforms such as the MagnNa PURE or the KingFisher Flex. Alternatively, the process can be integrated into a single platform, the Genetic Signatures GS1, for DNA purification and PCR set up. The system is set to detect *Cryptosporidium* spp., *G. intestinalis*, *D. fragilis*, *E. histolytica* and *Blastocystis* spp. Independent analysis has shown a sensitivity of 92–100% for these species, with a specificity of 100% for these parasites when compared with microscopy and in-house real-time PCR.

12.12. Digital PCR

The previously discussed, real-time assays may be used in a semi quantitative manner to determine the starting quantity of template. This is calculated from the number of cycles required to generate a sufficiently intense fluorescent signal, the 'threshold signal'. This cycle time has a log–linear relationship with the starting template concentration. When this data is integrated into a standard curve produced from a known dilution series of target nucleic acids (either whole parasite DNA or RNA or synthetically produced plasmid vectors) the number of parasite genomes in the PCR reaction may be determined. This is, of course, completely dependent upon the lack of inhibitors in the nucleic acid sample. Droplet PCR works on a serial dilution of sample and encapsulation of the target loci into microdroplets containing the amplification reaction components in such a way that there is, on average, less than one copy of the target locus per microdroplet. The number of positive microdroplets is then counted and the total number of loci copies in the reaction determined by this methodology. The quantitative diagnosis of malaria was shown by one study to vary by as much as sixfold between replicate assays, whereas the inter-assay variability with digital PCR was less than twofold. The accurate quantitation of malaria infection is necessary where alternative treatment pathways are triggered by different parasite densities in an infection. In those individuals with parasitaemias of greater than 3%, treatment with intravenous artesunate is recommended under UK guidelines, whereas other patients may be treated with oral schizontizides depending upon the patient's symptoms. A sixfold difference in parasite density caused by assay variability could therefore have a major effect on patient safety. The major drawback to digital PCR as a front line diagnostic technique is the requirement for dedicated equipment, sample preparation time and inability to batch analyse samples. Since a 3% parasitaemia is easily recognisable under the microscope there is little likelihood of digital PCR making it into mainstream diagnostic pathways for the detection of parasite infections in the near future, especially when many health providers are facing budgetary constraints on their services, including pathology-based specialties.

12.13. RNA-based Assays

Previously mentioned technologies have relied upon the amplification and detection of the parasite genome. The capacity to amplify parasite RNA offers the possibility to significantly increase the LOD of assays, whilst simultaneously differentiating between live and dead parasites. For malaria the LOD achievable using RNA-based methodologies has been listed as 20 parasites per ml for an assay replicated in several different research institutes, whilst other publications have claimed to achieve as little as two parasites per ml. Such levels are a significant improvement on the previously mentioned Snounou PCR. These assays are being challenged by DNA-based assays on multi-locus genes in telomeric regions of the *Plasmodium* genome, which claim to achieve 0.03–0.15 parasites per µl blood. Whilst these levels are useful in the detection of asymptomatic carriers of infection in endemic locations, such as sub-Saharan Africa, for a malaria eradication program, they are significantly over-spec for Western laboratories seeking to detect parasites in symptomatic patients. The inclusion of these assays in viral panels may allow for the commonality of the nucleic acid extraction and amplification pathway for infections such as HIV to be exploited without significant extra effort and the ability to determine live from dead parasites would be useful to screen returning travellers who have been treated abroad for malaria but still have residual symptoms.

12.14. Micro RNA

Micro RNAs (miRNAs) are short (often less than 22 bases) single-stranded non coding RNA molecules produced by eukaryotes and even some viruses. Helminths have been shown to excrete miRNAs into the host system, possibly in an attempt to affect the host immune system in a way that is preferential to their survival. A number of papers have shown this for *Heligmosomoides polygyrus* and *Brugia malayi* as well as *Onochocirca vulvulans* (the causative agent of river blindness). This parasite does not produce significant numbers of immature microfilariae in the blood stream, making its diagnosis reliant on sampling the parasite in onchocircic nodules, where the mature parasite resides, through skin snips and slit skin smears. The ability to detect the parasite remotely through its miRNA would, therefore, represent a significant improvement in diagnostic technology. Unfortunately, a recent publication stated that the levels of miRNA released by the helminth *O. volvulas* was insufficient to be detected by standard LNA-based RT-PCR, the technique commonly used to detect and quantify miRNA.

12.15. Isothermal Amplification Assays

PCR and real-time PCR both have drawbacks to their deployment in resource-poor locations and many of the technologies based upon these methodologies are suboptimal concerning the time following receipt of the sample at which a result may be deemed negative. Isothermal amplification uses different enzymes to *Taq* polymerase, which overcome may of the limitations imposed by this enzyme. Although extensively re-engineered many forms of the enzyme are highly susceptible to inhibition by haem (making direct amplification from blood difficult) as well as other inhibitors found commonly in other samples. In addition the enzyme requires temperature cycling in order to allow for template denaturation as well as primer binding and extension using the polymerase. Descriptions of a number of different technologies have been published including rolling circle amplification (RCA), strand displacement amplification (SDA), helicase-dependent amplification (HDA), nicking enzyme amplification reaction (NEAR), LPA probes and the following technologies that circumvent many of these problems.

12.15.1. Twist DX

The Twist DX assay is an isothermal assay using recombinase polymerase amplification (RPA) that works most usually between 37 and 42 °C (far lower temperatures than most other isothermal amplification assays). Recombinases bound to the oligonucleotide primers initiates the reaction by engineering the recombination of the oligonucleotide with its homologous sequence in the genome, in the process nicking the genomic strand and creating an identical overhanging DNA strand and a replication initiation site at the 3′ end of the integrated oligonucleotide. A single-stranded binding protein stabilizes the overhanging DNA strand whilst a polymerase with strand displacement initiates replication of the complimentary genomic strand. With two opposing recombinase-bound primers the system is capable of repeating this cycle to copy each amplicon many times over in a very short period of time. Descriptions of a number of assays for parasites have been published, including for malaria, *Schistosomiasis* and visceral *Leishmaniasis*. The latter is incorporated into a mobile suitcase platform suitable for diagnosing a number of viral diseases as well as visceral *Leishmaniasis* intended to be deployed in resource-poor locations.

12.15.2. NASBA

Nucleic acid sequence-based amplification (NASBA) is an isothermal amplification assay targeting RNA sequences, most commonly at around 41 °C. The reaction is initiated by an antisense primer binding to a complementary sequence at the 3′ end of the RNA molecule to initiate reverse transcription of a complementary (−) DNA by Avian myeloblastosis virus reverse transcriptase (AMV-RT). The RNA in the DNA–RNA hybrid is destroyed by RNAse H, leaving a single stranded cDNA that is bound by a second primer with a 5′ T7 promoter sequence to initiate its replication, utilising the capacity of AMV-RT to replicate single-stranded DNA sequences. The resultant double-stranded DNA has a functional T7 promoter site, allowing for transcription of the DNA by a T7 polymerase to produce multiple copies of antisense (−) RNA. Each resultant antisense RNA may be bound by the second primer, which primes the production of a (+) DNA (−) RNA duplex by AMV-RT with a T7 promotor site. The RNA in this duplex is degraded by RNAse H, resulting in a single-stranded (+) DNA, which is bound by the first primer and *via* the action of AMV-RT produces a double strand, resulting in multiple copies of the double-stranded sequence downstream of a functional T7 promoter. By this cycling action of AMV-RT, RNAse H and T7 polymerase multiple double-stranded DNA copies of the single-stranded RNA target may be produced. If a molecular beacon probe is designed to be complimentary to one of these strands, then reaction may be detected in real time. So far NASBA assays have been published as putative diagnostics for malaria, leishmaniasis, toxoplasmosis and human African trypanosomiasis (HAT).

12.15.3. Gen-Probe

The APTIMA® *Trichomonas vaginalis* (Hologic) assay is one of only a few amplification-based assays for *T. vaginalis* currently cleared by the US FDA. The assay uses the proprietary Hologic platforms for DNA purification by probe capture and amplification and detection. The amplification is drive from parasite RNA utilising a transcription-mediated amplification (TMA) and hybridization protection assay (HPA). TMA uses the same technology employed by NASBA, with the exception that the T7 promoter site is at the 5′ end of the antisense primer and the resultant RNA is a sense RNA, which is bound by the antisense primer to initiate reverse transcription. Detection is achieved using torch probes, involving the decoupling of quencher and fluorophore in the same way as for molecular beacon probes.

12.15.4. LAMP

Loop-mediated amplification of DNA (LAMP) utilises the strand-displacing polymerase, *Bst*, from *Bacillus stereothermophilus* to achieve isothermal amplification. The technology utilises four separate primers, two terminal primers (B1 and B2) and two composite primers (FIP and BIP) to produce a self-priming dumbbell structure by specific amplification of the target DNA. The 3′ region of the FIP and BIP primers contain regions complimentary to the target genome, whilst the 5′ sequence is complementary to the other strand of DNA. Replication begins when the 3′ region of either FIP or BIP binds to the target locus. This is achievable because at the amplification temperature the DNA duplex is in dynamic equilibrium, folding and unfolding at random. The BST enzyme is capable of displacing all double-stranded DNA 3′ to the primer binding site, thereby initiating synthesis of a new complementary strand of DNA. The terminal primers' only role is to release this as a single-stranded moiety through strand displacement when they initiate synthesis upstream of the original binding site.

Once displaced, loop-mediated amplification requires the other composite primer to bind to this single-stranded DNA and initiate amplification of it, including the complementary 5′ sequence and for the other terminal primer to release it through strand-displacement DNA amplification. The resultant molecule produces a dumbbell shape thanks to the complementary sequence at the 5′ and 3′ end of the molecule. It is from the 3′ end of the molecule that self-priming, template extension occurs. In addition, the loop contains the sequence necessary for the 3′ of the composite primers to bind, and thereby initiate new amplification and new self-priming template release. Each amplicon created and subsequently released has the ability to both self-prime to extend itself and to be bound by the composite primers to create new templates. The resultant reaction kinetics are significantly faster than PCR, quickly exhausting the reaction within a matter of minutes after self-priming amplicons are first created.

As well as speed and the ability to be performed in a simple water bath, LAMP has several other advantages over PCR. The *Bst* enzyme is far more resistant to inhibition than *Taq* polymerase, meaning a much cruder source of DNA may be used. This, in turn, allows for simpler and faster DNA preparation methods. Secondly the amount of amplicon produced is much greater than can be achieved with PCR. This, in turn, leads to the production of significant levels of orthophosphate ions, a by-product of DNA synthesis. When this combines with free magnesium ions in the buffer, insoluble magnesium orthophosphate is produced, allowing for the synthesis of DNA to be followed by a number of different methodologies, including visual interpretation of the reaction by eye for cloudiness (end-point monitoring). A more accurate way of monitoring the production of magnesium orthophosphate is turbidimetry, either real-time or end point. The strong affinity of orthophosphate ions for divalent metal ions means it binds very tightly to Mn^{2+} ions, to the extent that calcein molecules, which are quenched by bound Mn^{2+}, have their ions stripped from them and will fluoresce upon excitation with UV light. This may be monitored by real-time machinery or as a visual end-point assay using a simple low-energy black light. In addition, intercalating fluorophores may be used to detect DNA amplification or the products may be visualized using gel electrophoresis or by the use of lateral-flow immunoassays.

The requirements for LAMP primer design are far more exacting than those for either conventional or real-time PCR, due to the need for self-priming loop structures to be created by composite primers in sufficiently close proximity to the terminal primers. In addition the ability to include a third set of primers, so-called loop primers, in the reaction significantly increases the speed of the reaction and its sensitivity. Because of this the existence of high-quality genomic sequence covering a large number of naturally occurring isolates (where the locus in question is polymorphic) is essential to developing a product with high clinical and analytical sensitivity and specificity. The presence of a single-nucleotide polymorphism at the wrong position of a primer sequence can significantly impede the reaction, and with so many primers the potential for this is significantly raised in LAMP compared with PCR.

LAMP assays have been published for the detection number of parasitic infection including malaria, HAT, schistosomiasis, hook worm (*N. americanus*), filariasis, leishmaniasis, clonorchiasis, amoebic dysentery (*E. histolytica*), eosinophilic meningitis (*Angiostrongylus cantonesis*), *Enterocytozoon bieneusi*-induced microsporidiosis, giardiasis, ascariasis and toxoplasmosis and differentiation between *T. solium* and *T. saginatum* for human taenaisis in order to determine if a patient should be screened for cysticercosis and differentiation of *Opisthorchis viverrini* from *C. sinesis*. Of all the papers published, the majority have focused on malaria, and it was not surprising therefore that the first CE-marked LAMP-based diagnostic was developed against malaria. A collaborative project between the Hospital for Tropical Diseases, the Foundation for New and Improved Diagnostics (FIND) and Eiken Chemical Company Ltd, resulted in the commercial launch of a CE-marked product capable of detecting less that 0.5 parasites per microlitre within one hour of sample receipt using either a proprietary rapid DNA extraction method that did not need any centrifuges, or a simple boil-and-spin reaction. This product was followed by a similar product from Meridian Biosciences for the diagnosis of malaria using LAMP. In addition to malaria, a commercial product has been released for HAT, whilst products for additional infections such as Chagas disease (*Trypanosoma cruzi*) are in development.

The ability to incorporate intercalating fluorophores into the LAMP reaction opens up the ability to multiplex different LAMP reactions, such that future syndromic products may be achievable, or alternatively the reaction may incorporate an internal control to evaluate DNA extraction and amplification to reduce the likelihood of false negatives occurring.

12.16. Lab on a Chip

The capacity to integrate DNA extraction, amplification and analysis into a single device (or even do away with one of these steps) offers the possibility to simplify and speed up parasite diagnosis. So-called lab on a chip devices seek to replicate the performance of large complex DNA extraction and amplification platforms into a single disposable device, with much lower start up costs. Lab on a chip has been shown to be an achievable reality with various systems, including most successfully the gene expert system. Alternative technologies have been showcased for malaria in the guise of a gel-based real time PCR reaction developed by Aquilla (Canada) that can amplify down to 0.5 parasites per microlitre from diluted blood added straight to these reaction blocks on a disposable chip. Interestingly many lab-on-a-chip-based technologies being investigated for the detection of parasites are not based on molecular technologies but instead use more traditional methodologies, such as microscopy or nanoparticle sorting.

12.16.1. Gene Expert

The Gene Expert system is a plug-and-play system for the detection of human pathogens. Whilst most well known for its TB assays, the system can detect a number of pathogens in a range of samples, including blood, sputum, stool and nasal and skin swabs. The utility of this system is that it combines sample extraction, amplification and detection in a single-use disposable cartridge for an assay that can be run sequentially as samples become available, rather than having to batch samples, as occurs with most other systems. The largest platform for running Xpert assays is the Cephid Infinity, capable of running 48 or even 80 tests simultaneously. Currently the only parasite detectable by this system *T. vaginalis*. Expert TV is an FDA-cleared molecular *Trichomonas* test for both men and woman, providing an answer within as little as 40 minutes of sample addition. The test is marketed as being highly sensitive, detecting up to five times more positives than microscopy, and the rapid turn-around time allows patients to be treated the same day they are seen. *T. vaginalis* is the most common sexually transmitted infection that is not viral in the USA, with up to two and half million American women and one and a half million American men infected. The capacity for this system to detect bacterial pathogens, such as *Clostridium difficile*, in stools opens the possibility that gastrointestinal parasites may also be detectable if the correct amplification chemistry is achieved. It is worth noting, however, that bacterial pathogens are often present in stool samples at copy numbers that are orders of magnitude more than are seen with parasite pathogens.

12.17. How Good are Genome Based Assays in Reality?

A great number of assays originate from academic institutions, which may lack access to true diagnostic samples. Assays may be published therefore solely on the basis of cultured organisms, or even synthetic DNA constructs. Similar restrictions may also affect commercial bodies, even those with access to clinical samples. There may be no true standard upon which the actual assay limit of detection may be based. Where such standards are lacking then determining the LOD is entirely dependent upon the quality of those techniques for the counting of parasites in cultured or clinical samples. In order that molecular techniques can be universally and reliably evaluated by other institutes, universal standards must be created that are externally validated for use in the development of these assays.

The National Institute for Biological Standards and Calibrants (NIBSC), is a UK-based organisation that is charged with the production of such standards. Whereas NIBSC currently produces six different standards for serological assays, they only produce DNA standards for *Toxoplasma gondii* and *Plasmodium falciparum*. The

Minimum Information for Publication of Quantitative Real-Time PCR Experiments (MIQE) guidelines discuss the use of internationally recognized standards, the utility of such was shown very clearly by an independent evaluation of seven published malaria real-time PCR techniques using the MIQE guidelines, which showed an almost tenfold difference in the published and actual LOD of some of these assay.

12.18. If You Don't Look for it You Won't Find it

Possibly the most apparent limitation with the allele-specific detection techniques mentioned above, is that the repertoire of parasites detected is fixed by the primer and probe combinations in the assay. No other organism will be detected, and even a mutation in a sub-species or group of isolates may prevent their detection, as shown previously when *P. ovale wallikeri* failed to be detected by arguably the most commonly used malaria PCR. The limitation of commercial systems for enteric pathogens to six different parasite pathogens (even ones that include the major causes of parasite-induced gastro-enteritis in the developed world) will mean that any other parasites are missed where molecular testing replaces stool microscopy. Such a panel would therefore miss any helminth-based infections. This would have serious implications for anyone infected, who would be at risk of cysticerocosis through autoinfection (as would all the other members of their household). The increase in hookworm infections in Northern America re-enforces the need to supplement current molecular technologies with alternative methodologies for detection of parasite infection, until molecular methodologies are deployed that overcome this limitation.

12.19. Next-generation Sequencing

Next-generation sequencing is a term used to describe a number of different technologies for the determination of gene sequences using massively parallel approaches (see Chapter 3). The major advantage of these techniques is the creation of clonally amplified libraries of fragmented sequences from the input genome(s). NGS is being developed to identify bacterial pathogens and their resistance profiles. Initially, this required the culturing of organisms in order to produce sufficient DNA for sequence analysis, although some headway is being made towards sequencing directly from clinical samples. Since many parasites are difficult to culture and will occur in a background of competing DNA, especially in stool samples, it is difficult to see how this technology will transcend into the field of parasitic disease diagnosis, since any parasite sequences will be swamped by a multitude of sequences originating from human cells and commensal organisms.

12.19.1. Other Uses for Next-generation Sequencing

Resistance to anti-parasitics is well known across the globe, most notably with the evolution of drug-resistant isolates of *Plasmodium falciparum*, first with the widespread resistance to chloroquine, then sulphadoxine-permethrin (SP), and finally with reports of delayed clearance of artemisinin. The first drug-resistance mutation for chloroquine resistance was mapped utilizing the meiotic recombination that occurs during crosses between a sensitive and resistant mutant in a chimpanzee model. Using a series of microsatellite markers the resistance marker was mapped to a single region of chromosome 7. However, the locus thought to be responsible for the resistance was actually incorrect, and it was only after the advent of whole-genome sequencing that the chloroquine resistance gene was identified. The use of a chimpanzee model is not only significantly laborious but also has ethical concerns. Microsatellite analysis on a genome-wide scale in naturally occurring strains resistant and sensitive to SP allowed the identification of the resistance mutations without the use of a chimpanzee host.

Whole-genome sequencing allows for a much richer analysis of the genome including possible loci resulting in delayed response to artemisinin. As yet there is not the diversity of anti-malarial compounds to allow for tailored treatment of sensitive and resistant parasites, as can occur for prokaryotic pathogens, however, with the potential evolution and spread of artemisinin-resistant malaria isolates, it may conceivably be necessary to begin the screening of malaria-infected individuals to work out the most efficacious method of treatment. This could be done through whole-genome sequencing, or most likely by utilizing whole-genome sequencing to monitor the evolution and spread of resistance mutations to allow for sequence-specific PCR assays to be developed in order to test for them. Additional parasites for which evolution of drug-resistance is a problem include *G. intestinalis*. Analyses of the genomes of a susceptible *T. circumcincta* strain and of the human hookworm *N. americanus* are underway (funded by

NHGRI). Genomes of drug-resistant isolates will providc the world community of helminth researchers with the resources needed to create a comprehensive map of genetic diversity in a model parasite and allow the utilization of this information in similar studies involving parasites of importance to humans.

12.20. MALDI-TOF

The presence of high-quality genomes has allowed for the identification of organisms by the high-throughput analysis of their metabolomes using matrix-assisted laser desorption/ionization-time of flight (MALDI-TOF) mass spectrometry. Clinical samples (most often after organism enrichment through culture) are mixed with an energy-absorbent, organic compound called matrix. The energy absorbance of the matrix allows the ionization of the sample with laser excitation. The resultant protonated ions are then accelerated towards an analyser, where the time of flight is based upon their mass to charge ratio. Where a spectrum [peptide mass fingerprint (PMF)] is identified that correlates to one previously generated from reference samples, the corresponding organism may be identified on the basis of its PMF. Spectra have been elucidated for microsporidia (*Encephalitozoon cuniculi*, *Encephalitozoon hellem* and *Encephalitozoon intestinalis*), *E. histolytica*, *E. dispar*, *Leishmania* spp. *Acanthamoeba* spp., *T. vaginalis* and *Trichinella* spp. For the most part, these spectra have been created from purified parasites, for instance, from soluble extracts of oocytes isolated by gradient centrifugation from infected stools. The requirement for high concentration, purified material for MALDI-TOF limits its current utility to identify only those parasites which may be purified away from the clinical sample (especially in the case of stool pathogens) or amplified by culture. As such the potential use for the technology appears currently limited to identifying the species of *Leishmania* parasite in culture-positive samples (first reported in 2013) and potentially the identification of *T. solium* from *T. saginatum* in proglottids shed in the stool samples.

12.21. External Quality Assurance Schemes for Molecular Parasitology

An essential part of any diagnostic laboratory is quality assurance. Quality assurance is the process by which the diagnostic processes are proved fit for purpose. Laboratories can attempt this by the implementation of internal quality assurance schemes (IQA) where samples are reintroduced into testing having been previously recoded so that the tester is unaware of the intended results. Through reanalysis the tester can determine if repeatability is achievable in their assay repertoire. This will not address any bias in the system, such as a propensity to produce false positives due to cross reactivity. Such problems can only be addressed by laboratories joining an external quality assurance (EQA) scheme, provided by a registered scheme organizer. Such organisers will send out pre-distribution samples to a number of reference laboratories, which use a range of diagnostic technologies in order to identify samples which are consistently positive or negative by these different techniques. The organisers of such schemes are unlikely to repeatedly pick exceptionally strong positives, but instead will choose a range of samples to test the specificity and sensitivity of participating laboratories. Such samples are sent out blinded and reports collected before the intended results are published. For European providers, there is a requirement to be registered with ISO17148, the accreditation scheme for providers of proficiency testing. Such schemes will also have post-distribution samples returned to them for random testing by reference laboratories in order to ensure the samples have not degraded in transport.

There are a number of parasitology EQA scheme providers in Europe. Of these providers, UKNEQAS has by far the most comprehensive portfolio of test schemes, offering schemes for parasite serology, stool microscopy, blood microscopy and malaria antigen tests. It is currently the only scheme to offer a molecular testing scheme for parasitology, which currently caters only to those laboratories offering diagnostics based upon malaria genome amplification, although RNA-based schemes are intended to be released.

The product sent out is lyophilized blood, containing either no malaria parasites or malaria parasites of known density that have been titred against the international standard by reference laboratories using a number of different assays. The lyophilized blood has been shown to be a stable product, capable of being shipped without the use of dry ice, making it significantly easier and cheaper to transport by air freight. This has led to the including of laboratories in a wide number of countries in the scheme including reference centres in Africa and Asia. A number of other molecular malaria schemes have been started or trialled around the world including RNA-based schemes, and a meeting of a number of scheme developers and potential clients in London in 2015 chaired by the WHO lead to UKNEQAS providing a bespoke scheme to the WHO, thereby allowing the WHO to roll it out to a wider a wider number of participants, including those involved in drug trial monitoring, routine diagnostic procedures and epidemiological studies.

The current lack of scheme for genome-based diagnostic procedures to detect other parasites must surely be a difficulty faced by laboratories asked how they can prove that their diagnostic procedures are fit for purpose during accreditation inspections. The capacity of blood to be freeze-dried whilst preserving DNA indicates that other matrices, such as urine, stools, and liver aspirates, may also be suitable for processing to facilitate rolling out schemes against parasites found in these samples. Until then laboratories are going to have to rely on IQA or interlaboratory swap schemes to fulfil their quality assurance requirements.

12.22. What are the Clinical Implications of Genome-based Parasite Diagnosis? Improved LOD – Increased Detection

Numerous molecular assays have been published with significantly improved limits of detection compared with manual microscopy. Even accepting that some are somewhat fanciful in their claims, there is a comprehensive body of work that proves molecular assays can detect much lower levels of many parasites than are achievable using light microscopy, although, as stated, the repertoire of parasites detected is completely pre-set for targeted molecular strategies such as PCR. The improvement in LOD does not always translate to an improved detection rate if patients present with a parasite burden within the LOD of skilled microscopists. An example of this can be seen from the clinical trial of a LAMP diagnostic for malaria at a reference centre for parasitology. During the period of the trial all patient samples were subject to PCR analysis. Although PCR was able to detect parasites in a number of samples that were microscopy negative, over the seven months that the trial ran only one of these samples was taken during the initial patient presentation, the rest were all follow-up samples from patients admitted for treatment. Even the single primary sample that was PCR-positive but microscopy-negative was from a patient that had previously been treated for malaria and would have been referred for PCR following the hospital's diagnostic algorithm. In such a scenario routine screen of patient samples would have detected no additional cases of malaria in over 700 patients screened.

A similar situation was observed in a study on the implementation of helmith diagnostic PCRs at a second parasite reference centre. Over of several months that the study was run only a single patient was detected using the assay! Compare such data to the multiple studies published for enteric parasites, where PCR had produced up to a four-fold increase in the number of patients testing positive for pathogens, such as microsporidia and *Giardia intestinalis*. The impact of molecular diagnostics is therefore highly dependent upon the parasites to which they are targeted and the patient population in which they are deployed.[3,4]

12.23. Cost Implications for Testing

The high Capex cost and Opex costs associated with service contracts and some consumables has an obvious effect of raising the cost of this service provision. This must be balanced against the capacity to batch process samples with some systems and to automate systems, thereby reducing user time and error, which will affect cost in the opposite direction. For some laboratories staff time is a significant factor in assay prices, whilst the capacity to prevent repeat testing is another cost saver. For the clinical management, a higher positive and negative predictive value has a benefit in certain scenarios, in reducing the need for repeat testing, improving the accuracy of treatment regimens and improving the clinical management of patients. The knock-on effect of this is a lowering of health care costs, especially for gastrointestinal infections. The inconsistent shedding of cysts is well known to confound attempts to detect them by light microscopy, requiring multiple samples to be sent in by patients in order to exclude the presence of infection. Even with multiple tests, the perception that a test has a poor negative predictive value may still influence clinicians to undertake empirical testing where the clinical symptoms are indicative of a given infection. As syndromic tests become more readily accessible and volume and competition forces manufacturers to drive down costs the equation is skewed more in the favour of molecular testing, especially when larger panels of parasites may be tested simultaneously with bacterial and viral pathogen to exclude common causes of these syndromes (such as gastroenteritis). Where these tests can utilise cell-free DNA in samples such as urine and saliva, the patients' experience will also be improved where invasive techniques such as venipuncture are replaced.

References

1. J. J. Verweij and C. R. Stensvold, *Clin. Microbiol. Rev.*, 2014, **27**(2), 371–418.
2. K. Freeman, H. Mistry, A. Tsertsvadze, P. Royle, N. McCarthy, S. Taylor-Phillips, R. Manuel and J. Mason, *Health Technol. Assess.*, 2017, **21**(23), 1–188.
3. L. van Lieshout and M. Roestenberg, *Clin. Microbiol. Infect.*, 2015, **21**(6), 520–528.
4. S. Momčilović, C. Cantacessi, V. Arsić-Arsenijević, D. Otranto and S. Tasić-Otašević, *Clin. Microbiol. Infect.*, 2018, pii: S1198–743X(18)30395-1.
5. S. D. Polley, I. J. González, D. Mohamed, R. Daly, K. Bowers, J. Watson, E. Mewse, M. Armstrong, C. Gray, M. D. Perkins, D. Bell, H. Kanda, N. Tomita, Y. Kubota, Y. Mori, P. L. Chiodini and C. J. Sutherland, *J. Infect. Dis.*, 2013, **208**(4), 637–644.

13 Connective Tissue Disorders – Ehlers–Danlos Syndrome

P. J. Sawle*[a] and F. M. Pope[b]

[a]University of Hertfordshire, Department of Biological and Environmental Sciences, College Lane, Hatfield, AL10 9AB, UK; [b]Ehlers–Danlos Diagnostic Service, Northwick Park Hospital, Watford Road, Harrow, HA1 3UJ, UK
*E-mail: p.sawle@herts.ac.uk

13.1. What Is Connective Tissue

Connective tissue, as the name suggests, is tissue which supports or separates differing tissue types in the mammalian body. It can be found throughout the body and exists in a number of forms, mostly comprising collagen and elastin. The type of function depends on which particular form the connective tissue is found in. For example, when in a dense regular form it provides structure to ligaments, tendons and blood vessels. While when in an irregular loose or dense form connective tissue provides stretching and/or torsional resistance to organs.[1] Connective tissue, surprisingly to some, also includes bone. Many of these tissues contain cells of the immune system, such as macrophages and eosinophils.[1,2]

Hypermobility (often referred to as being 'double jointed') is an example of the variability in the general population of connective

Genomics and Clinical Diagnostics
Edited by David Whitehouse and Ralph Rapley

Published by the Royal Society of Chemistry, www.rsc.org

tissue, with some able to demonstrate extreme joint movement beyond normal physical parameters. It plays a pivotal role in so many aspects of the human body it is easy to understand why conditions associated with it can be so varied and debilitating.

13.1.1. Brief Overview of Connective Tissue Disorders

There exist a number of connective tissue disorders, owing to its variability in both composition and distribution. The incidence of connective tissue disorders depends on the condition, for example Sjögren's syndrome affects between 0.5 and 3% of a given population, while, systemic lupus erythematosus affects between 15 and 50 people per 100 000.[3] They are often divided in to autoimmune disorders and inherited conditions. Connective tissue disorders that involve the immune system are a group of conditions in which there initially appears to be no underlying condition and in which the immune system has been triggered to attack the body's own cells. The trigger for this is often attributed to environmental factors or genetic predisposition, these include rheumatoid arthritis and systemic lupus erythematosus.[4,5] Individuals who show signs and symptoms of multiple connective tissue disorders can be diagnosed with mixed connective tissue disease (MCTD).[6]

Scurvy is an example of a connective tissue disorder associated with a nutritional deficiency (vitamin C), individuals suffer from poor wound healing, bleeding and bone abnormalities due to impaired stabilisation of collagens.[7]

Congenital conditions of connective tissue include; Marfan, an autosomal dominant condition, in which individuals have a distinctive appearance and suffer from serious heart and aortic conditions (mitral valve prolapse and aortic aneurysms) as well as flexible joints and scoliosis; and Ehlers–Danlos Syndrome (EDS), which is associated with collagen defects.[8–10]

13.2. Collagen

Collagen is a structural protein which is a common component of connective tissues in mammals. The list of connective tissues which contain collagen includes fibrous tissues, such as skin, tendons and ligaments. It is also found in cartilage, blood vessels, the cornea, bones, the gut, intervertebral discs and dentin. It is estimated to make up to ~25% of the total protein mass in mammals.[11]

Table 13.1 Collagen types and their associated chains, genes and tissues.

Type	Chains	Genes	Distribution
I	α1(I)	COL1A1	Skin, tendons arteries, bone
	α2(I)	COL1A2	
II	α1(II)	COL2A1	Cartilage, vitreous
III	α1(III)	COL3A1	Skin, arteries, uterus
V	α1(V)	COL5A1	Skin, tendons
	α2(V)	COL5A2	Cornea, bone
	α3(V)	COL5A3	Ligaments
XI	α1(XI)	COL11A1	Cartilage
	α2(XI)	COL11A2	

13.2.1. Structure

Collagen consists of a long, uninterrupted triple-helix structure which is composed of three α-polypeptide chains. This structure is rich in glycine and proline with repeating Gly–X–Y triplets.[12,13] There is a large family of different α-chains, which results in a considerable diversity in composition of collagen trimers. A list of the key collagen types, associated α-chains, genes and tissue localisation can be seen in Table 13.1.

Initially formed as procollagen, the N and C propeptides are cleaved by procollagen peptidase to produce tropocollagen. Tropocollagen is then able to associate into a collagen fibre *via* the formation of aldol and aldol–histidine cross links (see Figure 13.1).[14] The small size of the glycine side chains facilitates the folding of the triple helix. It is common in collagen variants for the glycine to be altered to another, larger, side chain group thus affecting the ability of the triple helix to fold.[15]

13.2.2. Collagen Conditions

Collagen disorders can result from genetic defects, in which variants can arise in structural collagen genes or genes for enzymes involved in collagen biosynthesis. Disorders can also be associated with deficiencies that affect the biosynthesis, assembly, post-translational modification and secretion of collagen.[16] Conditions that affect collagen are a heterogeneous group of syndromes, which tend to be multi-systemic, have phenotypic variability and result in monogenic inheritance.[17,18] Collagen-associated diseases include; osteogenesis imperfecta (OI); Alport syndrome; collagenopathy and EDS.[19–23]

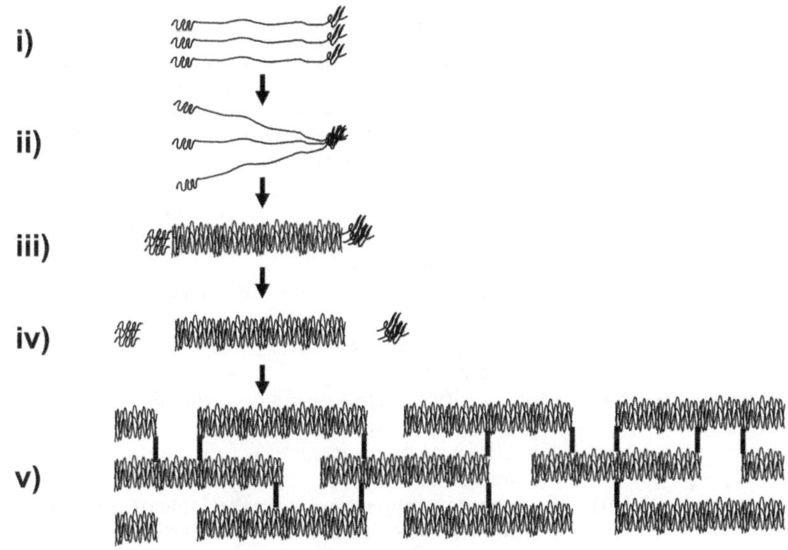

Figure 13.1 Fibrillar collagen processing pathway: (i) Synthesis and modification of procollagen chains. (ii) Chain association. (iii) Triple-helix formation. (iv) Cleavage of N and C propeptides. (v) Assembly and cross-linking of fibrils.

13.3. Ehlers–Danlos Syndrome

13.3.1. Overview

The Villefranche Nosology characterised and classified six types of EDS (see Table 13.2).[24] This classification was published in 1998 and, unsurprisingly, with the research and development into EDS in the years since this date many novel subtypes have been described with many feeling that reclassification or amendment of the subsets is required.[25] There exists a high degree of overlap among EDS, itself and other disorders which can be seen in Figure 13.2.

For the purpose of this review it is not necessary to discuss all of the subtypes and instead in this section we will consider a number of the prominent types of EDS with regard to their pathophysiology and genetic components.

13.3.2. Types of Ehlers–Danlos Syndrome

13.3.2.1. Classical Ehlers–Danlos

Classical EDS (cEDS) is an autosomal dominant condition for which the genetic basis is associated most commonly with mutations of COL5A1 and COL5A2, causing defects in Type V collagen.[25–28] It has

Table 13.2 Villefranche nosology of Ehlers–Danlos. The table details the type, major features and associated defect.[25]

Type	Major features	Defect
Classical EDS (EDS I and II)	Skin hyperextensibility; wide thin scars; joint hypermobility	Proα1(V) or proα2(V)
Hypermobility EDS (EDS III)	Generalised joint hypermobility; skin hyperextensibility	Unknown
Vascular EDS (EDS IV)	Arterial/intestinal/uterine fragility or rupture; easy bruising; characteristic facial appearance	Proα1(III)
Kyphoscoliosis (EDS VI)	Joint laxity; severe muscle hypotonia in infancy; scoliosis at birth; fragile sclera of the eye	Lysyl hydroxylase deficiency
Arthrochalasia (EDS VII)	Severe joint hypermobility with dislocations; congenital bilateral hip dislocations	Proα1(1) or proα2(I) deficiencies
Dermatosparaxis (EDS IV)	Severe skin fragility; sagging, redundant skin	Deficiency of procollagen 1 N-terminal peptidase in collagen type 1

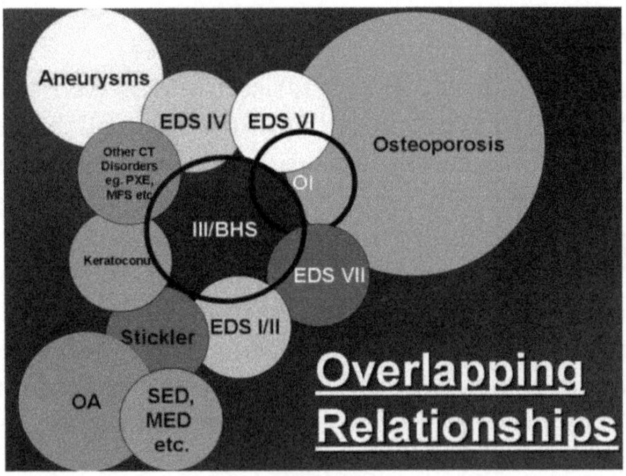

Figure 13.2 Overlap among types of EDS and other associated disorders.

also been associated, albeit in rare circumstances, with mutations of COL1A1, resulting in defects in Type I collagen.[26] Classical EDS can often be diagnosed from a clinical examination as well as some detail from an individual's medical history. The condition itself comprises a number of clinical criteria which can be categorised into the associated affected regions (see Figure 13.3).

Skin

- Atrophic scarring of the knees, shins, elbows, forehead and chin. This is due to the fragility of the skin. It is often highlighted in childhood as a result of the inquisitive and adventurous nature of children. Scarring tends to be wide with the scar tissue having a cigarette-paper-like appearance.
- Hyperextensible skin.
- Bruising, particularly of the shins.

Other

- Generalised joint hypermobility (GJH). This can result in dislocations or subluxations from innocuous incidents. As a consequences of this joint pain is often prevalent in those who suffer from cEDS.
- The fragility and extensibility of the skin is also found in other tissues, which can make hernias and cervical insufficiency more likely.
- Familial history of the condition.

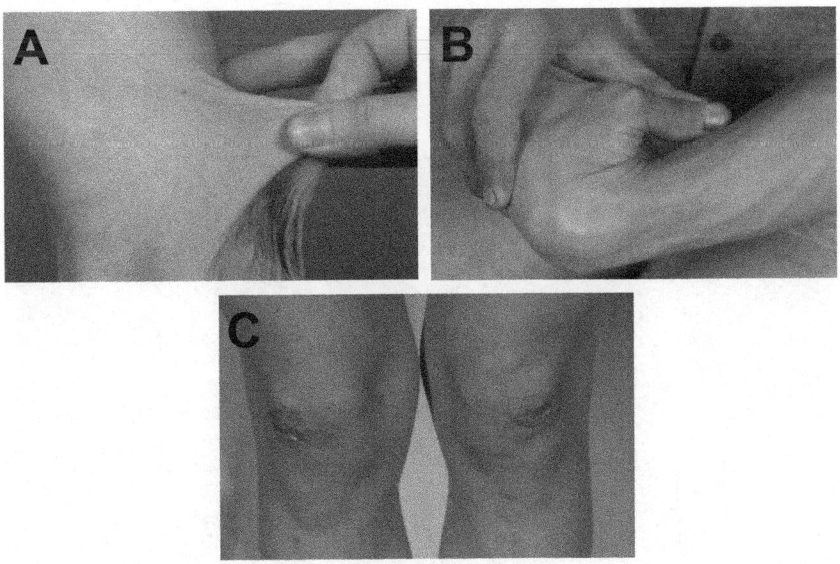

Figure 13.3 Classical EDS features: (A) Skin hyperextensibility. (B) Joint hypermobility. (C) Atrophic scarring of the knees.

13.3.2.2. Vascular Ehlers–Danlos

Vascular EDS (vEDS) is an autosomal dominant condition for which the genetic basis is associated most commonly with mutations of COL3A1, causing defects in Type III collagen.[25,26,29,30] It has also been associated, as with cEDS, in rare circumstances with mutations of COL1A1, resulting in defects in Type I collagen.[26] The condition itself comprises a number of clinical criteria which include aspects common to other types of EDS as well as aspects common to the general population, making definitive diagnosis difficult from purely clinical examination (see Figure 13.4).

Major criteria

- Familial history with a defined COL3A1 variant.
- Arterial rupture (presenting at a young age).
- Any of the following: spontaneous sigmoid colon perforation: uterine rupture: or carotid-cavernous sinus fistula (CCSF) formation in the absence of other associated causative factors.

Minor criteria

- History of easy and severe bruising. This may occur without any obvious trauma and occur in unusual places.
- Thin, translucent skin in which there is an increased venous visibility.
- Characteristic facial appearance often described as 'pixie' like. This constitutes thin lobeless ears, nose and lips and prominent eyes.

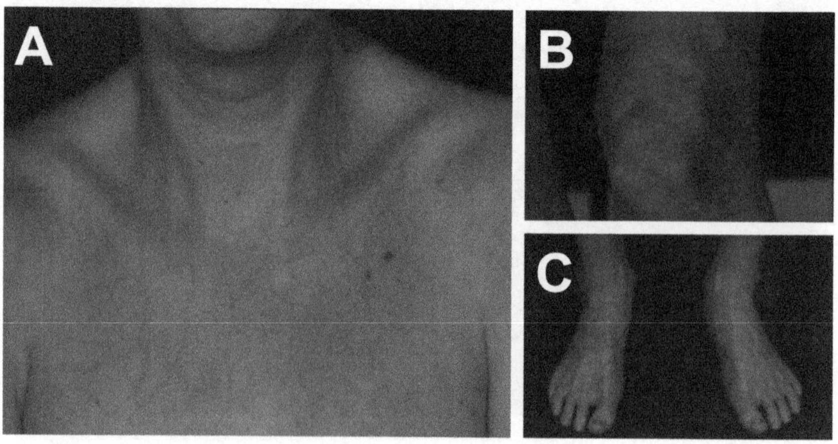

Figure 13.4 Vascular EDS features: (A) Note the thin translucent skin with easily visible vasculature. (B) Scarring of the knee. (C) Acrogeria of the feet.

- Acrogeria of both the hands and feet.
- Congenital hip dislocation.
- Hypermobility associated with the small joints.
- Early-onset varicose veins.

13.3.2.3. Hypermobility

Hypermobile EDS (hEDS) is an autosomal dominant condition for which the genetic basis remains unknown.[25,31] Hypermobile EDS diagnosis relies on thorough clinical examination as well as a detailed medical history. The condition itself comprises a number of clinical criteria which must be present for a definitive diagnosis to be made. The use of the Beighton score can give an indication as to whether an individual could potentially suffer from hypermobility and, therefore, be at risk of hEDS (see Figure 13.5).[32] The clinical diagnosis of hEDS needs the simultaneous presence of criteria 1, 2 and 3.[25] The general aspects of each criterion are listed below; however, this is not an exhaustive list.

Criterion 1 – GJH

Criterion 2 – Systemic manifestations of what could be considered a generalised connective tissue disorder. These are broken up into Feature A, B and C and can include:

Feature A

- Unusually soft or smooth skin.
- Mild hyperextensibility of the skin.
- Recurrent or multiple abdominal hernia(s) (*e.g.* umbilical, inguinal, crural).
- Atrophic scarring involving at least two sites.
- Dental crowding and high or narrow palate.
- Arm span-to-height ratio ≥ 1.05.

Feature B

- A positive family history for hEDS.

Feature C

- Musculoskeletal complications

Criterion 3

- The absence of skin fragility.
- Efforts to exclude other heritable and acquired connective tissue disorders.
- Efforts to exclude other alternative diagnoses that may also include joint hypermobility.

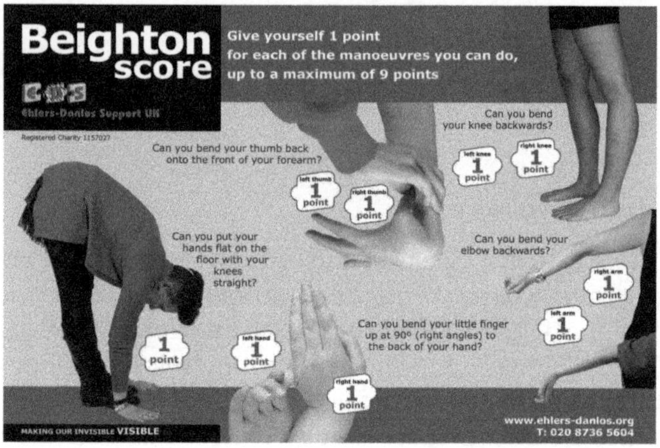

Figure 13.5 How to calculate Beighton Score. The Beighton Scale can be used to identify individuals with Hypermobility. A positive Beighton score does not necessarily mean you have EDS. (Reproduced from https://twitter.com/ehlersdanlosuk/status/799221824848621569 with permission from EDS UK).

13.3.2.4. Kyphoscoliotic Ehlers–Danlos

Kyphoscoliotic EDS (kEDS) is an autosomal recessive condition in which variants in either procollagen–lysine, 2-oxoglutarate 5-dioxygenase 1 (PLOD 1) or FKBP14 are causative.[25,26,33] Both are important in post translational modification of collagens, and their subsequent alteration means that collagens are not processed fully (under-hydroxylation, for example, in the case of kEDS PLOD1).[34,35] The condition is generally detected using urinary cross link analysis, absence of an abnormal urinary cross link analysis result excludes kEDS, whereas confirmatory genetic sequencing does not. There are a number of associated criteria which can help direct diagnosis of kEDS these are divided into major and minor criteria (see Figure 13.6).

Major Criteria

- GJH.
- Congenital muscle hypotonia.
- Kyphoscoliosis (congenital or early onset).

Minor Criteria

- Easily bruise.
- Hyperextensible skin.
- Blue sclerae.
- Medium sized aortic rupture or aneurysm.

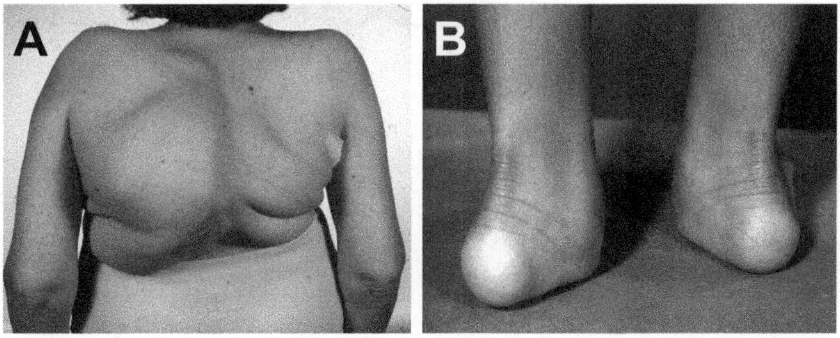

Figure 13.6 Kyphoscoliosis EDS Features: (A) scoliosis of the spine and (B) outward turned heals. Reproduced from ref. 56 with permission from Elsevier, Copyright 1995.

13.3.2.5. Arthrochalasia Ehlers–Danlos

Arthrochalasia EDS (aEDS) is an autosomal dominant condition in which variants in COL1A1 or COL1A2 lead to entire or partial deletion of exon 6 in the respective gene.[25,26] These variants result in the presence of pNα1(I) and/or pNα2(I) chains in which the C terminal but not the N terminal region has been cleaved.[36] The condition should be confirmed *via* sequencing, where absence of a complete or partial loss of exon 6 in COL1A1 or COL1A2 excludes aEDS.[25,37] There are a number of associated criteria which can be categorised into major and minor which can help indicate if an individual is a sufferer of aEDS (see Figure 13.7).

Major criteria

- Congenital bilateral hip dislocation.
- Severe GJH, with multiple dislocations and/or subluxations.
- Hyperextensible skin.

Minor criteria includes

- Tissue fragility.
- Muscle hypotonia.
- Easily bruises

13.3.3. Referral of Patients

Individuals for whom EDS testing is requested are referred from a number of sources. The range of these sources is varied, highlighting the diverse presentation of the condition. In terms of the EDS Diagnostic

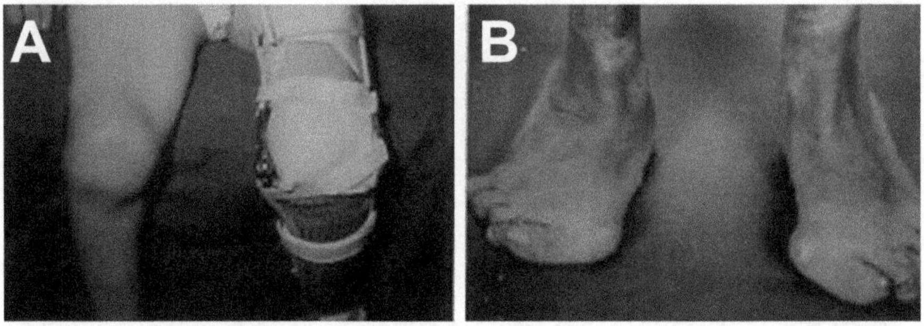

Figure 13.7 Atherosclerosis EDS Features: (A) bracing of joints to aid in support and (B) clubbed feet. Reproduced from ref. 56 with permission from Elsevier, Copyright 1995.

Service (specifically Northwick Park Hospital) in the UK, most referrals are received from rheumatology and genetics units, with more or less equal numbers from both. A large number of referrals are also received from cardiologists, surgical teams and neurologists. There are some referrals from gastroenterologists, paediatricians and haematologists and the occasional referral from other specialities (*e.g.* dermatologists).

13.4. Diagnostic Testing

Within the UK complex EDS diagnostic testing is carried out by the EDS Diagnostic Service which has units at Northwick Park Hospital in London and Sheffield Children's Hospital. The services provide diagnostic testing for individuals and families with suspected complex EDS. Testing can be carried out using a number of diagnostic protocols. Originally collagen protein biochemistry was the principal method of screening specific collagens, but with the wide availability of sequencing, including the rolling out of next-generation sequencing, alternative techniques are now available. Within the EDS diagnostic service a number of techniques were routinely conducted to help diagnose complex EDS cases. Each method provides information about a specific aspect of an individual's condition and together could be used to give a complete picture to what can be a multifaceted condition. The next sections contain descriptions of the various techniques, examples of common results and pros and cons for some of the protocols. There are no examples for hEDS as no specific test can be used to diagnose the condition.

13.4.1. Collagen Biochemistry

13.4.1.1. Overview of Protocol

To determine the collagen profile of an individual it is possible to perform collagen protein analysis based on the technique described by Nuytinck *et al.* in 1992.[38] The assay involves the culture of fibroblasts from skin biopsies taken from a patient's upper inner arm. Once confluent, patient fibroblasts are labelled with ^{14}C-proline and incubated for 24 hours, after which medium-associated procollagens and collagens are collected using ethanol precipitation. Samples are scintillated to prepare equal amounts for loading. Pepsin is used to partially digest medium-associated procollagens to get collagens. Proteins are separated on 5% sodium dodecyl sulphate–pulsed agarose gel electrophoresis (SDS–PAGE) gels, under reducing conditions. The SDS–PAGE gels are fixed and dried before exposing them to either photographic film or a phosphor screen. Photographic film is developed or the phosphor screen is read on a GE Typhoon Trio Variable Mode Imager to visualise the protein bands.[31]

Collagens from the cells and not just the medium can also be prepared and these are generally pepsinised to screen only the collagen aspect, as the procollagen pattern of cellular collagens is too complicated and not used diagnostically. Cellular collagen can provide a means to determine if collagen transport has been affected. Collagens that are retained in the cell would show up as a strong band on the cellular collagen gel and have a correspondingly weaker band on the medium collagen gel.

An example of a collagen gel can be seen in Figure 13.8 which shows both medium collagen and procollagen gels. The labelling of the bands is based on the classification of collagens which can be seen in Table 1.1.

13.4.1.2. Examples

13.4.1.2.1. EDS Biochemistry

In cEDS mutations of COL5A1 lead to defects in Type V collagen. In Figure 13.9 there is a doublet of the α1(V) band as a result of modifications to some of the α1(V) collagen.

13.4.1.2.2. vEDS Biochemistry

Vascular EDS is a result of mutations in COL3A1, resulting in defects in Type III collagen. This normally manifests as a reduction in Type

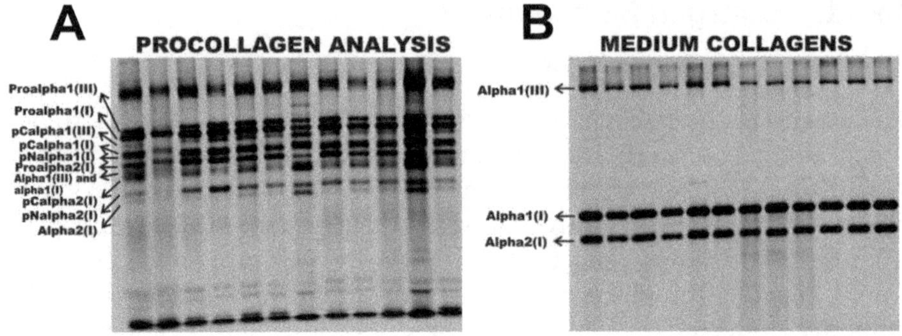

Figure 13.8 Example collagen and procollagen gel: (A) medium procollagen gel in which collagens with a mixture of N and C terminals are still present. (B) Medium collagen gel in which pepsin has been used to cleave all the N and C terminals thus greatly simplifying the gel. Reproduced with permission from Dr Sofie Symoens, Center for Medical Genetics Ghent (CMGG), Ghent University Hospital.

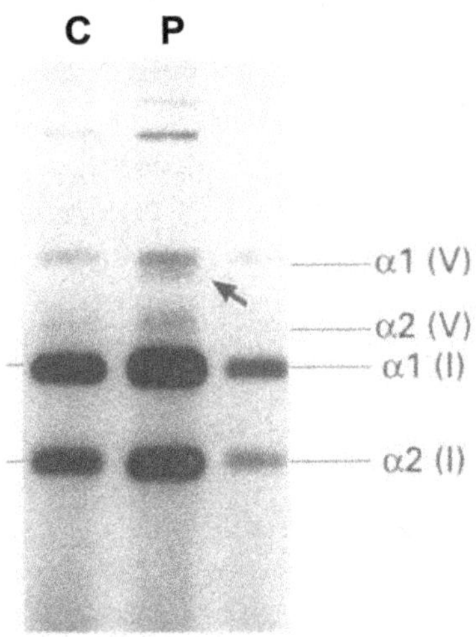

Figure 13.9 cEDS biochemistry example. A cellular collagen gel, note the doublet present in the patient sample (P) for α1(V) which is not present in the control (C). Reproduced from ref. 57 with permission from the BMJ Publishing Group Ltd, Copyright 1996.

III collagen, which is readily detected by collagen biochemistry. On a procollagen gel the amount of Proα1(III) and correspondingly on a collagen gel the amount of $\alpha1(III)_3$ are significantly reduced. This can be seen in Figure 13.10.

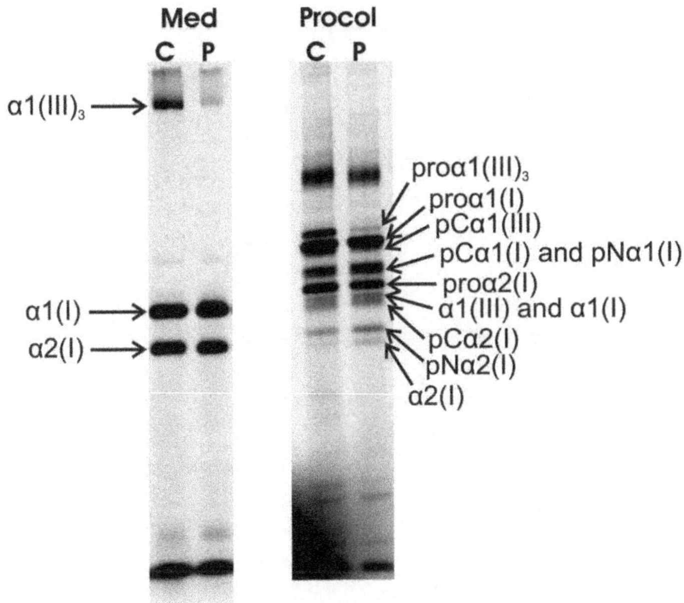

Figure 13.10 vEDS biochemistry example. Medium collagen and procollagen gels in which Type III collagen is significantly reduced (both α1(III)$_3$ and proα1(III)$_3$) in the patient sample (P) compared with the control (C). Reproduced with permission from Dr Sofie Symoens, Center for Medical Genetics Ghent (CMGG), Ghent University Hospital.

Often the reduction in Type III collagen can be striking, with almost no protein present, whilst on other occasions the reduction can be more subtle, requiring further investigation.

13.4.1.2.3. kEDS Biochemistry

Kyphoscoliotic EDS (PLOD1) can be indicated by under-hydroxylation of collagen chains α1(I) and α2(I). This can be seen in Figure 13.11 where both the α1(I) and α2(I) bands in lanes 3 and 4 run further than the same chains in the other lanes. This increase in run distance means a decrease in protein size and is associated with under-hydroxylation of the chains. These abnormalities are often subtle.

13.4.1.2.4. aEDS Biochemistry

In aEDS the retention of pNα1(I) or pNα2(I) (aEDS Type A/B) and the retention of both pNα1(I) and pNα2(I) (aEDS Type C) result in the condition. This can be detected on the collagen gel and Figure

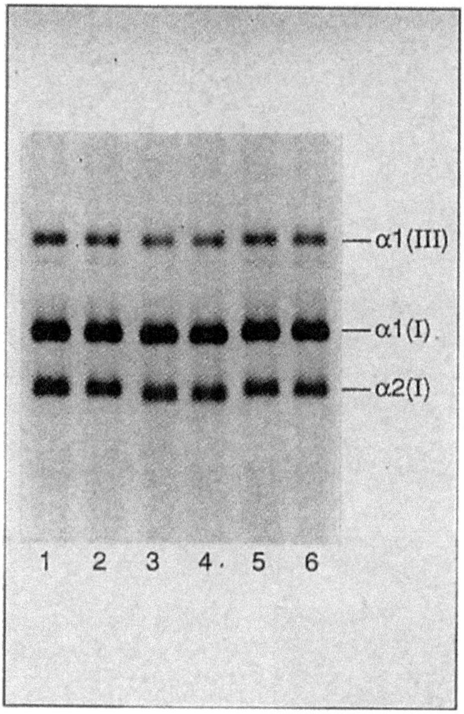

Figure 13.11 kEDS biochemistry example. Medium collagen gel in which the α1(I) and α2(I) of lanes 3 and 4 can be clearly seen to migrate further through the gel. Reproduced from ref. 56 with permission from Elsevier, Copyright 1995.

13.12 gives an example of aEDS Type B, where pNα2 (I) is retained, resulting in a doublet with the fully processed α2(I) and the pNα2(I) band.

13.4.1.3. Benefits

There are numerous benefits to performing collagen biochemistry to screen for EDS. One of the most prominent is the ability to screen all the selected collagens (Type I, Type III and Type V) in one process, allowing the collagen profile of an individual to be determined. Often with sequencing the most likely collagen candidate is chosen and sequenced. Running a collagen protein analysis would normally precede Sanger sequencing. By running the biochemistry first the likely collagen gene variant can be determined, saving time and money by determining which collagen should be sequenced. Next-generation sequencing has reduced the impact of this, as multiple collagens can be sequenced at once. Collagen biochemistry provides

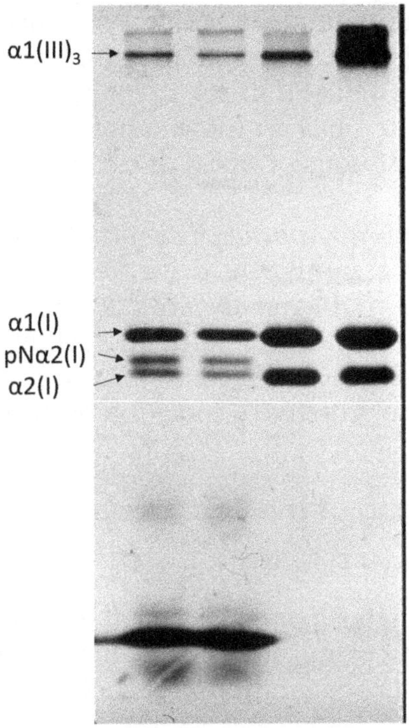

Figure 13.12 aEDS biochemistry example. Medium collagen gel in which there is reduction in α2(I) as a result of the retention of pNα2(I). Reproduced from ref. 56 with permission from Elsevier, Copyright 1995.

vital information on whether the protein is functional, by revealing if the collagens have been processed properly. Defectively processed collagens result in collagen deficiency. Collagen biochemistry can also highlight transport defects, in which collagen is produced but not exported from the cell effectively also resulting in collagen deficiency. Such issues cannot be detected by EDS sequencing, where the associated gene would be normal. As important as sequencing is, in a condition where other factors can potentially influence the outcome making sure the product of the gene is present, functional and where it needs to be is vitally important. Other benefits of collagen protein biochemistry include the low cost per individual and the relatively quick turn-around time.

13.4.1.4. Drawbacks

There are a number of drawbacks to the technique. The most significant of these would be the relatively complex interpretation of gels. Unlike certain other conditions where the presence or absence of a band can be used to determine the presence or absence of a condition, EDS biochemistry can have much more subtlety in its interpretation. Requiring specialist training to interpret, it is not uncommon for gels to yield little obvious information at first glance. This can complicate matters in today's world of relatively straightforward yes/no diagnostics.

The assay also still incorporates the use of radioactivity, albeit ^{14}C, but there has been a significant drive to get away from the use of radioactive isotopes in diagnostic work. There has been a switch to using fluorescence to detect medium and cellular collagens but this has not been able to be transferred to procollagen detection owing to the complexity of the protein composition of the sample.[39]

13.4.2. Sequencing of the COL Genes

13.4.2.1. Overview of Protocol

Sequencing of collagen genes can be carried out in a number of ways. Originally carried out using Sanger sequencing it is now being carried out more often by next-generation sequencing (commonly using Illumina technology).[40]

Sanger sequencing involves isolating total genomic DNA from 2–5 ml of peripheral blood using standardised extraction methods. DNA sequencing of polymerase chain reaction (PCR)-amplified collagen gene fragments (depending on the suspected EDS type - COL3A1, COL1A1, COL5A1, *etc.*) covering all exons and intron–exon boundaries is carried out using an automated sequencer (such as an ABI 3730) and Big Dye Terminator Sequencing protocol. Dosage analysis is then performed using multiplex ligation-dependent probe amplification.[31]

Next-generation sequencing can be carried out a number of ways: massively parallel signature sequencing (MPSS), Polony sequencing, 454 pyrosequencing, Illumina (Solexa) sequencing, SOLiD sequencing, Ion Torrent semiconductor sequencing, DNA nanoball sequencing, Heliscope single-molecule sequencing; for the purpose of this chapter the authors will refer to Illumina sequencing. During Illumina sequencing primers attach to the forward strands, after which fluorescently tagged nucleotides are added to the DNA strand of the sample. One base is added per round. Each nucleotide has a reversible terminator, which prevents multiple additions during each round. Each of the four nucleotide bases has a unique fluorescent emission, which after each round is recorded, meaning that the exact base added each round can be determined. Sequencing occurs for millions of clusters at once, with each cluster having ~1000 identical copies of the DNA. The sequence is then analysed by determining the various fragments and looking for overlapping areas, called contigs, which can be aligned. If a reference sequence is available, the contigs can be compared with it to identify mutations (variants). Illumina sequencing for EDS uses a panel of genes linked to EDS (COL1A1, COL3A1, COL5A1, PLOD1, FKBP14, *etc.*) and associated connective tissue disorders (such as OI). Thus, unlike Sanger sequencing, individuals can be sequenced for multiple genes during the same run, this saves both time and money.

13.4.2.2. Benefits

The most significant benefit sequencing has is that it can specifically identify the exact variation in the gene. This is important as it allows an individual to have a confirmed diagnosis, whereas with the other techniques there could be some level of interpretation associated with the result. Identifying a specific variant is also important as it means that familial testing of both immediate and extended family can take place. This is beneficial for two reasons; (1) family members can be excluded from suffering the condition, including for family planning (important as most subtypes of EDS are autosomal dominant) and (2) it also allows research into the genetics and syndrome development of the particular variant.

Since the switch to next-generation sequencing, panels of genes can now be run covering all the genes associated with EDS. This process is now much faster than screening with Sanger sequencing, where one gene would be screened at a time (see Chapter 3). The panel of genes has also helped to reduce the cost associated with screening.

13.4.2.3. Drawbacks

The main drawback still associated with sequencing is the cost, both the initial outlay, which can be in the hundreds of thousands of pounds, and the running costs, which can be up to £1000 per patient. This is coming down all the time as sequencing technology improves and also as it is used to diagnose further conditions.

There are specific drawbacks related to the type of sequencing undertaken. As alluded to previously, Sanger sequencing is used to sequence one specific gene. This gene is normally chosen on the basis of clinical examination and the results of the other techniques described in this chapter. As such, if nothing is found from sequencing that particular gene sometimes further genes are examined (due to the high degree of crossover within EDS). This wastes both time and money. With the introduction of next-generation sequencing and panels this is now no longer a problem since an individual can have all of their EDS-associated genes sequenced at once. But this does result in a new problem, which is the large amount of data associated with sequencing all these genes and understanding the natural variability *vs.* pathophysiological variability.[41]

Sequencing, it should also be reiterated, tells you whether you have a 'normal' gene but does not tell you whether the protein produced by that gene is functional. Is the protein processed properly? Is it exported to where it needs to be? Thus individuals can have no mutation but still suffer from EDS, highlighting the need for the biochemical functional assay.

13.4.3. Electron Microscopy

13.4.3.1. Overview of Protocol

Electron microscopy (EM) can be used to assess the collagen fibril structure. EM involves the use of a beam of electrons to visualize a sample. It is far more powerful than regular light microscopy and the increased magnification can see fine detail in much smaller objects. EM is used to look at the overall structure of the fibroblast, look for variations in the internal structure of the cell (such as the Golgi apparatus) and the overall organisation of collagen fibrils, which could indicate any problems.[42,43] There are various types of electron microscopes, commonly the transmission electron microscope (TEM), which has better resolution, is used in preference over the scanning electron microscope (SEM), which has better bulk imaging and depth of view. Skin biopsies of individuals (3 mm punch) are taken from the upper arm and added to a fixative (4% glutaraldehyde in 0.1 M phosphate buffer). Samples are then post-fixed in osmium tetroxide, dehydrated and embedded in epoxy resin. Thin sections are then stained with uranyl acetate and lead citrate prior to examination with the electron microscope.[31]

13.4.3.2. Examples

The following are examples of collagen fibril packing. The collagen fibrils are being viewed head on.

13.4.3.2.1. Normal EM

Within normal individuals collagen fibrils are of equal size and shape. They form a flower petal pattern (groups of seven fibres) which is consistent throughout (see Figure 13.13).

13.4.3.2.2. cEDS EM

Within cEDS the EM pattern changes, in this instance collagen fibrils are no longer equal in size or shape (although the variation is minimal) as a result of this the packing is no longer consistent and the flower petal pattern is lost (Figure 13.14). The changes in size and shape occur due to the variation in the makeup of the collagen. A common feature among cEDS individuals is rosettes, in which irregular collagen packing results in a deformed collagen fibril.

13.4.3.2.3. vEDS EM

The results of vEDS EM share some similarities with those for cEDS EM, however, collagen fibrils vary significantly in size and shape and the packing is loose in vEDS patients. There are no rosettes found

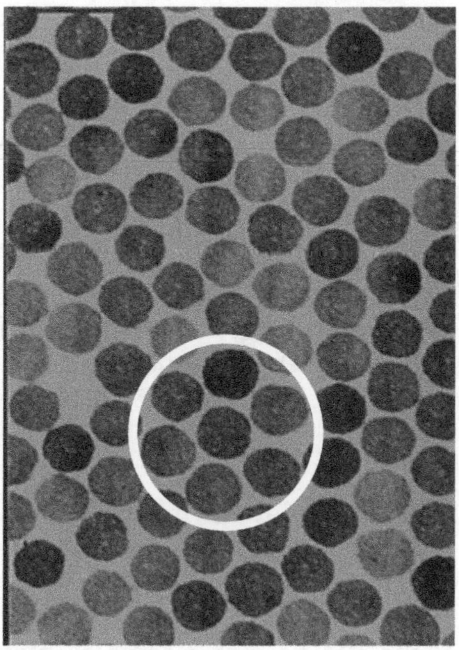

Figure 13.13 Normal EM example. Note the consistent shape, size and packing. Highlighted in the circle is an example of the normal flower petal (seven-fibril) shape that occurs.

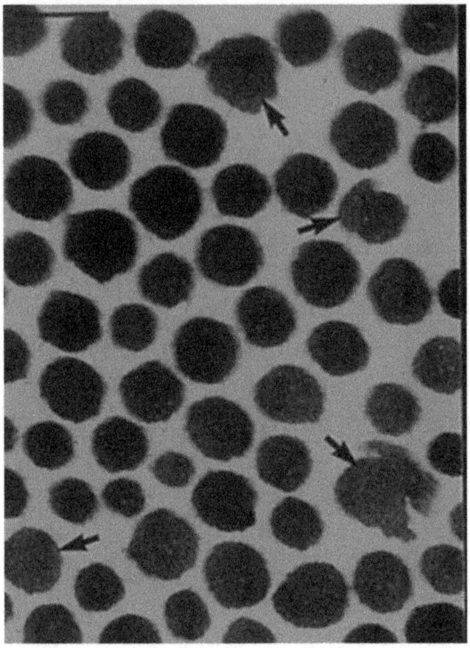

Figure 13.14 cEDS EM example. Note the inconsistency in size and shape. Also note, highlighted with arrows, the presence of rosettes indicative of cEDS.

within vEDS EM, as this is associated with Type 5 collagen. An example of vEDS EM can be seen in Figure 13.15. The degree of variation can be an indication as to the severity of the condition.

13.4.3.2.4. kEDS EM

The use of EM for kEDS shows up generic changes to the collagen fibrils (changes in size and shape) which cannot be used to diagnose kEDS but can be used to support it and help rule out other forms of EDS.

13.4.3.2.5. aEDS EM

Arthrochalasia EDS gives a truly distinctive EM result, which varies depending on the specific type of aEDS. In aEDS Type A/B the pattern is angular, almost jigsaw-like, as a result of the retention of pNα1(I) or pNα2(I) (Figure 13.16).

In aEDS Type C the EM pattern resembles hieroglyphics and is related to the retention of both pNα1(I) and pNα2(I) due to a peptidase deficiency (Figure 13.17).

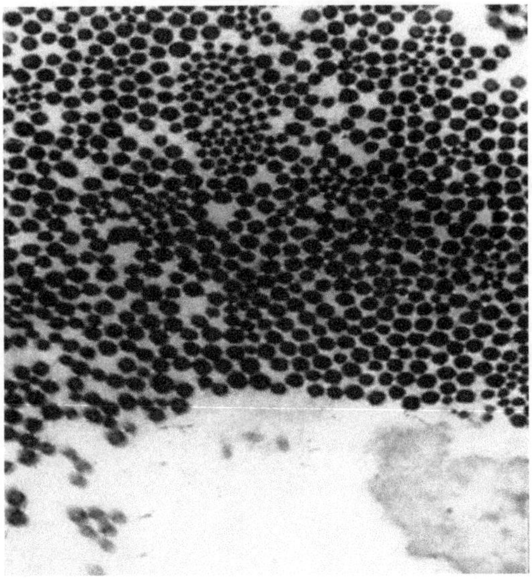

Figure 13.15 vEDS EM example. Note the significant variation in both size and shape and irregular packing. Reproduced from ref. 56 with permission from Elsevier, Copyright 1995.

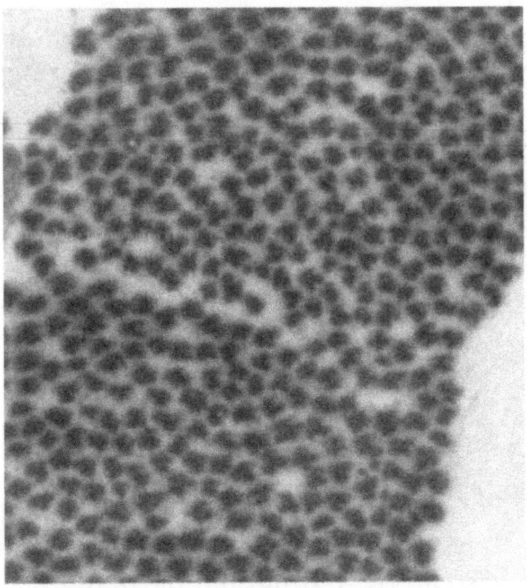

Figure 13.16 aEDS Type A/B EM example. Note that the fibrils are no longer rounded and instead have a 'jigsaw'-like pattern indicative of aEDS. Reproduced from ref. 36 with permission from the BMJ Publishing Group Ltd, Copyright 2000.

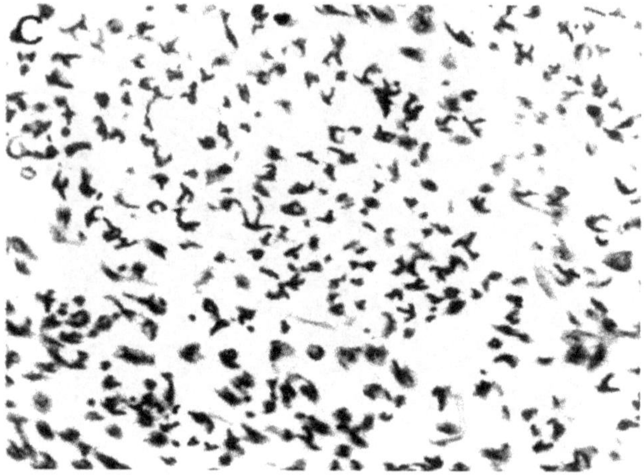

Figure 13.17 aEDS Type C EM example. The hieroglyphic pattern associated with aEDS, which is in stark contrast to the round consistently sized fibrils in a normal EM.

13.4.4. Light Microscopy

13.4.4.1. Overview of Protocol

Light microscopy (LM) is used to look at the amount of collagen in a sample, indicating if there is collagen deficiency.[44] Samples are collected in much the same way as for EM analysis. Skin biopsies of individuals (3 mm punch) are taken from the upper arm and added to a fixative (4% glutaraldehyde in 0.1 M phosphate buffer). Samples are then post-fixed in osmium tetroxide, dehydrated and embedded in epoxy resin. Thin sections are then stained with haematoxylin and eosin and elastin van Geison stains prior to examination using a light microscope.[31]

13.4.4.2. Examples

13.4.4.2.1. vEDS LM

Vascular EDS is characterized by a reduction in the quantity of collagen Type III. This is can be highlighted particularly well using LM in which collagen depletion is examined. Figure 13.18 shows a characteristic example of vEDS LM, collagen is visibly depleted in the vEDS sample compared with the control.

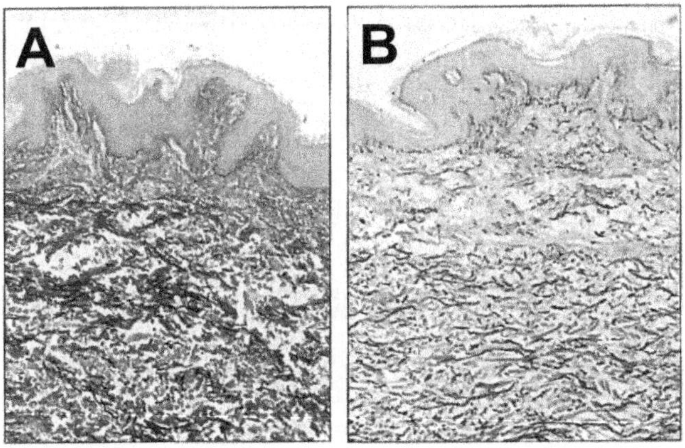

Figure 13.18 vEDS LM example: (A) LM from a control subject, in which a normal amount of collagen is present (the dark strands). (B) LM from an individual with vEDS, in which the amount of collagen is significantly reduced (note the increase in visible gaps). Reproduced from ref. 56 with permission from Elsevier, Copyright 1995.

13.4.5. Urinary Cross Link Analysis

13.4.5.1. Overview of Protocol

Pyridinolines, a product of collagen degradation, which cannot be metabolised, have been used to provide a measure of tissue collagen degradation. These pyridinoline cross-linking residues are excreted in urine and so are easy to collect and can be stored by freezing the sample to test at a later date.

More often associated with bone turnover Eyre *et al.* identified that type II collagen peptides also contributed to the urinary pyridinoline pool.[45] Tandem mass spectrometry has been used to profile the spectra of peptide degradation products derived from collagen domains, using enrichment by immunoaffinity chromatography. Thus the post-translational status of lysine and hydroxylysine residues within cross-linking sites in type I and II collagens of individuals suspected to have EDS VIA and other disorders caused by collagen defects can be examined non-invasively.[45]

13.4.5.2. Examples

13.4.5.2.1. kEDS Urinary Cross-link Analysis

Kyphoscoliotic EDS results from over-modification of α1(I) and α2(I). Kyphoscoliotic EDS can be tested for using urinary cross-link analysis, where a change in the ratio of deoxypyridinoline (Dpyr) *vs.* pyridinoline (Pyr) is indicative of the condition.[34] The normal ratio of Dpyr : Pyr crosslinks is approximately 0.2, whereas in kEDS the ratio can be significantly increased (roughly a 10–40-times increase).

This can be seen in Figure 13.19 where the level of the Dpyr : Pyr ratio is increased above the normal 0.2 ratio.

13.4.6. Overview of Relationship Between All Tests

Each test looks at a specific aspect of EDS. As such, combining tests can give a very detailed overview of a patient's condition. In recent years there has been a drive to increase sequencing for many conditions, to the detriment of a more holistic approach. This can be attributed to the fact that a number of conditions can be diagnosed based on the presence or absence of a variant, although this is not always the case. With regard to EDS, sequencing can provide

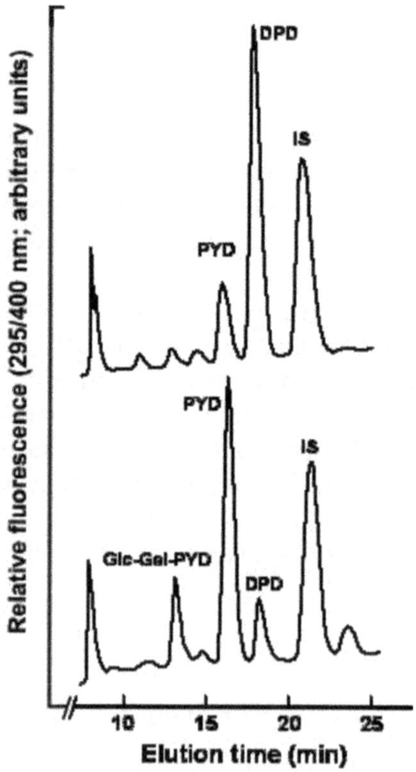

Figure 13.19 kEDS urinary cross-link analysis example. Two traces which highlight significant Dpyr:Pyr (on the figure DPD:PYD) ratios. Reproduced with permission from Dr Simon Robins.

definitive identification of a precise variant (if present) but does not provide information as to its effect. For a truly 'gold standard' diagnostic pathway one should consider a more holistic approach looking at all the aspects detailed in this chapter. From a clinician's perspective the order of importance, from most important to least important, with regards to each test would be as follows: LM, EM, collagen biochemistry, collagen sequencing and urinary cross-link analysis. Generally clinicians can make a confident assessment of EDS from a clinical examination which can be backed up by LM and EM.[46] Further testing with collagen biochemistry, collagen sequencing and urinary cross-link analysis helps with more subtle and/or complicated cases.

13.5. Clinical Aspects

13.5.1. The Importance of a Definitive Diagnosis

Individuals who suffer from EDS frequently have a complicated journey to diagnosis. Sufferers often experience being passed around various departments as clinicians struggle to diagnose their condition.[47] Misdiagnosis of not only the specific EDS subtype but the condition in general is common.[48] The nature of the presentation of symptoms can also lead to misdiagnosis as child abuse, since bruising and scarring are common.[49] It is, therefore, vitally important that these individuals are identified as early as possible and given the right guidance for the management of their specific conditions.

13.5.2. Treatments

There is currently no cure for EDS, and instead individuals who suffer from the syndrome are advised to manage their condition.[47,50] Precautionary measures and treatment in-general is dictated by the particular type of EDS that the individual suffers from and the presenting symptoms. Sufferers are often advised to avoid activities that could result in trauma, such as bruising, scarring, dislocations or fractures.[51] This is often a complex balance in which the individual should be allowed to experience 'life' but must be tempered with the knowledge that over-indulgence can result in detrimental short- and long-term side effects or damage.[47] Often simple protection, such as, padding of legs and arms is sufficient. Joints can be braced in less severe cases and physiotherapy can be used to help build muscle and joint strength.

Medical staff need to be aware of EDS and the risks associated with the treatment of EDS patients when serious injury may have occurred. Stitches and suturing requires great care and often changes to technique, as skin and tissue is fragile and hyperextensible, which can lead to tearing and reopening wounds.[52] This is especially important in vEDS individuals who may present with spontaneous ruptured aneurysms where surgical intervention can worsen the situation.[51]

Patients often suffer from psychological stress associated with the condition.[53,54] Sufferers often look healthy and fit but suffer from debilitating joint pain.[55] It may have taken a long time for diagnosis to take place and they are often misdiagnosed. Also in severe cases scarring may be severe, leading to disfigurement. It is, therefore, vital that the psychological health of EDS patients is taken into account and managed.

13.5.3. Ehlers–Danlos Support UK

Set up in 1987 EDS UK (https://www.ehlers-danlos.org/) is the only UK-based EDS charity that exclusively represents those with EDS. It seeks to advise and inform those who are living with EDS. It offers a support network to sufferers, running conferences and events to bring together people with the condition and the medical profession involved in looking after them. The charity also helps to educate the general medical community to promote and highlight the importance of continuity of care. They also provide money to help fund research into the condition.

Acknowledgements

The authors would like to acknowledge Dr Anthony Vandersteen in helping with the initial research required for this Chapter. Thanks also goes to Dr Simon Robins and Dr Sofie Symoens for generously allowing the reproduction of their work.

References

1. M. Ross and W. Pawlina, *Histology: A Text and Atlas, with Correlated Cell and Molecular Biology*, Lippincott Williams & Wilkins, 2011.
2. J. D. Bos and M. L. Kapsenberg, *Immunol. Today*, 1986, **7**, 235–240.
3. M. Gaubitz, *Rheumatology*, 2006, **45**(Suppl 3), iii3–iii4.
4. E. Bellucci, R. Terenzi, G. M. La Paglia, S. Gentileschi, A. Tripoli, C. Tani and A. Alunno, *Clin. Exp. Rheumatol.*, 2016, **34**, 793–801.

5. G. C. Tsokos, M. S. Lo, P. Costa Reis and K. E. Sullivan, *Nat. Rev. Rheumatol.*, 2016, **12**, 716–730.
6. G. C. Sharp, W. S. Irvin, E. M. Tan, R. G. Gould and H. R. Holman, *Am. J. Med.*, 1972, **52**, 148–159.
7. A. M. Swanson and L. C. Hughey, *Cutis*, 2010, **86**, 205–207.
8. S. J. Orlow, K. L. Watsky and J. L. Bolognia, *J. Am. Acad. Dermatol.*, 1991, **25**, 205–221.
9. S. J. Orlow, K. L. Watsky and J. L. Bolognia, *J. Am. Acad. Dermatol.*, 1991, **25**, 447–462.
10. H. C. Dietz, in *GeneReviews(R)*, ed. R. A. Pagon, M. P. Adam, H. H. Ardinger, S. E. Wallace, A. Amemiya, L. J. H. Bean, T. D. Bird, N. Ledbetter, H. C. Mefford, R. J. H. Smith and K. Stephens, Seattle (WA), 1993.
11. G. A. Di Lullo, S. M. Sweeney, J. Korkko, L. Ala-Kokko and J. D. San Antonio, *J. Biol. Chem.*, 2002, **277**, 4223–4231.
12. D. E. Gunson, *Equine Vet. J.*, 1979, **11**, 97–101.
13. K. Kuhn, *Essays Biochem.*, 1969, **5**, 59–87.
14. S. Ricard-Blum and G. Ville, *Int. J. Biochem.*, 1989, **21**, 1185–1189.
15. B. Steinmann, A. Superti-Furga and P. M. Royce, *Ann. N. Y. Acad. Sci.*, 1988, **543**, 47–61.
16. B. Peterkofsky, *Am. J. Clin. Nutr.*, 1991, **54**, 1135S–1140S.
17. F. M. Pope, *Baillieres Clin. Rheumatol.*, 1988, **2**, 673–702.
18. F. M. Pope, A. C. Nicholls, J. Dorling and J. Webb, *J. R. Soc. Med.*, 1983, **76**, 1050–1062.
19. R. Smith, *Clin. Rheum. Dis.*, 1986, **12**, 655–689.
20. F. M. Pope, *Lancet*, 1986, **2**, 341–342.
21. F. M. Pope, G. R. Martin, J. R. Lichtenstein, R. Penttinen, B. Gerson, D. W. Rowe and V. A. McKusick, *Proc. Natl. Acad. Sci. U. S. A.*, 1975, **72**, 1314–1316.
22. G. R. Martin, J. R. Lichtenstein, V. A. McKusick, R. Penttinen, D. W. Rowe, F. M. Pope and M. D. Sussman, *Birth Defects, Orig. Artic. Ser.*, 1975, **11**, 11–13.
23. A. J. Richards, J. Morgan, P. W. Bearcroft, E. Pickering, M. J. Owen, P. Holmans, N. Williams, C. Tysoe, F. M. Pope, M. P. Snead and H. Hughes, *J. Med. Genet.*, 2002, **39**, 661–665.
24. P. Beighton, A. De Paepe, B. Steinmann, P. Tsipouras and R. J. Wenstrup, *Am. J. Med. Genet.*, 1998, **77**, 31–37.
25. F. Malfait, C. Francomano, P. Byers, J. Belmont, B. Berglund, J. Black, L. Bloom, J. M. Bowen, A. F. Brady, N. P. Burrows, M. Castori, H. Cohen, M. Colombi, S. Demirdas, J. De Backer, A. De Paepe, S. Fournel-Gigleux, M. Frank, N. Ghali, C. Giunta, R. Grahame, A. Hakim, X. Jeunemaitre, D. Johnson, B. Juul-Kristensen, I. Kapferer-Seebacher, H. Kazkaz, T. Kosho, M. E. Lavallee, H. Levy, R. Mendoza-Londono, M. Pepin, F. M. Pope, E. Reinstein, L. Robert, M. Rohrbach, L. Sanders, G. J. Sobey, T. Van Damme, A. Vandersteen, C. van Mourik, N. Voermans, N. Wheeldon, J. Zschocke and B. Tinkle, *Am. J. Med. Genet. C Semin. Med. Genet.*, 2017, **175**, 8–26.
26. A. F. Brady, S. Demirdas, S. Fournel-Gigleux, N. Ghali, C. Giunta, I. Kapferer-Seebacher, T. Kosho, R. Mendoza-Londono, M. F. Pope, M. Rohrbach, T. Van Damme, A. Vandersteen, C. van Mourik, N. Voermans, J. Zschocke and F. Malfait, *Am. J. Med. Genet. C Semin. Med. Genet.*, 2017, **175**, 70–115.
27. N. P. Burrows, A. C. Nicholls, J. R. Yates, G. Gatward, P. Sarathachandra, A. Richards and F. M. Pope, *J. Invest. Dermatol.*, 1996, **106**, 1273–1276.
28. F. Malfait, P. Coucke, S. Symoens, B. Loeys, L. Nuytinck and A. De Paepe, *Hum. Mutat.*, 2005, **25**, 28–37.
29. F. M. Pope, A. C. Nicholls, P. Narcisi, A. Temple, Y. Chia, P. Fryer, A. De Paepe, W. P. De Groote, J. R. McEwan and D. A. Compston, *et al.*, *Clin. Exp. Dermatol.*, 1988, **13**, 285–302.

30. F. M. Pope, P. Narcisi, A. C. Nicholls, D. Germaine, G. Pals and A. J. Richards, *Br. J. Dermatol.*, 1996, **135**, 163–181.
31. N. S. Stembridge, A. M. Vandersteen, N. Ghali, P. Sawle, M. Nesbitt, R. C. Pollitt, D. J. Ferguson, S. Holden, F. Elmslie, A. Henderson, D. J. Hulmes and F. M. Pope, *Am. J. Med. Genet. A*, 2015, **167A**, 1763–1772.
32. A. D. Farmer, H. Douthwaite, S. Gardiner, Q. Aziz and R. Grahame, *J. Rheumatol.*, 2010, **37**, 1513–1518.
33. C. Giunta, M. Baumann, C. Fauth, U. Lindert, E. M. Abdalla, A. F. Brady, J. Collins, J. Dastgir, S. Donkervoort, N. Ghali, D. S. Johnson, A. Kariminejad, J. Koch, M. Kraenzlin, N. Lahiri, B. Lozic, A. Y. Manzur, J. E. V. Morton, J. Pilch, R. C. Pollitt, G. Schreiber, N. L. Shannon, G. Sobey, A. Vandersteen, F. S. van Dijk, M. Witsch-Baumgartner, J. Zschocke, F. M. Pope, C. G. Bonnemann and M. Rohrbach, *Genet. Med.*, 2018, **20**, 42–54.
34. D. Eyre, P. Shao, M. A. Weis and B. Steinmann, *Mol. Genet. Metab.*, 2002, **76**, 211–216.
35. H. N. Yeowell and L. C. Walker, *Mol. Genet. Metab.*, 2000, **71**, 212–224.
36. A. C. Nicholls, J. L. Sher, M. J. Wright, C. Oley, R. F. Mueller and F. M. Pope, *J. Med. Genet.*, 2000, **37**, E33.
37. M. Klaassens, E. Reinstein, Y. Hilhorst-Hofstee, J. J. Schrander, F. Malfait, H. Staal, L. C. ten Have, J. Blaauw, H. C. Roggeveen, D. Krakow, A. De Paepe, M. A. van Steensel, G. Pals, J. M. Graham Jr. and C. T. Schrander-Stumpel, *Clin. Genet.*, 2012, **82**, 121–130.
38. L. Nuytinck, P. Narcisi, A. Nicholls, J. P. Renard, F. M. Pope and A. De Paepe, *J. Med. Genet.*, 1992, **29**, 375–380.
39. Y. Li, D. Ho, H. Meng, T. R. Chan, B. An, H. Yu, B. Brodsky, A. S. Jun and S. Michael Yu, *Bioconjugate Chem.*, 2013, **24**, 9–16.
40. R. A. Weerakkody, J. Vandrovcova, C. Kanonidou, M. Mueller, P. Gampawar, Y. Ibrahim, P. Norsworthy, J. Biggs, A. Abdullah, D. Ross, H. A. Black, D. Ferguson, N. J. Cheshire, H. Kazkaz, R. Grahame, N. Ghali, A. Vandersteen, F. M. Pope and T. J. Aitman, *Genet. Med.*, 2016, **18**, 1119–1127.
41. M. Sayitoglu, *Turk. J. Haematol.*, 2016, **33**, 172–179.
42. H. Julkunen, P. Rokkanen and H. Inoue, *Ann. Med. Exp. Biol. Fenn.*, 1970, **48**, 201–204.
43. M. Sevenich, U. Schultz-Ehrenburg and C. E. Orfanos, *Arch. Dermatol. Res.*, 1980, **267**, 237–251.
44. L. C. Junqueira and J. T. Roscoe, *Histochem. J.*, 1985, **17**, 1197–1202.
45. D. R. Eyre, M. A. Weis and J. J. Wu, *Methods*, 2008, **45**, 65–74.
46. G. Sobey, *Arch. Dis. Child.*, 2015, **100**, 57–61.
47. L. K. Anderson, *Am. J. Nurs.*, 2015, **115**, , 38–46; quiz 47–38.
48. M. Jarmulowicz and W. G. Phillips, *J. R. Soc. Med.*, 2001, **94**, 28–30.
49. M. Castori, *Am. J. Med. Genet. A*, 2016, **170**, 1947.
50. F. Malfait and A. De Paepe, *Adv. Exp. Med. Biol.*, 2014, **802**, 129–143.
51. A. De Paepe and F. Malfait, *Br. J. Haematol.*, 2004, **127**, 491–500.
52. B. P. Kelley and L. Heller, *Hernia*, 2012, **16**, 369–372.
53. M. C. Scheper, B. Juul-Kristensen, L. Rombaut, E. A. Rameckers, J. Verbunt and R. H. Engelbert, *Arch. Phys. Med. Rehabil.*, 2016, **97**, 2174–2187.
54. B. Berglund, M. Anne-Cathrine and I. Randers, *Disabil. Rehabilit.*, 2010, **32**, 1–7.
55. L. Myerson, *Anthropol. Work Rev.*, 1994, **15**, 22–27.
56. F. M. Pope and R. Smith, *Color Atlas of Inherited Connective Tissue Disorders*, Mosby-Wolfe Turin, Italy, 1995.
57. A. Nicholls, J. Oliver, S. McCarron, J. Harrison, D. Greenspan and F. Pope, *J. Med. Genet.*, 1996, **33**, 940–946.

14 Antibody-based Diagnostics

John M. Davis

School of Life and Medical Sciences, University of Hertfordshire, College Lane, Hatfield, Hertfordshire, AL10 9AB, UK
*E-mail: j.m.davis@herts.ac.uk

14.1. Introduction

Antibodies have been used for diagnostic purposes since the agglutination, precipitin and complement fixation tests were first devised in the 1890s.[1] Landsteiner's description of blood groups and blood group testing in 1900 depended on the presence of naturally occurring antibodies in the plasma that caused erythrocyte agglutination when blood of different groups was mixed.[2] This antibody-based testing has permitted the safe transfusion of blood between different individuals for more than a century, and has saved an incalculable number of lives. Over the intervening years a huge number of antibody-based tests have been developed and applied to the diagnosis, definition and prognosis of disease. Despite the advent of nucleic-acid-based tests – many of which are described elsewhere in this book – antibody-based diagnostics remain a vital part of the armamentarium of methods that can be deployed by modern medicine to protect our health and help fight disease, and these are the subject of this chapter.

Genomics and Clinical Diagnostics
Edited by David Whitehouse and Ralph Rapley

Published by the Royal Society of Chemistry, www.rsc.org

There is not the space here to survey all the antibody-based diagnostic tests in use, and for a more comprehensive overview of immunoassay technology the reader should consult some of the excellent books on the subject.[3–5] Moreover, what may currently be used in one country may not be employed in another due to issues of cost and the availability of relevant technology and trained staff. Conversely, some tests that were once heavily dependent on antibodies have been replaced by DNA-based techniques; most notably HLA typing for transplant matching. Thus, this chapter will focus only on some of the most widely used antibody-based tests, the principles underlying them, and their clinical applications, as well as a brief look at what the future may hold.

14.2. Antibody Structure and Function

In order to fully grasp the way in which antibodies (also termed immunoglobulins) can be employed in diagnostic tests it is first necessary to understand their molecular structure. This was initially described in 1959 by Porter[6] and Edelman[7] in work that earned them the 1972 Nobel Prize in Physiology or Medicine. Further research by them and others then expanded and refined the details.[8–12]

The IgG molecule (Figure 14.1) is the most abundant type of antibody molecule in human blood. It is a protein that consists of four polypeptide chains, two heavy chains (each with a molecular mass of approximately 50 kDa) and two light chains (each of approximately 25 kDa) joined by inter-chain disulphide bonds as well as other non-covalent forces. There are also intra-chain disulphide bonds that help form domains within the polypeptide chains. Each light chain has two domains, and each heavy chain four. The N-terminal domains of the heavy and light chain are each termed variable domains. Each variable domain is made up of framework regions interspersed with hypervariable (also known as complementarity-determining) regions. The hypervariable regions on these two domains largely define both the specificity of the antibody molecule – what antigenic site (epitope) it binds to – and its affinity for that epitope. The other domains are constant domains that vary little between antibodies of different antigen-binding specificity, and these are involved in the effector and transport functions of the antibody. Another way to divide up the structure is in terms of its two Fab regions and the Fc region (Figure 14.1). It is important to note that the IgG molecule can potentially bind to (and thus cross-link) two identical epitopes.

In humans, the IgG class of antibodies can be subdivided into four subclasses; IgG_1, IgG_2, IgG_3 and IgG_4. In addition, there are four other classes of antibody; IgM, IgA, IgE and IgD. These each have a structure

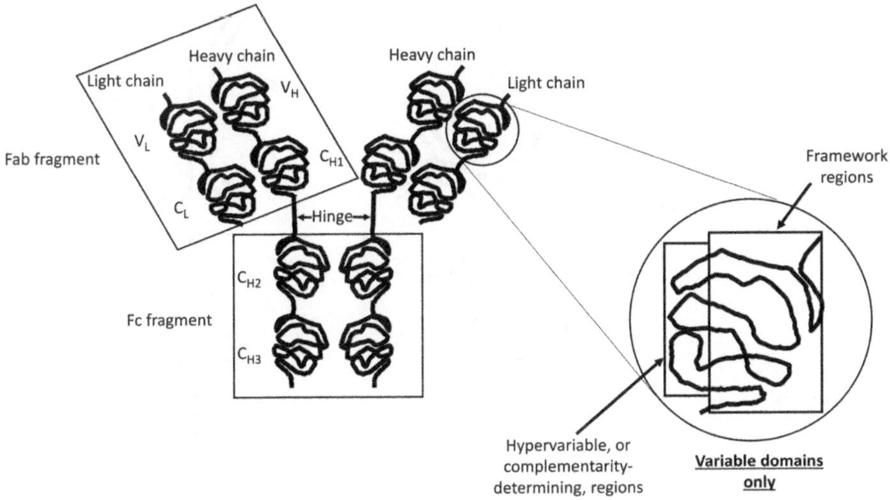

Figure 14.1 Structure of the IgG molecule. The two heavy and two light polypeptide chains are folded into a number of globular domains, four for each heavy chain and two for each light chain. The name of each domain is indicated (*e.g.* V_L, C_{H_2} *etc.*) – V_H and V_L domains are variable domains, whilst the other domains are constant domains. The areas comprising the Fab, Fc and hinge regions are indicated. The enlargement of a variable domain shows how the hypervariable regions, also called complementarity-determining regions, are discontinuous, being interspersed with framework regions that have a less-variable amino acid sequence.

that is related to, but different from, IgG (Figure 14.2). Most strikingly (and importantly in terms of antibody-based tests) IgM has a 'pentameric' structure compared with 'monomeric' IgG, a correspondingly large molecular mass and the potential to bind to ten identical epitopes. It appears early in the immune response to an antigen, and is particularly effective at binding and activating complement (a protein cascade that helps the immune system attack and clear microbes and damaged cells).

For a more detailed discussion of antibody structure and function, readers can refer to any of the many excellent general textbooks on Immunology.[8–12]

14.2.1. Polyclonal and Monoclonal Antibodies

Until 1975, all antibody preparations were polyclonal, *i.e.* they were derived directly from living organisms, usually from the blood serum of mammals, after exposure to an immunogen. (Note that naturally occurring human anti-A and anti-B blood group antibodies are an

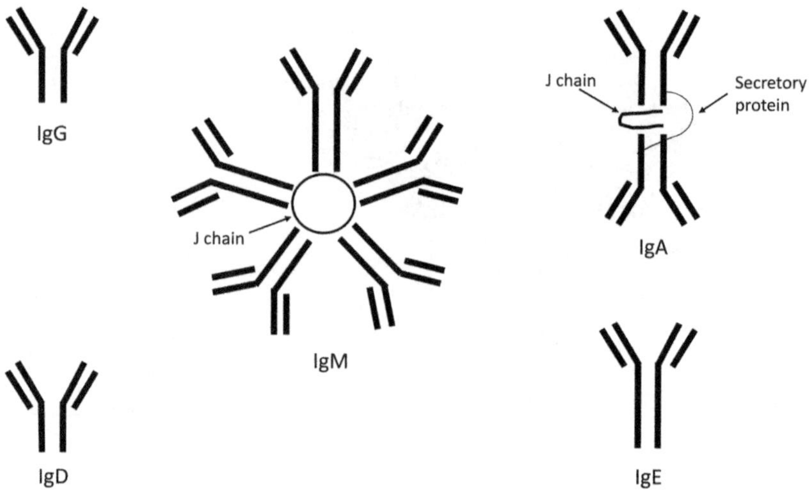

Figure 14.2 Comparison of the structures of the various immunoglobulin classes. IgD is similar in structure to IgG, having four domains on each heavy chain, although the molecular mass of IgD (approximately 175 kDa) is slightly greater than that of IgG (approximately 150–160 kDa). IgE is larger still, with five domains on each heavy chain and a molecular mass nearer 190 kDa. If one considers IgG, IgD and IgE as having a 'monomeric' structure, then IgA is most commonly found as a 'dimer' with the two 'monomers' linked by a J-chain and secretory protein, giving a molecular mass of around 320 kDa. Each IgA heavy chain has four domains. IgM is a much larger molecule than all the others. There are five domains on each heavy chain, and overall the molecule has a 'pentameric' structure stabilised by a J-chain. The molecular mass of IgM is around 900 kDa.

important exception to this latter point, as they arise without exposure to the blood group antigen of interest.) Such antibody preparations are termed polyclonal, as they are formed from antibodies secreted by the progeny of multiple different naturally occurring plasma (B-lymphocyte) cell clones, each making an antibody with a different primary amino acid structure. The consequence of this is that antigen-specific antibodies in a polyclonal preparation may be of a range of classes and/or subclasses, and will vary both in their fine specificity (*i.e.* the exact epitope to which they bind) and in their affinity for the epitope. Moreover, in the serum obtained from an individual immunised animal, less than 1% of the antibody molecules will actually bind to the antigen of interest, even if the animal has been heavily immunised. Thus 99% (or more) of the antibody molecules in any polyclonal preparation will potentially bind to irrelevant antigens. In many cases this led to the necessity to absorb the antibody preparations with preparations of antigens that had been identified as possibly causing unwanted reactions in the particular test being used. This was onerous and frequently had limited success. In addition, polyclonal antibody preparations had other drawbacks (see Table 14.1).

In 1975, Köhler and Milstein[13] devised the technology by which monoclonal antibodies could be obtained. By taking spleen cells from immunised mice and fusing them with myeloma cells, hybrids (hybridomas) could be isolated that secreted the unique antibody from the B-lymphocyte parent and also retained the ability to grow indefinitely in culture from the myeloma parent. The technology has been further improved over the intervening decades,[14,15] and monoclonal antibodies now offer many advantages over polyclonal antibodies for use in both diagnosis and therapy – see Table 14.1. However, it should be noted that monoclonal antibodies are not superior to polyclonal antibodies for all applications, and polyclonals still have an important role to play.

Table 14.1 Comparison of the properties of polyclonal and monoclonal antibody preparations.

Polyclonal antibodies:	Monoclonal antibodies:
Are relatively quick, cheap and easy to prepare – immunise animal, bleed, prepare serum	Are time-consuming and expensive to prepare, requiring skilled staff and a well-equipped laboratory
Are not highly specific – may bind to multiple antigens and epitopes, and are heavily contaminated with antibodies of unknown specificity	Are highly specific due to their single amino acid sequence (although it should be noted that occasional unexpected cross-reactions can still occur).
Contain a range of antibodies of different affinities for the antigen of interest	Have uniform affinity
Are not uniform within a preparation – as well as the specificity issues listed above, preparations may be contaminated with other blood-derived proteins and with antibodies of various isotypes	Are uniform within a preparation as they have a single amino acid sequence (although glycosylation will not be identical on every molecule).
Vary between preparations – even from the same animal, sequential antibody preparations will not be identical due to environmental influences (*e.g.* health, diet *etc.*)	Are identical between preparations if carefully prepared
Can only be raised against something that is immunogenic – antibodies cannot usually be raised against a normal component of the animal being immunised	Can be raised against almost anything – including non-immunogenic and toxic substances – by the use of modern technology, such as phage-display
Can bind to multiple epitopes on a single molecule, and multiple antigens within an impure preparation	Can be used to make a highly specific antibody against a single component (epitope) of a complex (impure) mixture
Are extremely difficult to use to identify previously unknown antigens	Because of the above point, they can be used routinely to identify previously unknown antigens and have been widely used for this purpose. (All the CD antigens have been identified in this way)

14.3. Antibody-based Diagnostic Technologies

All of the technologies described here utilise the exquisite specificity of antibodies for their cognate epitopes in order to detect the presence of, and in many cases quantify, either the relevant antibody or the antigen. Some technologies, such as enzyme-linked immunosorbent assay (ELISA), can be formatted to detect antigen or antibody as required, while others, such as immunohistochemistry, are normally only used for the detection of antigen.

We will first describe some of the techniques that were developed in the early days of immunoassay and that are still in use, before moving on to more recent technologies. The earlier techniques rely almost exclusively on the formation of immune complexes by unlabelled immune reagents, whereas more recently formats using labelled reagents have come to dominate. Among these, the ELISA assay will be described in some detail, as once the underlying principles are understood these can be extended to most of the labelled-antibody-based technologies described subsequently.

14.3.1. Agglutination Assays – Blood Typing and Other Applications

As already mentioned, this method of blood typing has been in use for more than a century but is still in daily use around the world. For ABO typing it depends on the presence or absence of the A and B antigens on human erythrocytes, and the complementary presence of naturally-occurring IgM antibodies reactive with those antigens not present in the individual. Thus a person with blood group A also has anti-B antibodies in their blood serum, a person with blood group B has anti-A, a person with blood group AB has neither anti-A nor anti-B, and a person of blood group O has neither A nor B antigens but both anti-A and anti-B antibodies. Because these antibodies are of the IgM class, they are sufficiently large (Figure 14.3A) that they are able to cross-link erythrocytes. Thus, for example, when serum from a subject of blood group A – containing anti-B antibodies – is mixed with erythrocytes from a person of blood group B, the anti-B antibodies will cross-link the group B erythrocytes into clusters. This can be seen with the naked eye and is termed agglutination.

It must be stressed that this ABO blood typing depends on naturally-occurring antibodies that are of the IgM class. Other blood group antigens, such as Rhesus-D, may be present on erythrocytes but individuals do not normally have anti-Rhesus-D antibodies in their blood. Only if

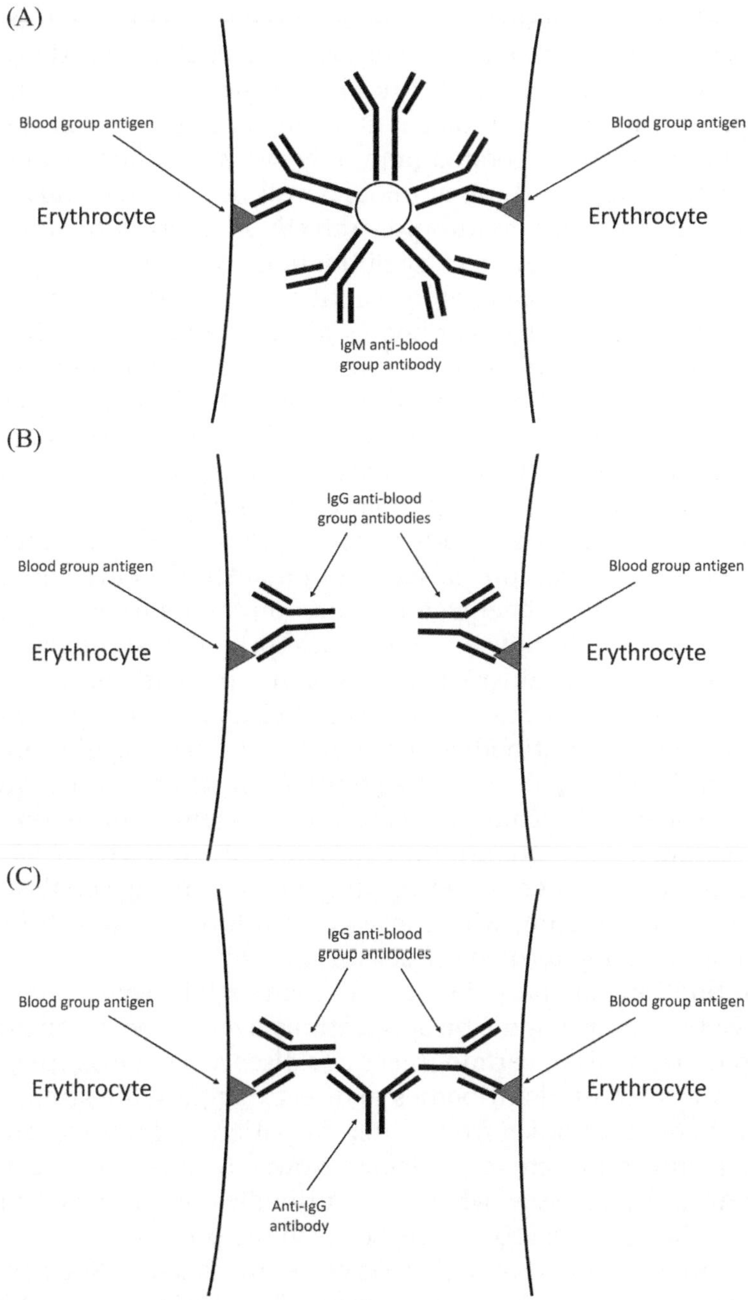

Figure 14.3 Erythrocyte agglutination. An IgM molecule against a blood group antigen is sufficiently large that it can mediate the direct cross-linking of erythrocytes (A) whereas an IgG molecule of the same specificity cannot (B). However, the addition of an anti-IgG antibody to the latter system can permit indirect cross-linking (C).

a Rhesus-D-negative individual has been exposed to Rhesus-D-positive erythrocytes, for example through a transplacental haemorrhage from a Rhesus-D-positive foetus to its Rhesus-D-negative mother, or similarly through the transfusion of Rhesus-D-mismatched blood, will anti-Rhesus-D antibodies be generated (and only ever in a Rhesus-D-negative individual). Moreover, these antibodies will be IgG, and thus will be too small to cross-link erythrocytes directly in an agglutination assay (Figure 14.3B). Thus indirect agglutination assays were developed[16] (Figure 14.3C) in which an anti-IgG antibody was added to the test system to cross-link the IgG antibodies already bound to erythrocytes. Whilst this system is still used for some purposes and in some settings, the indirect agglutination method for Rhesus-D blood typing became largely obsolete in the 1980s with the introduction of IgM monoclonal antibodies to Rhesus-D that were able to mediate direct agglutination.[17]

It should be noted that agglutination of erythrocytes (haemagglutination) can be caused by other factors. Certain viruses, including influenza, mumps and rubella, can agglutinate red blood cells because they bear molecules – haemagglutinins – on their surfaces. This phenomenon can be employed to detect the presence and measure the level of antibodies against such viruses in a patient's or immunised individual's serum. A known amount of virus is incubated with the serum under test. Antibodies that bind to the haemagglutinin molecules on the virus will inhibit agglutination when erythrocytes are added subsequently. Thus this is termed a haemagglutination inhibition assay. A semi-quantitative measure of the level of antibody present can be obtained by performing the assay using serial two-fold dilutions of the serum, with the highest dilution that still inhibits agglutination being termed the serum titre.[18,19]

Agglutination of particles other than erythrocytes has been employed in many diagnostic applications over the years, such as for the rapid diagnosis of certain infectious diseases. For example, in the 1920s Lancefield used antibodies to directly agglutinate streptococci, and by using antibodies from different patients identified different strains (serovars) of the organism.[20] Nowadays it is more common to use an indirect assay, where the antibodies are attached to latex beads, as the agglutination of the beads in the presence of antigen (in this example, the microorganism) is easier to visualise. Such an assay can be used to detect smaller, non-particulate, antigens, or by binding specific antigens to the beads, can be used to detect relevant antibodies.[21] As might be expected from the discussion above regarding blood typing, this latter assay would normally be employed when looking for IgM antibodies. Two common applications of this sort of test are:

In the detection of rheumatoid factor, in which non-specific IgG – here acting as an antigen – is attached to the beads in order to detect IgM anti-IgG antibodies, *i.e.* rheumatoid factor.[22,23]
In the diagnosis of infectious mononucleosis, which is caused by Epstein–Barr virus (EBV) infection. During this infection, heterophile antibodies are produced by the host, *i.e.* antibodies that react with antigens on cells from another species. Originally, these were detected by agglutination of sheep erythrocytes in the Paul–Bunnell test.[24] These were later replaced by horse and then bovine erythrocytes.[25] Nowadays, the bovine erythrocyte antigens are purified and coated on latex spheres, and these will agglutinate within two minutes in the presence of heterophile antibodies.[26]
A modern incarnation of this type of test is the QMS (Quantitative Microsphere System) from ThermoFisher™, an agglutination inhibition test in which microparticles bound to the analyte of interest (frequently a drug) can be cross-linked by the addition of antibodies specific for the analyte. Free analyte in a sample, mixed with the microparticle–analyte conjugate before addition of the agglutinating antibodies, will inhibit agglutination, and this can be detected photometrically as an inhibition in the rate of absorbance change (which is proportional to the rate of agglutination). This assay format is currently routinely used in clinical analysers to measure levels of both drugs of abuse and therapeutic drugs, as well as certain endocrine hormones.[27]

14.3.2. Precipitation-based Tests

The structure of antibody molecules (described above) is such that they each have the potential to bind simultaneously to at least two (identical) antigenic sites, with the IgM molecule theoretically able to bind up to ten. If an antigen molecule bears two or more antigenic sites then, under the right conditions, multiple antigen and antibody molecules can cross-link to form large aggregates that can precipitate out of solution. However, this only occurs when the concentrations of antibody and antigen are equivalent; if either is present in excess, no precipitation occurs.

This phenomenon has formed the basis for a range of antibody-based precipitation assays. Some can be carried out quantitatively in tubes or similar containers, in which case the formation of precipitate is measured by turbidimetry or nephelometry.[28] However, many are carried out in agar or agarose gels. These have the advantages that: (1) soluble antigen and antibody can move freely in the matrix, either by diffusion or when subjected to electrophoresis, and form a concentration gradient in the process; (2) precipitates (occurring at what is termed the zone of equivalence) are stabilised and immobilised; and (3) the translucent gel allows the precipitates to be visualised.

Some of the gel-based techniques developed over the years include:

Single radial immunodiffusion (SRID), – sometimes also called the Mancini test – where an analyte (usually the antigen to be assayed) is placed in a gel containing a uniform concentration of antibody. As the analyte diffuses outwards its concentration decreases to the zone of equivalence, where precipitation occurs. The diameter of the ring formed is proportional to the square of the analyte concentration. The technique has been employed for a variety of purposes, including in the diagnosis of certain immunodeficiencies and gammopathies[29] and for the detection of tuberculous meningitis and quantitation of the corresponding mycobacterial antigen in cerebrospinal fluid.[30]

Double immunodiffusion (Ouchterlony technique), where antigen and antibody are each placed in wells and diffuse towards each other. Again, precipitation occurs at the zone of equivalence. This is a strictly qualitative method, but allows the detection and analysis of multiple species at the same time, and can indicate non-identity, identity, or partial identity of molecular species – see Figure 14.4. Examples of diagnostic use include the identification of paraproteins in serum[31] and the characterisation of fungal and parasitic infections.[32,33]

Immunoelectrophoresis, where an antigen source (*e.g.* blood serum) is subjected to electrophoresis, then a trough and/or wells are cut in the gel and filled with an appropriate antibody, which diffuses into the gel to form precipitin lines. Again, this technique is predominantly qualitative. Examples of diagnostic use are the characterisation of primary immunodeficiencies,[34] and the identification of proteins in blood and urine, including the identification of Bence-Jones proteins in multiple myeloma.[35,36]

Rocket immunoelectrophoresis, which is a cross between immunoelectrophoresis and SRID. A source of antigen is placed in a well and electrophoresed into a gel containing a uniform concentration of a relevant antibody. The electrophoresis speeds up migration of the antigen through the gel (compared with SRID), and again precipitin lines form – in the shape of a rocket – at the zone of equivalence. This is a quantitative technique in which the distance of a precipitin line from the antigen well is related to the antigen concentration. Examples of diagnostic use include the measurement

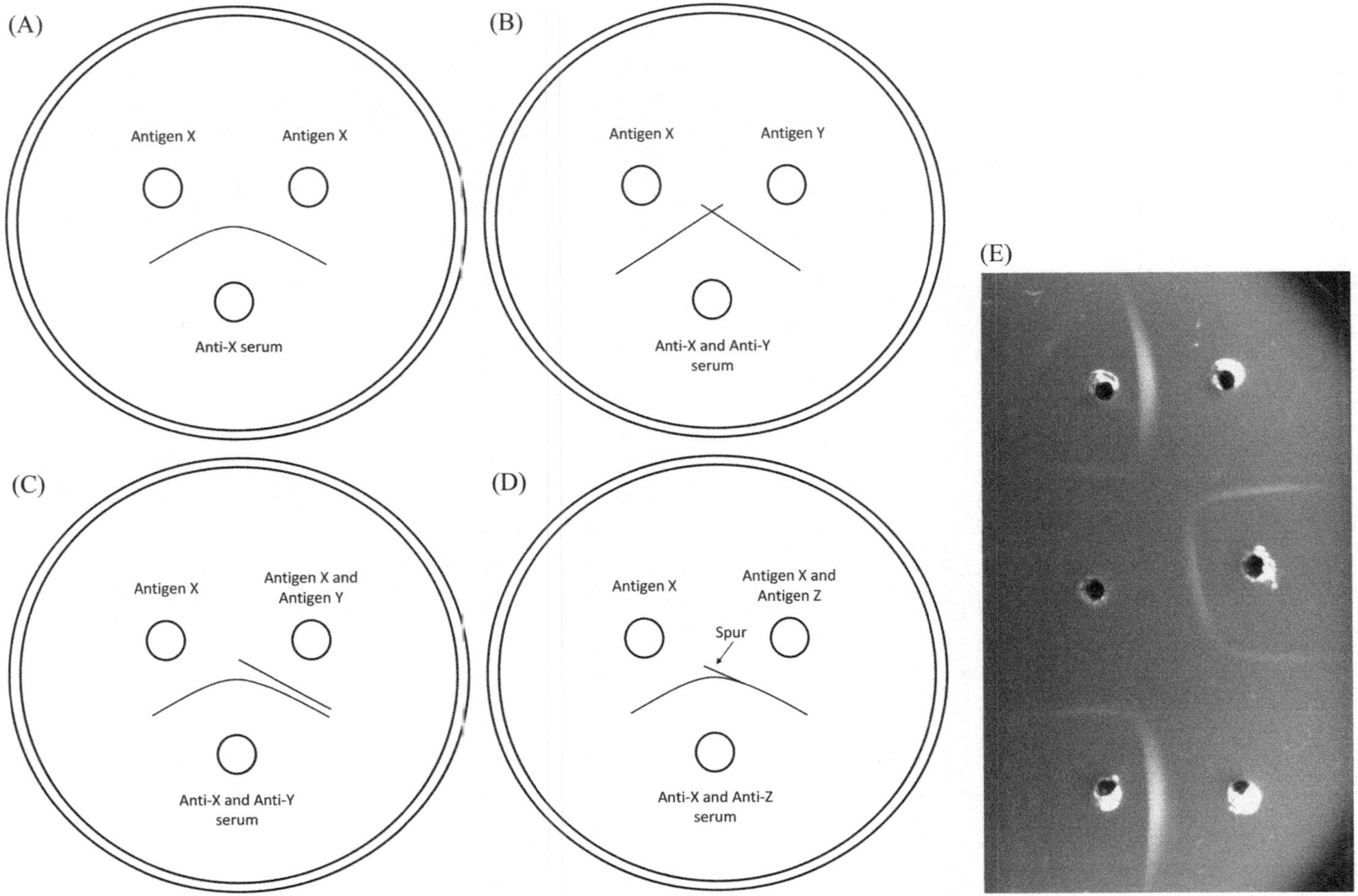

Figure 14.4 The double immunodiffusion (Ouchterlony) technique. Wells are made in a gel (usually agar) and antigen- and antibody-containing solutions are added. After incubation (frequently at 4 °C for 1–2 days) the gel is examined for the formation

of complement component C3d in the plasma and synovial fluids of patients with rheumatoid arthritis or systemic lupus erythematosus,[37] and the simultaneous measurement of α_1-antitrypsin and α_2-macroglobulin levels in serum.[38]

Whilst the above precipitation-based techniques are both cheap and low-tech, they are increasingly being replaced for routine diagnostic purposes by more sophisticated, higher-throughput, more precise and quantitative techniques such as ELISA (see Section 14.3.4).

14.3.3. Complement Fixation

The complement system is a protein cascade present in the blood that acts by a range of mechanisms to protect the host against invading organisms.[12] It can be activated (*via* what is known as the classical pathway) by antibody bound to a suitable antigen. Consumption (fixation) of complement can be used to measure antigen, antibody or antigen–antibody complex levels depending on how the assay is configured. The assay starts with a known, pre-determined amount of complement in the reaction mixture that is just sufficient to lyse all the pre-sensitized erythrocytes to be used (see below). It works on the principle that, when antibody binds to a multivalent antigen, immune complexes are formed to which the first component of the complement cascade (C1q) will bind. Thus, in the assay, antigen and antibody are mixed to allow the formation of complexes, and these bind a corresponding amount of C1q from the complement preparation in the reaction mixture. Pre-sensitised erythrocytes (usually sheep red blood cells coated with a haemolysin – an antibody against

---

of precipitates. (A–D): Diagrams of the precipitin lines formed in the presence of different combinations of antibody and antigen: (A) Both upper wells contain the same antigen. A continuous, smooth, boomerang-shaped precipitin line is formed. (B) The upper wells contain two different antigens with no shared epitopes. The precipitin lines cross. (C) One antigen (X) is present in both antigen preparations, leading to a boomerang-shaped precipitin line as in A, but antigen Y shares no epitopes with antigen X and forms a separate line with no boomerang shape as it is not present in the left-hand well. (D) Antigen X and antigen Z share epitopes, leading to the formation of a boomerang-shaped precipitin line, but Z also has another epitope not shared with X and this leads to the formation of a spur. (E) Photograph of the precipitin lines formed on an agar-coated slide (Image credit to Nathan Davies). The top right, middle left and bottom right wells contained a solution of bovine serum albumin, whilst the other wells contained a rabbit antibody preparation against bovine serum albumin. Compare this with (A).

sheep red blood cells capable of lysing them in the presence of complement) are then added, and these erythrocytes lyse in proportion to the amount of residual C1q that has not already been bound. After centrifugation, or allowing time for intact erythrocytes to settle, the assay can be read. If set up in microtitre plates, the degree of lysis can be estimated by eye. Alternatively, the level of free haemoglobin can be measured spectrophotometrically after removal of intact erythrocytes. The degree of lysis will be inversely related to the amount of analyte being measured. Normally, a series of dilutions of the analyte will have to be made to ensure that the concentration causing partial lysis is encompassed. In systems that are read by eye, this allows the result to be reported as a titre, *i.e.* the maximum dilution at which a pre-defined level of lysis occurs.[39]

It should be noted that only certain immunoglobulin isotypes (classes) mediate complement fixation, *i.e.* IgM, IgG_1 and IgG_3 (and IgG_2 weakly). Thus the assay cannot be employed with antibodies of other isotypes (although there may be a limited amount of activation *via* the alternative pathway). In addition, it requires a range of controls and is quite labour-intensive. As with many of the other assays described above, complement fixation was widely used in the past for purposes such as the diagnosis of syphilis[40] and other bacterial, viral, fungal and parasitic diseases[41–43] and more recently has been used to demonstrate patient response, as demonstrated by an increase in specific antibody levels, to a range of infectious diseases, including influenza, chlamydia, mycoplasma, mumps and measles.[39] However, it has increasingly been replaced in routine diagnosis by more modern serological assays, such as ELISA, or by nucleic-acid-based tests,[44] but is still employed in some places and for certain applications in microbiology[45,46] particularly where it is important to demonstrate the ability of an antibody to mediate a biological reaction.

14.3.4. ELISA (Enzyme-linked Immunosorbent Assay)

This is perhaps the most widely used antibody-based test system in use in medicine today. It was first described by Engvall and Perlmann in 1971 and has its roots in radioimmunoassay.[47] However, because of its many advantages, including the absence of radioactive hazard and the simpler equipment needed for measurement, it has essentially completely replaced radioimmunoassay.

As with many other antibody-based techniques, the ELISA test can be formatted in a number of different ways to quantify antigen or antibody, and these can be described by one or more terms including direct, indirect, competitive, non-competitive and sandwich. An indirect, non-competitive form will be described first, followed by the modifications needed to yield the other formats.

14.3.4.1. Indirect, Non-competitive ELISA for the Quantitation of Serum Antibodies

Most ELISAs today are performed in multi-well (frequently 96-well) plastic plates. For the quantitation of a specific antibody, the relevant antigen is first adsorbed onto the inner surface of the wells. The plastic surface and the buffer containing the antigen are designed such that the antigen adsorbs non-specifically to the surface. Excess antigen is then washed away, and any remaining sites on the plastic that could bind non-specifically to other components of the test system are blocked using a proteinaceous substance that will not interfere with the assay. Wells are then incubated with the serum under test (primary antibody), then washed to remove unbound antibody before incubation with an enzyme-linked detection reagent. This is frequently an antibody (termed a secondary antibody) against immunoglobulin from the species under test, but labelled antigen is used in some tests.[48] For clinical diagnostic purposes, human serum will normally be tested, so an enzyme-linked rabbit (or goat or sheep *etc.*) anti-human immunoglobulin would be used. After incubation, excess (unbound) enzyme-linked reagent would be washed away before introduction of a substrate. Enzyme molecules linked to the bound secondary antibody will react with the substrate to generate either a coloured substance (chromogenic assay), a fluorescent substance (fluorimetric assay), or a light-emitting substance (luminescent assay) – see Section 14.3.4.6. As long as the secondary antibody was added in excess, the strength of the signal generated, *e.g.* the depth of colour (absorbance), will be proportional to the amount of antigen-specific primary antibody bound in the well – see Figure 14.5. In order to ensure that a measurable (sub-maximal) absorbance is obtained, serial dilutions, frequently twofold, of the serum under test are made across the plate. Results can be described in semi-quantitative terms as a titre, *i.e.* the dilution at which a serum gives a certain pre-defined sub-maximal response, or in quantitative terms by the inclusion of suitable standards.

This type of ELISA is used diagnostically to measure the level of antibody generated against a pathogen or vaccine. Diagnosis of HIV is currently performed (in combination with other tests – see Sections 14.3.4.4. and 14.3.6.) by using an ELISA assay that detects antibody, in a person's serum or other bodily fluid, that is reactive against HIV viral lysate, recombinant

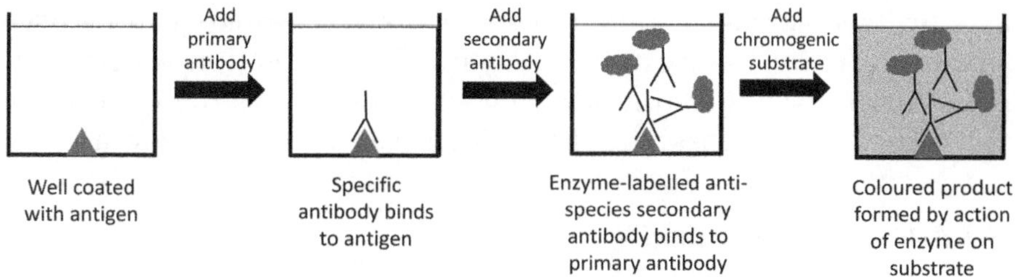

Figure 14.5 Indirect non-competitive ELISA assay. Plate wells are coated with antigen. The antiserum or antibody preparation to be assayed is then added, and after incubation unbound antibody is washed away. Enzyme-labelled anti-species antibody (secondary antibody) is added, and after incubation excess secondary antibody washed away. Chromogenic substrate is then added, and after a suitable incubation period the intensity of the colour formed is measured spectrophotometrically.

HIV proteins or synthetic peptides.[49,50] Similarly, prior infection with hepatitis B, or the success of immunisation with a hepatitis B vaccine, is assessed by measuring the level of antibody to recombinant hepatitis B surface antigen (HbSAg) in an individual's serum. If, following three injections, the level in a vaccinee's serum is below that defined as demonstrating successful vaccination, a fourth dose will be given.[51]

14.3.4.2. Direct, Non-competitive ELISA

Indirect ELISAs are so termed because the analyte itself (in the above examples, the primary antibody) is not labelled, but a labelled secondary antibody bears the enzyme that makes detection possible, so the measurement is indirect. A direct ELISA requires labelling of the primary antibody, and this is not usually feasible or economic in the diagnostic setting, as it would require isolation and enzyme labelling of antibody from every individual patient's sample. Instead, using an indirect ELISA, no handling of a patient's sample (other than pipetting) is required, and only a single enzyme-labelled reagent (which can be purchased ready to use) is needed. The indirect assay also has the advantage that the secondary reagent can bind to several sites on the primary antibody (see Figure 14.5), and this acts to amplify the signal.

14.3.4.3. Competitive ELISA

This format is less widely used than the non-competitive format, and relies on an analyte of interest in an added sample competing with a quantity of analyte that is either already bound to the wells of the plate or itself carries a suitable label. Different formats can be used depending on the nature of the analyte and the detection system to be used, but all have the common characteristic that as the quantity of added analyte increases, the signal decreases in a non-linear fashion.

14.3.4.4. Sandwich ELISA

This is a format used to measure antigen, in which the antigen forms the filling between two 'slices' of antibody 'bread'.

The first step in a sandwich ELISA is to immobilise on the assay plate's plastic wells a 'capture' antibody (*i.e.*, one that will bind – capture from the sample – the antigen of interest) – see Figure 14.6. It should be noted that whilst the formats described above, particularly the indirect, non-competitive ELISA, can successfully use polyclonal antibodies, the first stage of a sandwich ELISA really requires the use of a monoclonal antibody. This is because, as already described, only 1% or less of a polyclonal antibody preparation will be antibodies against the antigen of interest, whereas in a monoclonal preparation theoretically 100% will bind the relevant antigen. Thus use of a monoclonal capture antibody should yield a signal that is around two orders of magnitude greater than that attainable with a polyclonal preparation.

In the second step a sample containing the antigen of interest is added, and those antigen molecules will be bound by the capture antibody. After washing, another antibody against the antigen of interest is added. This antibody may be monoclonal or polyclonal, but must have one of the following two properties: either it must bind to a

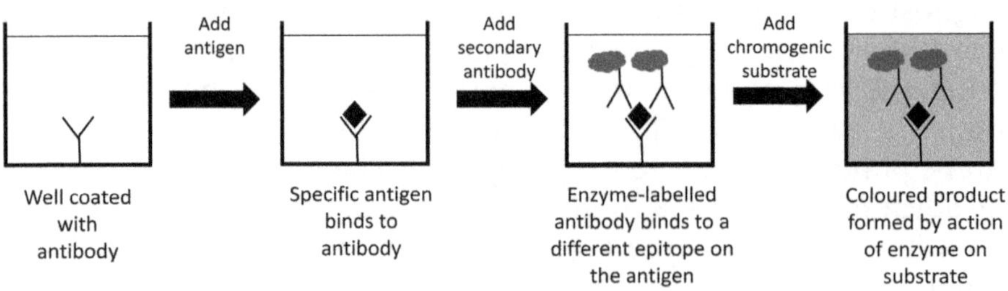

Figure 14.6 Sandwich ELISA assay. Plate wells are coated with monoclonal antibody against the relevant antigen. The putative source of antigen is then added, and after incubation unbound material is washed away. A second, enzyme-labelled antibody to the antigen of interest is added, and after incubation excess antibody is washed away. Chromogenic substrate is then added, and after a suitable incubation period the intensity of the colour formed is measured spectrophotometrically.

different epitope on the antigen from the one bound by the capture antibody, or it must bind to an identical epitope that is present in multiple copies. This second antibody may itself be enzyme-labelled, or may be followed by another (usually anti-species Ig) labelled antibody as in the indirect, non-competitive ELISA described above.

Clinical applications of the sandwich ELISA technique include the detection of p24 antigen in the bodily fluids of HIV-infected individuals,[49] as well as the identification of many other infections by the detection of their relevant antigens;[52] the measurement of cytokine levels in a range of immune-related diseases;[53] and the measurement of tumour-associated antigens in cancer.[54–56]

14.3.4.5. Signal Amplification

Depending on the application, the sensitivity (amount of analyte that can be detected) of the ELISA assay as described above may or may not be adequate for the intended purpose. For example, in the early days (*i.e.* the mid-1980s) of applying ELISAs to the detection of circulating anti-HIV antibodies, lack of sensitivity was a major issue that had to be overcome. Many approaches have been taken to increase the sensitivity of ELISA assays, and these have frequently proven useful in the other techniques described below. However, a common feature to all these approaches was that simple amplification of the signal achieved nothing if the background signal ("noise") was also amplified. Such "noise" can come from spontaneous reactions (*e.g.* non-enzymatic substrate cleavage), background signals in the detector equipment, the assay components themselves, or from non-specific binding of reagents. Thus special approaches had to be taken in order to increase the signal-to-noise ratio.

It has already been mentioned in Section 14.3.4.2 that using a labelled secondary antibody that can bind to several sites on the primary detection antibody offers a degree of amplification. The highly specific and very high-affinity binding of biotin to avidin (having a binding constant in the region of 10^{15} L mol^{-1}) has been used in a similar fashion. Quite high levels of biotin can be covalently linked to detection antibodies without loss of antibody activity. Labelled avidin can then be added, giving significant amplification without similarly increased "noise" levels, due to the highly specific nature of the biotin–avidin binding. A further option for even greater sensitivity uses the fact that each avidin molecule can bind four biotin molecules, so instead of using labelled avidin, unlabelled avidin can be used followed by biotinylated label.

14.3.4.6. Choice of Label

Several different enzymes are in common use as labels in ELISA assays. Of these, both horseradish peroxidase (HRP) and alkaline phosphatase (AP) are widely used in chromogenic assays (*i.e.* those to be read by photometry). The detection limit of HRP has been estimated to be 2×10^6 zeptomoles (abbreviated zmol, 1 zmol = 10^{-21} mol) whereas that of AP is rather better at 5×10^4 zmol.[28] However, in order to lower the detection limit further, techniques other than photometry must be employed, and this usually means the use of more-expensive equipment. Alkaline phosphatase, β-D-galactosidase and certain non-enzymatic labels can be utilised in several different fluorescence-based techniques that can yield sensitivities down to the 100–1000 zmol level. Beyond this, chemiluminescence or electrochemiluminescence come into play, again with both enzymatic and non-enzymatic [*e.g.* ruthenium(II) tris(bipyridyl)] labels. At their best, some of these are claimed to have sensitivities down to the 1 zmol level.[28]

It should be noted that many of the above amplification strategies may also be applicable to the western blotting and immunochemistry techniques described below, although consideration must be given to their applicability on a case-by-case basis. For example, it would be inappropriate to use a very sensitive fluorescence detection method for immunohistochemistry on a tissue with significant endogenous fluorescence.

14.3.5. Other Types of Enzyme-linked Immunoassays

14.3.5.1. Cloned Enzyme Donor Immunoassay (CEDIA®)

This assay from ThermoFisher™ combines the use of two complementary genetically engineered fragments of β-galactosidase with antibodies against the analyte of interest. One of the enzyme fragments is termed the enzyme donor (ED) and the other the enzyme acceptor (EA); only when the two fragments can combine with each other is enzyme activity reconstituted.

ED is first labelled with the analyte (antigen) of interest. This must be small enough that it does not interfere with the ability of the ED to spontaneously recombine with the EA to produce active enzyme. In the assay, ED–antigen conjugate is mixed with the sample under test, the analyte-specific antibody and EA. In the absence of free antigen in the sample, the (large) antibody binds to the ED–antigen conjugate and prevents it from binding to the added EA to form active enzyme. In the presence of free antigen, binding of antibody to the ED–antigen conjugate is inhibited, so the EA–antigen is free to combine with ED to yield enzymatic activity. Thus the activity measured increases with increasing analyte concentration.[57] This assay finds application in measuring levels of drugs of abuse and therapeutic drugs, as well as certain endocrine hormones.[27]

14.3.5.2. Enzyme-multiplied Immunoassay Technique (EMIT)

EMIT is similar in many ways to CEDIA, but does not rely on the use of engineered enzyme fragments. The analyte (antigen) of interest is labelled with an enzyme. Again, in the absence of free analyte antibody specific for the analyte inhibits the enzymatic activity of the enzyme–analyte conjugate. However, in the presence of free analyte (in the sample being assayed) this inhibition is itself inhibited, so the level of enzyme activity is directly related to the level of free analyte in the sample.[58] Applications are similar to those of CEDIA.[27]

14.3.6. Western Blotting

This test is generally more difficult, time-consuming and costly to perform than an ELISA, but has the potential to demonstrate greater specificity at the molecular level. It was used for many years as a confirmatory test for the presence of anti-HIV antibodies following positive ELISA or other tests for anti-HIV antibodies. However, it is now being replaced by nucleic-acid-based tests.[49]

A source of the relevant protein antigen(s) is denatured in a suitable buffer containing sodium dodecyl sulphate and a reducing agent (to break both inter- and intra-chain disulphide bonds), and is then subjected to polyacrylamide gel electrophoresis. Polypeptide chains migrate distances that are inversely related to their molecular mass. These polypeptide bands are then transferred (blotted) onto a nitrocellulose filter, which is then blocked – as in an ELISA test – to prevent subsequent non-specific binding to the filter. Commercial tests may be packaged at this stage for subsequent use in the purchaser's laboratory. A source (such as serum) of antibodies to be tested for reactivity with the polypeptides on the blot is then added, followed (after washing) with an enzyme-linked anti-species Ig and substrate as in the ELISA assay. Further amplification methods may be used, again as in an ELISA – see Section 14.3.4.5. Positive results show stained bands at positions that correspond to the molecular mass(es) of known relevant antigen polypeptides. So, for example, anti-HIV antibodies might be demonstrated by reaction with p17, p24, p31, gp41, p51, p55, p66, gp120 and gp160 from an HIV protein preparation (p = protein; gp = glycoprotein; number = molecular mass in kDa).[59,60] The method can also be used to detect the presence of an antigen or antigens in a preparation by using an antibody known to be reactive with it/them in the western blotting system – see Figure 14.7.

It should be noted that, in order to bind to a polypeptide on the blot, an antibody must be able to bind to the polypeptide's denatured structure, as secondary and higher-level structures will have been disrupted prior to electrophoresis. It should also be noted that this technique is primarily qualitative, and is not generally employed for quantitation of an antibody under test. However, in the presence of excess antibody the depth of staining can give an indication of the quantity of antigen present in a particular band.

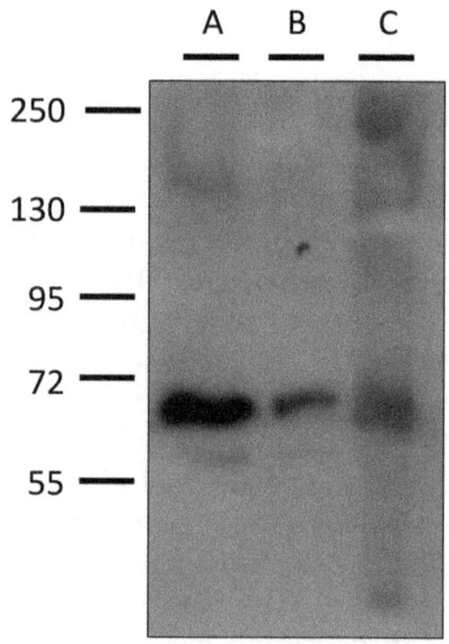

Figure 14.7 Example of a western blot. In this example, three cell extracts (lanes A, B and C) have been subjected to sodium dodecyl sulphate-polyacrylamide gel electrophoresis, transferred (blotted) onto a nitrocellulose filter, then incubated with a rabbit anti-NFκB p65 subunit antibody followed (after washing) by a horseradish peroxidase-conjugated goat anti-rabbit IgG antibody. This was then visualised by the addition of a chemiluminescent substrate. The positions of molecular mass markers (in kDa) are indicated on the left. The presence of NFκB p65 subunit (molecular mass approximately 65 kDa) can be clearly seen – at different concentrations – in all three extracts. (Image credit to Noelia Perez Diaz).

14.3.7. Immunochemistry

Immunochemistry is a term that covers techniques where substances (normally antigens) are located in or on cells or tissues by means of an immunological interaction, usually binding of an antibody. It can be divided into two related techniques that are very similar in principle: immunocytochemistry, which is applied to cell suspensions, and immunohistochemistry, which is applied to tissues (usually sections) on a microscope slide. The principle is essentially the same as that used in ELISA assays, in that the (potentially) antigen-containing material is incubated with an antibody that is labelled with, or can be indirectly labelled with, an enzyme or other chemical that permits bound antibody to be detected in the test system. However, in contrast to an ELISA, where a coloured product is normally obtained that is soluble in, and diffuses into, the surrounding milieu, in these techniques it is important that the signal being generated remains co-located with the antibody. Thus the combination of enzyme label and substrate to be used in immunohistochemical techniques will be critical in order to yield an insoluble product. Immunocytochemical techniques, by contrast, frequently employ fluorescent labels, although these can be used in immunohistochemistry too (although background tissue autofluorescence can cause problems). It should be noted that, by judicious choice of antibodies and labels, two or more antigens can be detected at the same time in the same sample.

14.3.7.1. Immunocytochemistry

An illustrative example of the use of immunocytochemistry in diagnosis is the detection and enumeration of circulating lymphocytes displaying the CD4 cell surface antigen, *i.e.* $CD4^+$ cells. (It should be noted that – as mentioned in Table 14.1 – all the CD [Cluster of Differentiation] antigens were identified by using the ability of monoclonal antibodies to identify single components in an impure preparation, frequently whole cells.) $CD4^+$ cell counts are important in monitoring patients who are HIV-positive, as they permit both detection of the progress of the disease towards AIDS and the assessment of the efficacy of anti-retroviral therapy (ART). Counts in an individual not infected with HIV would be expected to be between 500 and 1500 $CD4^+$ cells μl^{-1} of blood. In an HIV-positive person, ART would usually be started when $CD4^+$ counts fall below 350 μl^{-1}, and a count below 200 μl^{-1} would be diagnostic of AIDS.[61]

There are essentially two methods of enumerating $CD4^+$ cells. In less resource-limited countries where there is a high throughput of samples, it is usual to stain the cells using a fluorochrome-conjugated monoclonal antibody – thus this is an immunofluorescence technique. The stained cell preparation is then passed through a flow cytometer which counts both the stained and unstained cells. This method is considered to be the 'Gold Standard' by the World Health Organisation.[62]

In more resource-limited settings, or where there is only a small throughput of samples, bead-based systems combined with microscopy have been used. In one system (from Dynal), after lysis of erythrocytes the blood sample is incubated with magnetic beads coupled to an anti-CD14 antibody. This antibody binds to monocytes, and a magnet is placed by the tube. This holds the monocytes within the tube while unlabelled cells are decanted. These cells are then incubated with anti-CD4-coupled magnetic beads. A magnet is again used, this time to retain the $CD4^+$ cells in the tube whilst unlabelled cells are removed. The $CD4^+$ cells can then be counted.[63] In another system (from Coulter), monoclonals with the same specificity are used, but bound to latex beads. The anti-CD14 antibody is bound to small (0.2 μm diameter) beads and acts as a blocking reagent for monocytes, while the anti-CD4 antibodies are bound to larger (2 μm) beads, which form rosettes with the $CD4^+$ cells.[64] However, with the advent of smaller, cheaper, more-portable and battery-powered cytometers, the move is towards the type of cytometric assay described in the previous paragraph.

14.3.7.2. Immunohistochemistry (IHC)

This technique is used to look for the presence and location of antigens in tissue samples and other cell preparations (both intracellular and external plasma-membrane-bound antigens can be detected). Thus it is dependent on the retention both of antigenic activity and cell and/or tissue morphology. Samples to be tested may be presented in a variety of forms including paraffin-embedded sections, frozen sections, free-floating sections or cytological specimens.

An important aspect of this technique is that, apart from frozen sections, cytological smears and cytocentrifuge preparations, the other forms of sample will all have been subjected to chemical fixation and possibly other processes, such as decalcification, wax embedding and subsequent dewaxing protocols, that may adversely affect antigen structure. Indeed, a process termed "antigen retrieval" is frequently used on aldehyde-fixed samples (as well as samples fixed by other techniques) to reveal antigens that would otherwise not be detected. However, the conditions used in this process may include heating, proteolysis and use of a variety of chemicals and pH conditions. Thus it can be seen that what works as antigen retrieval for one antigen may destroy the antigenicity of another. Indeed, the true mechanisms of antigen retrieval remain unclear, and there is no such thing as a universal antigen retrieval protocol.[65,66] Because of the above considerations, it cannot be assumed that antibodies that detect a specific antigen in a different test system, *e.g.* ELISA, will do so in the IHC setting, and indeed they may not react in the same way in different IHC protocols. Consequently, individual antibodies must be validated in the actual protocol in use.

IHC is extensively used in diagnostic surgical pathology, and particularly in tumour diagnosis where the expression or upregulation of tumour-associated antigens can be both diagnostically and prognostically important. It can also be used in the diagnosis of certain infectious (particularly viral) diseases, and in the classification of certain muscular and neurodegenerative diseases.[67–69] Well over 200 different antigens are currently used as diagnostic targets in IHC, and that number is growing constantly.[66]

A simple light microscope is all that is normally needed to examine and evaluate stained slides in the diagnostic context, but for research the technique can also be used with samples intended for examination by electron microscopy.[28]

Some examples of tissue sections stained using IHC can be seen in Figure 14.8.

14.3.8. Other Non-isotopic Immunoassays

Above we have described the use of the most common non-isotopic techniques currently used in diagnostic immunoassays. However, there are a range of other non-isotopic assays – using different labels or approaches – that are either already in use or are under investigation, and these are described briefly in Table 14.2. Each has potential as well as drawbacks, and combinations of some of these techniques (*e.g.* bioluminescence and nanoparticles[85] or quantum dots and FRET[86]) are being investigated. The extent to which they will find wide application in the diagnostic immunoassay market remains to be seen.

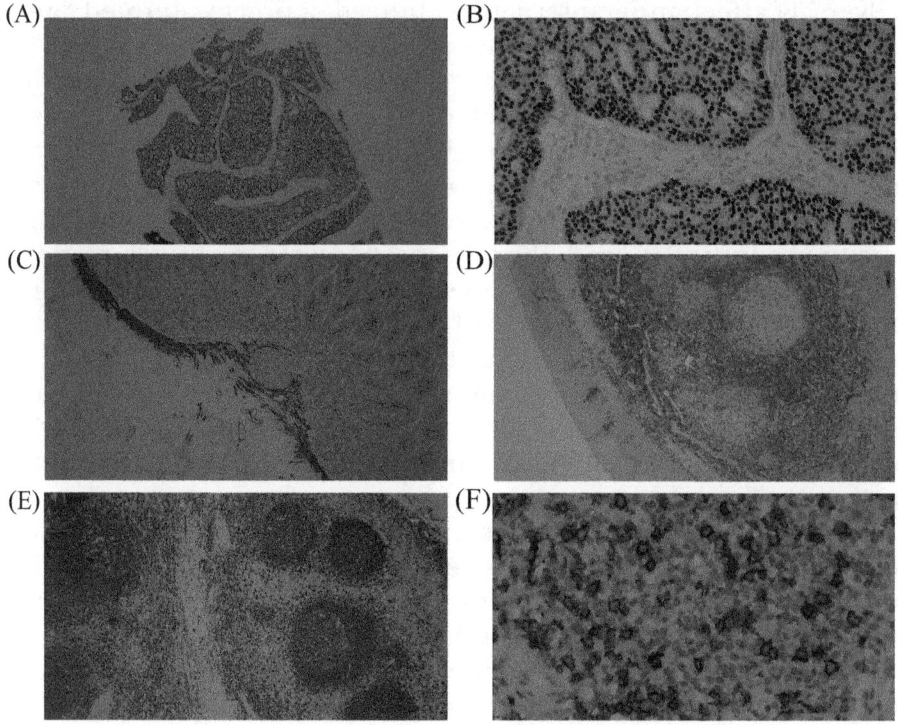

Figure 14.8 Examples of tissue sections stained by immunohistochemistry. All original sections were stained using the relevant DAKO kit, yielding a brown precipitate in the presence of antigen, then counterstained with haematoxylin. These are reproduced here in black and white. (A) Section through a breast tumour stained with antibodies to the oestrogen receptor, which is heavily over-expressed in some breast tumours. Strong (dark) staining of tumour cells can be seen. Tumour tissue is interspersed with unstained normal breast tissue. (Magnification ×40) (B) A close-up of the tumour shown in A, showing the nuclear location of the oestrogen receptor. Again, stained tumour tissue can be seen interspersed with unstained normal tissue. (Magnification ×200) (C) Section through human colon tissue stained with antibodies to the muscle marker desmin. To the bottom left is the connective tissue, running top left to bottom right is the muscularis mucosae, which has stained strongly for desmin, and to the top right are villi. The circular structure in the centre is a Peyer's patch. Note the desmin staining around the arteries in the connective tissue. (Magnification ×100) (D) Section through a lymph node, stained for the T-cell marker CD3. (Magnification ×40) (E) Section through the same lymph node as in D, this time stained for the B-cell marker CD20. Compare with D and note the different spatial distribution of the stained cells. (Magnification ×40) (F) A close-up of the section shown in E, demonstrating the location of CD20 on the cells' plasma membranes. (Magnification ×400) (Image credit to Rachel Smith).

Table 14.2 Some other non-isotopic immunoassays.

Bioluminescence immunoassays[70] – similar to other luminescence assays (see Section 14.3.4.6) except that no enzymatic activity is involved. The label used is apoaequorin (a jellyfish protein), which emits light in the presence of Ca^{2+} ions
Erenna® immunoassay[71] – similar in principle to a sandwich ELISA, but the fluorophore-labelled secondary antibody is eluted (*i.e.* the capture antibody–antigen–secondary antibody complex is chemically broken apart) and detected using a capillary-flow-laser-induced fluorescence system. Sensitivity in the range of femtograms ml^{-1} is claimed.[72]
Fluorescence excitation transfer (FRET) immunoassays[73,74] – these assays depend on the molecule absorbing the excitation photon (the donor) having been brought so close to an acceptor molecule that dipole–dipole coupling can occur. The acceptor then radiates the energy at a longer wavelength. This mechanism can thus detect inter-molecular interactions between a molecule labelled with a donor and another labelled with an acceptor, *e.g.* an antigen and an antibody. In the absence of an interaction, no energy transfer occurs. Measurement can be made of the emission from the acceptor, or (more commonly) the quenching of emission from the donor, both of which increase with the proximity of the molecules. Thus the technology is suitable for assays, particularly competitive assays, with an unlabelled analyte.
Immuno-PCR[75] – the secondary antibody in a sandwich assay is labelled with a segment of DNA. Bound DNA can then be amplified using the PCR reaction. It is claimed to be highly sensitive, but the necessity to separate the product by gel electrophoresis and quantitate it by densitometry means that in its present form it is not suitable for large numbers of samples.
Nanotechnology-based immunoassays – a wide variety of approaches have been taken using a range of different nanoparticle types, mostly in order to increase assay sensitivity.[76–80]
Phosphor immunoassays[81] – the labels used are nano-particulate up-converting phosphors based on rare-earth elements, that absorb two or more photons before emitting light at a shorter wavelength. (This is the opposite of what fluorescent labels do; they absorb a single photon before emitting one at a longer wavelength.) The background signal is low because biological samples do not generate an upconverted signal. Multiplexing can be achieved by the use of nanoparticles of different chemistries, with consequently different absorption and emission properties.
Quantum dot immunoassays[82,83] – these assays use highly-fluorescent nanometre-sized quantum dots as labels. The emission wavelength of these dots can be tuned to a desired wavelength, consequently facilitating the simultaneous measurement of multiple analytes.
Surface effect immunoassays[84] – an antibody is immobilised on a suitable surface, and binding is measured by techniques such as total internal reflection, interferometry or surface plasmon resonance.

14.3.9. Rapid/Point-of-care Testing

Traditionally, the techniques described above have had to be carried out in a laboratory by trained personnel usually using expensive equipment. This has time and cost implications – samples have to be collected, stored and transported to a central laboratory under appropriate conditions, tested in batches, reported and the results communicated to the subject. This can mean that results are not received quickly – which can delay treatment and/or allow the spread of disease – or can mean that testing is not performed soon enough or frequently enough. Thus there has been a move in many areas to devise cheaper and more rapid tests that can be performed by the patient or the healthcare worker either at home, at the bedside or in the clinic.

Possibly the most widely used antibody-based test that falls into this category is the home pregnancy-testing kit, versions of which have been available since the 1970s. It detects the presence of human chorionic gonadotrophin (HCG) in a pregnant woman's urine and illustrates a principle – very similar to that of the sandwich ELISA – that can be adapted for detecting other antigens. It uses lateral flow technology, in which an absorptive matrix acts like a wick. A liquid sample to be tested is placed at one end of the matrix and spreads along its length, carrying the analyte (if present) with it and encountering along the way the antibodies that will permit its detection. Antibody labels that permit easy visualisation are usually latex beads, which give a blue colour, or colloidal gold, which gives a red colour, but other labels can be used.[87] In the pregnancy testing kit (Figure 14.9) urine placed at one end of the testing "stick" flows along it, dissolving the labelled monoclonal anti-HCG antibody (adsorbed to the matrix during manufacturing) which will bind any HCG present. Both HCG-bound and -unbound antibody will continue to flow along, eventually encountering an area where an unlabelled anti-HCG antibody, specific for a different HCG epitope from the labelled antibody, has been covalently immobilised to the matrix. The labelled antibody will be captured at this point only if already bound to HCG, and will give a visible readout indicating pregnancy. In the absence of HCG, the labelled antibody will flow past the immobilised antibody yielding no visible signal. The test usually incorporates a control to check for correct functioning. This frequently takes the form of a line of material, such a protein A, placed downstream of the capture antibody, which will bind excess labelled antibody. This confirms adequate liquid flow through the device, dissolution of the labelled antibody from the matrix, and the maintenance of a degree of integrity of the labelled antibody but does not guarantee the ability of the device to detect HCG.[87]

A very similar test has become important in the diagnosis of malaria in settings where good-quality microscopy services (for the visualisation of the parasite) are not readily available. These tests come in dipstick, cassette or card form, and require lysis of the blood sample before application to the device.[88] More and more

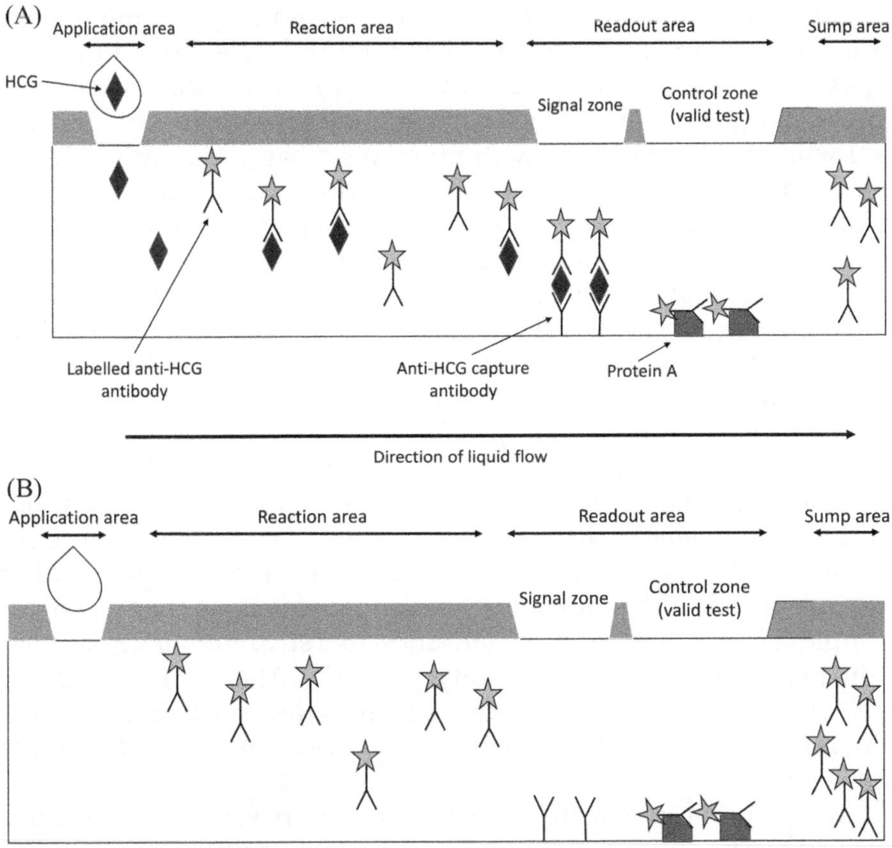

Figure 14.9 The technology behind a home pregnancy-testing kit. (A) Urine from a pregnant female contains human chorionic gonadotrophin (HCG). When this urine is added to the test cassette in the application area it is absorbed by the wicking material within and flows along it toward the sump area. As it does so it dissolves the labelled monoclonal anti-HCG antibody absorbed to the wicking material. The HCG and antibody combine and subsequently bind to the second anti-HCG antibody covalently bound in the signal zone. Excess antibody binds to the agent bound in the control zone, in this example protein A which binds to IgG. Any further unbound material is drawn into the sump area. In a positive pregnancy test, label will be seen in both the signal and control areas. (B) Urine from an individual who is not pregnant contains no HCG. Thus the labelled antibody is not retained in the signal zone, and whist some labelled antibody will be bound in the control zone the rest will flow into the sump area. In a negative pregnancy test, label will only be seen in the control zone, with none in the signal zone. Should there be label in neither the signal nor control zone, then the test is invalid. Note that these diagrams are to demonstrate the principles, and do not necessarily represent any particular test on the market.

antigen-detection tests of this type are becoming available for use in healthcare, including for the detection of Ebola virus, influenza types A and B, and *Streptococcus*.[89] Beyond healthcare they have spread to areas as diverse as environmental monitoring, biodefence and wine production.[90] As with many tests based on antigen–antibody binding, the antigen-detection technology can also be adapted to detect the presence of a specific antibody. Devices similar to those described above are available to detect antibodies to, for example, HIV and hepatitis C virus in saliva.[89] In this form of test the adsorbed detection reagent is a labelled substance, such as colloidal-gold-labelled protein A, that will bind to antibodies, and the covalently bound substance in the signal zone is a synthetic or recombinant HIV or hepatitis C peptide or protein.[91]

There are many variations on the detection systems outlined above. For example, a device for the detection of troponin T (released into the blood following damage to cardiac muscle) uses two soluble antibodies to detect the antigen, one of them gold-labelled and the other biotin-labelled. If the antigen is present it cross-links them, and the complex is immobilised for visualisation by a strip containing streptavidin (which has a very high affinity for biotin). Many other variations also exist, and such systems can be formatted such that multiple different tests can be performed in one device from a single sample application.[92]

Rapid or point-of-care devices can be qualitative, semi-quantitative or quantitative, may be designed for single or multiple use and may be read by eye or by the use of portable or benchtop equipment.[92] Their speed, convenience and relative cheapness make it inevitable that ever more of them will come to the market in the years to come, limited only by the ingenuity of the scientists designing them and the presence of a sufficiently large market to support a particular application.

14.3.10. Microfluidics

This class of automated device – often termed a lab-on-a-chip – is usually designed to require much smaller sample volumes (<1 μl) than ELISAs or the point-of-care devices described above. Initially they were fabricated in either silicon or glass, but the range of materials that can be used has expanded to include polymers[93] and even paper.[94]

Their potential advantages for antibody-based diagnostics go beyond the small sample volumes required, including offering much (theoretically >1000-fold) shorter analysis times due to shorter molecular diffusion distances,[95] greater sensitivity[96] and reduced reagent costs. Some of the immunodiagnostic applications that have been successfully piloted include the measurement of secretory IgA,[95] the detection of viruses,[96] the measurement of α-fetoprotein in blood[97] and the detection of matrix metalloproteinase-8 in saliva in periodontal disease.[98]

Disappointingly, although the potential of microfluidic devices to provide cheap, simple and effective diagnostics has been recognised for at least two decades, this promise has still not been fully realised, with few devices so far reaching the market.[99]

14.3.11. *In Vivo* Diagnostics

All of the techniques described so far are for use *in vitro*, *i.e.* they are performed outside of the patient's body. However, a small group of specialised monoclonal-antibody-based diagnostics have been developed for use within the body. Unfortunately, these have not lived up to the financial predictions made for them in the 1980s and 1990s, and almost all of them are no longer available for *in vivo* use. Myoscint®, for example, an ^{111}In-labelled mouse monoclonal IgG_{2a} Fab fragment against human myosin (which is exposed in necrotic myocardial cells) was developed to determine the existence and location of myocardial necrosis following myocardial infarction. It was licensed in Europe in 1989 and in the USA in 1996, but in 1996 was said to generate less than $1m in European sales[100] and was withdrawn for *in vivo* use in 1999.[101] Similarly OncoScint®, an ^{111}In-labelled monoclonal antibody against carcinoembryonic antigen was licensed in 1992 for the detection of metastases in colorectal cancer,[102] but is no longer commercially available. Verluma®, and CEA-Scan®, also designed for imaging cancer, met a similar fate.[103]

More interesting is the case of the monoclonal antibody Zevalin® (ibritumomab tiuxitan), which binds to CD20 and is used in the treatment of non-Hodgkin's lymphoma. Between 2002 (when it was first approved for therapeutic use) and 2011, it could be labelled with the radioactive isotopes ^{111}In or ^{90}Y. The ^{111}In-labelled version was injected into the patient to determine the biodistribution of the antibody. Imaging scans of the patient's body were then performed using a gamma-camera to detect the distribution of the γ-emitting isotope. Only if the biodistribution was deemed satisfactory would the therapeutic dose of β-emitting ^{90}Y-Zevalin be administered. However, in 2011 the US FDA decided that the ^{111}In imaging step was unnecessary, so it is no longer performed.[104]

The only *in vivo* diagnostic that seems to have survived is ProstaScint®, an ^{111}In-labelled monoclonal antibody against prostate-specific membrane antigen, which was licensed in 1996 for the detection of metastases in prostate cancer. It too has had a troubled history, having changed ownership several times and been available only intermittently over the last 21 years. Now (July 2017) marketed by AYTU Bioscience, it booked more than $1.5 million in revenue in 2016 and the company is hoping to leverage modern imaging technologies, such as fused single-photon emission computed tomography (SPECT-CT) imaging, and the ProxiScan™ gamma camera, which is small enough for trans-rectal prostate cancer diagnosis, to increase sales further.[105,106]

Despite this disappointing history, there has recently been a resurgence in interest in *in vivo* antibody-based diagnostics, due in part to the improved imaging possibilities made available by the use of positron-emission tomography (PET) cameras, the development of related labelling technologies and their combination with (X-ray) computed tomography. Currently, the main utility of these tests is in trials relating to the therapeutic use of monoclonal antibodies; whether any will achieve regulatory approval for routine use in patients remains to be seen. However, the use of such materials will always be limited by the radiological hazard to patients and caregivers (hospital staff, family *etc.*) and the high costs relating to the use of radionuclides.[107]

14.3.12. The Future of Antibody-based Diagnostics

It is clear from the above what some of the drivers will be in the future development of antibody-based diagnostics. These will include:

- Diagnostic need.
- Speed of readout.
- Reduction of cost.
- Reduction of the requirement for trained operatives.
- Reduction in the requirement for sophisticated laboratory facilities.
- Improvements in technology (*e.g.* greater sensitivity) or the advent of new technologies.
- New diagnostic targets (the identification of new molecular markers for known diseases, or the emergence of new diseases).

Much has already been achieved in this area with visible-readout rapid diagnostic tests as described above. The range of these and their sensitivity will continue to increase, and they will be augmented by "lab-on-a-chip" devices that can perform more complex tests automatically and/or perform multiple analyses at the same time. Wearables and integration with smart phones[77] may offer further advances.

Microarrays offer potential in applications that require the detection of many analytes simultaneously. Hundreds, or even thousands, of nanolitre-volume droplets of antibody or an antigen preparation can be spotted or printed onto a glass or plastic surface. Typically, a sandwich-type assay is then performed with an enzymatic or fluorescent readout that is measured by a scanning device. This approach may prove useful for high-throughput drug screening, for disease biomarker identification[108,109] or the detection of known markers, or for detecting changing patterns of analytes during disease development, treatment, relapse and recovery. It has shown particular promise in the classification of cancers and the stratification of patients.[110] However, whilst much exciting research and development work is being performed on this technology,[111] including interfacing it with microfluidics,[112] it has yet to progress to the routine diagnostic laboratory due to a range of problems that have yet to be solved.[113]

Any new (immuno-)diagnostic can only be implemented for routine use once the technology and the applications to which it is being applied reach a sufficient level of reliability and maturity. Quality Assurance programmes are in place in many countries to help ensure this is the case. For example, in the UK the United Kingdom Accreditation Service accredits medical laboratory compliance with ISO 15189.[114] It seems that most "cutting edge" techniques (other than rapid/point-of-care tests) will continue to initially require expensive facilities and staff working in a centralised laboratory for their implementation.

Sadly, it will always be the case that some techniques that appear promising may take far longer than expected to reach their potential, or may never do so. Surface plasmon resonance is a good example of such a technology. It was touted as a technique for use in rapid point-of-care diagnostics in the mid-1980s, but has yet to escape the confines of the research laboratory.[115,116]

Acknowledgements

I would like to thank all my colleagues in the Department of Biological and Environmental Sciences at the University of Hertfordshire for their suggestions, feedback and other forms of assistance with this chapter. I would also like to thank my wife Carole for her support and forbearance while I worked on this chapter.

References

1. A. M. Silverstein, *A History of Immunology*, Academic Press, London, 1989.
2. K. Landsteiner, *Centralbl. Bakt. Parasitenk. Infectionskr.*, 1900, **27**, 357.
3. C. P. Price and D. J. Newman, *Principles and Practice of Immunoassay*, McMillan, New York, 2nd edn, 1997.
4. *The Immunoassay Handbook: Theory and Applications of Ligand Binding, ELISA and Related Techniques*, ed. D. Wild, Elsevier, Amsterdam, Netherlands, 4th edn, 2013.
5. *Teitz Textbook of Clinical Chemistry and Molecular Diagnostics*, ed. C. A. Burtis, E. R. Ashwood and D. E. Bruns, Elsevier Saunders, St Louis, USA, 5th edn, 2012.
6. R. R. Porter, *Biochem. J.*, 1959, **73**, 119.
7. G. M. Edelman, *J. Am. Chem. Soc.*, 1959, **81**, 3155.
8. D. Ribatti, *Immunol. Lett.*, 2015, **164**, 72.
9. A. K. Abbas, A. H. Lichtman and S. Pillai, *Cellular and Molecular Immunology*, Elsevier, Philadelphia, PA, USA, 9th edn, 2017.
10. D. Male, J. Brostoff, D. B. Roth and I. Roitt, *Immunology*, Elsevier, Philadelphia, PA, USA, 8th edn, 2012.
11. P. J. Delves, S. J. Martin, D. R. Burton and I. M. Roitt, *Roitt's Essential Immunology*, Wiley Blackwell, Chichester, UK, 13th edn, 2017.
12. T. W. Mak, M. E. Saunder and B. D. Jett, *Primer to the Immune Response*, Elsevier, London, UK, 2nd edn, 2014.
13. G. Köhler and C. Milstein, *Nature*, 1975, **256**, 495.
14. *Monoclonal Antibodies – Methods and Protocols*, ed. V. Ossipow and N. Fischer, Springer, New York, 2nd edn, 2014.
15. J. D. Marks, *Molecular Biology and Biotechnology*, ed. R. Rapley and D. Whitehouse, Royal Society of Chemistry Publishing, Cambridge, UK, 6th edn, 2015, pp. 167–194.
16. R. R. Coombs, A. E. Mourant and R. R. Race, *Br. J. Exp. Pathol.*, 1945, **26**, 255.
17. K. M. Thompson, *et al.*, *Immunology*, 1986, **58**, 157.
18. D. L. Noah, H. Hill, D. Hines, E. L. White and M. C. Wolff, *Clin. Vaccine Immunol.*, 2009, **16**, 558.
19. https://www.virology.ws/2009/05/27/influenza-hemagglutination-inhibition-assay/, last accessed October 2017.
20. R. C. Lancefield, *J. Exp. Med.*, 1928, **47**(1), 91.
21. J. A. Molina-Bolivar and F. Galisteo-Gonzalez, *J. Macromol. Sci., Part C: Polym. Rev.*, 2005, **45**, 59.
22. C. M. Plotz and J. M. Singer, *Am. J. Med.*, 1956, **21**, 888.
23. N. Banchuin, K. Janyapoon, S. Sarntivijai and L. Parivisutt, *Asian Pac. J. Allergy Immunol.*, 1992, **10**, 47.
24. J. R. Paul and W. W. Bunnell, *Am. J. Med. Sci.*, 1932, **1932**(183), 90.
25. B. Seitanidis, *J. Clin. Pathol.*, 1969, **22**, 321.
26. https://www.oxoid.com/UK/blue/prod_detail/prod_detail.asp?pr=DR0680-&minfo=Y, last accessed July 2017.
27. https://tools.thermofisher.com/content/sfs/brochures/10017521MTL_Technology%20Brochure%20NS.pdf, last accessed August 2017.
28. L. J. Kricke and J. Y. Park, *Teitz Textbook of Clinical Chemistry and Molecular Diagnostics*, ed. C. A. Burtis, E. R. Ashwood and D. E. Bruns, Elsevier Saunders, St Louis, USA, 5th edn, 2012, pp. 379–399.
29. G. Virella and H. H. Fudenberg, *Clin. Chem.*, 1977, **23**, 1925.
30. R. S. Kashvap, *et al.*, *Med. Sci. Monit.*, 2002, **8**, MT95.
31. D. Burnett, *The Science of Laboratory Diagnosis*, ed. D. Burnett and J. Crocker, CRC Press, Boca Raton, FL, USA, 1999, pp. 461–470.

32. J. Stanley, *Essentials of Immunology & Serology*, Delmar, Albany, NY, USA, 2002.
33. V. Kumar, *Trematode Infections and Diseases of Man and Animals*, Springer Science + Business Media, Dordrecht, Netherlands, 1999.
34. A. Cantani, *Pediatric Allergy, Asthma and Immunology*, Springer, Berlin, 2008.
35. D. F. Keren, *Protein Electrophoresis in Clinical Diagnosis*, Hodder Arnold, London, 2003.
36. J. Wallach, *Interpretation of Diagnostic Tests*, Lippincott, Williams & Wilkins, Philadelphia, USA, 8th edn, 2007.
37. B. E. Bourke, I. K. Moss and R. N. Maini, *J. Immunol. Methods*, 1982, **48**(1), 97.
38. S. A. Misbah, *The Science of Laboratory Diagnosis*, ed. D. Burnett and J. Crocker, CRC Press, Boca Raton, FL, USA, 1999, pp. 485–504.
39. https://www.gov.uk/government/uploads/system/uploads/attachment_data/file/344179/V_18i3.2.pdf, last accessed August 2017.
40. A. Wadsworth, N. M. Harris and R. Gilbert, *Am. J. Public Health*, 1934, **24**(7), 727.
41. C. E. Rice, *Can. J. Comp. Med. Vet. Sci.*, 1960, **24**(12), 352.
42. C. E. Rice, *Can. J. Comp. Med. Vet. Sci.*, 1961, **25**(2), 35.
43. J. E. Blair, B. Coakley, A. C. Santelli, J. G. Hentz and N. L. Wengenack, *Mycopathologia*, 2006, **162**(5), 317.
44. https://www.gov.uk/government/publications/smi-v-18-complement-fixation-tests, last accessed August 2017.
45. B. D. Foy, *et al.*, *Emerging Infect. Dis.*, 2011, **17**(5), 880.
46. G. Turabelidze, R. K. Aggu-Sher, E. Jahanpou and C. J. Hinkle, *Morb. Mortal. Wkly. Rep.*, 2015, **64**(23), 636.
47. E. Engvall and P. Perlmann, *Immunochemistry*, 1971, **8**, 871.
48. https://www.fda.gov/BiologicsBloodVaccines/BloodBloodProducts/ApprovedProducts/LicensedProductsBLAs/BloodDonorScreening/InfectiousDisease/UCM080466#HumanImmunodeficiencyVirusType1HIV1Assay, last accessed October 2017.
49. T. S. Alexander, *Clin. Vaccine Immunol.*, 2016, **23**, 249.
50. https://www.fda.gov/downloads/BiologicsBloodVaccines/BloodBloodProducts/ApprovedProducts/LicensedProductsBLAs/BloodDonorScreening/InfectiousDisease/UCM091700.pdf, last accessed July 2017.
51. https://www.gov.uk/government/uploads/system/uploads/attachment_data/file/503768/2905115_Green_Book_Chapter_18_v3_0W.PDF, last accessed July 2017.
52. L. J. Stockman, *et al.*, *J. Clin. Microbiol.*, 2008, **16**, 1842.
53. T. L. Whiteside, *BioTechniques*, 2002, **33**, S4.
54. http://www.isogen-lifescience.com/cancer-antigen-elisa-kits, last accessed October 2017.
55. H. Saadi, *et al.*, *Results Immunol.*, 2013, **3**, 51.
56. https://www.sinobiological.com/cancer-biomarker-ELISA-kit-r051.html, last accessed October 2017.
57. D. R. Henderson, S. B. Friedman, J. D. Harris, W. B. Manning and M. A. Zoccoli, *Clin. Chem.*, 1986, **32**, 1637.
58. K. E. Rubenstein, R. S. Schneider and E. F. Ullman, *Biochem. Biophys. Res. Commun.*, 1972, **47**, 846.
59. Association of State and Territorial Public Health Laboratory Directors and AIDS Program, Center for Infectious Diseases, Public Health Practice Program Office and Centers for Disease Control, *Morb. Mortal. Wkly. Rep.*, 1989, **38**(S7), 1.
60. World Health Organisation, *Wkly. Epidem. Rec*, 1990, **37**, 281.
61. https://www.hiv.va.gov/patient/diagnosis/labs-CD4-count.asp, last accessed July 2017.
62. https://www.who.int/hiv/amds/LaboratoryGuideEnumeratingCD4TLymphocytes.pdf, last accessed July 2017.

63. https://www.who.int/hiv/amds/LaboratoryGuideEnumeratingCD4TLymphocytes.pdf, last accessed July 2017.
64. H. Karcher, D. Böhning, R. Downing, S. Mashate and G. Harms, *Cytometry*, 2006, **70B**(3), 163.
65. S. Renshaw, in *Immunohistochemistry*, ed. S. Renshaw, Scion Publishing Ltd., Bloxham, Oxfordshire, UK, 2007, ch. 4, pp. 45–96.
66. R. Chetty, K. Cooper and A. M. Gown, *Leong's Manual of Diagnostic Antibodies for Immunohistology*, Cambridge University Press, Cambridge, UK, 3rd edn, 2016.
67. J. Duraiyan, R. Govindarajan, K. Kaliyappan and M. Palanisamy, *J. Pharm. BioAllied Sci.*, 2012, **4**(suppl. 2), S307.
68. J. L. Hornick, *Mod. Pathol.*, 2014, **27**, S47.
69. C. Suriyonplengsaeng, *et al.*, *Diagn. Pathol.*, 2017, **12**, 19.
70. S. Zenno and S. Inouye, *Biochem. Biophys. Res. Commun.*, 1990, **171**, 169.
71. J. Todd, *et al.*, *Clin. Chem.*, 2007, **53**, 1990.
72. https://www.merckmillipore.com/GB/en/life-science-research/protein-detection-quantification/Immunoassay-Platform-Solutions/smc-erenna-immunoassay-platform/CIeb.qB.KJUAAAFOhkA1lU_l,nav?ReferrerURL=https%3A%2F%2F, last accessed August 2017.
73. E. F. Ullman, M. Schwarzberg and K. E. Rubenstein, *J. Biol. Chem.*, 1976, **251**, 4172.
74. E. F. Ullman and P. L. Khanna, *Methods Enzymol.*, 1981, **74**, 28.
75. T. Sano, C. L. Smith and C. R. Cantor, *Science*, 1992, **258**, 120.
76. J. W. Kim, W. Heu, S. Joeng and H. S. Kim, *Anal. Chim. Acta*, 2017, **988**, 81.
77. N. Cheng, Y. Song, M. M. A. Zeinhom, Y. C. Cheng, L. Sheng and H. Li, *et al.*, *ACS Appl. Mater. Interfaces*, 2017, **9**, 40671.
78. A. Kaminska, *et al.*, *Sci. Rep.*, 2017, **7**, 10656.
79. S. H. Liao, H. S. Huang, J. J. Chieh, Y. K. Su, Y. F. Tong and K. W. Huang, *Sensors (Basel)*, 2017, **17**, 2018.
80. S. S. Nanda, M. J. Kim, K. Kim, G. C. Palaefthymiou, S. T. Selvan and D. K. Yi, *Colloids Surf., B*, 2017, **159**, 644.
81. R. S. Niedbala, *et al.*, *Anal. Biochem.*, 2001, **293**, 22.
82. A. M. Sobolev, M. V. Pozharov, N. V. Beloglazova and I. Y. Goryacheva, *2016 International Conference Laser Optics*, DOI: 10.1109/LO.2016.7549788.
83. W. C. Chan, D. J. Maxwell, X. Gao, R. E. Bailey, M. Han and S. Nie, *Curr. Opin. Biotechnol.*, 2002, **13**, 40.
84. D. R. Purvis, D. Pollard-Knight and C. R. Lowe, *Principles and Practice of Immunoassay*, ed. C. P. Price and D. J. Newman, Macmillan, New York, 2nd edn, 1997, ch. 20, pp. 511–543.
85. Y. Ikeda, Y. Mashimo, M. Mie and E. Kobatake, *Analyst*, 2016, **141**, 6557.
86. Q. Zeng, Q. Li, W. Ji, X. Bin and J. Song, *Sci. Rep.*, 2016, **6**, 26534.
87. J. R. L. Ehrenkranz, *Epidemiology*, 2002, **13**(38), S15.
88. World Health Organisation, *The Use of Malaria Rapid Diagnostic Tests*, WHO, Geneva, 2nd edn, 2006.
89. https://www.orasure.com/products/products.asp, last accessed July 2017.
90. https://www.rapidmicrobiology.com/test-method/lateral-flow-immunoassays/, last accessed July 2017.
91. https://www.fda.gov/downloads/BiologicsBloodVaccines/BloodBloodProducts/ApprovedProducts/PremarketApprovalsPMAs/UCM091861.pdf, last accessed July 2017.
92. C. P. Price and A. St. John, *Teitz Textbook of Clinical Chemistry and Molecular Diagnostics*, ed. C. A. Burtis, E. R. Ashwood and D. E. Bruns, Elsevier Saunders, St Louis, USA, 5th edn, 2012, pp. 487–505.
93. L. A. L. Bayzydlo and J. P. Landers, *Teitz Textbook of Clinical Chemistry and Molecular Diagnostics*, ed. C. A. Burtis, E. R. Ashwood and D. E. Bruns, Elsevier Saunders, St Louis, USA, 5th edn, 2012, pp. 443–468.

94. https://dfa.org/, last accessed July 2017.
95. K. Sato, *et al.*, *Anal. Chem.*, 2000, **72**(6), 1144.
96. W. T. Liu, *et al.*, *Lab Chip*, 2005, **5**(11), 1327.
97. Y. Liu, *et al.*, *Anal. Chim. Acta*, 2009, **650**, 77.
98. A. E. Herr, *et al.*, *Proc. Natl. Acad. Sci. U. S. A.*, 2007, **104**, 5268.
99. X. Mao and T. J. Huang, *Lab Chip*, 2012, **12**(8), 1412.
100. https://www.diagnosticimaging.com/articles/centocors-myoscint-returnswith-positive-fda-panel-review, last accessed July 2017.
101. G. Walsh, *Proteins: Biochemistry and Biotechnology*, Wiley Blackwell, Chichester, UK, 2014.
102. https://www.diagnosticimaging.com/articles/oncoscint-nod-opens-door-usmoab-imaging-procedures, last accessed July 2017.
103. C. A. Boswell and M. W. Brechbiel, *Nucl. Med. Biol.*, 2007, **34**, 757.
104. https://www.zevalin.com/H501-faqs.aspx, last accessed July 2017.
105. https://aytubio.com/aytu-bioscience-provides-third-fiscal-quarter-2016-business-update/, last accessed July 2017.
106. https://aytubio.com/downloads/Aytu-Corporate-Overview.pdf, last accessed July 2017.
107. L. E. Lamberts, *et al.*, *J. Clin. Oncol.*, 2015, **33**, 1491.
108. J. C. Miller, H. Zhou, J. Kwekel, R. Cavallo, J. Burke and E. B. Butler, *et al.*, *Proteomics*, 2003, **3**(1), 56.
109. S. Chen, T. LaRoche, D. Hamelinck, D. Bergsma, D. Brenner and D. Simeone, *et al.*, *Nat. Methods*, 2007, **4**(5), 437.
110. C. P. Paweletz, L. Charboneau, V. E. Bichsel, N. L. Simone, T. Chen and J. W. Gillespie, *et al.*, *Oncogene*, 2001, **20**(16), 1981.
111. L. Barrade, A. E. Garcia and J. A. Camerero, *Pharm. Res.*, 2011, **28**(7), 1480.
112. C. K. Dixit and G. R. Aguirre, *Microarrays*, 2014, **3**, 180.
113. U. Sauer, *Sensors*, 2017, **17**, 256.
114. https://www.ukas.com/services/accreditation-services/medical-laboratory-accreditation-iso-15189/, last accessed October 2017.
115. H. H. Nguyen, J. J. Park, S. Kang and M. Kim, *Sensors*, 2015, **15**, 10481.
116. J.-F. Masson, *ACS Sens.*, 2017, **2**, 16.

15 Molecular Biomarkers: Overview, Technologies and Strategies[†]

Mukesh Verma[a], Debmalya Barh*[b], and Vasco Azevedo[c]

[a]Epidemiology and Genomics Research Program, Division of Cancer Control and Population Sciences, National Cancer Institute (NCI), National Institutes of Health (NIH), 9609 Medical Center Drive, Rockville, MD 20850, USA; [b]Centre for Genomics and Applied Gene Technology, Institute of Integrative Omics and Applied Biotechnology (IIOAB), Nonakuri, Purba Medinipur, West Bengal-721172, India; [c]Departamento de Biologia Geral, Instituto de Ciências Biológicas (ICB), Universidade Federal de Minas Gerais, Pampulha, Belo Horizonte, Minas Gerais, Brazil
*E-mail: dr.barh@gmail.com

15.1. Introduction

Biomarkers can be specific cells, molecules, images, genes, gene products, enzymes, hormones, or receptors.[1] The term "biomarkers" is sometimes replaced by "actionable biomarkers" to describe biomarkers that can inform that clinical practice can be applied for disease diagnosis, prognosis and treatment. Most of such actionable biomarkers exist in cancer although in other diseases also progress has been made in the last few years. Sometimes entirely different kinds of biomarkers exist, for example, temperature is a biomarker

[†]This chapter has been reproduced from *Molecular Biology and Biotechnology*, ed. R. Rapley and D. Whitehouse, 2014, 6th edition, copyright The Royal Society of Chemistry.

Genomics and Clinical Diagnostics
Edited by David Whitehouse and Ralph Rapley

Published by the Royal Society of Chemistry, www.rsc.org

of abnormal physiology of the body (fever), possibly due to infection, and blood pressure is a biomarker for stroke. To evaluate the current status of the biomarker field in disease diagnosis and therapy, we analysed published literature starting from 1985 until April 2013 and summarized in Table 15.1. The total number of publications in the biomarker field was 601 401. The number of publications in the biomarker diagnosis field was almost double that of those in therapeutics. This seems reasonable because more biomarkers exists which can be used for disease diagnosis but it takes time to develop biomarkers for therapies. Metabolic disease biomarkers are fewer compared with cardiovascular disease markers.[2] Another important point in such meta-analysis is the proper use of terms for the analysis. For example, when we used "infectious agent biomarker", the number of publications was much less than using the term "bacterial biomarkers" or "viral biomarker" (see Table 15.1). In some diseases, a single biomarker is sufficient to diagnose a disease, whereas in other diseases multiple biomarkers are needed for diagnosis. The utility of multiple biomarkers is determined based on their combined sensitivity and specificity, which should be superior to those of a single marker. In this chapter, biomarkers of immune diseases, cardiovascular diseases (CVDs), metabolic diseases, infectious diseases, neurological diseases and cancer are discussed (see Tables 15.2–15.7).

Table 15.1 Publications of biomarker studies indicating investigator interest in the field.[a]

Topic	Number of Publications
Biomarker	601 401
Biomarker and diagnosis	373 653
Biomarker and therapy	173 756
Biomarker and cancer	215 529
Biomarker and neurological disorders	46 330
Biomarker and cardiovascular disorders	57 744
Biomarker and metabolic disorders	521
Biomarker and immunologic disorders	73 417
Biomarker and infectious agents related diseases	159
Biomarkers of viruses in diseases	11 445
Biomarkers of bacteria in diseases	14 538
Biomarker and breast cancer and epigenetics	45
Biomarker and breast cancer and methylation	381
Biomarker and breast cancer and histone	247
Biomarker and breast cancer and microRNA	209
Biomarker and breast cancer and proteomics	419
Biomarker and breast cancer and imaging	1094

[a]The analysis was PubMed-based, and references up to April 2013 were considered.

15.2. Overview of Biomarkers

Biomarkers are often used in clinics, where they are derived from biological fluids that are easily available. Biomarkers show characteristic biological properties that can be detected and measured in parts of the body, such as organ or body fluid (blood, urine, saliva, nipple aspirate, pleural lavage, pancreatic juice).[1,3] They represent normal or disease state of the body and can be used to follow up treatment in a disease state. A number of assays have been developed to identify biomarkers. These assays include immunohistochemistry, imaging, gene constitution (amplification, mutation and rearrangement), gene and protein expression analysis, such as single gene or protein expression, methylation and histone profiling, micro RNA (miRNA) polymorphism and miRNA profiling.[3–6] Biomarkers play major roles in medicinal biology.[4–7] In medicine, a biomarker generally refers to a specific protein concentration in blood that may, in turn, reflect the presence, progression or severity of a disease and guide the treatment.[8] Biomarkers have been used for more than half a century but their application has increased remarkably in the 21st century. A diseased state indicates a structural change in the proteins or enzymes. The physiological changes between the normal and diseased state are compared to look out for biomarkers. This is because the gene and protein expression profiles, along with the metabolic expression profiles, change in a diseased state. The researchers study the upregulated and downregulated genes, proteins and metabolites and understand the genetic patterns closely associated with a particular type of disease that leads to the discovery of a biomarker. Once identified, the biomarkers are validated in a large number of samples and possibly at two different laboratories. Microarrays – RNA, DNA, protein or antibody – play an important role in analysing a biomarker. The comparison of plasma protein concentration levels in normal and diseased states is often used in biomarker analysis. For example, various new forms of glycoproteins are formed during glycosylation wherein polysaccharides or sugars are added to the polypeptides (proteins). Any abnormal concentration of glycoproteins can act as a biomarker for various diseases, such as muscular dystrophy and acute or chronic inflammation. In the case of pancreatic cancer, RNAase-1 is used as a biomarker for disease diagnosis. An obvious alteration in pattern of glycosylation of the enzyme was observed in the urine and blood serum in tumorous pancreatic cells. In molecular terms, a biomarker is the subset of markers that can be discovered using various omics or imaging technologies. We have selected biomarkers of cancer, cardiovascular disease, neurological disorders, infectious-agents-related diseases, metabolic diseases and immune diseases to cover a wide-spectrum of diseases and provide up-to-date information about biomarkers associated with these diseases and disorders.

15.3. Types of Biomarkers

15.3.1. Based on Utility

Physicians and scientists use various types of biomarkers to study human diseases, either to track the progression of a disease, or to detect the effects of a drug. The use of biomarkers in the diagnosis of infections, genetic disorders and cancer is well known. The expansion of laboratory technology and growth of molecular biology increases the feasibility of biomarkers that are technically advanced. Biomarkers play a very important role in personalized medicine as they facilitate the combination of therapeutics with diagnostics. In contrast, big pharmaceutical and biotechnology companies use biomarkers as a drug discovery tool. Biomarkers are important in drug development as they help to determine the pharmacodynamic effects of the drug being developed and to assess the safety and the efficacy of the drug. Safer drugs with better efficacy can be developed in a cost-effective manner by using biomarkers.

Biomarkers include tools and technologies that can be used in prediction of progression and outcome (survival, recurrence, multiple diseases) of a disease. Biomarkers can be classified on the basis of detection techniques (imaging and non-imaging), properties (molecular, cellular and nuclear) and utility (diagnosis, prognosis and therapy). Because of the latest high-throughput omics technologies in genomics, epigenomics, proteomics, metabolomics and transcriptomics, a large number of potential biomarkers have been identified that can be utilized in various ways for disease management and patient care.[2,3] Biomarkers can be classified as diagnostic, prognostic and therapeutic biomarkers on the basis of their utility.[8] Early detection biomarkers are used to detect a particular disease in its early stages. Diagnostic biomarkers are those that are used to identify the presence or absence of a disease or to diagnose a disease. Predictive biomarkers are those that are present prior to an event occurring and that predict the outcome and the efficacy of the drug in a treatment. Prognostic biomarkers give valid information about the outcome without using a therapy. They are used to determine how such a disease may develop (aetiology of the disease) in an individual and increase survival of patients. Disease prognosis biomarkers are related to measures of a disease state. Efficacy biomarkers reflect the effects of a particular treatment using a specific drug. Since they correlate with the desired clinical outcomes, they can be used to obtain provisional regulatory approval of a drug. Surrogate biomarkers are used to measure the clinical outcomes. Toxicity biomarkers, as their name reflects, measure the toxicity of drugs or interventions. Target biomarkers reflect the presence of a specific drug target and indicate the drug–target interaction and outcome. Lastly, pharmacodynamic biomarkers (pharmacogenomics and pharmacoepigenomics) belong to the category of biomarkers that are used in drug development for intervention and/or treatment.

15.3.2. Based on Diagnosis Approaches

On the basis of diagnostic approaches, biomarkers can also be classified as imaging biomarkers or non-imaging biomarkers. Imaging biomarkers are those biomarkers which are detectable by using imaging techniques, such as X-ray (mammograms), electrocardiogram (ECG), ultrasound imaging, computed tomography (CT) and CT scanning, magnetic resonance imaging (MRI) and quantum dots.[9] Imaging biomarkers give reproducible results and have tremendous potential in diagnosis as well as therapy and prognosis.

Non-imaging biomarkers can be molecular biomarkers with biophysical properties. They may include nucleic-acid-based biomarkers, such as gene mutations or polymorphisms, alterations in copy number (copy number variance or CNVs), changes in the mitochondrial genome, chromosomal aberrations, gene and protein expression profile, microRNA, metabolites, enzymes, antigens, hormone *etc.*[1,4–7] Serum levels of interleukin (IL)-27, IL-29 and IL-31 in bronchoalveolar lavage fluid (BALF) were high in lung cancer patients.[10] Higher levels of microRNA miR-21 and lower levels of miR-451 and miR-485 were observed in lung cancer.[11]

If a combination of biomarkers (biomarker profiles) can detect a disease early, it provides an opportunity for developing intervention and therapeutic approaches.[3] For example, if biomarkers can discriminate pre-rheumatoid arthritis subjects from normal subjects, intervention and therapeutic approaches can be applied in high-risk individuals because drugs exist which can treat rheumatoid arthritis. Biomarkers are useful in detecting secondary cancers (recurrence) also.

15.3.3. Based on Therapy

Biomarkers are used to follow up therapy and their levels indicate the host's response to therapy. Due to recent developments in genomics, it is possible to make therapy personalized. The identification of biomarkers represents a fundamental medical advance that can lead to an improved understanding of a disease, and holds the potential to define surrogate diagnostic and prognostic end points.

Understanding biomarkers in a complex system, such as the central nervous system (CNS), is extremely valuable for developing therapeutics in brain and neurological disorders. Brain parenchyma is a highly complex microvascular structure and it undergoes a variety of changes during tumour formation. Neo-angiogenesis starts in tumours and if anti-angiogenesis therapy is planned for such brain tumours, biomarkers should be known, which can be used to follow therapy. Some biomarkers in this category are vascular endothelial growth factor (VEGF), phosphatidylinositol glycan biosynthesis class F protein (PIGF), angiopoitin-1 and integrin.[12] In Alzheimer's disease, glutamate levels are increased (in the hippocampus region of the brain) after treatment with galantamine.[13] Therefore, glutamate can be considered a therapeutic biomarker of Alzheimer's disease. Galantamine is a cholinesterase inhibitor that is generally used in the treatment of this disease.[14]

An ideal biomarker should indicate whether a treatment is non-toxic and effective. Molecular biomarkers are required to predict the likelihood of an individual tumour's responsiveness or of toxicity in normal organs and to indicate optimized treatments with improved efficacy at reduced side effects for each cancer patient. Biomarkers with prognostic value concerning treatment response and patient survival can then be used as targets to develop optimized drugs. For example, (1) chemo-selective treatment of tumours with 9p21 deletion by L-alanosine, (2) treatment of multidrug-resistant P-glycoprotein-expressing tumour cells by non-cross-resistant natural products or by inhibitors of P-glycoprotein to overcome multidrug resistance, and (3) natural products that inhibit the epidermal growth factor receptor (EGFR) in EGFR-overexpressing tumour cells.[15]

In cancer cells, cyclic-AMP-dependent protein kinase (PKA) is secreted into the conditioned medium. This PKA, designated as extracellular protein kinase A (ECPKA), is markedly upregulated in the sera of patients with cancer. The currently available tumour biomarkers are based on the antigen determination method and lack specificity and sensitivity. An ECPKA autoantibody detection method for a universal biomarker that detects cancer of various cell types has been reported. The receiver-operating characteristic (ROC) plot indicated that autoantibody enzyme immunoassay exhibited 90% sensitivity and 88% specificity, whereas the enzymatic assay exhibited 83% sensitivity and 80% specificity. These results indicated that the autoantibody method distinguished between patients with cancer and controls better than the antigen method could. Serum biomarker measurement in body fluid immunoassays has been the most widely used approach, generally of established tumour-associated markers, such as carcinoembryonic antigen (CEA), alpha fetoprotein (AFP), human chorionic gonadotrophin (hCG), prostate specific antigen (PSA), and carcinoembryonic antigen 125 (CA125). These biomarkers have low specificity and sensitivity; therefore, they are either not used in screening and diagnosis or used in combination with other biomarkers. The limitations of the presently available serum tumour biomarkers, based on the antigen determination method, indicate the need for other means of screening.

Plasma and urinary concentrations of two members of the VEGF family and their receptors as potential response and toxicity biomarkers of bevacizumab with neoadjuvant chemoradiation in patients with localized rectal cancer were evaluated by different groups of investigators.[16] Of all biomarkers, pre-treatment plasma soluble VEGF receptor 1 (sVEGFR-1), an endogenous blocker of VEGF and placental growth factor (PlGF), a factor linked with vascular normalization, was associated with both primary tumour regression and the development of adverse events after neoadjuvant bevacizumab and chemoradiation. Plasma sVEGFR-1 should be further evaluated to establish it as a potential biomarker to stratify patients in future studies of bevacizumab and/or cytotoxics in the neoadjuvant setting.

15.3.4. Imaging Biomarkers and Non-imaging Biomarkers

Mammography is the process of using low-energy X-rays to examine human breast cancer and is used as a screening and diagnostic tool. Mammography is applied in clinics for breast cancer screening and has helped in reducing breast cancer mortality.[17] We should, however, keep in mind that the radiation exposure associated with mammography has a potential health risk.[18,19]

Imaging is an enabling scientific discipline combining advanced technology and complex computational and analytical methods to provide a unique ability to extract spatially and temporally defined information from humans.[20,21] It allows us to investigate intact biological systems (without isolating samples or taking biopsies) across the spectrum from sub-cellular to macroscopic and from discovery to clinical decision making. By this technology early breast cancer is detected *via* characteristic masses and/or microcalcification. Thus, mammography is considered to be a non-invasive biomarker of cancer diagnosis. For the average woman between the ages of 50 and 74 years, mammography is recommended every two years. This helps in avoiding unnecessary surgery, treatment and anxiety. On a cautionary note, mammography has a false-negative rate of approximately 10% due to dense tissues obscuring the cancer and also due to the fact that the appearance of cancer on mammogram has a large overlap with the appearance of normal tissues.[22] Quantum dots technology (nanotechnology) is also based on imaging and has been successfully used for cancer diagnosis.[23]

Another technology called a positron emission tomography (PET) scan was used to evaluate the treatment response in breast cancer patients.[24] Although reasonable success could be achieved, the main problem with imaging technologies (along with health hazards of multiple exposures) is that tumour heterogeneity interferes with interpretation of results and a combination of other biomarkers and patient-related information is needed to infer any clinical value. It is re-emphasized here that multiple biomarkers should be used to achieve high sensitivity and specificity (discussed in other sections of this chapter).

15.3.5. Invasive and Non-invasive Biomarkers

Tissues are the best source of material to assay early detection cancer biomarkers because they represent true expression of biomarkers during disease development.[9,10] However, tissue collection is an invasive procedure and it is difficult to get healthy tissue for comparison. Therefore, biomarkers that can be assayed in samples collected noninvasively are preferred.[10] Biofluids (urine, blood, sputum) and exfoliated cells are good examples of non-invasive sources of biomarkers for early diagnosis of the disease. After identifying biomarkers, the assay and the biomarker have to be approved by the Food and Drug Administration (FDA) so that these biomarkers can be assayed in clinical samples.[25] The FDA has provided guidelines in this direction (www.fda.gov). If biomarkers, assays or devices are planned for clinical use in patient samples, they should be reviewed by the FDA's Center for Devices and Radiological Health (CDRH) for their ability to analytically measure the biomarker.[26] Biomarkers and devices for quantification are expected to yield equivalent results. Biomarkers should have passed analytical and clinical validation tests specified by the FDA. Analytical validity in this context is defined as the ability of an assay to accurately and reliably measure the analyte in the laboratory as well as in the clinical sample. Clinical validation requires the detection or prediction of the associated disease in specimens from targeted patients. Biomarker qualification by the FDA enables collaboration among stakeholders, reduces costs for individual stakeholders and provides biomarkers that are useful for the general public and private parties.

Sometimes biomarkers fail to show reasonable sensitivity and specificity when validation is conducted.[27] Although non-invasive biomarkers are the best for breast cancer detection, when these biomarkers are validated in large number of samples, some of them do not show reasonable sensitivity and specificity.[28] The traditional treatment options for breast cancer are radiation, chemicals and surgery (lumpectomy, quadrantectomy, mastectomy). Surgery is usually combined with adjuvant therapy (hormonal and/or chemical therapy). Chemicals used for therapy have considerable toxicity and hormonal treatments also have long lasting adverse effects. Surviving patients generally have poor quality of life. Furthermore, resistance to chemotherapy is another problem observed in breast cancer patients.[29,30] Pharmacogenomics is an area of research which may provide some useful information in these cases.[31] By applying diagnostic tests and knowing the genetic background of an individual, personalized treatments are possible.

Non-invasive biomarkers may also help in guiding the type of therapy that is more beneficial for patients. For example, in the case of breast cancer, in women with ductal carcinoma *in situ* (DCIS) the treatment is with tamoxifen instead of an aromatase inhibitor.[32] However, at the early invasive stage, as judged by a panel of biomarkers, aromatase inhibitor proved better for treatment than tamoxifen. The use of hormonal therapy changed with the age of patient and tumour characteristics. Most of these characteristics correlated better with early stage than DCIS. This research led to development of prevention strategies. Now endocrine therapy is used for preventing new primary breast cancers and invasive recurrence for women with DCIS or early invasive breast cancer. The dose used was higher at an early stage but decreased with age of the patient.

Epigenomic biomarkers have enormous potential clinical implications in cancer diagnosis and prognosis.[3] Because of the availability of genome-wide methylation, histone and miRNA analysis technologies, and our rapidly accumulating knowledge regarding the epigenome, the translation of findings discussed in this article may be possible in the near future. Epigenetic biomarkers may also help in identifying patients who will benefit from the therapy and will not develop resistance to drugs, and ultimately increase survival.[33] Recently developed drugs for disease treatment are based on specific pathways and may be useful for those individuals where those pathways are altered. This approach can be designed for personalized medicine and precision medicine. Epigenetic biomarkers may help in such approaches. All potential treatment biomarkers for diseases discussed in this chapter have by no means been exhausted, and it is expected that additional high-penetrance biomarkers will be identified.

15.4. Omics Approaches in Biomarker Discovery

Omics is an emerging and exciting area in the field of science and medicine.[4,5,9,34] In the post-genome era, efforts are focused on biomarker discovery and the early diagnosis of different diseases through the application of various omics technologies.[3] Numerous promising developments have been elucidated using omics (transcriptomics, proteomics, metabolomics, peptidomics, glycomics, phosphoproteomics or lipidomics – on tissue samples and body fluids) in different diseases.[32,34] The development of high-throughput technologies that permit the solution of deciphering a disease from higher dimensionality may provide a knowledge base, which changes the face of the disease biology, aetiology and therapeutics. The omics technology that has driven these new areas of research consists of DNA and protein microarrays, mass spectrometry and a number of other instruments that enable high-throughput analyses at relatively low cost.[1,34] High-throughput omics technologies are being established and validated to create better predictive models for diagnosis, prognosis and therapy to identify and characterize key signalling networks and to find new targets for drug development.[34]

15.4.1. Immune Diseases

Various applications of omics technology include study of changes in tissue-specific protein expression in normal and disease samples, evaluating changes in immune response of the proteins in disease states, translating results from the laboratory to bedside delivery of patient care and developing an artificial neural network (ANN) to distinguish high-risk individuals from normal subjects. Interferons and cytokines were identified as a big group of biomarkers in immune diseases.[35] Interferons belong to three families: I, II and III, with multiple subtypes. In infections and chronic inflammatory diseases, interferons and their stimulating genes could be utilized to follow up outcome and survival. Multiple sclerosis is an immune disease with characteristic feature of deregulated T lymphocyte apoptosis.[36] Biomarkers identified during multiple sclerosis development were CASP8AP2, IL-23, CD36, ITGAL, OLR1, RNASEL, RTN4RL2 and THBS1.[36] Selected biomarkers are shown in Table 15.2.

15.4.2. Cardiovascular Diseases (CVDs)

Cardiovascular diseases may arise due to psychosocial stress.[37] CVD biomarkers, adhesion and proinflammatory molecules (IL-6, other cytokines, C-reactive proteins and fibrinogen), and pathogens were evaluated to assess their contribution in socioeconomic position (SEP) and CVDs.[37] Selected biomarkers are shown in Table 15.3.

15.4.3. Metabolic Diseases

A multitude of factors should be considered in selecting the specific technology to be adopted for metabolomics studies.[2] Two major technologies, mass spectrometry (MS) and nuclear magnetic resonance spectroscopy (NMR), are generally considered, as they can measure hundreds to thousands of unique chemical entities. MS is highly sensitive and has the capacity to detect metabolites with concentrations in the picomole range and above, requires small biospecimen volumes, enables metabolites to be individually identified and quantified and is suitable for use in a high-throughput mode. However, MS requires expensive consumables, has relatively lower analytical reproducibility, poorly represents highly polar metabolites when using standard chromatography protocols and requires more complex software and algorithms for routine data analysis.[2] In contrast, NMR allows for the comprehensive generation of metabolite profiles by a single non-destructive method, is fully automated with high-throughput capacity, inherently quantitative, highly suitable for metabolite structure elucidation and has very high analytical reproducibility with a well-established mathematical and statistical tool box. The disadvantages to using NMR include its relative insensitivity in detecting metabolites

Table 15.2 Biomarkers and their characteristics in immune diseases.[a]

Disease	Biomarker	Characteristics
Autoimmune disease (rheumatoid arthritis)	Antigen arrays	Can be used for disease stratification and planning treatment strategies.[73]
Autoimmune disease (rheumatoid arthritis, systemic and cutaneous lupus erythematosus, (SLE))	Annexin-I	Annexin-1 exerts its anti-inflammatory effect by suppressing the generation of inflammatory mediators and anti-annexin I antibodies are present in rheumatoid arthritis and SLE patients.[74]
Autoimmune disease (systemic sclerosis, systemic and cutaneous lupus erythematosus, (SLE)), polymyositis (PM), dermatomyositis (DM), multiple sclerosis	Prolactin (PRL), ferritin, vitamin D, tumour marker tissue polypeptide antigen (TPA)	High levels of PRL, ferritin, vitamin D, and TPA were observed in SLE, PM, DM and multiple sclerosis although levels of these markers were different in different diseases.[75]
Rheumatoid arthritis (RA)	IL-17 and Wnt/β-catenin	In rheumatoid arthritis (RA) bone loss occurs.[97] Transcriptomics approaches were applied to identify genes and pathways that contributed to bone loss in RA patients. Results indicated IL-17 and Wnt/β-catenin as transcriptional biomarkers of RA.[98]
Multiple sclerosis	CASP8AP2, IL-23, CD36, ITGAL, OLR1, RNASEL, RTN4RL2, and THBS1	In infections and chronic inflammatory diseases, interferons and their stimulating genes can be utilized to follow up outcome and survival. Multiple sclerosis is an immune disease with characteristic feature of deregulated T lymphocyte apoptosis. Biomarkers identified during multiple sclerosis development were CASP8AP2, IL-23, CD36, ITGAL, OLR1, RNASEL, RTN4RL2, and THBS1.[36]
Asthma	Nitric oxide	Measurements of fractional excretion of nitric oxide can be used as a biomarker for diagnosing asthma.[162]
Rheumatoid arthritis (RA)	IL-17 and Wnt/beta catenin	In rheumatoid arthritis (RA) bone loss occurs. Transcriptomics approaches were applied to identify genes and pathways, which contributed to bone loss in RA patients. Results indicated IL-17 and Wnt/beta catenin to be transcriptional biomarkers of RA.[98]
Autism	Increased number of mast cells and serotonin in urine	High number of mast cells and serotonin in urine are observed in autism patients. Mast cells produce inflammatory cytokines in large amounts in autism patients and the possibility of autoimmunity has been proposed.[108]

(continued)

Table 15.2 *(continued)*

Disease	Biomarker	Characteristics
Multiple sclerosis	HLA DRB1-1*5 haplotype, SHP-1 hypermethylation	Multiple sclerosis is an inflammatory disease in which the fatty myelin sheaths around the axons of the brain and spinal cord are damaged. Female-specific multiple sclerosis is associated with HLA DRB1-1*5 haplotype.[163] SHP-1 hypermethylation is a biomarker for multiple sclerosis.[145]
Asthma	Arginine kinase, sarcoplasmic calcium binding proteins, and tropomyosin	Proteomics approaches identified allergenic proteins, based on their reactivity to patient's sera, using tandem mass spectrometry.[161] The most significant allergens identified were arginine kinase, sarcoplasmic calcium binding proteins, and tropomyosin that can be used for screening.
Rheumatoid arthritis (RA)	Aggrecan fragments, C-propeptide of type II collagen	In rheumatoid arthritis (RA) inflammation of joints synovium is persistent, which leads to erosion of cartilage and bone. A biomarker, aggrecan fragments, was identified which was detected in synovial fluid (SF). Levels of aggrecan fragments (structural components of cartilage) were higher in RA patients than in healthy individuals. Another biomarker was C-propeptide of type II collagen and its levels were directly proportional to the rate of collagen synthesis. Cartilage oligomeric matrix protein (COMP) is also a component of cartilage and it can be measured in serum and SF.
Multiple sclerosis	Neurofilament light protein (NFL), Glial fibrillary acidic protein (GFAP)	Multiple sclerosis is associated with autoimmune-mediated inflammation of the central nervous system and may lead to demyelination and axonal damage. Biomarker neurofilament light protein (NFL) is a cytoskeleton component in large myelinated axons and it is released into the cerebrospinal fluid (CSF) and can be used to determine the level of axonal damage. Glial fibrillary acidic protein (GFAP) is another biomarker for multiple sclerosis.
Rheumatoid arthritis and systemic and cutaneous lupus erythematosus (SLE)	Annexin I, II and V	Annexin-I exerted its anti-inflammatory effect by suppressing the generation of inflammatory mediators and anti-annexin I antibodies were present in rheumatoid arthritis and systemic and cutaneous lupus erythematosus (SLE) patients.[74] Annexin II and V were involved in the coagulation cascade because they had affinity towards phospholipids.

Autoimmune disease (systemic sclerosis, systemic and cutaneous lupus erythematosus, SLE), polymyositis (PM), dermatomyositis (DM), multiple sclerosis	Prolactin (PRL), ferritin, vitamin D	Prolactin (PRL), ferritin, vitamin D and the tumour biomarker tissue polypeptide antigen (TPA) levels were higher in patients with autoimmune disease (systemic sclerosis, systemic and cutaneous lupus erythematosus, (SLE)), polymyositis (PM), dermatomyositis (DM), multiple sclerosis.[75] The granin family is a group of acidic proteins present in the secretory granules of a wide variety of endocrine, neuronal, and neuroendocrine cells. A few important granins, chromogranin A and B, secretogrannin II, HISL-19 antigen, NESP55, ProSAAS, should be studied further for their clinical implication.[76]
Autism	Serotonin, inflammatory cytokines	High number of mast cells and serotonin in urine were observed in autism patients. Mast cells produce inflammatory cytokines in large amounts in autism patients and the possibility of autoimmunity was proposed.[108]
Asthma	Glutathione S transferase M1 (GSTM1)	In one meta-analysis glutathione S transferase M1 (GSTM1) biomarker was identified which helped in screening a cohort of children who were at high risk of developing asthma.[164] In a population-based study C-reactive protein, fibrinogen and interleukin-6 were found to be useful biomarkers for screening.[165]
RA	CTX-1, CTX-II, C-telopeptide I and II	Biomarkers of bone and cartilage turnover are collagen C-telopeptides I and II, which are predictors of structural damage in RA patients. C-terminal cross-linked telopeptide of type I collagen (CTX-I) could be measured in serum or urine after it was released during bone resorption. Bone erosions and osteoporosis both occurred as a consequence of RA, and CTX-I levels could be used for prognosis. On the other hand, increased CTX-II levels were associated with rapid progression of joint damage. For clinical implication, investigators of these studies recommended that new prognostic biomarkers should supply information beyond that provided by risk factors in the prediction of disease outcome.[166]

[a]Note: The same diseases are mentioned in more than one row because biomarkers shown in column 2 were identified in different studies.

Table 15.3 Biomarkers and their characteristics in cardiovascular disease (CVD).[a]

Disease	Biomarker	Characteristics
Cardiovascular disease (CVD)	C-reactive protein	In a risk prediction model of cardiovascular disease (CVD) c-reactive protein levels were implemented.[8]
Cardiovascular disease (CVD)	Blood pressure	Biomarker studies of CVD prediction in elderly indicate that mortality and cardiovascular events are dependent on low peripheral pulse pressure not on high blood pressure.[167]
Cardiovascular disease (CVD)	Lipoprotein-associated phospholipase A2 and antiphosphorylcholine IgM	In cardiovascular disease (CVD), the relative contribution of genetic and environmental effect was studied for two inflammatory biomarkers, lipoprotein-associated phospholipase A2 and antiphosphorylcholine IgM, in a Swedish population.[168]
Cardiovascular disease (CVD)	TNFα, IL-6, plasma vitamin E concentrations, total and LDL cholesterol, and antioxidant profiles	Biomarkers discussed in the diagnostic section can be used for follow up of the treatment. Dietary intervention of CVD by fish oil (salmon, herring, and pompano) and other nutrients has been demonstrated in a number of studies. Some of the participants had higher levels of triacylglycerolaemia. Biomarkers TNFα and IL-6 were reduced and level of adiponectin increased in the treated arm. Thus TNFα, IL-6 and adiponectin were used as therapeutic biomarkers. In another study, argon oil supplement reduced plasma levels of lipids and antioxidant status. Therapeutic biomarkers used in this study were plasma vitamin E concentrations, total and LDL cholesterol and antioxidant profiles.[169]
Cardiovascular disease (CVD)	Atherosclerosis	A very well characterized biomarker of CVD.
Cardiovascular disease (CVD)	IL-6, other cytokines, C-reactive proteins, and fibrinogen	CVD biomarkers, adhesion and proinflammatory molecules (IL-6, other cytokines, C-reactive proteins, and fibrinogen) and pathogens were evaluated to assess their contribution in socioeconomic position (SEP) and CVD.[37]
Atherosclerosis	Osteoprotegerin (OPG)	Osteoprotegerin (OPG) is a marker, which is independently associated not only with risk factors of atherosclerosis but also with subclinical peripheral atherosclerosis and clinical atherosclerosis and is recommended as a prognostic biomarker for ischemic heart disease and ischemic stroke.[77]
Hypertension	Methylated ADD1 gene	In hypertension, global methylation profiling and individual gene methylation status (ADD1 gene) have been used for diagnosis and outcome.[147]

Atherosclerosis	Hypermethylation of monoamine oxidase A (MACA)	Atherosclerosis can be diagnosed when other biomarkers are combined with epigenetic biomarkers, such as hypermethylation of monoamine oxide A (MAOA).[146]
Atherosclerosis (early diagnosis)	Carotid intima media thickness (IMT), circulating oxidized low-density lipoprotein (LDL), and flow-mediated dilation (FMD)	For early diagnosis of carotid atherosclerosis, for which obesity is a risk factor, the early biomarkers are carotid intima media thickness (IMT) and circulating oxidized low-density lipoprotein.[170] To evaluate endothelium status, flow-mediated dilation (FMD) and IMT have been used as early markers of atherosclerosis in patients with non-alcoholic fatty liver disease (NAFLD).[171]
Atherosclerosis	CARD8, Ephs, ephrins	Atherosclerosis is a chronic inflammatory disease of the vessel wall. The gene CARD8, which encodes proteins involved in innate immunity in atherosclerosis patients, is an excellent biomarker for atherosclerosis. Inflammatory markers also overexpressed in this disease. Inflammosomes produce interleukin 1β in response to cholesterol crystal accumulation in macrophages. Ephs and ephrins also have been proposed as biomarkers of atherosclerosis.[109]
Atherosclerosis	Hypermethylation of monoamine oxidase A (MAOA) and ADD1	Atherosclerosis can be diagnosed when other biomarkers are combined with epigenetic biomarkers, such as hypermethylation of monoamine oxidase A (MAOA). Along with gene-specific methylation biomarkers, global DNA methylation biomarkers were identified for atherosclerosis.[146] Similarly, in other CVDs, such as hypertension, global methylation profiling and individual gene methylation status (ADD1 gene) were used for diagnosis and outcome.[147,148]

[a]Note: The same diseases are mentioned in more than one row because biomarkers shown in column 2 were identified in different studies.

with concentrations in the micromole range and below. Furthermore, its validity is dependent on the quality of sample collection and handling, as well as the available metadata. Before applying either of these technologies to population-based studies, investigators must consider the advantages and disadvantages in relation to the design and aims of the study.

Quantitative proteomics can be used for the identification of disease biomarkers that could be used for early detection, serve as therapeutic targets, or monitor response to treatment. Several quantitative proteomics tools are currently available to study differential expression of proteins in a variety of types of samples. Two-dimensional gel electrophoresis (2-DE), which was classically used for proteomic profiling, has been coupled to fluorescence labelling for differential proteomics. Isotope labelling methods, such as stable isotope labelling with amino acids in cell culture (SILAC), isotope-coded affinity tagging (ICAT), isobaric tags for relative and absolute quantitation (iTRAQ), and ^{18}O labelling have all been used in quantitative approaches for identification of biomarkers. In addition, heavy-isotope-labelled peptides can be used to obtain absolute quantitative data. Label-free methods for quantitative proteomics, which have the potential of replacing isotope-labelling strategies, are becoming popular. Other emerging technologies, such as protein microarrays, have the potential for providing additional opportunities for biomarker identification. Selected biomarkers are shown in Table 15.4.

15.4.4. Infectious Diseases

Bacterial infection occurs in pneumonia. Community-acquired pneumonia is very common and for proper diagnosis and treatment the biomarker which showed promise was procalcitonin.[38] C-reactive protein and inflammatory biomarkers could be used in combination with procalcitonin to make intelligent clinical decisions. Selected biomarkers are shown in Table 15.5.

15.4.5. Neurological Diseases

For the diagnosis of seizures, carbohydrate-deficient transferrin (CDT) was evaluated and results indicated that this biomarker alone could not diagnose the disease but contributed to better diagnosis when combined with other biomarkers, such as gamma glutamyl transferase (GGT), aspartate aminotransferase (ASAT), alanine aminotransferase

Table 15.4 Biomarkers and their characteristics in metabolic diseases.[a]

Disease	Biomarker	Characteristics
Diabetes	Alanine amino transferase (ALT), gamma glutamyl transferase (GGT), triglyceride, plasminogen activator inhibitor (PAI-1) antigen, ferritin, C-reactive protein (CRP), sex-hormone binding globulin (SHBG)	Biomarkers of risk for diabetes.
Diabetes	IL-6	Inflammation marker of diabetes (genetic studies on association of IL-6 with diabetes have not been completed yet).
Diabetes	Adiponectin	Some groups of investigators consider adiponectin as an excellent biomarker of diabetes pathogenesis.[172]
Diabetes	Insulin resistance (IR) and blood glucose levels	The most studied biomarkers, which are used in clinics routinely
Diabetes	Haemoglobin A1c	For the management of diabetes haemoglobin A1c (HbA1c) is used, which is considered as a reliable indicator of glycaemic control. In most of the clinical studies in diabetes, HBA1c biomarker is used to determine the glucose control.
Diabetes	1,5-Anhydroglucitol (1.5 A)	Circulating biomarker used to measure hyperglycaemic condition
Metabolic diseases	Alpha-hydroxybutyrate (αHB) and linolyl-glycerophosphocholine (L-GPC)	Alpha-hydroxybutyrate (αHB) and linolyl-glycerophosphocholine (L-GPC) were identified as biomarkers of insulin resistance and glucose intolerance in a large population study of more than 1000 participants from the relationship between insulin sensitivity and cardiovascular disease (RISC) study.[2]
Chronic kidney disease	Proteomic profiling	Differentially expressed peaks in the spectra are future biomarkers for chronic kidney disease.[173]
Diabetes	Proteomic profiling	Isobaric tags for relative and absolute quantitation (iTRAQ) in combination with two-dimensional gel electrophoresis technologies were used to identify a group of proteins, which were associated with diabetes, pancreatitis and/or pancreatic cancer.[79]
Diabetes	RCAN1, perilipin A and G0/G1 switch gene, G0S2	Glucose response gene, RCAN1, was overexpressed, whereas perilipin A and G0/G1 switch gene, G0S2, were underexpressed in diabetes-2.[100,101]
Diabetes (early diagnosis)	Dysglycemia, alphahydroxybutyrate, linoleyol-glycerophosphocholine, advanced glycation end-products (age), albumin excretion rate (AR) and single-nucleotide polymorphisms (SNPs) in epidermal growth factor gene intron 2	A number of markers have been described for early diagnosis of diabetes. A few of them are dysglycaemia, alphahydroxybutyrate, linoleyolglycerophosphocholine, advanced glycation end-products (age), albumin excretion rate (AR) and SNPs in epidermal growth factor gene intron 2.[2,174,175]

[a]Note: The same diseases are mentioned in more than one row because biomarkers shown in column 2 were identified in different studies.

Table 15.5 Biomarkers and their characteristics in infection diseases.[a]

Disease	Biomarker	Characteristics
Infectious diarrhoea	Calprotectin	In cases of infectious diarrhoea, faecal calprotectin is a good prediction marker.
Fungal infection	1, 3-β-D-Glucan (BG)	1,3-β-D-Glucan (BG) can be used as a biomarker in invasive fungal infections (especially infections involving *Candida*) in patients undergoing treatment of candedaemia with anidulafungin
Community acquired pneumonia (CAP)	Procalcitonin, c-reactive protein, inflammatory cytokines, and bacterial infection	Bacterial infection occurs in pneumonia. Community-acquired pneumonia (CAP) is very common and for proper diagnosis and treatment the biomarker which showed promise was procalcitonin.[38] C-reactive protein and inflammatory biomarkers can be used in combination with procalcitonin to make intelligent clinical decisions.
Ulcer diseases and gastritis (and gastric cancer)	*Helicobacter pylori*	*H. pylori* is the most common chronic bacterial infection in humans. This bacterium is involved not only in gastric cancer but in ulcer diseases and gastritis also. Its synergistic gastrotoxic interaction with non-steroidal anti-inflammatory drugs, and association with atherosclerotic events is a matter of concern.
Diarrhoea in AIDS patients	Tubuloreticulin inclusions (TRIs)	Diarrhoea in AIDS patients was treated with specific medications and therapeutic response was measured by levels of tubuloreticulin inclusions (TRIs).
Pneumonia	Procalcitonin (PCT)	For pneumonia therapy on more than 100 patients, procalcitonin (PCT) levels were very useful and this biomarker has been recommended for future prognosis. Sometimes detecting bacteria alone is not sufficient to design therapy.[176]
HIV/AIDS	CRC5-Δ2 mutation	In one study conducted in Georgia, where the prevalence of HIV/AIDS is high, polymorphism of the CCR5 gene was studied.[50] More than 100 subjects were enrolled in this study. Results identified CRC5-Δ32 mutation as a marker of the disease in this population.
HIV/AIDS	Mitochondrial DNA mutations	Mitochondrial DNA mutations are also biomarkers for AIDS, especially in those individuals who are undergoing therapy.[51]
Liver cancer (HCC)	p16 and c-myc hypermethylation	In early liver cancer, where risk factors are infection by HBV and HCV, clustered DNA methylation changes in polycomb repressor target genes were observed. P16 and c-myc hypermethylation also suggested initiation of hepatocellular carcinoma.[127,128]

Tuberculosis	Vitamin D receptor (VDR) methylation	In tuberculosis, vitamin D receptor (VDR) methylation indicated high risk of developing the disease.[150] The VDR gene encodes a transcription factor that alters calcium homeostasis and immune function.
Tuberculosis (early diagnosis)	Region-of-difference-1 (RD-1) gene product and sputum cytokine levels	Region-of-difference-1 (RD-1) gene product and sputum cytokine levels are considered a biomarker for early detection of tuberculosis.[177,178]
Hepatocellular carcinoma (early diagnosis)	HSP70, CAP2, glypican 3 and glutamine synthase	Hepatocellular carcinoma (HCC) involves infectious agents and the early diagnostic markers for HCC are HSP70, CAP2, glypican 3 and glutamine synthase.[179] These markers were identified based on gene expression analysis of HCC samples.
Hepatocellular carcinoma (HCC)	HBV, HCV, methylation of polycomb repressor, hypermethylation of p16 and c-myc,	In early liver cancer, where risk factors are infection by HBV and HCV, clustered DNA methylation changes in polycomb repressor target genes were observed. P16 and c-myc hypermethylation also indicated initiation of hepatocellular carcinoma.[127,128]
Tuberculosis	Vitamin D receptor (VDR) methylation	In tuberculosis, vitamin D receptor (VDR) methylation indicated a high risk of developing the disease.[150] The VDR gene encodes a transcription factor, which alters calcium homeostasis and immune function.

[a]Note: For the same diseases are mentioned in more than one row because biomarkers shown in column 2 were identified in different studies.

(ALAT), and ASAT : ALAT ratio.[39] A novel missense mutation at position 134 T to A of the Norrie disease protein (NDP) gene, resulting in an amino acid change at codon V45E, was identified as a biomarker for Norrie disease (ND) which is a rare X-linked disorder characterized by congenital blindness and sometimes mental retardation and deafness.[40] A novel mutation in the ATP7B gene was used as a diagnostic biomarker for neurological impairment in Wilson's disease.[41] Selected biomarkers are shown in Table 15.6.

15.4.6. Cancer

Proteomic profiling of samples from individuals with cancer and healthy individuals has identified disease-associated profiles, especially in matrix-associated laser desorption/ionization time-of-flight mass spectrometry (MALDI TOF MS) in colorectal and other cancers, which can be used for diagnosis, either alone or in combination with other biomarkers.[42] Methylation biomarkers were useful in diagnosis of almost all major cancers.[7,9,43] Defects in the mismatch repair (MMR) genes, such as MLH1, MSH2 or MSH6, or methylation of the MLH1 promoter led to erroneous replication of segments of simple nucleotide repeats, which contributed to microsatellite instability (MSI). MSI increases the risk of cancer occurrence. However, MSI is uncommon in cancers of the breast as compared with some other cancers, such as colorectal carcinoma (CRC). It was observed that the prognosis in MSI-positive breast cancer patients was worse than that of patients with MSI-negative tumours. Selected biomarkers are shown in Table 15.7.

15.5. Technologies and Strategies for Molecular Biomarker Discovery

Technologies and strategies developed for cancer diagnosis can be applied to other diseases also.[4–6] The detection and treatment of cancer is greatly facilitated by the omics technologies. For example, genomics analysis provides clues about gene regulation and gene knockdown for cancer management. The approval of Mammaprint and Oncotype DX indicates that multiplex diagnostic biomarker sets are becoming feasible. The microRNA field in human cancers has opened a new avenue for cancer researchers. Some therapeutic drugs targeting DNA methylation and histone deacetylation are currently undergoing intense studies. Proteomics also plays an important role

Table 15.6 Biomarkers and their characteristics in neurological diseases.[a]

Disease	Biomarker	Characteristics
Neurological disorders	Grannins	The granin family is a group of acidic proteins present in the secretory granules of a wide variety of endocrine, neuronal, and neuroendocrine cells. A few important granins, chromogranin A and B, secretogrannin II, HISL-19 antigen, NESP55 and ProSAAS should be studied further for their clinical implications.[76]
Neurological disorders	Microvesicles	Microvesicles (MVs) are used as biomarkers of neurological disorders because their release is increased in these diseases.[81]
Neurological disorders (Alzheimer's disease)	Glutamate	In Alzheimer's disease, glutamate levels increased (in the hippocampus region of the brain) after treatment with galantamine.[13] Therefore, glutamate can be considered a therapeutic biomarker of Alzheimer's disease.[14]
Neurological disorders (seizures)	Carbohydrate-deficient transferrin (CDT)	For the diagnosis of seizures, carbohydrate-deficient transferrin (CDT) was evaluated and results indicated that this marker alone couldn't diagnose the disease but could contribute to better diagnosis when combined with other biomarkers, such as GGT, ASAT, ALAT and ASAT: ALAT ratio.[39]
Norrie disease (ND) (X-linked disorder)	Norrie disease protein gene (*NDP*) missense mutation at position 134 T to A	A novel missense mutation at position 134 T to A of the Norrie disease protein gene (*NDP*) resulting in an amino acid change at codon V45E was identified as a biomarker for Norrie disease (ND) which is a rare X-linked disorder characterized by congenital blindness and sometimes mental retardation, and deafness.[40]
Wilson's disease	Mutation in ATP7B gene	A novel mutation in ATP7B gene can be used as a diagnostic marker for neurological impairment in Wilson's disease.[41]
Alzheimer's disease	CXR4 and CCR3	Genes involved in inflammation and immune system regulatory pathways were identified when profiling of Alzheimer's and non-Alzheimer's samples was conducted.[114] Results also indicated a role for chemokines and their receptors (CXR4 and CCR3) in patient samples and these two receptors may be considered biomarkers for Alzheimer's disease.
Alzheimer's disease	Altered methylation of Alu, Line-1, and SAT-α sequences	In Alzheimer's disease altered methylation levels of repeat sequences (Alu, Line-1 and SAT-α) were observed. This modification induces genomic instability, which contributes to disease initiation and progression.[151]
Autism	MECP2 hypermethylation	Epigenetic regulation in autism was also studied and locus-specific hypermethylation of MECP2 was observed.[152]

[a]Note: The same disease is mentioned in more than one row because biomarkers shown in column 2 were identified in different studies.

Table 15.7 Biomarkers and their characteristics in cancer.[a]

Disease	Biomarker	Characteristics
Apoptosis (cancer)	Annexin XI	Anti-annexin XI has been reported in different cancers.[74]
Oral cancer	Mutation in exon 4, codon 63 of the p53 gene	Antibodies against abnormal p53 were found in saliva and serum of oral cancer patients.[53]
Gastric cancer	Metabolites of glycolysis, fatty-acid β oxidation, and cholesterol and amino acid metabolism	In gastric cancer, alterations in metabolites of glycolysis, fatty-acid β-oxidation, and cholesterol and amino acid metabolism were observed.[121] These metabolites can be used to follow gastric cancer development and treatment response.
Pancreatic cancer	Serum levels of antibodies against periodontal bacteria *Poryphromonas gingivalis*	In pancreatic cancer the incidence and mortality rates are the same and the survival is only five years after the diagnosis of the disease. Plasma and serum of pancreatic cancer contains antibodies against periodontal bacteria *P. gingivalis* and their level is associated with pancreatic cancer.[82]
Pancreatic cancer	C-reactive proteins, interleukin-6 (IL-6), and soluble receptor of tumour necrosis factor α	Inflammatory biomarkers, C-reactive proteins, interleukin-6 (IL-6), and soluble receptor of tumour necrosis factor α are being used in identifying pancreatic cancer patients.[83]
Ovarian cancer	CA 125, osteopontin, kllikrein 6, B7–H4, spondin 2, and DcR3	B7–H4, spondin 2, and DcR3 were identified as early biomarkers in ovarian cancer.[180]
Prostate cancer	Prostate-specific antigen, α methyl CoA-racemase	Results have not been validated for α methyl CoA-racemase.[181]
Colon cancer	APC, CDKNA	Expression analysis was used to identify these biomarkers.[182]
Lung cancer	EGFR, KRAS	Multiple investigators have identified these biomarkers.[183,184]
Breast cancer	BRCA-1, BRCA-2, Let-7	These biomarkers have been used in clinical samples.[185]
Acute myeloid leukaemia, chemotherapy-resistant AML, acute lymphoblastic leukaemia, acute non-promyelocytic leukaemia	FLT3, DAPK1, hPer3, DNMT3A, repeat sequences LINE-1	Several investigators have identified these markers. Several methylation markers of AML have also been reported.[186–188]
Renal cell carcinoma (Kidney cancer)	APAF-1,DAPK-1	Hypermethylation of APAF-1 and DAPK-1

Breast cancer	SNAPs in in 1p11.2, 2q35, 3p, 5p12, 8q24, 10q23, 13, 14q24.1, and 16q regions	In GWAS, SNPs identified were primary located in 1p11.2, 2q35, 3p, 5p12, 8q24, 10q23, 13, 14q24.1, and 16q regions.
Pancreatic cancer	*P. gingivalis*	Plasma and serum of pancreatic cancer contained antibodies against periodontal bacteria *P. gingivalis* and their level was associated with pancreatic cancer.[82]
Pancreatic cancer	Inflammatory biomarkers, C-reactive proteins, interleukin-6 (IL-6), and soluble receptor of tumour necrosis factor α, cell cycle regulatory proteins (cyclin D1 and Ki67), glycolytic enzyme lactate dehydrogenase (LDH), matrix metalloproteinases (MMPs)	Inflammatory biomarkers, C-reactive proteins, interleukin-6 (IL-6), and soluble receptor of tumour necrosis factor α also were used in identifying pancreatic cancer patients.[83] Some biomarkers identified include increased salivary levels of cell cycle regulatory proteins (cyclin D1 and Ki67), glycolytic enzyme lactate dehydrogenase (LDH), matrix metalloproteinases (MMPs) and reduction in DNA repair enzyme (8-oxoquanine DNA glycosylase) and mapsin in oral cancer patients.[84]
Hepatocellular carcinoma (HCC)	hTERT, α-fetoprotein (AFP), des-γ-carboxy prothrombin (DCP)	A highly sensitive method for serum human telomerase reverse transcriptase (hTERT) mRNA for hepatocellular carcinoma (HCC) was reported.[85] α-fetoprotein (AFP) and des-γ-carboxy prothrombin (DCP) were found to be good markers for HCC.
Retinoblastoma	Trimethylation of H4K20	Retinoblastoma levels were lower whenever trimethylation of H4K20 was present. A correlation with the tumour stage and grade was also established based on these histone biomarkers.[189]
Breast cancer	HDAC1	Another interesting study reported quantitative expression of HDAC1 and its correlation with breast cancer patient's age, lymph node status, tumour size and her2/neu-negative, ER- and PR-positive status.[147]
Kidney cancer	VHL, MET, FLCN, fumarate hydratase, succinate dehydrogenase, TSC1, TSC2, TFE3	Altered expression of VHL, MET, FLCN, fumarate hydratase, succinate dehydrogenase, TSC1, TSC2 and TFE3 genes in kidney cancer was observed.[190]
Lung cancer	IL-27, IL-29, IL-31 in BALF	Serum levels of IL-27, IL-29 and IL-31 in BALF are high in lung cancer patients.[10]
Lung cancer	miR-21, miR-485, miR-451	Higher levels of miR-21 and lower levels of miR-451 and miR-485 were observed in lung cancer.[11]
Lung cancer	CYFRA 21-1	Increased CYFRA 21-1 levels were observed in samples from lung cancer patients.[191]

[a]Note: The same diseases are mentioned in more than one row because biomarkers shown in column 2 were identified in different studies.

in cancer biomarker discovery and quantitative proteome–disease relationships provide a means for connectivity analysis. Fluorescent dye enables a more reliable and quantitative analysis and it facilitates the progress of biochips and cytomics. The huge amount of information collected by multiparameter single-cell flow or slide-based cytometry measurements serves to investigate the molecular behaviour of cancer cell populations. Metabolite profiling is such a field, which can be applied to cancer and other diseases.

15.5.1. Genomics-based Biomarkers in Immune Diseases, Cardiovascular Diseases (CVD), Metabolic Diseases, Infectious Diseases, Neurological Diseases and Cancer

Discovery and validation of novel disease-associated biomarkers remain a crucial goal of future patient care. Advanced genomic technologies, such as single nucleotide polymorphism (SNP) array and next-generation sequencing, help shape the landscapes of genomics and epigenomics. Genome-wide association studies (GWAS), which act as a powerful approach to identify common, low-penetrance disease loci, have been conducted in several types of diseases, such as diabetes and cancer, and have identified many novel associated loci, confirming that susceptibility to these diseases is polygenic. Though the creation of risk profiles from combinations of susceptible SNPs is not yet clinically applicable, future, large-scale GWAS holds great promise for the individualized screening and prevention. Epigenomic biomarkers, like DNA methylation, have emerged as highly promising biomarkers and are being actively studied for multiple diseases. Validated as being associated with disease risk or drug response, some DNA methylation biomarkers are being transferred into clinical use.[34] Discovery of the genes and pathways mutated in diseases states, especially through large-scale genome-wide sequencing, has provided key insights into the mechanisms underlying disease processes and indicated new candidate biomarkers for diagnosis and clinical intervention as well as prognosis. The comprehensive landscapes of cancer genomes indicate the convergence of mutations onto pathways that govern the course of disease development and indicate that rather than seeking genomics and epigenomics alterations of specific mutated genes, the combination with dynamic transcriptomics, proteomics and metabolomics of the downstream mediators or key nodal points may be preferable for future disease biomarker discovery.

Most of the genomic biomarkers are mutations in genes or small-nucleotide polymorphisms (generally present in the noncoding regions).

15.5.1.1. Immune Diseases

Among immune diseases, autoimmune diseases are well characterized. These diseases result when the immune system goes awry and recognizes self-tissues as foreign. The major contribution to these diseases is by autoantibodies and autoreactive cellular responses, which ultimately contribute to the ongoing autoimmune disease process. During the process, inflammatory enzymes are recruited to the affected organ and tissue-degrading proteolytic enzymes are released. Therefore, identifying biomarkers that may detect the process early is very significant. An ideal biomarker for these diseases should reach abnormal levels (either higher or lower levels compared with controls) in conjunction with disease development, should fluctuate in relation to disease severity and should normalize after treatment. Three groups of biomarkers are being characterized in these diseases. First, degradation products arising from destruction of the affected tissue; second, enzymes that play a role in tissue degradation, and; third, cytokines and other proteins associated with immune system activation and the inflammatory response. In rheumatoid arthritis (RA) inflammation of joints' synovium is persistent, which leads to erosion of cartilage and bone. A biomarker, aggrecan fragments, was identified which was detected in synovial fluid (SF). Levels of aggrecan fragments (structural components of cartilage) were higher in RA patients than in healthy individuals. Another biomarker was C-propeptide of type II collagen and its levels were directly proportional to the rate of collagen synthesis. Cartilage oligomeric matrix protein (COMP) is also a component of cartilage and it can be measured in serum and SF. Multiple sclerosis is associated with autoimmune-mediated inflammation of the CNS and may lead to demyelination and axonal damage. The biomarker neurofilament light protein (NFL) is a cytoskeleton component in large myelinated axons and it is released into the cerebrospinal fluid (CSF) and can be used to determine the level of axonal damage. Glial fibrillary acidic protein (GFAP) is another biomarker for multiple sclerosis.

A genetic link with HLA-DR4 and related allotypes of MHC class II and T-cell-associated protein PTPN22 with rheumatoid arthritis was established.[44] The presence of autoantibodies to IgGFc (rheumatoid factor) and antibodies to citrullinated peptide (ACPA) also indicated the presence of disease.

In rheumatoid arthritis, SNPS were reported in several genes which were used as biomarkers for disease diagnosis. These genes include PTPN22, IL23R, TRAF1, CTLA4, IRF5, STAT4, CCR6 and PAD14.[45]

15.5.1.2. Cardiovascular Diseases (CVDs)

Pulmonary arterial hypertension (PAH) is influenced by genetic background and may be useful in guiding therapy of the disease.[46] In this disease an increase in blood pressure in the pulmonary artery and pulmonary vein occurs, which leads to shortness of breath, dizziness and fainting. GWAS studies indicated that other factors (rare exonic mutations, epigenetic phenomena and interaction with environmental factors) might also contribute to the development of this disease.[47]

15.5.1.3. Metabolic Diseases

Like other omics approaches, metabolomics also takes advantage of non-targeted approach for identifying disease-associated biomarkers. The technology has the advantage of using the minimum amount of sample and needing no prior knowledge of the substances to be analysed by nuclear magnetic resonance (NMR) or mass spectrometry (MS).[34] The complete process includes the acquisition of the experimental data, multivariate statistical analysis and the projection of the profiles, which is the acquired information, to construct the patient map (or phenotype). The main diseases where metabolomics has been applied are rheumatoid arthritis, spondyloarthritis, systemic lupus erythematosus and osteoarthritis.[48] In spondyloarthritis, association of HLA-A, B and HLA-DR gene expression was studied and population-specific alterations were reported.[49]

15.5.1.4. Infectious Diseases

Regarding genomic biomarkers in infectious diseases, we have selected AIDS (although the topic is so big that we cannot cover it in this article). In one study conducted in Georgia, where the prevalence of HIV/AIDS is high, polymorphism of CCR5 gene was studied.[50] More than 100 subjects were enrolled in this study. Results identified the CRC5-Δ32 mutation as a biomarker of the disease in this population. Mitochondrial mutations could also be used as biomarkers for AIDS, especially in those individuals undergoing therapy.[51]

15.5.1.5. Neurological Diseases

Because of the anatomical location of the nervous system, it is difficult to obtain biomarkers for neurological diseases. The accessibility of the affected organ is a challenge and therefore surrogate biomarkers are generally used to identify and follow these diseases. Gene expression analysis, mutations and SNPs were the common approaches to identify neurological diseases. Compared with healthy subjects, altered gene expression profiling was observed in schizophrenia and Alzheimer's patient samples.[52]

15.5.1.6. Cancer

Here we described the specific example of breast cancer, with reference to genomic approaches in population science for screening, which may result in early detection of the disease and ultimately longer survival. Mortality from breast cancer is very high worldwide. More than half of breast cancer cases occur in Western countries. The cost of treatment is higher when breast cancer is detected late in its development; therefore, detecting this cancer early is the key to success. Mammography has been successful in reducing mortality from this cancer, but it is an expensive technique. The occurrence of breast cancer in the general population can be explained by inherited genetic susceptibility, somatic changes, effects of endogenous and exogenous environments and interaction of these factors (especially gene–environment interactions). In the case of oral cancer, exon 4, codon 63 of the p53 gene was mutated in salivary DNA in patients.[53] Autoantibodies against abnormal p53 were reported in saliva and serum of these patients.

Inherited genes for breast cancer susceptibility can be low- or high-penetrance genes; the few genes with allelic variants that confer a high degree of risk to the individual are known as high-penetrance genes. Other genes confer a small to moderate degree of breast cancer risk to the individual and are known as low-penetrance genes. Relatively few individuals in the population carry risk-increasing genotypes at the loci where high-penetrance genes act; therefore, the population-attributable risk is low. On the other hand, the low-penetrance genes are not associated with syndromic or Mendelian patterns but are associated with sporadic breast cancer. The allelic variation of low-penetrance genes is relatively high, and large breast cancer populations carry low-penetrance genes. Different investigators identified low- and high-penetrance genes in breast cancer in a number of populations. The BRCA1 gene was the first gene identified to represent susceptibility to hereditary breast cancer and later on BRCA2 (located on chromosome 17q21) was also confirmed for breast cancer and ovarian cancer.[54] To identify breast-cancer-associated genetic biomarkers a number of cohorts with exposure and life-style data and other details of the participants were used.[55] One of those cohorts, the Collaborative Oncological Gene Environment Study (COGS), which is a large-scale genotyping cohort funded by European Commission was utilized to identify disease-associated biomarkers. More than 150 000 samples were genotyped in this study. Familial-based high-penetrance susceptibility genes were identified first and then low-penetrance genes by association studies.[56,57] Carriers of such genes and SNPs are predisposed to breast cancer. A panel of 70 genes was able to predict breast cancer prognosis.[58] Genomic markers include SNPs, mutations, additions, deletions, recombinations and changes in copy number (altered CNVs).[59,60]

GWAS were conducted by different groups in different cohorts to identify breast-cancer-susceptibility genes which may be useful for breast cancer screening of high-risk populations.[61,62] In one such study, genotyping of 2702 women of European ancestry with invasive breast cancer and 5726 controls was conducted.[61] SNPs identified in this study were primary located in the 1p11.2, 2q35, 3p, 5p12, 8q24, 10q23, 13, 14q24.1 and 16q regions. Genes affected by these SNPs are involved in regulation of actin cytoskeleton, glycan degradation, alpha linoleic acid metabolism, circadian rhythm and drug metabolism.

15.5.2. Proteomics-based Biomarkers in Immune Diseases, Cardiovascular Diseases (CVD), Metabolic Diseases, Infectious Diseases, Neurological Diseases and Cancer

Proteomics technologies have emerged as a useful tool in the discovery of diagnostic biomarkers and substantial technological advances in proteomics and related computational science have been made in the last decade.[1,63] These advances overcome, in part, the complexity and heterogeneity of the human proteome, permitting the quantitative analysis and identification of protein changes associated with tumour development. With the advent of new and improved proteomic technologies, it is possible to discover new biomarkers for the early detection and treatment monitoring of different diseases. The contribution of the Human Proteomic Organization is valuable in this regard because this organization has provided a dataset of normal healthy human proteomic profiles which can be used as a reference dataset for identifying disease-associated proteomic changes (http://www.hupo.org/). Protein biomarkers are more related to disease phenotype and are more targetable for therapy in comparison with transcriptomic or genomic biomarkers. Proteomics provides a powerful tool to investigate potential biomarkers in diseases due to its high sensitivity, precise characterization of protein interactions and ability to detect functionally significant post-translational modifications. Proteomic biomarkers have been identified in blood (serum and plasma) as well as in tissue samples by applying approaches such as nuclear magnetic resonance spectroscopy (NMR), mass spectrometry (MS), two-dimensional gel electrophoresis and immunoprecipitation. In one study, investigators identified circulating proteomic biomarkers from different stages of breast cancer using an innovative strategy employing high-sensitivity label-free proteomics. The approach was MS-based and provided semi-quantitative results and could be applied in preclinical and clinical studies. Furthermore, breast cancer patient serum was analysed by bidimensional nano ultra performance liquid chromatography (nanoUPLC) tandem nano electrospray ionization–mass spectrometry (ESI-MS) to identify breast cancer biomarkers that were differentially expressed at early stages of cancer development.[64] Higher grainyhead-like 3 (GRHL3) expression and lower levels of tumour necrosis factor alpha (TNFα) were reported during early stage of the diseases whereas post-mitotic segregation increased homologue 2 (PMS2) expression was high in advanced stages of the disease. These results were validated in a different set of patients although the number of participants was low. These investigators plan to evaluate the effects of such markers in determining survival rates of patients and recurrence of breast cancer or other cancers.

Proteomics with the recent advances in mass spectrometry is considered a powerful analytical method for deciphering protein expression alterations as a function of disease progression.[63,65] Proteomics-based analyses of breast serum and tissue lysates have resulted in the finding of a number of potential tumour biomarkers, providing, therefore, a basis for a better understanding of breast-cancer development and progression, and eventually serving as diagnostic and prognostic markers.[64] Probably the most widely used proteomic technology is the identification of alterations in protein expression between two different samples through comparative two-dimensional gel electrophoresis (2-DE), which provides high-resolution separation of proteins and offers a powerful method for their identification and characterization.[66]

Proteomic analysis is an essential component to explain the information contained in genomic sequences in terms of the structure, function and control of biological processes and pathways.[67] The proteome reflects the cellular state or the external conditions encountered by a cell. In addition, proteomic analysis is a genome-wide assay to differentiate distinct cellular states and to determine the molecular mechanisms that control them.[68] Infection-associated proteomic biomarkers have also been characterized.[69] High-throughput proteomic methodologies have the potential to revolutionize protein biomarker discovery and to allow for multiple protein biomarkers to be assayed simultaneously. With the significant advances in 2-DE and mass spectrometry (MS), protein biomarker discovery has become one of the central applications of proteomics.[67]

15.5.2.1. *Immune Diseases*

Antigen arrays are valuable for profiling autoantibodies in diverse rheumatic autoimmune diseases and can be composed of proteins, peptides, protein complexes, glycoproteins, sugar nucleic acids and lipids. T and B cells could be isolated from disease subjects and used for disease stratification, which ultimately helped in designing treatment and disease management of autoimmune diseases (such as rheumatoid arthritis (RA)).[70] The proteomic profile might indicate whether the disease was aggressive or not. Treatment was selected based on the aggressiveness of the disease.[71] Less than two-thirds of all individuals with RA had an adequate response to anti-TNF therapy. Biomarkers could identify those individuals who were less likely to respond to this therapy.[70,72] This would not only reduce the cost associated with therapy but also avoid unwanted adverse reaction due to anti-TNF therapy among non-responders. Technologies such as mass cytometry, peptide and protein arrays and BCR (B cell receptor) and TCR (T cell receptor) sequencing might prove useful to improve the management of autoimmune diseases and represent the state of art in analysing cells, soluble proteins and genes.[73]

Another group of biomarkers that was characterized for autoimmune diseases was annexins, a group of 12 highly conserved proteins which regulate the cell cycle. Their abnormal expression was associated with disease development.[74] Annexin-I exerted its anti-inflammatory effect by suppressing the generation of inflammatory mediators, and anti-annexin I antibodies were present in RA and systemic and cutaneous lupus erythematosus (SLE) patients.[117]

Annexin II and V were involved in the coagulation cascade because they had affinity towards phospholipids. Levels of prolactin (PRL), ferritin, vitamin D and the tumour biomarker tissue polypeptide antigen (TPA) were higher in patients with autoimmune disease (systemic sclerosis and SLE), polymyositis (PM), dermatomyositis (DM) and multiple sclerosis.[75] The granin family is a group of acidic proteins present in the secretory granules of a wide variety of endocrine, neuronal and neuroendocrine cells. Few important granins, chromogranin A and B, secretogrannin II, HISL-19 antigen, NESP55 and ProSAAS, should be studied further for their clinical implication.[76]

15.5.2.2. Cardiovascular Diseases (CVDs)

Atherosclerosis is the main cause of cardiovascular diseases. Diagnosis of subclinical atherosclerosis is a clinical challenge. Furthermore, determining the extent of atherosclerosis aggressiveness in individual patient is a challenge too. Plasma osteoprotegerin (OPG) turned out to be an excellent proteomic biomarker that could be used to detect preclinical atherosclerosis, as validated in a case–control study.[77] Proteomic profiling in circulating cells and plasma extracellular vesicles could distinguish populations at high risk of developing atherosclerosis.[78]

15.5.2.3. Metabolic Diseases

Isobaric tags for relative and absolute quantitation (iTRAQ) in combination with two-dimensional gel electrophoresis technologies identified a group of proteins which were associated with diabetes, pancreatitis and/or pancreatic cancer.[79]

15.5.2.4. Infectious Diseases

HIV-associated neurodegenerative disease progression and treatment response could be measured by following levels of the biomarkers complement C3, soluble superoxide dismutase and prostaglandin synthase.[123] Although studies of immune system disorders have focused more on HIV-related diseases, HIV infection has effects on the CNS as this lentivirus could infect brain cells. CNS dysfunction then led to a group of cognitive and behavioural changes (called HIV-associated neurocognitive disorders or neuroAIDS) which served as biomarkers.[80]

15.5.2.5. Neurological Diseases

Because of the anatomical location of the nervous system, it is difficult to obtain biomarkers for the neurological disorders. The accessibility of the affected organ is a challenge and therefore surrogate biomarkers are generally used to identify and follow neurological disorders. Microvesicles (MVs) were used as biomarkers of neurological disorders because their release was increased in these diseases.[81] MVs originated from exosomes (which are derived from endothelial cells) and could be found in plasma or serum. MVs were linked to neurological pathologies with a vascular or ischemic pathogenic component (sometimes used in diagnosis and follow up of strokes). In another disease, multiple sclerosis, MVs of oligodendroglial origin were reported in the cerebrospinal fluid (CSF).[81] Detection of MVs should be explored further to gain pathogenic information, identify therapeutic targets and select specific biomarkers for neurological diseases.

15.5.2.6. Cancer

In pancreatic cancer, the incidence and mortality rates are the same and the survival is only five years after the diagnosis of the disease. Plasma and serum of pancreatic cancer contained antibodies against periodontal bacteria *Poryphromonas gingivalis* and their level was associated with pancreatic cancer.[82] Inflammatory biomarkers, C-reactive proteins, IL-6, and soluble receptor of tumour necrosis factor alpha also were used in identifying pancreatic cancer patients.[83] Some biomarkers identified include increased salivary levels of cell cycle regulatory proteins (cyclin D1 and Ki67), glycolytic enzyme lactate dehydrogenase (LDH) and matrix metalloproteinases (MMPs) and a reduction in DNA repair enzyme (8-oxoquanine DNA glycosylase) and mapsin in oral cancer patients.[84]

A highly sensitive method for serum human telomerase reverse transcriptase (hTERT) mRNA for hepatocellular carcinoma (HCC) has been reported.[85] Alpha-fetoprotein (AFP) and des-gamma-carboxy prothrombin (DCP) were found to be good markers for HCC. This group also verified the significance of hTERT mRNA in a large scale multi-centred trial. hTERT mRNA was demonstrated to be independently correlated with clinical parameters (tumour size and tumour differentiation). hTERT mRNA proved to be superior to AFP, AFP-L3 and DCP in the diagnosis and underwent an indisputable change in response to therapy. The detection rate of small HCC by hTERT mRNA was superior to those for the other biomarkers.[85]

To identify the potential biomarkers involved in hepatocellular carcinoma (HCC) carcinogenesis, a comparative proteomics approach was utilized to identify the differentially expressed proteins in the serum of ten HCC patients and ten controls. A total of 12 significantly altered proteins were identified by mass spectrometry. Of the 12 proteins identified, HSP90 was one of the most significantly altered proteins and its overexpression in the serum of 20 HCC patients was confirmed using enzyme-linked immunosorbent analysis (ELISA). The observations indicate that HSP90 might be a potential biomarker for early diagnosis, prognosis and monitoring in the therapy of HCC.[86]

Proteomic analysis with 2-DE and MS was used to identify other potential serum markers for breast cancer.[87,88] Protein extracts expressed in the serum of breast cancer patients after depletion of high-abundance proteins were compared with sera from healthy women using proteomic approaches. By comparing 2-DE profiles between tumour and non-tumour samples and using MALDI-TOF mass spectrometry of their trypsinized fragments, the identification of two proteins of interest, haptoglobin precursor and alpha-1-antitrypsin precursor, were observed.[130] Separation and analysis of proteins from cells, tissue samples and breast tumour biopsies proved very successful in identifying novel biomarkers.[89] Using proteomic approaches, 26 immunoreactive proteins (antigens), against which sera from newly diagnosed patients with infiltrating ductal carcinomas exhibited reactivity, were detected.[87,90] Among these antigens, peroxiredoxin-2 (Prx-2) belongs to a family of thiol-specific antioxidant proteins that control intracellular H_2O_2 by reducing reactive oxygen species (ROS) issued from free radicals. Such proteins might have an important role and protect the breast tumour cells against oxidative injury and modulate cell proliferation and apoptosis of malignant cells.[91]

Proteomic approaches are useful to identify protein–protein interaction and in one study oestrogen receptor alpha and its interaction with a number of transcription factors was characterized, which resulted in clinically useful information about breast cancer therapeutics.[92] Laser-capture microdissected breast cancer and normal tissue cells were analysed by mass spectrometry to identify proteomic profiles associated with breast cancer.[93] In another study, glyoxalase-1 was found to be expressed in breast cancer.[87] This protein was involved in detoxification of methylglyoxal, which is a cytotoxic product of glycolysis. Further analysis of tissue microarray indicated correlation of glyoxalase-1 with tumour grade. Based on results from a reversed-phase protein array, a model was created to predict pathological response in patients receiving neoadjuvant taxane and anthracyclin taxane-based systemic therapy, thus indicating the translational significance of proteomic biomarkers in breast cancer.[94]

15.5.3. Transcriptomics-based Biomarkers in Immune Diseases, Cardiovascular Diseases (CVD), Metabolic Diseases, Infectious Diseases, Neurological Diseases and Cancer

Transcriptomics is the study of how our genes are regulated and expressed in different biological settings. All expressed genes can be quantitatively measured in a tissue at a given time (this science is termed transcriptomics).[95] Over the last decade, microarray-technology-based transcriptomic analysis has contributed enormously to our understanding of the molecular basis of a number of diseases. Gene expression profiling offers an unparalleled opportunity to develop biomarkers that are useful in diagnosis and prognosis and in helping to achieve the goal of individualized treatment. However, the limitations of the technology and the danger of inappropriate experimental processes should not be underestimated. In clinical settings, blood transcriptomics of Alzheimer's patients undergoing treatment with EHT 0202 has been conducted.[96] Transcriptomic analysis indicated activation of pathways associated with CNS disorders, diabetes, inflammation and autoimmunity. Treatment resulted in deactivation of these pathways in patients. Thus, transcriptomic biomarker profiling could be used in disease prognosis. Such studies would help us identify those patients' populations who would respond to treatment; and also identify those individuals who would be likely to have recurrence of the disease.

15.5.3.1. Immune Diseases

In RA bone loss occurs.[97] Transcriptomics approaches were applied to identify genes and pathways which contributed to bone loss in RA patients. Results indicated IL-17 and Wnt/β catenin to be transcriptional biomarkers of RA.[98]

15.5.3.2. Cardiovascular Diseases (CVDs)

Atherosclerosis is a pathological process in which the walls of large arteries thicken and lose elasticity as a result of the growth of atheromatous lesions. Transcriptomics-based gene expression analysis did not identify specific genes but pathways (especially inflammation-related pathways) were identified, which were associated with atherosclerosis development.[99] The analysis was conducted on monocytes and macrophages.

15.5.3.3. Metabolic Diseases

Most gene expression profiling identified diabetes but in a few cases individual gene expression (also called the candidate gene expression approach) also provided information about disease-associated transcriptomics. The glucose response gene, RCAN1, was overexpressed, whereas perilipin A and the G0/G1 switch gene, G0S2, were underexpressed in diabetes-2.[100,101]

15.5.3.4. Infectious Diseases

Based on transcriptomics, a regression model was developed which could predict the onset of ventilator-associated pneumonia.[102] In another study, transcriptomics identified a gene-expression pattern which helped in identifying patients who were at high risk of developing tracheobronchitis or pneumonia.[103]

15.5.3.5. Neurological Diseases

In the neurobiology field, neuronal development, function and the subsequent degeneration of the brain are still serious problems. There is a need to find better targets for developing therapeutic intervention by identifying new biomarkers by applying transcriptomics and other approaches.[104] A high-throughput high-content screening approach is needed to go from genes to gene-networks.

15.5.3.6. Cancer

To implicate a combination of omics technologies, one investigator performed transcriptomics and metabolomics in breast cancer samples and identified a disease-associated profile, which could be used for diagnosis purposes.[105] Transcriptomics identified genes and pathways whereas the functional significance was evaluated by metabolite characterization.

15.5.4. Immunomics-based Biomarkers in Immune Diseases, Cardiovascular Diseases (CVD), Metabolic Diseases, Infectious Diseases, Neurological Diseases and Cancer

The huge amount of immunological information hidden in the plasma could be better revealed by combining the characterization of antibody-binding target epitomes with improved estimation of effector functions triggered by these binding events. Functional immune profiles can be generated, characterizing general immune responsiveness, by designing arrays with information about epitope collections from different antibody targets. Immunomics was implicated in searching and identifying proteins of interest in the case of breast or colorectal cancers.[106] Two approaches were developed during these studies: the top-down serological proteome analysis (SERPA) and the bottom-up multiple affinity protein profiling (MAPPing) strategies. The first one relied on two-dimensional electrophoresis (2-DE), immunoblotting, image analysis and mass spectrometry. The second approach dealt with the use of two-dimensional immunoaffinity chromatography, enzymatic digestion of the antigens and analyses by tandem mass spectrometry. Using an immunoinformatics approach, putative T- and B- cell epitopes of capsid proteins were identified that were conserved in existing serotypes.[107]

15.5.4.1. Immune Diseases

Although autism is a neurological disorder its regulation involves autoimmunity. This disease is characterized by impaired social interaction and verbal and non-verbal communication. High numbers of mast cells and serotonin in urine were observed in autism patients. Mast cells produce inflammatory cytokines in large amounts in autism patients and the possibility of autoimmunity was proposed.[108]

15.5.4.2. Cardiovascular Diseases (CVDs)

Atherosclerosis is a chronic inflammatory disease of the vessel wall. The gene CARD8, which encodes proteins involved in innate immunity in atherosclerosis patients, is an excellent biomarker for atherosclerosis. Inflammatory markers were also overexpressed in this disease. Inflammosomes produce interleukin 1β in response to cholesterol crystal accumulation in macrophages. Ephs and ephrins also were proposed as biomarkers of atherosclerosis.[109]

15.5.4.3. Metabolic Diseases

Chronic low-grade inflammation contributes to the pathogenesis of insulin resistance. In diabetes innate immune cells accumulate in metabolic tissues and release inflammatory cytokines, especially IL-1β and TNF. One investigator proposed cell-mediated immunity in diabetes. Regulation of type 1 and type 2 diabetes by the immune system occurred differently in diabetic patients under study.[110,111]

15.5.4.4. Infectious Diseases

Tuberculosis (TB) is a common and lethal infectious disease caused by various strains of mycobacteria. Lung is the primary organ that is infected by the bacteria although any part of the body can be infected. For several years antibiotics were able to treat TB but in the past few years antibiotic-resistant mycobacteria have been reported. To address this point, the use of anti-mycobacteria antibodies, enhancing the Th1 protective responses by using mycobacterial antigen, or increasing Th1 cytokines, was evaluated, and promising results were obtained.[112] Childhood deficiency of vitamin D was proposed to be a susceptibility biomarker for TB.[113]

15.5.4.5. Neurological Diseases

Genes involved in inflammation and immune system regulatory pathways were identified when profiling of Alzheimer's and non-Alzheimer's samples was conducted.[114] The results also indicated a role for chemokines and their receptors (CXR4 and CCR3) in patient samples; and these two receptors may be considered to be biomarkers for Alzheimer's disease. Another biomarker for the same disease was amyloid beta, which was expressed at higher levels in patients compared with age- and sex-matched controls.

15.5.4.6. Cancer

Genetic variations in innate immunity and inflammation pathways associated with lung cancer risk were proposed in different studies. Some investigators proposed T-cell-mediated immunity in lung cancer using human-leukocyte-antigen D-related (HLA-D) telomerase-derived epitopes. In cervical cancer, cell-mediated immunity is the main mechanism of cancer development. Infection-associated cancers, infectious agents and their epitopes were considered excellent biomarkers, which could be used for diagnostic purposes.[115–117]

15.5.5. Metabolomics-based Biomarkers in Immune Diseases, Cardiovascular Diseases (CVD), Metabolic Diseases, Infectious Diseases, Neurological Diseases and Cancer

Metabolomics is the study of small molecules of both endogenous and exogenous origin, such as metabolic substrates and their products, lipids, small peptides, vitamins and other protein cofactors, generated by metabolism,[2,118] which are downstream from genes. This approach has received more attention in recent years as an ideal methodology to unravel signals closer to the culmination of the disease process. The compounds identified through metabolomic profiling represent a range of intermediate metabolic pathways that may serve as biomarkers of exposure, susceptibility or disease.[119] Therefore, it is a valuable approach for deciphering metabolic outcomes with a phenotypic change. Inherited metabolic disorders can be measured by following a set of metabolomic biomarkers by directly measuring metabolites using liquid chromatography–mass spectrometry (LC-MS) or NMR. A few examples of these diseases are: amino acid metabolism disorders (phenylketonuria, maple syrup urine disease, tyrosinaemia I, arginineaemia, homocystineuria, ornithine transcarbamylase deficiency and non-ketotic hyperglycaemia), organic acidurias and mitochondrial defects.[120]

Risk prediction for diabetes and CVD is difficult although a number of biomarkers have been identified (see Table 15.4). So far insulin resistance (IR) and atherosclerosis are the only promising biomarkers.

In gastric cancer, alterations in metabolites of glycolysis, fatty-acid beta-oxidation, and cholesterol and amino acid metabolism were observed.[121] These metabolites can be used to follow gastric cancer development and treatment response.

15.5.5.1. Immune Diseases

Diagnosis of asthma is sometimes difficult and respiratory symptoms alone are not sufficient for diagnosing this disease. Measurements of fractional excretion of nitric oxide can be used as a biomarker for diagnosing asthma. There is a considerable effect of the environment on health from pollution, climate change and epigenetic influences, underlining the importance of understanding gene–environment interactions in the pathogenesis of asthma and its response to treatment. Metabolomic and proteomic approaches can be applied to determine levels of nitric oxide.[122,123]

15.5.5.2. Cardiovascular Diseases (CVDs)

Metabolites characterized from biofluids of CVD patients serve as biomarkers for CVDs, such as atherosclerosis. Dyslipidaemia in HIV patients put them at high risk of developing CVDs. In another study, intestinal microbiodata and metabolites were suggested to contribute to CVD development.[124,125]

15.5.5.3. Metabolic Diseases

Metabolomic profiling represents a snapshot of the health of an individual. Serum metabolites from healthy individuals and diabetes-2 patients showed different profiles, and disease-associated profiles indicated biomarkers for diabetes diagnosis. In a large population study diabetes-associated metabolites were identified, which were generated by seven genes.[126,127]

15.5.5.4. Infectious Diseases

Antibiotic resistance is a common feature of TB, where mycobacteria become resistant to antibiotic treatment. Gas chromatography–mass spectrometry (GC-MS)-based technologies were used to identify metabolites, especially metabolites from fatty acid metabolism, from wild-type mycobacteria-infected and mutant-infected samples. These observations laid the groundwork for developing therapeutics for TB treatment. Metabolomic adaptation of the bacteria in the host also was evaluated.[128]

15.5.5.5. Neurological Diseases

An epileptic seizure is a transient symptom of abnormal excessive or synchronous neuronal activity in the brain. Errors of metabolism result in abnormal levels of metabolites at the neonatal stage of development, resulting in neonatal seizures. One group of investigators demonstrated altered metabolism of astrocytes contributing to seizures.[129,130]

15.5.5.6. Cancer

GC–MS-based metabolomics in serum identified colorectal-cancer-associated biomarkers. Metabolites were successfully used to detect bladder, breast and pancreatic cancer. Here, it is emphasized that pancreatic cancer is a disease where early detection markers are sought because the incidence and mortality rates are almost the same, therefore, metabolomic biomarkers should be explored for this fatal disease.[131,132]

15.5.6. Epigenomics- and miRNOMICS-based Biomarkers in Immune Diseases, Cardiovascular Diseases (CVD), Metabolic Diseases, Infectious Diseases, Neurological Diseases and Cancer

In addition to the genetic code, human cells contain another regulatory level this is called epigenetic code.[34] This involves altered gene expression without changing the genomic structure.[33] Due to different chromatin status, condensed or relaxed, the same genetic variants might be associated with different phenotypes, depending on the environment, life-style and exposure. A rapidly growing number of genes with epigenetic regulation altering their expression by chromatin-remodelling (condensation and relaxation) have been identified.[7,34,43,133] Methylation of cytosines in DNA, histone modifications and alterations of non-coding RNAs (especially microRNAs) are the mechanisms involved in chromatin remodelling.

The term epigenome is used to define a cell's overall epigenetic state. Basic biological properties of DNA-segments, such as gene density, replication timing and recombination, are linked to their GC content. Promoter regions are rich in CpG content. A genomic region of about 0.4 kb with more than 50% GC content is called a CpG island. In mammals, CpG islands typically are 200–300 bp long. Promoters of tissue-specific genes that are situated within CpG islands are generally unmethylated, but during disease development these CpG sites start getting methylated. Cytosine methylation can regulate gene expression by hindering the association of some transcriptional factors with their cognate DNA recognition sequences, or methyl CpG-binding protein (MBP) can bind to methylated cytosines, mediating a repressive signal, or MBPs can interact with chromatin-forming proteins modifying the surrounding chromatin, linking

DNA methylation with chromatin modification. DNA methylation at position five of cytosine is conducted by DNA methyltransferases (DNMTs). These enzymes are needed for methylation initiation and methylation maintenance.[34]

Alterations due to epigenetic mechanisms can be stably passed on over numerous cycles of cell division. Selected epigenetic alterations can be inherited from one generation to another generation.[134] Cancer-specific methylation alterations are a hallmark of different cancers. Alteration in methylation may cause genomic instability, genomic alterations and changes in gene expression.[34,135] A systematic approach to determine epigenetic changes in tumour development may lead to identification of biomarkers needed for cancer diagnosis. It has been suggested that an integration of the genome and hypermethylome might provide insight into major pathways of cancer development, which, in turn, might help us identify new biomarkers of cancer diagnosis and prognosis.[136] Methylation and microRNA alterations are the main biomarkers which can be assayed easily in samples non-invasively.[137] The finding that monozygotic twins are epigenetically indistinguishable early in life but with age exhibit substantial difference in the epigenome, indicates that environmentally determined alterations in a cell's epigenetic marks are responsible for altered epigenome.[138] Examples exist to show that environmental factors influence disease initiation, progression and development.[139,140]

MicroRNAs (miRNAs), small RNA molecules of approximately 22 nucleotides, have been shown to be upregulated or downregulated in specific cell types and disease states. These molecules have become recognized as one of the major regulatory gatekeepers of coding genes in the human genome. The structure, nomenclature, mechanism of action, technologies used for miRNA detection and associations of miRNAs with human cancer has been evaluated by a number of investigators.[141,142] miRNAs are produced in a tissue-specific manner, and changes in miRNA within a tissue type can be correlated with disease status.[143] miRNAs regulate mRNA translation and degradation. Among a number of regulators of gene expression, miRNAs are the key regulators. Tissue-specific miRNAs have been reported by different groups.[28] These RNAs are of small size with distinct stem-and-loop structure.[144] A number of miRNAs can be isolated in circulation. Because of their small size and stability (due to secondary structure), these circulating microRNAs provide a rich source of diagnostic biomarkers.

15.5.6.1. Immune Diseases

Genetic and environmental factors contribute to multiple sclerosis. However, recent research indicates a role of epigenetics in multiple sclerosis development. Multiple sclerosis is an inflammatory disease in which the fatty myelin sheaths around the axons of the brain and spinal cord are damaged. Female-specific multiple sclerosis is associated with the HLA DRB1-1*5 haplotype. SHP-1 hypermethylation is a biomarker for multiple sclerosis.[145]

15.5.6.2. Cardiovascular Diseases (CVDs)

Atherosclerosis can be diagnosed when other biomarkers are combined with epigenetic biomarkers, such as hypermethylation of monoamine oxide A (MAOA). Along with gene-specific methylation biomarkers, global DNA methylation biomarkers were identified for atherosclerosis.[146] Similarly, in other CVDs, such as hypertension, global methylation profiling and individual gene methylation status (ADD1 gene) were used for diagnosis and prediction of outcome.[147]

15.5.6.3. Metabolic Diseases

A combination of biomarkers, genetics and epigenetics was used for diabetes diagnosis using mutations in duodenal homeobox-1 (PDX-1) and its methylation levels. Hypermethylation sites were reported at multiple sites distributed throughout the genome. At least one investigator reported a combination of histone and methylation biomarkers for diabetes diagnosis.[148,149]

15.5.6.4. Infectious Diseases

In early liver cancer, where risk factors are infection by HBV and HCV, clustered DNA methylation changes in polycomb repressor target genes were observed. P16 and c-myc hypermethylation also indicated initiation of hepatocellular carcinoma. In tuberculosis, vitamin D receptor (VDR) methylation indicated a high risk of developing the disease. The VDR gene encodes a transcription factor, which alters calcium homeostasis and immune function.[150]

15.5.6.5. Neurological Diseases

In Alzheimer's disease, altered methylation levels of repeat sequences (Alu, Line-1 and SAT-alpha) were observed. These modifications induced genomic instability that contributed to disease imitation and progression.[151] Epigenetic regulation in autism was also studied and locus-specific hypermethylation of MECP2 was observed.[152]

15.5.6.6. *Cancer*

Both histone modifications and DNA methylation were observed in different cancers, especially breast cancer. When epigenetic profiling of MCF-7, MDA-MB-231 and MDA-MB-231(S30) was followed, decreased trimethylation of H4K20 and hyperacetylation of H4 was observed. Concomitant to this decrease in trimethylation, lower levels of Suv4-20h2 histone methyl transferase were also observed. The effect was greater in MDA-MB-231 compared with other cells, which indicated that differential expression of histone modifications could represent aggressiveness of the disease. In another study, HDAC6 (one of the histone deacetyl transferases) responded to oestrogen treatment.[153] Retinoblastoma levels were lower whenever trimethylation of H4K20 was present. A correlation with the tumour stage and grade was also established, based on these histone biomarkers. Another interesting study reported quantitative expression of HDAC1 and its correlation with breast cancer patient's age, lymph node status, tumour size and human epidermal growth factor receptor 2/neural tumour-derived (her2/neu)-negative, oestrogen receptor (ER) and progesterone receptor (PR)-positive status.[154]

Cancer cells accumulate abnormal DNA methylation patterns that result in malignant phenotypes. The genomic distribution of methylation is not well understood. In this direction, a number of genome-wide association studies have been conducted so that cancer-risk-associated biomarkers can be identified. Using methylated DNA immunoprecipitation combined with high-throughput sequencing (MeDIP-seq), levels of methylation were compared in samples from normal and cancer cells and global hypomethylation was observed in cancer samples, especially in the CpG-rich regions. The location of these CpG-rich regions was not related to the transcription start sites of various genes. Using this approach, the methylation patterns during epithelial to mesenchymal transition also were evaluated and used for disease stratification.[155]

In breast cancer, methyl acceptance capacity in malignant breast tissues was approximately twofold to threefold greater compared with matched controls. However, the variation in methyl acceptance capacity among patients varied a lot.[156] Quantitative analysis for 5-methyl cytosine levels showed a substantial decrease compared with normal tissues. Levels of hypomethylation in BRCA1 and BRCA2 cancers were slightly lower but significant.[157] Genome-wide hypomethylation correlated with satellite sequence hypomethylation. Specific regions (Sa2-coding) on chromosome 1 and sat-alpha were specifically hypomethylated.[158] On chromosome 5, the region containing the coding sequence of SATr-1 also showed hypomethylation.

The tissue concentrations of specific miRNAs have been associated with tumour invasiveness, metastatic potential and other clinical characteristics for several types of cancers, including chronic lymphocytic leukaemia and breast, colorectal, hepatic, lung, pancreatic and prostate cancers. By targeting and controlling the expression of mRNA, miRNAs can control highly complex signal-transduction pathways and other biological pathways. The biological roles of miRNAs in cancer indicate a correlation with prognosis and therapeutic outcome. Further investigation of these roles may lead to new approaches for the categorization, diagnosis and treatment of human cancers. Frequent dysregulation of miRNA in malignancy highlights the study of molecular factors upstream of gene expression following the extensive investigation on elucidating the important role of miRNA in carcinogenesis. For example, oesophageal carcinogenesis is a multi-stage process, involving a variety of changes in gene expression and physiological structural changes. Recent innovation in miRNAs profiling technology have shed new light on the pathology of oesophageal carcinoma (EC), and also showed great potential for exploring novel biomarkers for both EC diagnosis and treatment. A thorough review of the role of miRNAs in EC, addressing miRNA functions, their putative role as oncogenes or tumour suppressors and their potential target genes has been performed by different investigators.[159]

In inflammatory breast cancer cells, more than 300 miRNAs were evaluated for their association with the disease.[160] The most promising miRNAs were miR-29a, miR-30b, miR-342-5p and miR-520a-5p. The functional analysis of these miRNAs revealed their role in cell proliferation and signal transduction pathways. These markers could be useful to identify inflammatory breast cancer cells. The promoter regions of miRNA coding region were evaluated and several miRNA promoters were found to be hypermethylated, especially those of miR-31, miR-130a, miR-let7a-3/let 7-b, miR-155 and miR-137.[143] In one example, an advantage of using miRNAs for detecting cancer was demonstrated due to their stability even in fixed tissues.[144] miR-155 predicted prognosis of triple-negative breast cancer (higher miR-155 expression correlated with higher angiogenesis and aggressiveness).[161]

Abbreviations

CMTC Clinico Molecular Triad Classification
DCIS Ductal carcinoma *in situ*
DNMT DNA methyltransferases
ER Oestrogen receptor
GC-MS Gas chromatography–mass spectrometry
GWAS Genome-wide association studies
HDAC Histone deacetylase
her2/neu Human epidermal growth factor receptor 2/neural tumour-derived
LC-MS Liquid chromatography–mass spectrometry
MBP Methyl CpG binding protein
miRNA microRNA
NMR MS Nuclear magnetic resonance mass spectroscopy
PR Progesterone receptor
SNP Single-nucleotide polymorphism

References

1. M. Verma, J. Kagan, D. Sidransky and S. Srivastava, Nature reviews, *Cancer*, 2003, **3**, 789–795.
2. E. Ferrannini, A. Natali, S. Camastra, M. Nannipieri, A. Mari, K. P. Adam, M. V. Milburn, G. Kastenmuller, J. Adamski, T. Tuomi, V. Lyssenko, L. Groop and W. E. Gall, *Diabetes*, 2013, **62**, 1730–1737.
3. M. Verma, *Curr. Genomics*, 2012, **13**, 308–313.
4. M. Verma and S. Srivastava, *Recent Results Cancer Res.*, 2003, **163**, 72–84discussion 264–266.
5. M. Verma, G. L. Wright Jr., S. M. Hanash, R. Gopal-Srivastava and S. Srivastava, *Ann. N. Y. Acad. Sci.*, 2001, **945**, 103–115.
6. S. Srivastava, M. Verma and D. E. Henson, *Clin. Cancer Res.*, 2001, **7**, 1118–1126.
7. M. Verma, *Methods Mol. Biol.*, 2012, **863**, 467–480.
8. S. A. Peters, F. L. Visseren and D. E. Grobbee, Nature reviews, *Cardiology*, 2013, **10**, 12–14.
9. C. W. Peng, Q. Tian, G. F. Yang, M. Fang, Z. L. Zhang, J. Peng, Y. Li and D. W. Pang, *Biomaterials*, 2012, **33**, 5742–5752.
10. W. Naumnik, B. Naumnik, K. Niewiarowska, M. Ossolinska and E. Chyczewska, *Exp. Oncol.*, 2012, **34**, 348–353.
11. C. C. Solomides, B. J. Evans, J. M. Navenot, R. Vadigepalli, S. C. Peiper and Z. X. Wang, *Acta Cytol.*, 2012, **56**, 645–654.
12. A. Di Ieva, *Microvasc. Res.*, 2010, **80**, 522–533.
13. J. Penner, R. Rupsingh, M. Smith, J. L. Wells, M. J. Borrie and R. Bartha, *Prog. Neuro-psychopharmacol. Biol. Psychiatry*, 2010, **34**, 104–110.
14. R. Rupsingh, M. Borrie, M. Smith, J. L. Wells and R. Bartha, *Neurobiol Aging*, 2011, **32**, 802–810.
15. T. Efferth, *Planta Med.*, 2010, **76**, 1143–1154.

16. D. G. Duda, C. G. Willett, M. Ancukiewicz, E. di Tomaso, M. Shah, B. G. Czito, R. Bentley, M. Poleski, G. Y. Lauwers, M. Carroll, D. Tyler, C. Mantyh, P. Shellito, J. W. Clark and R. K. Jain, *Oncologist*, 2010, **15**, 577–583.
17. H. Xu, C. Chen, C. M. Liu, J. Peng, Y. Li, Z. L. Zhang and H. W. Tang, *Guangpuxue Yu Guangpu Fenxi*, 2009, **29**, 3216–3219.
18. S. J. Otto, J. Fracheboud, A. L. Verbeek, R. Boer, J. C. Reijerink-Verheij, J. D. Otten, M. J. Broeders and H. J. de Koning, *Cancer Epidemiol., Biomarkers Prev.*, 2012, **21**, 66–73.
19. H. Allahverdipour, M. Asghari-Jafarabadi and A. Emami, *Women Health*, 2011, **51**, 204–219.
20. M. Heijblom, J. M. Klaase, F. M. van den Engh, T. G. van Leeuwen, W. Steenbergen and S. Manohar, *Technol. Cancer Res. Treat.*, 2011, **10**, 607–623.
21. M. Sentis, *Breast Cancer Res. Treat.*, 2010, **123**(Suppl 1), 11–13.
22. R. Luckmann, *ACP J. Club*, 2005, **142**, 23.
23. H. Zhang, D. Yee and C. Wang, *Nanomedicine*, 2008, **3**, 83–91.
24. L. M. Kenny, A. Al-Nahhas and E. O. Aboagye, *Nucl. Med. Commun.*, 2011, **32**, 333–335.
25. R. Lieberman, *Am. J. Ther.*, 2012, **19**, 395–396.
26. J. L. Parker, N. Lushina, P. S. Bal, T. Petrella, R. Dent and G. Lopes, *Breast Cancer Res. Treat.*, 2012, **136**, 179–185.
27. C. S. Zhu, P. F. Pinsky, D. W. Cramer, D. F. Ransohoff, P. Hartge, R. M. Pfeiffer, N. Urban, G. Mor, R. C. Bast Jr., L. E. Moore, A. E. Lokshin, M. W. McIntosh, S. J. Skates, A. Vitonis, Z. Zhang, D. C. Ward, J. T. Symanowski, A. Lomakin, E. T. Fung, P. M. Sluss, N. Scholler, K. H. Lu, A. M. Marrangoni, C. Patriotis, S. Srivastava, S. S. Buys and C. D. Berg, *Cancer Prev. Res.*, 2011, **4**, 375–383.
28. K. Tjensvoll, K. N. Svendsen, J. M. Reuben, S. Oltedal, B. Gilje, R. Smaaland and O. Nordgard, *Biomarkers*, 2012, **17**, 463–470.
29. S. Knappskog, R. Chrisanthar, E. Lokkevik, G. Anker, B. Ostenstad, S. Lundgren, T. Risberg, I. Mjaaland, B. Leirvaag, H. Miletic and P. E. Lonning, *Breast Cancer Res.*, 2012, **14**, R47.
30. E. Rivera and H. Gomez, *Breast Cancer Res.*, 2010, **12**(Suppl 2), S2.
31. W. G. Newman and D. Flockhart, *Pharmacogenomics*, 2012, **13**, 629–631.
32. B. A. Virnig, M. T. Torchia, S. L. Jarosek, S. Durham and T. M. Tuttle, in *Data Points Publication Series*, Rockville (MD), 2011.
33. M. Verma, *Curr. Opin. Clin. Nutr. Metab. Care*, 2013, **16**, 376–384.
34. M. Verma, M. J. Khoury and J. P. Ioannidis, *Cancer Epidemiol., Biomarkers Prev.*, 2013, **22**, 189–200.
35. M. Lucas and S. Gaudieri, *Biomarkers Med.*, 2012, **6**, 133–135.
36. C. Ferrandi, F. Richard, P. Tavano, E. Hauben, V. Barbie, J. P. Gotteland, B. Greco, M. Fortunato, M. F. Mariani, R. Furlan, G. Comi, G. Martino and P. F. Zaratin, *Mult. Scler.*, 2011, **17**, 43–56.
37. A. E. Aiello and G. A. Kaplan, *Biodemography Soc. Biol.*, 2009, **55**, 178–205.
38. G. Lippi, T. Meschi and G. Cervellin, *Eur. J. Intern. Med.*, 2011, **22**, 460–465.
39. G. Brathen, K. S. Bjerve, E. Brodtkorb and G. Bovim, *J. Neurol., Neurosurg., Psychiatry*, 2000, **68**, 342–348.
40. D. Lev, Y. Weigl, M. Hasan, E. Gak, M. Davidovich, C. Vinkler, E. Leshinsky-Silver, T. Lerman-Sagie and N. Watemberg, *Am. J. Med. Genet., Part A*, 2007, **143A**, 921–924.
41. N. Elleuch, I. Feki, E. Turki, M. I. Miladi, A. Boukhris, M. Damak, C. Mhiri, E. Chappuis and F. Woimant, *Rev. Neurol.*, 2010, **166**, 550–552.
42. D. Zhu, J. Wang, L. Ren, Y. Li, B. Xu, Y. Wei, Y. Zhong, X. Yu, S. Zhai, J. Xu and X. Qin, *J. Cell. Biochem.*, 2013, **114**, 448–455.
43. M. Verma, *Antioxid. Redox Signaling*, 2012, **17**, 355–364.
44. S. Viatte, D. Plant and S. Raychaudhuri, *Nat. Rev. Rheumatol.*, 2013, **9**, 141–153.

45. J. Kurkó, T. Besenyei, J. Laki, T. T. Glant, K. Mikecz and Z. Szekanecz, *Clin. Rev. Allergy Immunol.*, 2013, **45**, 170–179.
46. B. P. Smith, D. H. Best and C. G. Elliott, *Heart Failure Clin.*, 2012, **8**, 319–330.
47. J. Simino, D. C. Rao and B. I. Freedman, *Curr. Opin. Nephrol. Hypertens.*, 2012, **21**, 500–507.
48. G. Massawi, P. Hickling, D. Hilton and C. Patterson, *Rheumatology*, 2003, **42**, 1012–1014.
49. N. Mahfoudh, M. Siala, M. Rihl, A. Kammoun, F. Frikha, H. Fourati, M. Younes, R. Gdoura, L. Gaddour, F. Hakim, Z. Bahloul, S. Baklouti, N. Bargaoui, S. Sellami, A. Hammami and H. Makni, *Clin. Rheumatol.*, 2011, **30**, 1069–1073.
50. G. Kamkamidze, C. Capoulade-Metay, M. Butsashvili, Y. Dudoit, O. Chubinishvili, P. Debre and L. Theodorou, *Georgian Med. News*, 2005, 74–79.
51. B. J. Grady, D. C. Samuels, G. K. Robbins, D. Selph, J. A. Canter, R. B. Pollard, D. W. Haas, R. Shafer, S. A. Kalams, D. G. Murdock, M. D. Ritchie and T. Hulgan, *J. Acquired Immune Defic. Syndr.*, 2011, **58**, 363–370.
52. J. C. Mar, N. A. Matigian, A. Mackay-Sim, G. D. Mellick, C. M. Sue, P. A. Silburn, J. J. McGrath, J. Quackenbush and C. A. Wells, *PLoS Genet.*, 2011, **7**, e1002207.
53. F. D. Shah, R. Begum, B. N. Vajaria, K. R. Patel, J. B. Patel, S. N. Shukla and P. S. Patel, *Indian J. Clin. Biochem.*, 2011, **26**, 326–334.
54. R. Kirk, Nature reviews, *Clin. Oncol.*, 2011, **8**, 383.
55. P. Hall, *J. Intern. Med.*, 2012, **271**, 318–320.
56. A. C. Antoniou, A. B. Spurdle, O. M. Sinilnikova, S. Healey, K. A. Pooley, R. K. Schmutzler, B. Versmold, C. Engel, A. Meindl, N. Arnold, W. Hofmann, C. Sutter, D. Niederacher, H. Deissler, T. Caldes, K. Kampjarvi, H. Nevanlinna, J. Simard, J. Beesley, X. Chen, S. L. Neuhausen, T. R. Rebbeck, T. Wagner, H. T. Lynch, C. Isaacs, J. Weitzel, P. A. Ganz, M. B. Daly, G. Tomlinson, O. I. Olopade, J. L. Blum, F. J. Couch, P. Peterlongo, S. Manoukian, M. Barile, P. Radice, C. I. Szabo, L. H. Pereira, M. H. Greene, G. Rennert, F. Lejbkowicz, O. Barnett-Griness, I. L. Andrulis, H. Ozcelik, A. M. Gerdes, M. A. Caligo, Y. Laitman, B. Kaufman, R. Milgrom, E. Friedman, S. M. Domchek, K. L. Nathanson, A. Osorio, G. Llort, R. L. Milne, J. Benitez, U. Hamann, F. B. Hogervorst, P. Manders, M. J. Ligtenberg, A. M. van den Ouweland, S. Peock, M. Cook, R. Platte, D. G. Evans, R. Eeles, G. Pichert, C. Chu, D. Eccles, R. Davidson, F. Douglas, A. K. Godwin, L. Barjhoux, S. Mazoyer, H. Sobol, V. Bourdon, F. Eisinger, A. Chompret, C. Capoulade, B. Bressac-de Paillerets, G. M. Lenoir, M. Gauthier-Villars, C. Houdayer, D. Stoppa-Lyonnet, G. Chenevix-Trench and D. F. Easton, *Am. J. Hum. Genet.*, 2008, **82**, 937–948.
57. S. A. Gayther, H. Song, S. J. Ramus, S. K. Kjaer, A. S. Whittemore, L. Quaye, J. Tyrer, D. Shadforth, E. Hogdall, C. Hogdall, J. Blaeker, R. DiCioccio, V. McGuire, P. M. Webb, J. Beesley, A. C. Green, D. C. Whiteman, M. T. Goodman, G. Lurie, M. E. Carney, F. Modugno, R. B. Ness, R. P. Edwards, K. B. Moysich, E. L. Goode, F. J. Couch, J. M. Cunningham, T. A. Sellers, A. H. Wu, M. C. Pike, E. S. Iversen, J. R. Marks, M. Garcia-Closas, L. Brinton, J. Lissowska, B. Peplonska, D. F. Easton, I. Jacobs, B. A. Ponder, J. Schildkraut, C. L. Pearce, G. Chenevix-Trench, A. Berchuck and P. D. Pharoah, *Cancer Res.*, 2007, **67**, 3027–3035.
58. P. D. Pharoah and C. Caldas, Nature reviews, *Clin. Oncol.*, 2010, **7**, 615–616.
59. A. A. Ponomareva, E. Rykova, N. V. Cherdyntseva, E. L. Choinzonov, P. P. Laktionov and V. V. Vlasov, *Molekuliarnaia Biologiia*, 2011, **45**, 203–217.
60. V. R. Adams and R. D. Harvey, *Am. J. Health-Syst. Pharm.*, 2010, **67**, S3–S9 quiz S15–16.
61. J. Li, K. Humphreys, T. Heikkinen, K. Aittomaki, C. Blomqvist, P. D. Pharoah, A. M. Dunning, S. Ahmed, M. J. Hooning, J. W. Martens, A. M. van den Ouweland, L. Alfredsson, A. Palotie, L. Peltonen-Palotie, A. Irwanto, H. Q. Low, G. H. Teoh, A. Thalamuthu, D. F. Easton, H. Nevanlinna, J. Liu, K. Czene and P. Hall, *Breast Cancer Res. Treat.*, 2011, **126**, 717–727.

62. C. Turnbull, S. Ahmed, J. Morrison, D. Pernet, A. Renwick, M. Maranian, S. Seal, M. Ghoussaini, S. Hines, C. S. Healey, D. Hughes, M. Warren-Perry, W. Tapper, D. Eccles, D. G. Evans, M. Hooning, M. Schutte, A. van den Ouweland, R. Houlston, G. Ross, C. Langford, P. D. Pharoah, M. R. Stratton, A. M. Dunning, N. Rahman and D. F. Easton, *Nat. Genet.*, 2010, **42**, 504–507.
63. P. R. Srinivas, M. Verma, Y. Zhao and S. Srivastava, *Clin. Chem.*, 2002, **48**, 1160–1169.
64. C. Panis, L. Pizzatti, A. C. Herrera, R. Cecchini and E. Abdelhay, *Cancer Lett.*, 2013, **330**, 57–66.
65. M. Verma, *Methods Mol. Biol.*, 2009, **471**, 197–215.
66. K. Na, M. J. Lee, H. J. Jeong, H. Kim and Y. K. Paik, *Methods Mol. Biol.*, 2012, **854**, 223–237.
67. E. S. Baker, T. Liu, V. A. Petyuk, K. E. Burnum-Johnson, Y. M. Ibrahim, G. A. Anderson and R. D. Smith, *Genome Med.*, 2012, **4**, 63.
68. N. G. Anderson, *Clin. Chem.*, 2010, **56**, 154–160.
69. W. Rozek, J. Horning, J. Anderson and P. Ciborowski, *Proteomics: Clin. Appl.*, 2008, **2**, 1498–1507.
70. E. J. Toonen, C. Gilissen, B. Franke, W. Kievit, A. M. Eijsbouts, A. A. den Broeder, S. V. van Reijmersdal, J. A. Veltman, H. Scheffer, T. R. Radstake, P. L. van Riel, P. Barrera and M. J. Coenen, *PLoS One*, 2012, **7**, e33199.
71. P. Emery, *Rheumatology*, 2012, **51**(Suppl 5), v22–30.
72. E. Solau-Gervais, C. Prudhomme, P. Philippe, A. Duhamel, C. Dupont-Creteur, J. L. Legrand, E. Houvenagel and R. M. Flipo, *Jt., Bone, Spine*, 2012, **79**, 281–284.
73. H. T. Maecker, T. M. Lindstrom, W. H. Robinson, P. J. Utz, M. Hale, S. D. Boyd, S. S. Shen-Orr and C. G. Fathman, *Nat. Rev. Rheumatol.*, 2012, **8**, 317–328.
74. L. Iaccarino, A. Ghirardello, M. Canova, M. Zen, S. Bettio, L. Nalotto, L. Punzi and A. Doria, *Autoimmun. Rev.*, 2011, **10**, 553–558.
75. H. Orbach, G. Zandman-Goddard, H. Amital, V. Barak, Z. Szekanecz, G. Szucs, K. Danko, E. Nagy, T. Csepany, J. F. Carvalho, A. Doria and Y. Shoenfeld, *Ann. N. Y. Acad. Sci.*, 2007, **1109**, 385–400.
76. A. Bartolomucci, G. M. Pasinetti and S. R. Salton, *Neuroscience*, 2010, **170**, 289–297.
77. R. Mogelvang, S. H. Pedersen, A. Flyvbjerg, M. Bjerre, A. Z. Iversen, S. Galatius, J. Frystyk and J. S. Jensen, *Am. J. Cardiol.*, 2012, **109**, 515–520.
78. O. B. Bleijerveld, Y. N. Zhang, S. Beldar, I. E. Hoefer, S. K. Sze, G. Pasterkamp and D. P. de Kleijn, *Proteomics: Clin. Appl.*, 2013, **7**, 490–503.
79. W. S. Wang, X. H. Liu, L. X. Liu, D. Y. Jin, P. Y. Yang and X. L. Wang, *J. Proteomics*, 2013, **84C**, 52–60.
80. G. Pendyala and H. S. Fox, *Genome Med.*, 2010, **2**, 22.
81. E. Colombo, B. Borgiani, C. Verderio and R. Furlan, *Front. Physiol.*, 2012, **3**, 63.
82. D. S. Michaud, J. Izard, C. S. Wilhelm-Benartzi, D. H. You, V. A. Grote, A. Tjønneland, C. C. Dahm, K. Overvad, M. Jenab, V. Fedirko, M. C. Boutron-Ruault, F. Clavel-Chapelon, A. Racine, R. Kaaks, H. Boeing, J. Foerster, A. Trichopoulou, P. Lagiou, D. Trichopoulos, C. Sacerdote, S. Sieri, D. Palli, R. Tumino, S. Panico, P. D. Siersema, P. H. Peeters, E. Lund, A. Barricarte, J. M. Huerta, E. Molina-Montes, M. Dorronsoro, J. R. Quirós, E. J. Duell, W. Ye, M. Sund, B. Lindkvist, D. Johansen, K. T. Khaw, N. Wareham, R. C. Travis, P. Vineis, H. B. Bueno-de-Mesquita and E. Riboli, *Gut*, 2013, **62**, 1764–1770.
83. V. A. Grote, R. Kaaks, A. Nieters, A. Tjonneland, J. Halkjaer, K. Overvad, M. R. Skjelbo Nielsen, M. C. Boutron-Ruault, F. Clavel-Chapelon, A. Racine, B. Teucher, S. Becker, T. Pischon, H. Boeing, A. Trichopoulou, C. Cassapa, V. Stratigakou, D. Palli, V. Krogh, R. Tumino, P. Vineis, S. Panico, L. Rodriguez, E.

J. Duell, M. J. Sanchez, M. Dorronsoro, C. Navarro, A. B. Gurrea, P. D. Siersema, P. H. Peeters, W. Ye, M. Sund, B. Lindkvist, D. Johansen, K. T. Khaw, N. Wareham, N. E. Allen, R. C. Travis, V. Fedirko, M. Jenab, D. S. Michaud, S. C. Chuang, D. Romaguera, H. B. Bueno-de-Mesquita and S. Rohrmann, *Br. J. Cancer*, 2012, **106**, 1866–1874.

84. S. Tejpar, M. Bertagnolli, F. Bosman, H. J. Lenz, L. Garraway, F. Waldman, R. Warren, A. Bild, D. Collins-Brennan, H. Hahn, D. P. Harkin, R. Kennedy, M. Ilyas, H. Morreau, V. Proutski, C. Swanton, I. Tomlinson, M. Delorenzi, R. Fiocca, E. Van Cutsem and A. Roth, *Oncologist*, 2010, **15**, 390–404.
85. N. Miura, Y. Osaki, M. Nagashima, M. Kohno, K. Yorozu, K. Shomori, T. Kanbe, K. Oyama, Y. Kishimoto, S. Maruyama, E. Noma, Y. Horie, M. Kudo, S. Sakaguchi, Y. Hirooka, H. Ito, H. Kawasaki, J. Hasegawa and G. Shiota, *BMC Gastroenterol.*, 2010, **10**, 46.
86. Y. Sun, Z. Zang, X. Xu, Z. Zhang, L. Zhong, W. Zan, Y. Zhao and L. Sun, *Int. J. Mol. Sci.*, 2010, **11**, 1423–1433.
87. B. Hamrita, K. Chahed, M. Trimeche, C. L. Guillier, P. Hammann, A. Chaieb, S. Korbi and L. Chouchane, *Clin. Chim. Acta*, 2009, **404**, 111–118.
88. M. A. Fonseca-Sanchez, S. Rodriguez Cuevas, G. Mendoza-Hernandez, V. Bautista-Pina, E. Arechaga Ocampo, A. Hidalgo Miranda, V. Quintanar Jurado, L. A. Marchat, E. Alvarez-Sanchez, C. Perez Plasencia and C. Lopez-Camarillo, *Int. J. Oncol.*, 2012, **41**, 670–680.
89. B. Hamrita, H. Ben Nasr, P. Hammann, L. Kuhn, A. Ben Anes, S. Dimassi, A. Chaieb, H. Khairi and K. Chahed, *Ann. Biol. Clin.*, 2012, **70**, 553–565.
90. B. Hamrita, H. B. Nasr, K. Chahed and L. Chouchane, *Gulf J. Oncol.*, 2011, **1**, 36–44.
91. Y. M. Chung, Y. D. Yoo, J. K. Park, Y. T. Kim and H. J. Kim, *Anticancer Res.*, 2001, **21**, 1129–1133.
92. F. Cirillo, G. Nassa, R. Tarallo, C. Stellato, M. R. De Filippo, C. Ambrosino, M. Baumann, T. A. Nyman and A. Weisz, *J. Proteome Res.*, 2013, **12**, 421–431.
93. N. Q. Liu, R. B. Braakman, C. Stingl, T. M. Luider, J. W. Martens, J. A. Foekens and A. Umar, *J. Mammary Gland Biol. Neoplasia*, 2012, **17**, 155–164.
94. A. M. Gonzalez-Angulo, B. T. Hennessy, F. Meric-Bernstam, A. Sahin, W. Liu, Z. Ju, M. S. Carey, S. Myhre, C. Speers, L. Deng, R. Broaddus, A. Lluch, S. Aparicio, P. Brown, L. Pusztai, W. F. Symmans, J. Alsner, J. Overgaard, A. L. Borresen-Dale, G. N. Hortobagyi, K. R. Coombes and G. B. Mills, *Clin. Proteomics*, 2011, **8**, 11.
95. D. M. Pedrotty, M. P. Morley and T. P. Cappola, *Prog. Cardiovasc. Dis.*, 2012, **55**, 64–69.
96. L. Desire, E. Blondiaux, J. Carriere, R. Haddad, O. Sol, P. Fehlbaum-Beurdeley, R. Einstein, W. Zhou and M. P. Pando, *J. Alzheimer's Dis.*, 2013, **34**, 469–483.
97. C. Q. Yi, C. H. Ma, Z. P. Xie, Y. Cao, G. Q. Zhang, X. K. Zhou and Z. Q. Liu, *Genet. Mol. Res.*, 2013, 12.
98. J. Caetano-Lopes, A. Rodrigues, A. Lopes, A. C. Vale, M. A. Pitts-Kiefer, B. Vidal, I. P. Perpetuo, J. Monteiro, Y. T. Konttinen, M. F. Vaz, A. Nazarian, H. Canhão and J. E. Fonseca, *Clin. Rev. Allergy Immunol.*, 2014, **47**, 38–45.
99. J. C. Laguna and M. Alegret, *Pharmacogenomics*, 2012, **13**, 477–495.
100. H. Peiris, R. Raghupathi, C. F. Jessup, M. P. Zanin, D. Mohanasundaram, K. D. Mackenzie, T. Chataway, J. N. Clarke, J. Brealey, P. T. Coates, M. A. Pritchard and D. J. Keating, *Endocrinology*, 2012, **153**, 5212–5221.
101. T. S. Nielsen, U. Kampmann, R. R. Nielsen, N. Jessen, L. Orskov, S. B. Pedersen, J. O. Jorgensen, S. Lund and N. Moller, *J. Clin. Endocrinol. Metab.*, 2012, **97**, E1348–E1352.

102. J. M. Swanson, G. C. Wood, L. Xu, L. E. Tang, B. Meibohm, R. Homayouni, M. A. Croce and T. C. Fabian, *PLoS One*, 2012, **7**, e42065.
103. I. Martin-Loeches, E. Papiol, R. Almansa, G. Lopez-Campos, J. F. Bermejo-Martin and J. Rello, *Med. Intensiva*, 2012, **36**, 257–263.
104. S. Jain and P. Heutink, *Neuron*, 2010, **68**, 207–217.
105. E. Borgan, B. Sitter, O. C. Lingjaerde, H. Johnsen, S. Lundgren, T. F. Bathen, T. Sorlie, A. L. Borresen-Dale and I. S. Gribbestad, *BMC Cancer*, 2010, **10**, 628.
106. J. Hardouin, J. P. Lasserre, L. Sylvius, R. Joubert-Caron and M. Caron, *Ann. N. Y. Acad. Sci.*, 2007, **1107**, 223–230.
107. M. R. Amin, M. S. Siddiqui, D. Ahmed, F. Ahmed and A. Hossain, *Int. J. Bioinf. Res. Appl.*, 2011, **7**, 287–298.
108. M. Careaga and P. Ashwood, *Methods Mol. Biol.*, 2012, **934**, 219–240.
109. S. D. Funk and A. W. Orr, *Pharmacol. Res.*, 2013, **67**, 42–52.
110. J. I. Odegaard and A. Chawla, *Cold Spring Harbor Perspect. Med.*, 2012, **2**, a007724.
111. H. S. Kim and M. S. Lee, *Curr. Mol. Med.*, 2009, **9**, 30–44.
112. M. Gonzalez-Juarrero, *Immunotherapy*, 2012, **4**, 187–199.
113. A. J. Battersby, B. Kampmann and S. Burl, *Clin. Dev. Immunol.*, 2012, **2012**, 430972.
114. A. T. Weeraratna, A. Kalehua, I. Deleon, D. Bertak, G. Maher, M. S. Wade, A. Lustig, K. G. Becker, W. Wood, III, D. G. Walker, T. G. Beach and D. D. Taub, *Exp. Cell Res.*, 2007, **313**, 450–461.
115. R. P. Young and R. J. Hopkins, *Cancer*, 2013, **119**, 1761.
116. Y. Godet, E. Fabre, M. Dosset, M. Lamuraglia, E. Levionnois, P. Ravel, N. Benhamouda, A. Cazes, F. Le Pimpec-Barthes, B. Gaugler, P. Langlade-Demoyen, X. Pivot, P. Saas, B. Maillere, E. Tartour, C. Borg and O. Adotevi, *Clin. Cancer Res.*, 2012, **18**, 2943–2953.
117. N. Leung, *Med. J. Malays.*, 2005, **60**(Suppl B), 63–66.
118. H. Wang, V. K. Tso, C. M. Slupsky and R. N. Fedorak, *Future Oncol.*, 2010, **6**, 1395–1406.
119. Y. He, Z. Yu, I. Giegling, L. Xie, A. M. Hartmann, C. Prehn, J. Adamski, R. Kahn, Y. Li, T. Illig, R. Wang-Sattler and D. Rujescu, *Transl. Psychiatry*, 2012, **2**, e149.
120. H. Janeckova, K. Hron, P. Wojtowicz, E. Hlidkova, A. Baresova, D. Friedecky, L. Zidkova, P. Hornik, D. Behulova, D. Prochazkova, H. Vinohradska, K. Peskova, P. Bruheim, V. Smolka, S. Stastna and T. Adam, *J. Chromatogr. A*, 2012, **1226**, 11–17.
121. H. Song, L. Wang, H. L. Liu, X. B. Wu, H. S. Wang, Z. H. Liu, Y. Li, D. C. Diao, H. L. Chen and J. S. Peng, *Oncol. Rep.*, 2011, **26**, 431–438.
122. L. Pedersen, J. Elers and V. Backer, *Physician Sports Med.*, 2011, **39**, 163–171.
123. A. J. Apter, *J. Allergy Clin. Immunol.*, 2010, **125**, 79–84.
124. H. Rose, H. Low, E. Dewar, M. Bukrinsky, J. Hoy, A. Dart and D. Sviridov, *Atherosclerosis*, 2013, **229**, 206–211.
125. R. A. Koeth, Z. Wang, B. S. Levison, J. A. Buffa, E. Org, B. T. Sheehy, E. B. Britt, X. Fu, Y. Wu, L. Li, J. D. Smith, J. A. Didonato, J. Chen, H. Li, G. D. Wu, J. D. Lewis, M. Warrier, J. M. Brown, R. M. Krauss, W. H. Tang, F. D. Bushman, A. J. Lusis and S. L. Hazen, *Nat. Med.*, 2013, **19**, 576–585.
126. N. Friedrich, *J. Endocrinol.*, 2012, **215**, 29–42.
127. R. Wang-Sattler, Z. Yu, C. Herder, A. C. Messias, A. Floegel, Y. He, K. Heim, M. Campillos, C. Holzapfel, B. Thorand, H. Grallert, T. Xu, E. Bader, C. Huth, K. Mittelstrass, A. Doring, C. Meisinger, C. Gieger, C. Prehn, W. Roemisch-Margl, M. Carstensen, L. Xie, H. Yamanaka-Okumura, G. Xing, U. Ceglarek, J. Thiery, G. Giani, H. Lickert, X. Lin, Y. Li, H. Boeing, H. G. Joost, M. H. de Angelis, W. Rathmann, K. Suhre, H. Prokisch, A. Peters, T. Meitinger, M. Roden, H. E. Wichmann, T. Pischon, J. Adamski and T. Illig, *Mol. Syst. Biol.*, 2012, **8**, 615.

128. I. du Preez and T. Loots du, *OMICS*, 2012, **16**, 596–603.
129. C. Ficicioglu and D. Bearden, *Pediatr. Neurol.*, 2011, **45**, 283–291.
130. M. Broberg, K. J. Pope, T. Lewis, T. Olsson, M. Nilsson and J. O. Willoughby, *Epilepsy Res.*, 2008, **80**, 132–141.
131. T. Kobayashi, S. Nishiumi, A. Ikeda, T. Yoshie, A. Sakai, A. Matsubara, Y. Izumi, H. Tsumura, M. Tsuda, H. Nishisaki, N. Hayashi, S. Kawano, Y. Fujiwara, H. Minami, T. Takenawa, T. Azuma and M. Yoshida, *Cancer Epidemiol., Biomarkers Prev.*, 2013, **22**, 571–579.
132. M. Verma, *Technol. Cancer Res. Treat.*, 2005, **4**, 295–301.
133. S. Khare and M. Verma, *Methods Mol. Biol.*, 2012, **863**, 177–185.
134. M. D. Anway and M. K. Skinner, *Endocrinology*, 2006, **147**, S43–S49.
135. F. Fang, S. Turcan, A. Rimner, A. Kaufman, D. Giri, L. G. Morris, R. Shen, V. Seshan, Q. Mo, A. Heguy, S. B. Baylin, N. Ahuja, A. Viale, J. Massague, L. Norton, L. T. Vahdat, M. E. Moynahan and T. A. Chan, *Sci. Transl. Med.*, 2011, **3**, 75ra25.
136. J. M. Yi, M. Dhir, L. Van Neste, S. R. Downing, J. Jeschke, S. C. Glockner, M. de Freitas Calmon, C. M. Hooker, J. M. Funes, C. Boshoff, K. M. Smits, M. van Engeland, M. P. Weijenberg, C. A. Iacobuzio-Donahue, J. G. Herman, K. E. Schuebel, S. B. Baylin and N. Ahuja, *Clin. Cancer Res.*, 2011, **17**, 1535–1545.
137. L. Yu, N. W. Todd, L. Xing, Y. Xie, H. Zhang, Z. Liu, H. Fang, J. Zhang, R. L. Katz and F. Jiang, *Int. J. Cancer*, 2010, 127.
138. A. Harder, S. Titze, L. Herbst, T. Harder, K. Guse, S. Tinschert, D. Kaufmann, T. Rosenbaum, V. F. Mautner, E. Windt, U. Wahllander-Danek, K. Wimmer, S. Mundlos and H. Peters, *Twin Res. Hum. Genet.*, 2010, **13**, 582–594.
139. J. Ashley-Martin, J. VanLeeuwen, A. Cribb, P. Andreou and J. R. Guernsey, *Int. J. Environ. Res. Public Health*, 2012, **9**, 1846–1858.
140. J. Qiu, R. Yang, Y. Rao, Y. Du and F. W. Kalembo, *PLoS One*, 2012, **7**, e36497.
141. M. Alshalalfa, *Adv. Bioinf.*, 2012, **2012**, 839837.
142. K. W. Chang, S. Y. Kao, Y. H. Wu, M. M. Tsai, H. F. Tu, C. J. Liu, M. T. Lui and S. C. Lin, *Oral Oncol.*, 2013, **49**, 27–33.
143. N. Nishida, M. Nagahara, T. Sato, K. Mimori, T. Sudo, F. Tanaka, K. Shibata, H. Ishii, K. Sugihara, Y. Doki and M. Mori, *Clin. Cancer Res.*, 2012, **18**, 3054–3070.
144. P. S. Mitchell, R. K. Parkin, E. M. Kroh, B. R. Fritz, S. K. Wyman, E. L. Pogosova-Agadjanyan, A. Peterson, J. Noteboom, K. C. O'Briant, A. Allen, D. W. Lin, N. Urban, C. W. Drescher, B. S. Knudsen, D. L. Stirewalt, R. Gentleman, R. L. Vessella, P. S. Nelson, D. B. Martin and M. Tewari, *Proc. Natl. Acad. Sci. U. S. A.*, 2008, **105**, 10513–10518.
145. C. Kumagai, B. Kalman, F. A. Middleton, T. Vyshkina and P. T. Massa, *J. Neuroimmunol.*, 2012, **246**, 51–57.
146. J. Zhao, C. W. Forsberg, J. Goldberg, N. L. Smith and V. Vaccarino, *BMC Med. Genet.*, 2012, **13**, 100.
147. L. N. Zhang, P. P. Liu, L. Wang, F. Yuan, L. Xu, Y. Xin, L. J. Fei, Q. L. Zhong, Y. Huang, L. M. Hao, X. J. Qiu, Y. Le, M. Ye and S. Duan, *PLoS One*, 2013, **8**, e63455.
148. V. K. Rakyan, H. Beyan, T. A. Down, M. I. Hawa, S. Maslau, D. Aden, A. Daunay, F. Busato, C. A. Mein, B. Manfras, K. R. Dias, C. G. Bell, J. Tost, B. O. Boehm, S. Beck and R. D. Leslie, *PLoS Genet.*, 2011, **7**, e1002300.
149. L. M. Villeneuve, M. A. Reddy, L. L. Lanting, M. Wang, L. Meng and R. Natarajan, *Proc. Natl. Acad. Sci. U. S. A.*, 2008, **105**, 9047–9052.
150. C. Andraos, G. Koorsen, J. C. Knight and L. Bornman, *Human Immunol.*, 2011, **72**, 262–268.
151. V. Bollati, D. Galimberti, L. Pergoli, E. Dalla Valle, F. Barretta, F. Cortini, E. Scarpini, P. A. Bertazzi and A. Baccarelli, *Brain, Behav., Immun.*, 2011, **25**, 1078–1083.

152. R. P. Nagarajan, K. A. Patzel, M. Martin, D. H. Yasui, S. E. Swanberg, I. Hertz-Picciotto, R. L. Hansen, J. Van de Water, I. N. Pessah, R. Jiang, W. P. Robinson and J. M. LaSalle, *Autism Res.*, 2008, **1**, 169–178.
153. S. Saji, M. Kawakami, S. Hayashi, N. Yoshida, M. Hirose, S. Horiguchi, A. Itoh, N. Funata, S. L. Schreiber, M. Yoshida and M. Toi, *Oncogene*, 2005, **24**, 4531–4539.
154. Z. Zhang, H. Yamashita, T. Toyama, H. Sugiura, Y. Ando, K. Mita, M. Hamaguchi, Y. Hara, S. Kobayashi and H. Iwase, *Breast Cancer Res. Treat.*, 2005, **94**, 11–16.
155. Y. Ruike, Y. Imanaka, F. Sato, K. Shimizu and G. Tsujimoto, Genome-wide analysis of aberrant methylation in human breast cancer cells using methyl-DNA immunoprecipitation combined with high-throughput sequencing, *BMC Genomics*, 2010, **11**, 137.
156. J. Soares, A. E. Pinto, C. V. Cunha, S. Andre, I. Barao, J. M. Sousa and M. Cravo, *Cancer*, 1999, **85**, 112–118.
157. K. Jackson, M. C. Yu, K. Arakawa, E. Fiala, B. Youn, H. Fiegl, E. Muller-Holzner, M. Widschwendter and M. Ehrlich, *Cancer Biol. Ther.*, 2004, **3**, 1225–1231.
158. A. Narayan, W. Ji, X. Y. Zhang, A. Marrogi, J. R. Graff, S. B. Baylin and M. Ehrlich, *Int. J. Cancer*, 1998, **77**, 833–838.
159. S. L. Zhou and L. D. Wang, *World J. Gastroenterol.*, 2010, **16**, 2348–2354.
160. I. Van der Auwera, R. Limame, P. van Dam, P. B. Vermeulen, L. Y. Dirix and S. J. Van Laere, *Br. J. Cancer*, 2010, **103**, 532–541.
161. A. M. Abdel Rahman, S. D. Kamath, S. Gagne, A. L. Lopata and R. Helleur, *J. Proteome Res.*, 2013, **12**, 647–656.
162. T. L. Mertz, *J. Am. Osteopath. Assoc.*, 2011, **111**, S27–S29S32.
163. M. J. Chao, S. V. Ramagopalan, B. M. Herrera, M. R. Lincoln, D. A. Dyment, A. D. Sadovnick and G. C. Ebers, *Hum. Mol. Genet.*, 2009, **18**, 261–266.
164. F. Li, S. Li, H. Chang, Y. Nie, L. Zeng, X. Zhang and Y. Wang, *Genet. Test. Mol. Biomarkers*, 2013, **17**, 656–661.
165. D. R. Taylor, *Thorax*, 2009, **64**, 261–264.
166. M. G. Tektonidou and M. M. Ward, *Nat. Rev. Rheumatol.*, 2011, **7**, 708–717.
167. G. B. Lim, *Nat. Rev. Cardiol.*, 2012, **9**, 672.
168. I. Rahman, R. Atout, N. L. Pedersen, U. de Faire, J. Frostegard, E. Ninio, A. M. Bennet and P. K. Magnusson, *Atherosclerosis*, 2011, **218**, 117–122.
169. S. Sour, M. Belarbi, D. Khaldi, N. Benmansour, N. Sari, A. Nani, F. Chemat and F. Visioli, *Br. J. Nutr.*, 2012, **107**, 1800–1805.
170. I. Okur, L. Tumer, F. S. Ezgu, E. Yesilkaya, A. Aral, S. O. Oktar, A. Bideci and A. Hasanoglu, *J. Pediatr. Endocrinol. Metab.*, 2013, 1–6.
171. M. Kucukazman, N. Ata, B. Yavuz, K. Dal, O. Sen, O. S. Deveci, K. Agladioglu, A. O. Yeniova, Y. Nazligul and D. T. Ertugrul, *Eur. J. Gastroenterol. Hepatol.*, 2013, **25**, 147–151.
172. S. G. Wannamethee, P. H. Whincup, L. Lennon and N. Sattar, *Arch. Inter. Med.*, 2007, **167**, 1510–1517.
173. W. Mullen, C. Delles and H. Mischak, *Curr. Opin. Nephrol. Hypertens.*, 2011, **20**, 654–661.
174. *Duke Med. Health News*, 2012, **18**, 4–5.
175. K. Nakanishi and C. Watanabe, *Clin. Chim. Acta*, 2009, **402**, 171–175.
176. A. Maisel, S. X. Neath, J. Landsberg, C. Mueller, R. M. Nowak, W. F. Peacock, P. Ponikowski, M. Mockel, C. Hogan, A. H. Wu, M. Richards, P. Clopton, G. S. Filippatos, S. Di Somma, I. Anand, L. L. Ng, L. B. Daniels, R. H. Christenson, M. Potocki, J. McCord, G. Terracciano, O. Hartmann, A. Bergmann, N. G. Morgenthaler and S. D. Anker, *Eur. J. Heart Failure*, 2012, **14**, 278–286.
177. D. P. Dosanjh, M. Bakir, K. A. Millington, A. Soysal, Y. Aslan, S. Efee, J. J. Deeks and A. Lalvani, *PLoS One*, 2011, **6**, e28754.
178. R. Ribeiro-Rodrigues, T. Resende Co, J. L. Johnson, F. Ribeiro, M. Palaci, R. T. Sa, E. L. Maciel, F. E. Pereira Lima, V. Dettoni, Z. Toossi, W. H. Boom, R. Dietze, J. J. Ellner and C. S. Hirsch, *Clin. Diagn. Lab. Immunol.*, 2002, **9**, 818–823.

179. M. Sakamoto, *J. Gastroenterol.*, 2009, **44**(Suppl 19), 108–111.
180. I. Simon, Y. Liu, K. L. Krall, N. Urban, R. L. Wolfert, N. W. Kim and M. W. McIntosh, *Gynecol. Oncol.*, 2007, **106**, 112–118.
181. J. N. Mubiru, A. J. Valente and D. A. Troyer, *Prostate*, 2005, **65**, 117–123.
182. D. J. Birnbaum, S. Laibe, A. Ferrari, A. Lagarde, A. J. Fabre, G. Monges, D. Birnbaum and S. Olschwang, *Transl. Oncol.*, 2012, **5**, 72–76.
183. F. Al Dayel, *J. Infect. Public Health*, 2012, **5**(Suppl 1), S31–S34.
184. H. Takeda, N. Takigawa, K. Ohashi, D. Minami, I. Kataoka, E. Ichihara, N. Ochi, M. Tanimoto and K. Kiura, *Exp. Cell Res.*, 2013, **319**, 417–423.
185. S. Bernholtz, Y. Laitman, B. Kaufman, S. Shimon-Paluch and E. Friedman, *Breast Cancer Res. Treat.*, 2012, **132**, 669–673.
186. M. Gonen, Z. Sun, M. E. Figueroa, J. P. Patel, O. Abdel-Wahab, J. Racevskis, R. P. Ketterling, H. Fernandez, J. M. Rowe, M. S. Tallman, A. Melnick, R. L. Levine and E. Paietta, *Blood*, 2012, **120**, 2297–2306.
187. R. Claus, B. Hackanson, A. R. Poetsch, M. Zucknick, M. Sonnet, N. Blagitko-Dorfs, J. Hiller, S. Wilop, T. H. Brummendorf, O. Galm, U. Platzbecker, J. C. Byrd, K. Dohner, H. Dohner, M. Lubbert and C. Plass, *Int. J. Cancer*, 2012, **131**, E138–E142.
188. P. A. Ho, M. A. Kutny, T. A. Alonzo, R. B. Gerbing, J. Joaquin, S. C. Raimondi, A. S. Gamis and S. Meshinchi, *Pediatr. Blood Cancer*, 2011, **57**, 204–209.
189. C. A. Krusche, P. Wulfing, C. Kersting, A. Vloet, W. Bocker, L. Kiesel, H. M. Beier and J. Alfer, *Breast Cancer Res. Treat.*, 2005, **90**, 15–23.
190. W. M. Linehan, *Genome Res.*, 2012, **22**, 2089–2100.
191. J. Hur, H. J. Lee, J. E. Nam, Y. J. Kim, Y. J. Hong, H. Y. Kim, S. K. Kim, J. Chang, J. H. Kim, K. Y. Chung, H. S. Lee and B. W. Choi, *BMC Cancer*, 2012, **12**, 392.

16 Next-Generation Molecular Markers: Challenges, Applications and Future Perspectives†

Mukesh Verma[a], Debmalya Barh*[b], and Syed Shah Hassan[c]

[a]Epidemiology and Genomics Research Program, Division of Cancer Control and Population Sciences, National Cancer Institute (NCI), National Institutes of Health (NIH), 9609 Medical Center Drive, Rockville, MD 20850, USA; [b]Centre for Genomics and Applied Gene Technology, Institute of Integrative Omics and Applied Biotechnology (IIOAB), Nonakuri, Purba Medinipur, West Bengal-721172, India; [c]Departamento de Biologia Geral, Instituto de Ciências Biológicas (ICB), Universidade Federal de Minas Gerais, Pampulha, Belo Horizonte, Minas Gerais, Brazil
*E-mail: dr.barh@gmail.com

16.1. Introduction

16.1.1. Biological Limitations in Cancer Biomarker Discovery

Metabolomics represents the phenotypic state of a disease.[1] Here we have described challenges in translational research when metabolomic biomarkers are used for disease diagnosis, risk assessment and prognosis. One barrier is a paucity of commercially available

†This chapter has been reproduced from *Molecular Biology and Biotechnology*, ed. R. Rapley and D. Whitehouse, 2014, 6th edition, copyright The Royal Society of Chemistry.

Genomics and Clinical Diagnostics
Edited by David Whitehouse and Ralph Rapley

Published by the Royal Society of Chemistry, www.rsc.org

standards for identification and quantification of metabolites for humans, as well as data comparisons across studies.[2] Particular consideration should be given to the quantity needed and which compounds should be included. Additionally, the lack of carefully selected, well-annotated and easily accessible reference samples greatly limits investigations. Although the standard National Institute of Standards and Technology (NIST) pooled plasma reference sets are available, they are of limited value in determining the individual metabolite variation. Standards for other biological media, such as urine, are needed for investigating the most physiologically plausible pathways that best reflect the aetiology of disease. In the case of epidemiologic consortia, different laboratories may be involved in sample analyses; and in turn, inter-laboratory comparisons are problematic without reference samples. An opportunity surrounds the use of archived samples from population-based studies to gain insights into optimizing collection and storage protocols for different media for sample integrity. In addition to high-quality samples, quantitative robustness requires provision for quality controls, pooled references and standard reference materials to control for instrumentation variability drift and allow for comparisons across laboratories. Well-standardized protocols for sample collection, storage and analysis are also needed.

Cancer cells show progressive heterogeneity at the cellular and molecular levels. Transient expression of various intermediate components (proteins, genes, metabolites *etc.*) during intermediate stages of carcinogenesis affects detection of biomarkers, resulting in false positives or negatives. Similarly, biomarkers may get affected in response to therapy and other internal physiological and pathological factors, including age of the subject. Other influential factors in biomarker variation include food habit, nutrition and life style.

16.1.2. Clinical and Pathological Factors in Cancer Biomarker Discovery

Although a considerable amount of knowledge has been obtained in understanding disease biology and identification of biomarkers that can detect a disease, utilization of that information in clinical settings is still challenging. Clinical validation is the main hurdle in the process. Just to make our point clear we describe an example in cancer. In one case–control study of the European Prospective Investigation into Cancer and nutrition (EPIC) where more than 300 breast cancer patients and matched controls were tested for breast cancer over a period of three years using a panel of eight serum markers (osteopontin, haptoglobin, cancer antigen 15-3, carcinoembryonic antigen, cancer antigen-125, prolactin, cancer antigen 19-9 and alpha-fetoprotein), very low specificity (50%) and sensitivity (50%) was observed.[3] This may be due to the different subtypes of breast cancer present in the collected samples. Such epidemiological studies should select a broader target set of potential biomarkers, which could be enabled by antibody array technologies where profiles of up to 100 antibodies can be followed simultaneously. Making different groups, based on subtypes of cancer on the basis of the status of hormone receptors (oestrogen and progesterone), might also be helpful.

Although current biomarkers can differentiate normal *versus* cancer conditions; more accurate detection of each stages of cancer, such as initiation, progression, metastasis, early and aggressive stages and recurrence, are required to be developed.

16.1.3. Analytical Limitations in Cancer Biomarker Discovery

Analytical limitations exist for different biomarkers mainly because they have not been validated. Another area that needs progress and attention is the cost and high throughput. Systematic and adequate progress has not been made in the application of biomarkers in clinical settings. Proper analytical and clinical validation of early markers has not been achieved and the number of biomarkers approved by the Food and Drug Administration (FDA) is very small for clinical use.[4] Clinical validation of identified biomarkers is especially the key challenge in the field. Here is one example to emphasize our point. The National Cancer Institutes have developed guidelines for the analytical and clinical validation of biomarkers but none of the biomarkers has been validated to date.[5] Integration of genomic and proteomic markers with epigenetic markers may help us in subtyping different cancers and cancer stages (and this is true for other diseases also).[6] Values of biomarkers may be different in different tissues. For example, results of methylation profiling from blood and tissues are different. Koestler *et al.* conducted a systematic epigenome-wide methylation analysis and demonstrated that shifts in leukocyte subpopulations might account for a considerable proportion of variability in these patterns.[7] Multiplexing of biomarkers may reduce false positive results in screening studies where the intention is to identify populations which are at high risk of developing a disease. Quantitative imaging data storage and maintenance have their own challenges as we discussed above. Whether expression of microRNAs is localized in a specific part of the diseased tissue has to be carefully evaluated.[2] In a tissue biopsy the local concentration (number of microRNAs) may be low or high. Determining the accurate level of microRNAs is very critical.

Although opportunities exist for the use of metabolomic profiling in population-based research, multiple challenges prevent the proper integration of these data into epidemiological studies for meaningful interpretation.[1,2] Therefore, epidemiologists should strive to understand the principles of metabolomics to better determine and apply their appropriate uses. In addition, it is essential for metabolite profiles to be validated, both for analytical purposes and clinical use as biomarkers. Some of the primary challenges to population-based studies include the incorporation of this technology in the initial study design, identifying appropriate sample collection protocols and quality control methods and the selection of analytical approaches and quantification techniques. New strategies will probably be necessary for combining data from different analytical platforms to allow generalizability of data and interpretation. Furthermore, the need to develop better methods for analysing large amounts of data, including improvement of statistical and bioinformatics methods for data analysis, remains a critical barrier.

Unfortunately, metabolomic profiling is used far too frequently *post hoc* in epidemiology studies. Investigators should strive for well-designed, prospective studies that would establish causal effect, as well as temporal changes. In any case, causality is difficult to establish with a clear biological explanation in association studies, even with particularly well-designed studies. Therefore, it is critical to improve methodologies for integration of metabolomics data with other data, such as genomic and proteomic data, to help understand the functional and temporal relationship between a biomarker and an effect.

In general the concentration of biomarkers is very much less in specimens collected from patients, therefore quantification of markers requires highly sensitive assays.[2] Standard procedure, reference materials and quality control need to be strictly followed to ensure accuracy as well as reproducibility of the assay for a given biomarker. But, unfortunately, such good manufacturing/laboratory practice (GMP/GLP) and quality control are not followed in most cases in biomarker discovery.

16.1.4. Intellectual Property in Cancer Biomarker Discovery

The National Institutes of Health (NIH) provides an opportunity for investigators to keep intellectual property (IP) rights although the NIH supports the project. Leaving regulatory and financial, intellectual property and cultural issues aside, developing a diagnostic biomarker often requires expertise or patients that its discoverer may not possess. We very well know that discovery, development and validation of biomarkers are costly and there is no assurance that a project will succeed. Therefore, investors are often not interested investing in such projects. However, investors may agree if they are assured good return on their investment by means of patent protection of novel biomarkers of high sensitivity, efficacy and clinical utility.

16.1.5. Health Economy Factors in Cancer Biomarker Discovery

For a better outcome of treatment, the procedures should be inexpensive. Most of the novel and useful biomarkers and assay methods are patent protected. Therefore, the cost of diagnostic and prognostic assays using such biomarkers is very high and usually not affordable by common people. Cost effective assays need to be developed to meet the needs of common people.

16.2. Next-Generation Molecular Markers and Their Applications

Affinity reagents are considered the most significant tools for biomarker discovery. Affimers are a type of affinity reagents, which are relatively small (13 kDa), non-post-translationally modified, biophysically stable protein scaffolds.[8] They contain three variable regions into which distinct peptides are inserted. This structure then binds to proteins and other molecules the way antigen and antibody bind. So far 20 000 affimer arrays have been tested in serum of rheumatoid arthritis patients, and elevated levels of biomarkers, such as C-reactive protein, have been identified along with more than 20 new biomarkers. The detection and treatment of different diseases is greatly facilitated by the omics technologies. For example, in cancer, genomics analysis provides clues about gene regulation and gene knockdown for cancer management. Discovery of the involvement of microRNAs in human cancers has opened a new page for cancer researchers.[9,10]

The latest are Mammaprint and Oncotype DX, which, are being used in clinics, confirming the feasibility of multiplexing of techniques.[11,12] Some therapeutic drugs, called epigenetic drugs, target DNA methylation and histone deacetylation in solid tumours and leukaemia. These drugs work well with poly ADP ribose polymerase (PARP) inhibitors.[13] Proteomics also plays an important role in cancer biomarker discovery, and quantitative proteome–disease relationships provide a means for connectivity analysis.[14] Fluorescent dye enables a more reliable analysis, and it facilitates the progress of biochips and cytomics. The huge amount of information collected by multiparameter single-cell flow or slide-based cytometry measurements serves to investigate the molecular behaviour of cancer cell populations. Diseases selected in this article are very appropriate for metabolite profiling owing to their unique biochemical properties.

In Figure 16.1, we have presented a schematic approach to demonstrate factors contributing to disease development and approaches to isolate clinical samples and analyse biomarkers for disease detection, diagnosis, prognosis, and drug response. Table 16.1 highlights key biomarkers in different diseases and their application in detection, diagnosis, screening, prognosis and treatment response.

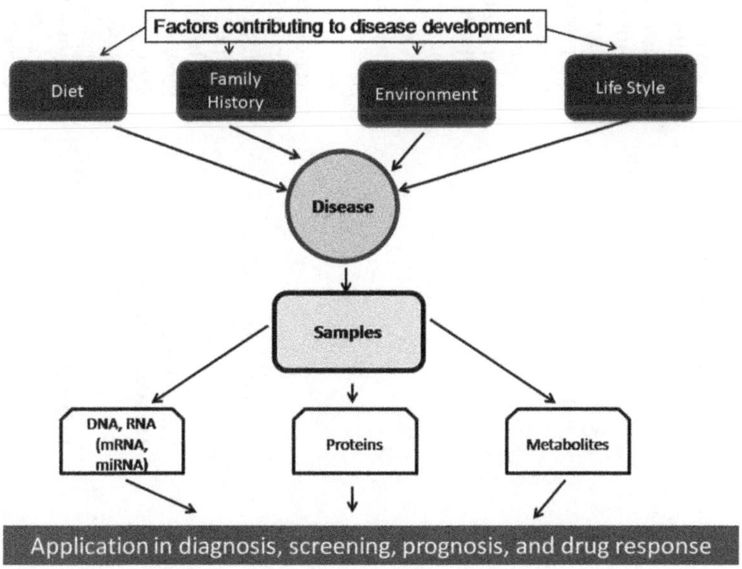

Figure 16.1 Schematic diagram to demonstrate factors contributing to disease development and approaches to isolate clinical samples and analyze biomarkers for disease detection, diagnosis, prognosis and drug response.

Table 16.1 Next-generation molecular markers for diagnostic screening, early diagnosis, risk assessment and prognosis.

Category	Diseases	Markers
Diagnostic screening	Immunological diseases	Arginine kinase, sarcoplasmic calcium binding proteins, tropomyosin.[15]
	Cardiovascular disease (CVD)	Lipoprotein-associated phospholipase A2, antiphosphorylcholine IgM.[16]
	Metabolic diseases	Glycine, glutamine, glycerophosphatidyl choline.[18]
	Infectious diseases	Hepatitis B virus (HBV), hepatitis C virus (HCV), certain strains of the human papillomavirus (HPV), Epstein–Barr virus (EBV), human immunodeficiency virus type 1 (HIV-1), human T-cell lymphotropic virus type-1 (HTLV-1) and the Gram-negative bacterium *Helicobacter pylori.*[24]
	Neurological diseases	Proteomic profiling of serum (multiple biomarkers).[29]
	Cancer	Gene promoter hypermethylation in death-associated protein kinase 1 (*DAPK1*), *p16* and *RASSF1A*1 could be used as biomarkers for detection of head and neck squamous cell carcinomas (HNSCCs); prostate-specific antigen (PSA) for prostate cancer; serum markers (PAP, tPSA, fPSA, proPSA, PSAD, PSAV, PSADT, EPCA and EPCA-2), tissue markers (AMACR, methylated GSTP1 and the TMPRSS2–ETS gene rearrangement) and a urine marker (DD3PCA3/UPM-3) for prostate cancer.[35–42] Serum tumour markers like carcinoembryonic antigen (CEA) and cancer antigen (CA) CA 15.3, CA 27.29 for breast and ovarian cancer.[44]
		Using saliva as the starting material, mRNAs specific for oesophageal, head and neck, oral and lung cancer also were identified.[48–50] MSI biomarkers were reported in a number of cancers, including bladder, colon, oesophageal and skin cancer.[54–56]
Early diagnosis	Immunological diseases	Glutathione S transferase M1 (GSTM1) biomarker was identified which helped in screening a cohort of children who were at high risk of developing asthma;[73] C-reactive protein, fibrinogen and interleukin-6 for screening.[74]
	Cardiovascular disease (CVD)	Flow-mediated dilatation (FMD) and IMT for atherosclerosis[60]
	Metabolic diseases	Dysglycaemia, alpha-hydroxybutyrate, linoleoyl glycerophosphocholine, advanced glycation end products (AGE), albumin excretion rate (AR) and SNPS in epidermal growth factor gene intron 2 (ref. 1, 62 and 63).
	Infectious diseases	Region-of-difference-1 (RD-1) gene product and sputum cytokine levels for early detection of tuberculosis;[64] HSP70, CAP2, glypican 3 and glutamine synthase[65] for HCC.
	Neurological diseases	Beta amyloid for early detection of Alzheimer's disease[66]
	Cancer	B7-H4, spondin 2 and DcR3 were identified as early biomarkers in ovarian cancer.[68–70]

Risk assessment (screening)	Immunological diseases	C-reactive protein, fibrinogen and interleukin-6 (ref. 74).
	Cardiovascular disease (CVD)	C-reactive protein.[75]
	Metabolic diseases	Alanine amino transferase (ALT), gamma glutamyl transferase (GGT), triglyceride, plasminogen activator inhibitor (PAI-1) antigen, ferritin, C-reactive protein (CRP), sex-hormone binding globulin (SHBG), ALT and GGT for diabetes.[78]
	Infectious diseases	HIV infection for AIDS;[80,81] gamma interferon, p2X7 polymorphism and mycobacteria antigens for tuberculosis.[82,83]
	Neurologic diseases	ABCA1 gene polymorphism and apolipoprotein E genotyping for Alzheimer's disease.[84–86]
	Cancer	Methylation levels of genes NKX-25, CLSTN1, SPOCK2, SLC16A12, DPYS and NSE1 for screening prostate, colon and breast cancer[88]
Prognostic markers	Immunological diseases	C-telopeptides I and II (CTX-1 and CTX-2) for rheumatoid arthritis (RA)[92]
	Cardiovascular disease (CVD)	Osteoprotegerin (OPG) for atherosclerosis.[95]
	Metabolic diseases	Alpha-hydroxybutyrate (Alpha-HB) and linoleoyl-glycerophosphocholine (L-GPC) for insulin resistance (IR) and glucose intolerance (GI).[1]
	Infectious diseases	Faecal calprotectin in infectious diarrhoea.[96] In a case–control study, both bacterial and viral infection was involved. Main bacteria were *Salmonella*, *Campylobacter*, *Yersinia* and *Shigella* and viruses were rotavirus, norovirus and adenovirus.
	Neurological diseases	MRI and proteomic based biomarkers in spinal fluid for Alzheimer's disease.[97]
	Cancer	Comparative genomic hybridization (CHG) identified complex genetic variants associated with adverse prognosis in cancer;[98] cytogenetics biomarkers for breast, head and neck, lung, liver and ovarian cancers.[99–102]

16.2.1. Diagnostic Screening

16.2.1.1. Immunological Diseases

Asthma involves inflammation, hyperresponsiveness, bronchoconstriction and symptoms (episodic breathlessness, wheezing, coughing, tightness of the chest and shortness of the breadth). Proteomics approaches identified allergenic proteins, based on their reactivity to patient's sera, using tandem mass spectrometry.[15] The most significant allergens identified were arginine kinase, sarcoplasmic calcium binding proteins and tropomyosin, which can be used for screening. This study also emphasized that mass spectroscopy (MS) analysis is a sensitive and accurate tool in identifying and quantifying aerosolized allergens.

16.2.1.2. Cardiovascular Disease (CVD)

In cardiovascular disease (CVD), the relative contribution of genetic and environmental effects was studied for two inflammatory biomarkers, lipoprotein-associated phospholipase A2 and antiphosphorylcholine IgM, in a Swedish population.[16] The results indicated that lipoprotein-associated phospholipase A2 had low heritability and higher environmental regulation. Therefore, for diagnostic screening of CVD, biomarkers should be selected based on the context (family history, exposure history, life-style, *etc.*). Non-invasive technologies, transcranial Doppler, magnetic resonance and computed tomography, were used for the diagnosis of intracranial atherosclerosis.[17] In these cases images were considered as biomarkers.

16.2.1.3. Metabolic Diseases

Metabolomic profiles can be used as biomarkers and can indicate the development of obesity, which, in turn, affects diabetes, cardiovascular disease, liver disease, renal disease and selected cancers. Metabolomic analysis is a valid and powerful tool that helps us to understand the mechanisms underlying different diseases. Biomarkers identified in obesity included glycine, glutamine and glycerophosphatidyl choline in serum.[18]

16.2.1.4. Infectious Diseases

Potential applications of biomarkers in infectious diseases include distinguishing bacterial from non-bacterial infection, monitoring response to treatment and predicting survival (outcome). Periodontal disease, where aerobic and anaerobic bacteria infect gums, was proposed as a biomarker and risk factor for CVDs, especially abnormal function of progenitor endothelial cells.[19] Another group demonstrated an association of the oral microbiome with gastrointestinal cancer.[20] In such situations oral microbiome data was considered a biomarker for gastrointestinal cancer. Procalcitonin (PCT), a proinflammatory biomarker, is an excellent early detection biomarker of invasive bacterial infection in febrile children evaluated in an emergency department.[21]

Infectious agents are associated with at least 15% of cancers.[22] Chronic infections are the second most preventable cause of cancer,[23] and about 18% of the global cancer burden has been attributed to infectious agents.[24] The presence of an infectious agent in tumour tissue is not sufficient to establish it as a causal agent; but the agent is termed carcinogenic if evidence from epidemiological, clinical and biological studies indicates a strong cancer aetiology.[25] The International Agency for Research on Cancer (IARC) has identified seven major infectious agents as carcinogenic: hepatitis B virus (HBV), hepatitis C virus (HCV), certain strains of the human papillomavirus (HPV), Epstein–Barr virus (EBV), human immunodeficiency virus type1 (HIV-1), human T-cell lymphotropic virus type-1 (HTLV-1) and the Gram-negative bacterium *Helicobacter pylori*.[24] Aflatoxin B1 (AFB1), the product of the common mould *Aspergillus flavus* was also considered in this category because the IARC has identified AFB1 as a chemical liver carcinogen.[26] Hepatitis C-infected individuals can be distinguished from healthy people on the basis of metabolomic profiling (which can be considered a biomarker for screening).[27] It was demonstrated that interleukin (IL)-18 levels in serum could be used as a biomarker for acute Epstein–Barr virus (EBV) infection.[28]

16.2.1.5. Neurological Diseases

Metabolomics-based technologies (liquid chromatography quadrupole time-of-flight mass spectrometry) and chemometrics were used to identify Alzheimer's disease (AD)-specific profiles.[29] Samples from healthy age-matched controls were also used to identify new biomarkers (profiles). The prediction value of AD was 94–97% in this metabolomic study, which was considered quite remarkable for early detection of AD. The confirmation of these results was done in another set of patients and controls and 100% diagnosis was achieved.[29] Such studies are significant for understanding the aetiology, pathophysiology and treatment of degenerative brain disorders.[30]

16.2.1.6. Cancer

We discuss different tumour types in this section and the next generation of biomarkers identified in different tumour types. Cancer is a genetic and epigenetic disease, and methylation biomarkers have been used in diagnostic screening of cancer.[31,32] Methylation changes at the precarcinoma stage could be used as biomarkers for diagnosis in cancer if these epigenetic changes were not present in normal cells.[32,33] With the development of high-throughput next-generation sequencing, it has become easier to detect methylation changes qualitatively as well as quantitatively, and, recently, global methylation profiling has also been accomplished.[34] Gene promoter hypermethylation in death-associated protein kinase 1 (*DAPK1*), *p16* and *RASSF1A*1 could be used as biomarkers for detection of head and neck squamous cell carcinomas (HNSCCs). The development of biomarkers for prostate cancer screening, detection and prognostication revolutionized the management of this disease. During the progression of cancer, it was observed that the levels of certain proteins were elevated. These abnormally increased molecules could be used as biomarkers for gaining an insight into the course of the disease. Prostate-specific antigen (PSA) is a useful, though not specific, biomarker for detecting prostate cancer. Serum markers (PAP, tPSA, fPSA, proPSA, PSAD, PSAV, PSADT, EPCA and EPCA-2), tissue markers (AMACR, methylated GSTP1 and the TMPRSS2–ETS gene rearrangement), and a urine marker (DD3PCA3/UPM-3) of prostate cancer are very well characterized biomarkers.[35–42] Disease-specific protein biomarkers also were identified[43] and characterization of such biomarkers in bodily fluids could aid in the early detection of cancer and help in monitoring cancer progression. It has become easier to identify cancer-specific markers with the advancement in the available technologies. In the case of breast cancer, some of the serum tumour markers, like carcinoembryonic antigen (CEA) and cancer antigen (CA) CA 15.3 and CA 27.29, were not confirmed to be sensitive for early detection; however, their high levels did reflect disease progression and recurrence.[44] Mammaglobin and MASPIN (mammary serine protease inhibitor) were shown to be diagnostic biomarkers.[45] The early detection of circulating breast cancer cells by morphological methods is currently being challenged by ultrasensitive proteomic and PCR-based methods, often enhanced by immunomagnetic-bead-based cell capture.[46] The role of MASPIN in diagnosis and prognosis of a variety of cancers is being explored.[47]

It has been reported that low or absent MASPIN cytoplasmic expression was frequently observed in oral carcinomas with lymph node metastasis.

Using saliva as the starting material, mRNAs specific for oesophageal, head and neck, oral and lung cancer were also identified.[48–50] Seven biomarkers showed 3.5-fold increases in their levels in oral SCC patients compared with healthy individuals. Further validation of values is needed before implementation in clinics.

Microsatellites are repeated sequences of DNA.[51,53] When these repeats are shortened this is called microsatellite instability (MSI). This stage arises in regions where the mutation rate is very high, mainly due to defective DNA repair genes, such as mismatch repair gene. MSI biomarkers have been reported in a number of cancers, including bladder, colon, oesophageal and skin cancer.[54–56] As high as 30% of HNSCC patients had MSI,[52] indicating that MSI is a good biomarker for HNSCC. A number of studies evaluated the potential of detecting these markers in tumour samples as well as in other biological specimens and found these biomarkers to be useful for diagnosis and prognosis.[57]

16.2.2. Early Diagnosis

16.2.2.1. *Immune Diseases*

Airways inflammation starts soon after inception of exposure to allergen in asthma. Inflammatory cytokines, along with arginine kinase and sarcoplasmic calcium binding proteins, are the most practical biomarkers for early diagnosis of asthma.[15,58]

16.2.2.2. *Cardiovascular Diseases*

For early diagnosis of carotid atherosclerosis, for which obesity is the risk factor, the early biomarkers are carotid intima media thickness (IMT) and circulating oxidized low-density lipoprotein.[59] To evaluate endothelium status, flow-mediated dilatation (FMD) and IMT were used as early biomarkers of atherosclerosis in patients with non-alcoholic fatty liver disease (NAFLD).[60]

16.2.2.3. *Metabolic Diseases*

A number of markers have been described for early diagnosis of diabetes.[1,61] A few of them are dysglycaemia, alpha-hydroxybutyrate, linoleoyl glycerophosphocholine, advanced glycation end products

(AGE), albumin excretion rate (AR) and single nucleotide polymorphisms (SNPs) in epidermal growth factor gene intron 2.[1,62,63]

16.2.2.4. Infectious Diseases

Region-of-difference-1 (RD-1) gene product and sputum cytokine levels are considered biomarkers for early detection of tuberculosis.[64] Hepatocellular carcinoma (HCC) involves infectious agents and the early diagnostic biomarkers for HCC are HSP70, CAP2, glypican 3 and glutamine synthase.[65] These biomarkers were identified on the basis of the results from gene expression analysis of HCC samples.

16.2.2.5. Neurological Diseases

The bioinformatic approach has helped in identifying early diagnosis biomarkers for Alzheimer's disease. On the basis of data from patients and matched controls, an artificial neural network (ANN) identified beta amyloid cascade as a potential biomarker for early detection of Alzheimer's disease.[66]

16.2.2.6. Cancer

The key to successful application of biomarkers is to detect the disease early so that a variety of treatment approaches can be applied.[67] Initially it was thought that genetic changes arise first at the time of initiation of a disease but genome-wide profiling indicates that epigenetic changes occur much earlier than genomic changes.[33] Technologies exist to follow up these changes for early detection. B7-H4, spondin 2 and DcR3 were identified as early biomarkers in ovarian cancer although other investigators have proposed other groups of biomarkers.[68–70]

16.2.3. Risk Assessment (Screening)

Genome-wide association studies (GWAS) are extremely powerful in identifying low-penetrance new SNPs (biomarkers) which may have therapeutic implications.[71] Identification of common low-susceptibility alleles is useful because it provides possible insights into the mechanisms of tumour biology in the case of cancer and for identifying high-risk individuals.[72]

Associations do not necessarily mean causality; the potential for confounding and reverse causality should always be kept in mind. It will be beneficial to predict complications, in the case of diabetes in particular and other diseases in general, and to determine subgroups that may be responsive to therapy. Clinical implications should not be exaggerated while characterizing prediction biomarkers.

16.2.3.1. Immune Diseases

In one meta-analysis glutathione *S* transferase M1 (GSTM1) biomarker was identified, which helped in screening a cohort of children who were at high risk of developing asthma.[73] In a population-based study C-reactive protein, fibrinogen and interleukin-6 were found to be useful biomarkers for screening.[74]

16.2.3.2. Cardiovascular Diseases

In a risk-prediction model of cardiovascular disease, C-reactive protein (CRP) levels were investigated.[75] However, after the completion of studies, it was suggested that evaluation of the potential influence of CRP levels would require studies to quantify the effects of additional CRP assessment on medical decision making. Biomarkers studies of CVD prediction in the elderly indicated that mortality and cardiovascular events were dependent on low peripheral pulse pressure not on high blood pressure.[76] Genetic variants in CVD were studied to evaluate their association with the disease but with limited success.[77]

16.2.3.3. Metabolic Diseases

For diabetes, alanine amino transferase (ALT), gamma glutamyl transferase (GGT), triglyceride, plasminogen activator inhibitor (PAI-1) antigen, ferritin, C-reactive protein (CRP) and sex-hormone binding globulin (SHBG) were identified as prediction markers after a large study "West of Scotland Coronary Prevention Study (WOSCOPS)" was conducted. Results from this study were validated in a different set of populations, and two biomarkers, ALT and GGT, looked very promising for risk assessment.[78] CRP was not causally related to insulin resistance or obesity. IL-6 upregulation was also associated with diabetes. Early diagnosis of chronic kidney diseases and atherosclerosis in same subjects could be accomplished by carotid ultrasound technology.[79]

16.2.3.4. Infectious Diseases

HIV/AIDS screening involves behavioural biomarkers as well as molecular omics biomarkers and a number of investigators have reported the screening strategies and their outcome in different populations.[80,81] In tuberculosis, gamma interferon, p2X7 polymorphism and mycobacteria antigens were used as biomarkers for risk assessment and screening.[82,83]

16.2.3.5. Neurological Diseases

The most common biomarkers used for risk assessment of Alzheimer's disease were ABCA1 gene polymorphism and apolipoprotein E genotyping; although a combination of markers also was used, synergism was lacking.[84–86]

16.2.3.6. Cancer

Transcriptomic and micro RNA (miRNA) biomarkers were used for screening colorectal cancer.[87] For screening prostate, colon and breast cancer, methylation levels of the genes NKX-25, CLSTN1, SPOCK2, SLC16A12, DPYS and NSE1 were used.[88] Malignant melanoma is one of the most aggressive types of tumour. Because malignant melanoma is difficult to treat once it has metastasized, early detection and treatment are essential. The search for reliable biomarkers of early-stage melanoma, therefore, has received much attention. By using an approach of screening tumour antigens and their auto-antibodies, bullous pemphigoid antigen 1 (BPAG1) was identified as a melanoma antigen recognized by its auto-antibody when anti-BPAG1 auto-antibodies were detected in melanoma patients at both early and advanced stages of disease.[89]

16.2.4. Prognostic Markers

Prognostic biomarkers should be tightly linked to the outcome so that they can be used as surrogate measures of efficacy and treatment response. Prognostic markers can be defined as factors that can predict an outcome in the absence of systemic therapy or predict an outcome different from that for patients who are devoid of the biomarker, despite empiric therapy (initiating treatment before confirming diagnosis).[90] It is not known whether there is any association between ovarian cysts and oestrogen levels during tamoxifen use. A very well designed study was conducted where breast cancer prognostic markers were utilized to evaluate the effect of tamoxifen use in premenopausal women with ovarian cysts and results indicated an association.[91] Hence, prognostic markers can be utilized to classify patients into appropriate groups for treatment.

16.2.4.1. *Immune Diseases*

Biomarkers of bone and cartilage turnover are collagen C-telopeptides I and II, which are predictors of structural damage in rheumatoid arthritis (RA) patients. C-terminal cross-linked telopeptide of type I collagen (CTX-I) could be measured in serum or urine after it was released during bone resorption. Bone erosions and osteoporosis both occurred as a consequence of RA, and CTX-I levels could be used for prognosis. On the other hand, increased CTX-II levels were associated with rapid progression of joint damage. For clinical implications, the investigators who performed these studies recommended that new prognostic biomarkers should supply information beyond that provided by risk factors in the prediction of disease outcome.[92]

European countries restrict anti-tumour-necrosis-factor therapy for RA and prescription is provided based on evaluation of inflammatory markers, C-reactive protein levels and general health of a patient.[93] Genomic variants of C-reactive proteins play a major role in therapeutics of RA and can be used as biomarkers.[94]

16.2.4.2. *Cardiovascular Diseases*

Osteoprotegerin (OPG) is a biomarker, which is independently associated not only with risk factors for atherosclerosis but also with subclinical peripheral atherosclerosis and clinical atherosclerosis and is recommended as a prognostic biomarker for ischemic heart disease and ischemic stroke.[95]

16.2.4.3. *Metabolic Diseases*

Alpha-hydroxybutyrate (AHB) and linoleoyl-glycerophosphocholine (L-GPC) were identified as biomarkers of insulin resistance (IR) and glucose intolerance (GI) in a large population study of more than 1000 participants from the Relationship between Insulin Sensitivity and Cardiovascular Disease (RISC) study.[1] AHB correlated positively and L-GPC negatively with both diseases (IR and GI), indicating that AHB was a positive predictor and L-GPC a negative predictor, independent of family history of diabetes, sex, age, fasting glucose and body mass index (BMI).

16.2.4.4. Infectious Diseases

Diarrhoea in children may involve infection. In cases of infectious diarrhoea, faecal calprotectin is a good prediction marker.[96] In a case–control study, both bacterial and viral infections were involved. The main bacteria were *Salmonella*, *Campylobacter*, *Yersinia* and *Shigella*; and viruses were rotavirus, norovirus and adenovirus. The results of these studies indicated that faecal calprotectin could be used as a non-invasive biomarker in management of children with infectious diarrhoea.

16.2.4.5. Neurological Diseases

Few biomarkers of Alzheimer's diagnosis are suitable for prognosis also. One investigator combined MRI and spinal-fluid-based biomarkers and observed better results than using single biomarkers for prognosis.[97] This investigation was performed in different sites of the brain (for MRI) and spinal fluid biomarker values were used as reference standards.

16.2.4.6. Cancer

Despite huge efforts into research for studying new biological prognostic markers, only a few out of several hundreds have progressed to clinical use. Comparative genomic hybridization (CHG) identified complex genetic variants associated with adverse prognosis in cancer.[98] Cytogenetics biomarkers turned out to be very useful biomarkers for breast, head and neck, lung, liver and ovarian cancers.[99–102] In the case of HNSCCs, loss of heterozygosity (LOH) on the distal arm of 18q was reported to be associated with poorer survival.[103] About 70% of colorectal cancers showed allelic deletions in chromosomes 18q and 17p. The *p53* gene located on 17p is mutated in about 40–60% of colorectal cancers (CRCs) and demonstrated association with prognosis and prediction of the disease. CRC patients with chromosome 18q loss showed worse disease-free and overall survival. Markers like Ki-67 staining detecting cell proliferation were significantly correlated with breast cancer outcome. One-fifth of breast cancer patients showed amplification or overexpression of cyclin D1 (*PRAD1* or bcl-1). Germline prognostic markers of bladder cancer were identified and characterized.[104] Stage

II and III CRC patients with high microsatellite instability have been reported to show improved survival and better relapse-free survival as compared with microsatellite stable (MSS) patients.

16.2.5. Drug Response

Systems immunology, based on mathematical and computational models, has the potential to predict treatment response.[105] High throughput omics technologies generate vast amount of data, which may help in identifying immunological processes at high resolution in disease development. Using the systems immunology approach it is possible to construct causal relationships between complex molecular processes and specific disease-associated phenotypes.[106] Drug response and pharmacogenomics in the context of biomarkers in different diseases are shown in Table 16.2.

16.2.5.1. Immune Diseases

The prevalence of autism is 1 in 80 in the USA. Evidence-based treatments are practiced in autism[107] although sometimes drug treatment (metabolic glutamate receptor (mGluR) antagonists and gamma aminobutyric acid (GABA) agonists) is also recommended, based on the severity and stage of the disease.[108] The aetiology of autism is not completely understood. However, synaptic maturation and plasticity in the pathogenesis of autism spectrum disorder, resulting in imbalance of excitation and inhibition, has been observed. The drugs mentioned above control excessive excitement.

16.2.5.2. Cardiovascular Diseases

Nutritional foods and their biological food components alter cardiovascular disease status and there is very well established evidence of disease-modifying effects of these compounds with anti-inflammatory and antioxidant effects in CVD patients. Reduced total low-density lipoprotein (LDL) cholesterol levels were observed when polyphenolic compounds were given to patients because bioactive phytochemicals play an important therapeutic role in attenuating oxidative damage.

16.2.5.3. Metabolic Diseases

The key players in measuring drug response in type II diabetes mellitus (T2DM) are glycaemic index (GI), glucose response curves (GRCs) and

Table 16.2 Drug response and pharmacogenomics in the context of biomarkers.

Category	Diseases	Markers
Drug response	Immunological diseases	Evidence-based treatments are practiced in autism[107] although sometimes drug treatment (mGluR antagonist and GABA agonist) is also recommended based on the severity and stage of the disease.[108]
	Cardiovascular disease (CVD)	Reduced total LDL cholesterol levels were observed when polyphenolic compounds were given to patients because bioactive phytochemicals play an important therapeutic role in attenuating oxidative damage. Nutritional foods and their biological food components alter cardiovascular disease status and have anti-inflammatory and antioxidant effects in CVD patients.
	Metabolic diseases	In a trial, a variety of foods were supplied to participants undergoing treatment with oral antidiabetic drugs (OADs).[109] The key factors in measuring drug response in type II diabetes mellitus (T2DM) are glycaemic index (GI), glucose response curves (GRCs) and daily mean plasma glucose (DMPG).
	Infectious diseases	In those patients who were given anidulafungin followed by fluconazole/voriconazole therapy, decreasing 1,3-β-D-Glucan (BG) concentrations reflected success of the treatment. BG can be used as a biomarker in invasive fungal infections (especially infections involving *Candida*) in patients undergoing treatment of candidaemia with anidulafungin.[110]
	Neurological diseases	For the treatment of schizophrenia, ionotropic glutamate receptors are targeted.[112] Recent pharmacological strategies have focused on improving cognition with drugs.[113]
	Cancer	Biomarkers of drug response may be new or the same as diagnostic or prognostic biomarkers depending on the biology and aetiology of the type of cancer studied. In different kinds of cancers pharmacological response of drugs is different and a variety of factors (genetic background, lifestyle, other diseases in the same person, BMI *etc.*) contribute to drug response.[114] Development of epigenetic drugs showed promise in selected cancer treatments.[115]

Pharmacogenomics	Immunological diseases	The response to infliximab was favourable in a group of patients where genes regulated by TNF were active compared with other participants of the study, providing strong evidence of the role that genomic background plays in responding to pharmacological drugs.[120]
		Pharmacogenomics of rheumatoid arthritis (RA) has been studied by several investigators[120,121] although RA is a highly heterogeneous disease in all aspects including response to therapy.
	Cardiovascular disease (CVD)	Infliximab treatment was provided to RA patients to reduce levels of TNF and results were evaluated based on the genomic background of the participants.[120] Host response due to inflammatory diseases, such as arthritis, depends on the genomic background, and pharmacological agents give responses based on the genomic susceptibility and some other, as yet unidentified, factors.[120–122]
	Metabolic diseases	Genetic background was considered in treatment of type 2 diabetes and an association of ADIPOR2 gene variants with CVDs and type 2 diabetes risk in individuals with defective glucose tolerance (conducted in a Finnish population) was observed.[123] Additionally, genetic predisposition and non-genetic risk factors of thioazolidine-related oedema in type 2 individuals and the pharmacogenomics of metformin were also observed.[124]
	Infectious diseases	Genetic variants at least at three loci, NAT2, CYP2E1 and GSTM1, were involved in pharmacogenomics. Pharmacogenomics of anti-TB drug-related hepatotoxicity was studied to understand the involvement of genetic background in response to treatment in TB patients.[125]
	Neurologic diseases	Genetic variants were identified in Alzheimer's disease and demonstrated treatment response.[126,127] In a recent study of an Italian cohort, response to cholinesterase inhibitors was evaluated in Alzheimer's patients using 48 SNPs.[126] Results indicated association of two SNPs with the response to treatment.
	Cancer	Promising results were obtained in colon, gastric, breast, ovarian, GI tract and lung cancer.[129–131] Polymorphism in miRNA-encoding genes also was evaluated for its implication in pharmacogenomics.[128]

daily mean plasma glucose (DMPG). In a trial, a variety of foods were supplied to participants undergoing treatment with oral antidiabetic drugs (OADs).[109] Promising results were obtained in this pilot study and validation of these observations in large population is planned.

16.2.5.4. Infectious Diseases

1,3-β-D-Glucan (BG) can be used as a biomarker in invasive fungal infections (especially infections involving *Candida*) in patients undergoing treatment of candidaemia with anidulafungin.[110] In those patients who were given anidulafungin followed by fluconazole/voriconazole therapy, decreasing BG concentrations reflected success of the treatment.

In unipolar depression, transcriptomic biomarker approaches were used to identify responders to treatment.[111] The main player in the process was tumour necrosis factor in the inflammatory cytokine pathway. These investigators evaluated genetic background of participants and suggested that SNPs rs1126757 in IL-11and rs7801617 in IL-6 play major role in response to antidepressants. This was an excellent example where genomics and transcriptomics biomarkers were used to follow up treatment.

Helicobacter pylori is the most common chronic bacterial infection in humans. This bacteria is involved not only in gastric cancer but in ulcer diseases and gastritis also. Its synergistic gastrotoxic interaction with non-steroidal anti-inflammatory drugs and association with atherosclerotic events is a matter of concern. Transmission of *H. pylori* is through oral ingestion, mainly within families in early childhood. Therefore, treatment and prevention approaches are more appropriate for children. One-week proton-pump-inhibitor-based triple therapy with clarithromycin and either amoxicillin or metronidazole is the most common therapy.

16.2.5.5. Neurological Diseases

Schizophrenia is a chronic brain disorder and affects approximately 2.5 million Americans and more than 24 million people worldwide. For the treatment of schizophrenia, ionotropic glutamate receptors are targeted.[112] Recent pharmacologic strategies have focused on improving cognition with drugs.[113] The consensus among investigators in the schizophrenia field is that physiological and pharmacological approaches should be combined for this disease for better outcome of the treatment.

16.2.5.6. Cancer

In different kinds of cancers pharmacological response of drugs is different and a variety of factors (genetic background, life style, other diseases in the same person, BMI *etc.*) contribute to drug response.[114] Biomarkers of drug response may be new or the same as diagnostic or prognostic biomarkers, depending on the biology and aetiology of the type of cancer studied. Personalized medicine is recommended for cancer because cancer is a heterogeneous disease and the rate of recurrence after the treatment is high for some cancer types. Development of epigenetic drugs showed promise in selected cancer treatments.[115]

16.2.6. Pharmacogenomics

Pharmacogenomics is the study of inter-individual genetic variability that plays a significant role in treatment response and toxicity due to the drug. Based on a better understanding of biological variability, attempts are being made to customize treatment at a personal level.[116] In pharmacogenomic approaches the following points are considered critical: pharmacokinetics-related genes and phenotypes, pharmacodynamic targets, genes and products, risk of disease related with metabolic cycle, physiological variations and environment interaction. When targeted agents matched with tumour molecular aberrations were examined in a phase I clinical trial, encouraging results were observed.[117] Since pharmacogenomics is a new science, success is not guaranteed, as happened in case of gastric cancer, where efficacy of Trastuzumab was checked in patients with advanced gastric cancer and the drug was not effective.[118] Other investigators have identified challenges in the field, including poorly coordinated diagnostic testing and current models of implication in disease stratification.[119]

16.2.6.1. Immune Diseases

Compared with other diseases, pharmacogenomics of RA has been studied by several investigators,[120,121] although RA is a highly heterogeneous disease in all aspects, including response to therapy. The response to infliximab was favourable in a group of patients where genes regulated by tumour necrosis factor (TNF) were active compared with other participants in the study, indicating strong evidence of role that genomic background plays in responding to pharmacological drugs.[120]

16.2.6.2. Cardiovascular Diseases

Host response due to inflammatory diseases, such as arthritis, depends on the genomic background, and pharmacological agents give responses based on the genomic susceptibility and some other, as yet unidentified, factors.[120–122] Infliximab treatment was provided to rheumatoid arthritis patients to reduce levels of TNF, and results were evaluated based on the genomic background of the participants.[120] The pharmacogenomics field in CVDs is still in its infancy, and during the next few years research should focus on accurately identifying, from available therapies, which is the optimal therapy for any particular individual.[121]

16.2.6.3. Metabolic Diseases

In a few studies the role of genetic background in treatment of type 2 diabetes was demonstrated, such as association of ADIPOR2 gene variants with CVDs and type 2 diabetes risk in individuals with defective glucose tolerance (conducted in a Finnish population),[123] genetic predisposition and non-genetic risk factors of thioazolidine-related oedema in type 2 individuals and the pharmacogenomics of metformin.[124] Validation in a large number of participants is still awaited.

16.2.6.4. Infectious Diseases

Pharmacogenomics of anti-TB drug-related hepatotoxicity was studied to understand the involvement of genetic background in response to treatment in TB patients.[125] Genetic variants at least at three loci, NAT2, CYP2E1 and GSTM1, were involved in pharmacogenomics. This study was conducted only in one population and it should be validated in populations with different ethnic backgrounds.

16.2.6.5. Neurological Diseases

Due to its serious effects on day-to-day life, Alzheimer's pharmacogenomics has been studied by a number of groups in the last decade and genetic variants were identified which were associated with treatment response.[126,127] In a recent study of an Italian cohort, response to cholinesterase inhibitors was evaluated in Alzheimer's patients using 48 SNPs.[126] The results indicated association of two SNPs with the response to treatment.

16.2.6.6. Cancer

The National Cancer Institute has identified pharmacoepidemiology related to pharmaceutical use and cancer risk, recurrence and survival as one of the priority areas of research. Polymorphism in microRNA-encoding genes also was evaluated for its implications in pharmacogenomics.[128] Although a considerable amount of new-targeted agents have been designed, based on cancer biology, challenges and gaps exist between pharmacogenomics knowledge and clinical application. Promising results were obtained for colon, gastric, breast, ovarian, GI tract and lung cancer.[129–131]

16.2.7. Therapeutic Biomarkers in Immune Diseases, Cardiovascular Diseases, Metabolic Diseases, Infectious Diseases, Neurological Diseases and Cancer

Since there is no measurable disease while undergoing treatment, generally therapeutic biomarkers are difficult to study. Another key point is adverse reaction of patients due to treatment. Such parameters are not needed for prognostic biomarkers; therefore, it is easier to identify prognostic biomarkers than therapeutic biomarkers. In the following section, examples of therapeutic biomarkers in different diseases are described.

16.2.7.1. Immune Diseases

In the case of multiple sclerosis, treatment with disease modifying therapies (DMT) with ability to prevent axonal damage resulted in reduction of progression of the disease from clinically isolated syndrome (CIS).[132] In another study, inclusion of information about Myxovirus-resistance-protein A (MxA) RNA during interferon treatment of multiple sclerosis resulted in a better response than interferon only treatment.[133] Other investigators also observed similar results.[134,135]

16.2.7.2. Cardiovascular Diseases

Biomarkers discussed in the diagnostic section can be used for follow up of the treatment. Dietary intervention of CVD by fish oil (salmon, herring and pompano) and other nutrients was demonstrated in a number of studies.[136] Some of the participants had higher levels of triacylglycerolaemia. Biomarkers TNF-α and IL-6 were reduced and level of adiponectin increased in the treated arm. Thus TNF-α IL-6 and adiponectin were used as therapeutic biomarkers. In another study, argon oil supplement reduced plasma levels of lipids and antioxidant status.[137] Therapeutic biomarkers used in this study were plasma vitamin E concentrations, total and LDL cholesterol and antioxidant profiles.

16.2.7.3. Metabolic Diseases

For the management of diabetes haemoglobin A1c (HbA1c) is used, which is considered a reliable indicator of glycaemic control. In most of the clinical studies in diabetes, HBA1c biomarker is used to determine glucose control.

16.2.7.4. Infectious Diseases

Diarrhoea in AIDS patients was treated with specific medications and therapeutic response was measured by levels of tubuloreticulin inclusions (TRIs).[138] Although this study may be considered an isolated study, TRIs have potential to be a therapeutic biomarker after larger studies are conducted. For pneumonia therapy of more than 100 patients, the biomarker procalcitonin (PCT) levels were very useful and this biomarker has been recommended for future prognosis.[139]

16.2.7.5. Neurological Diseases

In Parkinson's disease, glial cell-line-derived neurotrophic factor (GDNF) and its family of ligands (GDFLs) were used as biomarkers to follow up therapy.[140]

16.2.7.6. Cancer

Several cancer biomarkers are currently used to develop targeted anti-cancer drugs. Topoisomerase I inhibition in colorectal cancer is one example of targeted therapeutics and utilization of biomarkers.[141] In leukaemia, an antibody therapeutics approach was applied, using rituximab, and therapeutic biomarkers Ki-67 and PIM1 were followed for their response.[142] Results indicated that Ki-67 was an independent biomarker. In another study CD20 biomarker was used to evaluate the therapeutic potential of rituximab in B-cell lymphoma.[143] In B-cell non-Hodgkin's lymphoma (NHL), treatment with the same agent, IL-6 plasma levels were followed to evaluate drug response.[144] Eradication of NHL was achieved with a monoclonal antibody therapy combining rituximab with a blocking anti-CD40 antibody.[145] Other examples of treatment are trastuzumab for breast cancer (target p185neu), gemtuzumab for acute myeloid leukaemia (AML) (target CD33), ibritumomab for AML (target CD2090Y), edrecolomab for colorectal cancer (target EpCAM), tositumomab for NHL (target CD20), and cetuximab for colorectal cancer (target EGFR). Cisplatin is a widely used chemotherapeutic in head and neck cancer patients. However, the effect of the drug is limited by the resistance observed in these patients; and it has been seen that hypermethylation in some of the genes responsible for cytotoxicity, for example the *S100P* gene, causes the resistance. The above description indicates that biological markers that can predict therapeutic outcome enable guidance of the choice of treatment; and depending on the levels of the biomarkers it could be estimated if the patient would respond to a particular therapy or not and finally an appropriate treatment regimen could be selected. In addition, prediction of drug response helps in reducing the treatment cost.

16.3. Recent Trends and Future Directions

Significant progress with a variety of biomarkers has been made in the last two decades. Metabolomic and epigenomic biomarkers are relatively new and have shown promising results in disease diagnosis and prognosis. The monitoring of relative changes in metabolomic profiles in predisposed *versus* healthy individuals may help identify unique metabolites involved in disease processes.[146] These profiles have been used to predict the risk of diabetes,[147] cardiovascular disease[148] and lung cancer,[149] to diagnose prostate cancer,[150] to differentiate benign and malignant ovarian tumours[151] and to identify biomarkers of Crohn's disease.[7] Such shifts also may identify diagnostic biomarkers, which could provide insights into strategies for disease prevention and be used to monitor the response to treatment. In recent years, metabolomics and other post-GWAS platforms, such as proteomics and transcriptomics, have undergone rapid improvement in both their reliability and throughput; as such, it may be an appropriate time for their use in epidemiologic studies.[152] Although metabolomic profiling has been used in some larger-scale population studies,[153] the number of published reports to date remains small. If the successes of genomics and transcriptomics in epidemiology are reliable indicators, there is a large, as yet unexplored, potential for metabolomics to contribute to public health research. So far the assessment of genotypes of candidate biomarkers in metabolic diseases and CVD in blood samples has not improved prediction of these diseases. Probably multiple markers (transcriptomics, epigenomics, metabolomics and genomics) should be tested for better prediction of these diseases. Prognostication is one of the promising areas in translation of experimental research into clinical practice. In this approach, patterns of altered gene expression in tumours are used to construct classifiers instead of standard indices such as Nottingham Prognostic Index, Adjuvant Online and Predict.[154]

Multiplexing of biomarkers may reduce false-positive results in screening studies where the intention is to identify populations that are at high risk of developing a disease.[2] Quantitative imaging data storage and maintenance have their own challenges, as discussed above. In the case of miRNA biomarkers, whether miRNAs expression is localized in a specific part of the tissue has to be carefully evaluated. In a tissue biopsy the local concentration (number of miRNAs) may be low or high. Determining the accurate level of miRNAs is very critical.

It has been seen that a combination of genes measured concurrently imparts better information about the clinical effect than a single gene. Oncotype DX™ is the first multigene predictive test of prognosis for breast cancer patients receiving anti-oestrogen therapy (Genomic Health). The FDA has also approved the test MammaPrint that predicts relapse in breast cancer patients by analysing the activity of 70 genes. In the case of prostate cancer, the prostate-specific antigen (PSA) test has been approved, where PSA levels help in detecting prostate cancer and also predict a recurrence in patients suffering from the disease. Alpha-fetoprotein (AFP) has been approved for the diagnosis and monitoring of patients with non-seminoma testicular cancer.

Significant challenges exist regarding intellectual property (IP) issues. At times tests are known but their associations with diseases are new. Sometime patents belong to smaller companies for new tests but their utility in clinical settings can only be developed with larger companies in large clinical trials. In general, every situation represents unique opportunities and IP-related issues can only be solved in a case-by-case manner.

The presence of oestrogen receptor (ER), progesterone receptor (PR), and human epidermal growth factor receptor 2 (Her2, or ERBB2) is used for the clinical and pathological classification of breast cancer.[3] Generally, ER-positive (+) and PR+ are indicators of good prognosis, and Her2+ is an indicator of bad prognosis. In addition, ER-negative (−), PR−, and Her2− (also called triple-negative) status is considered to be an indicator of poor prognosis. Basal cells exhibit triple-negative features. On the basis of oncological pathway activity analysis, up to 18 subtypes of breast cancer have been suggested.[4] However, the implications of this information for clinical practice remain to be determined. Furthermore, many prognostic gene expression signatures that dichotomize patient populations into treatment-responsive and non-responsive groups lack specificity.[155,156] Additional biomarkers are needed that are better prognostic indicators than hormone receptor status, and a better understanding of the genetic characteristics of patients is needed to improve current clinical practice. Ideally, a method for preoperative molecular profiling should be developed that can guide treatment strategies.

In recent years metabolites of biofluids have been analysed for their potential in cancer diagnosis and follow up of treatment. Especially, urine analysis for the routine monitoring of metabolic disorders has attracted a reasonable amount of interest among scientists because the procedure can be done easily, non-invasively and repeatedly for a large number of samples with high precision. Generally volatile organic metabolites (VOM) get enriched in urine and their analysis is easy.[157] The advantage of adopting the metabolomic approach lies in the fact that metabolites are much more stable than RNA and proteins and their levels predict those pathways that are affected during disease development. In one small study with urine from controls and breast cancer, VOMs were identified which were differentially expressed in patients only.[158] Higher levels of 4-carene, 3-heptanone, 1,2,4-trimethylbenzene, 2-methoxythiophene and phenol, and lower levels of dimethyl sulphides were observed in breast cancer patients. Urine metabolites have been used for diagnosis of colon, lung, liver and prostate cancer also.[157]

Characterizing metabolomic pathways help in making treatment decisions.[159] Circulating biomarkers, especially diabetes biomarkers discussed in this article, should be used for disease stratification of patients followed by completion of questionnaire such as: (1) How can

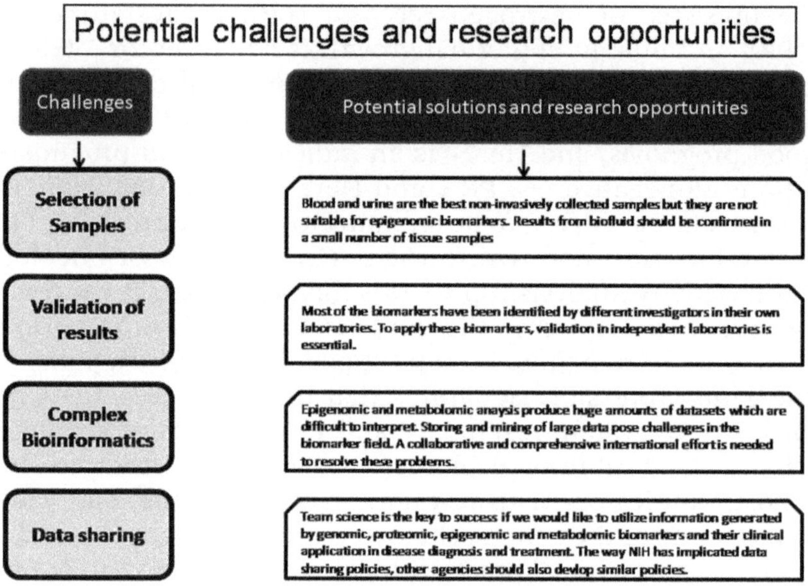

Figure 16.2 Potential challenges and research opportunities in utilizing biomarker information in disease detection, diagnosis, prognosis and drug response.

the group be best defined in terms of biomarker levels? (2) What is the cause of the unexpected results? (3) What is the underlying genetic trait? (4) Will one group respond better than other? This approach may lead us in the direction of personalized medicine. We have summarized potential challenges and research opportunities in utilizing biomarker information in disease detection, diagnosis, prognosis and drug response in Figure 16.2.

16.4. Conclusions

Biomarker screening tests face a challenge in transition of tests from the research level in the laboratory to beneficial use in clinical settings. The biomarkers need to be highly specific and sensitive for detection in clinical samples in order to avoid false-positive tests that could lead to misdiagnosis and incorrect therapeutic selection. An ideal biomarker should be used for screening in a large population study to validate the effectiveness of the screening and be proven effective in populations with different genetic backgrounds. It has been seen that many biomarkers that correlate with disease statistically may not be useful in clinical settings. It is advisable to study patient populations with diversity because the biomarkers might show different responses in different populations.

A considerable amount of knowledge has been obtained in understanding cancer biology and identification of biomarkers, which can detect cancer, but implementation of that information in clinical settings is still challenging.[160] Clinical validation is the main hurdle in the process. In one case control study of the Prospect-EPIC (European Prospective Investigation into Cancer and nutrition) where more than 300 breast cancer patients and matched controls were tested for breast cancer over a period of three years using a panel of eight serum biomarkers (osteopontin, haptoglobin, cancer antigen 15-3, carcinoembryonic antigen, cancer antigen-125, prolactin, cancer antigen 19-9 and alpha-fetoprotein), very low specificity (50%) and sensitivity (50%) was observed.[1] This may be due to different subtypes of breast cancer in collected samples. Such epidemiological studies should select a broader target set of potential biomarkers, which could be enabled by antibody array technologies, where profiles of up to 100 antibodies can be followed simultaneously. Making different groups, based on the status of hormone receptors (oestrogen and progesterone), might also be helpful.

In the case of cancer, the need for identification and characterization of early cancer diagnostic biomarkers is high because cancer is a heterogeneous disease and patient's individual molecular profiling due to tumour microenvironment determines the disease development and response to treatment.[2,5] Tumour microenvironment is affected by several factors, including epigenetic factors of the cell.

Reasonable progress has been made in clinical settings in some cases where biomarkers lead to better efficacy, less toxicity, better diagnosis and predictable prognosis. However, this has been possible only for a few selected disease conditions. We anticipate that such success will soon be replicated in several critical diseases. There is an apparent need for new biomarkers and the upcoming technologies promise the development of new biomarkers, which would change the course of disease detection and management. This will be a gain for the medical field, with improved patient care and better clinical outcome.

The main areas that need progress and attention are the cost and high throughput. Another area where scope for further progress remains is the application of biomarkers in clinical settings. Proper analytical and clinical validation of early biomarkers has not been achieved. Clinical validation of identified biomarkers is the key challenge in the field. The National Cancer Institutes have developed guidelines for the analytical and clinical validation of biomarkers, but none of the biomarkers has been validated to date.[161] Integration of genomic and proteomic biomarkers with epigenetic biomarkers may help us with subtyping disease stages.[6] Many times results of methylation profiling from blood and tissues are different. Koestler *et al.* conducted a systematic epigenome-wide methylation analysis and demonstrated that shifts in leukocyte subpopulations might account for a considerable proportion of variability in these patterns.[7] The location of miRNAs in any specific organ should be carefully determined. In a tissue biopsy the local concentration (number of miRNAs) may be low or high. Determining the level of miRNAs accurately is very critical.

Association studies are extremely powerful in identifying low-penetrance new SNPs (biomarkers) which may have therapeutic implications. Identification of common low-susceptibility alleles is useful because it provides possible insight into the mechanisms of tumour biology and allows identification of high-risk individuals. Since genotyping is not expensive these days, the information from such studies can be utilized in personalized medicine by targeted primary and secondary prevention.

Although individual omics techniques have generated a large number of potential biomarkers for any given cancer; due to heterogeneity of the carcinogenesis process, an integrated approach is required to gather all potential biomarkers and to identify key biomarkers for diagnosis, prognosis and therapy.

We emphasize that considerable progress has been made on disease-associated biomarkers, which can be used for a spectrum of different diseases and from risk assessment to follow-up survival. Information discussed in this article may be useful in developing new intervention and therapeutic targets.

References

1. E. Ferrannini, A. Natali, S. Camastra, M. Nannipieri, A. Mari, K. P. Adam, M. V. Milburn, G. Kastenmuller, J. Adamski, T. Tuomi, V. Lyssenko, L. Groop and W. E. Gall, *Diabetes*, 2013, **62**, 1730–1737.
2. M. Verma, M. J. Khoury and J. P. Ioannidis, *Cancer Epidemiol., Biomarkers Prev.*, 2013, **22**, 189–200.
3. A. W. Opstal-van Winden, W. Rodenburg, J. L. Pennings, C. T. van Oostrom, J. H. Beijnen, P. H. Peeters, C. H. van Gils and A. de Vries, *Int. J. Mol. Sci.*, 2012, **13**, 13587–13604.
4. C. S. Zhu, P. F. Pinsky, D. W. Cramer, D. F. Ransohoff, P. Hartge, R. M. Pfeiffer, N. Urban, G. Mor, R. C. Bast Jr, L. E. Moore, A. E. Lokshin, M. W. McIntosh, S.

J. Skates, A. Vitonis, Z. Zhang, D. C. Ward, J. T. Symanowski, A. Lomakin, E. T. Fung, P. M. Sluss, N. Scholler, K. H. Lu, A. M. Marrangoni, C. Patriotis, S. Srivastava, S. S. Buys and C. D. Berg, *Cancer Prev. Res.*, 2011, **4**, 375–383.
5. M. Verma, *Methods Mol. Biol.*, 2012, **863**, 467–480.
6. J. M. Yi, M. Dhir, L. Van Neste, S. R. Downing, J. Jeschke, S. C. Glockner, M. de Freitas Calmon, C. M. Hooker, J. M. Funes, C. Boshoff, K. M. Smits, M. van Engeland, M. P. Weijenberg, C. A. Iacobuzio-Donahue, J. G. Herman, K. E. Schuebel, S. B. Baylin and N. Ahuja, *Clin. Cancer Res.*, 2011, **17**, 1535–1545.
7. D. C. Koestler, C. J. Marsit, B. C. Christensen, W. Accomando, S. M. Langevin, E. A. Houseman, H. H. Nelson, M. R. Karagas, J. K. Wiencke and K. T. Kelsey, *Cancer Epidemiol., Biomarkers Prev.*, 2012, **21**, 1293–1302.
8. A. Johnson, Q. Song, P. Ko Ferrigno, P. R. Bueno and J. J. Davis, *Anal. Chem.*, 2012, **84**, 6553–6560.
9. W. Gao, J. Xu and Y. Q. Shu, *Expert Rev. Respir. Med.*, 2011, **5**, 699–709.
10. C. A. Castaneda, M. T. Agullo-Ortuno, J. A. Fresno Vara, H. Cortes-Funes, H. L. Gomez and E. Ciruelos, *Expert Rev. Anticancer Ther.*, 2011, **11**, 1265–1275.
11. J. E. Joh, N. N. Esposito, J. V. Kiluk, C. Laronga, M. C. Lee, L. Loftus, H. Soliman, J. C. Boughey, C. Reynolds, T. J. Lawton, P. I. Acs, L. Gordan and G. Acs, *Oncologist*, 2011, **16**, 1520–1526.
12. E. A. Slodkowska and J. S. Ross, *Expert Rev. Mol. Diagn.*, 2009, **9**, 417–422.
13. B. A. Teicher, *Curr. Opin. Pharmacol.*, 2010, **10**, 397–404.
14. J. Pang, W. P. Liu, X. P. Liu, L. Y. Li, Y. Q. Fang, Q. P. Sun, S. J. Liu, M. T. Li, Z. L. Su and X. Gao, *J. Proteome Res.*, 2010, **9**, 216–226.
15. A. M. Abdel Rahman, S. D. Kamath, S. Gagne, A. L. Lopata and R. Helleur, *J. Proteome Res.*, 2013, **12**, 647–656.
16. I. Rahman, R. Atout, N. L. Pedersen, U. de Faire, J. Frostegard, E. Ninio, A. M. Bennet and P. K. Magnusson, *Atherosclerosis*, 2011, **218**, 117–122.
17. G. Rebovich, E. J. Duffis and L. R. Caplan, *Expert Opin. Med. Diagn.*, 2010, **4**, 267–279.
18. A. Zhang, H. Sun and X. Wang, *Obes. Rev.*, 2013, **14**, 344–349.
19. X. Li, H. F. Tse and L. J. Jin, *J. Dent. Res.*, 2011, **90**, 1062–1069.
20. J. Ahn, C. Y. Chen and R. B. Hayes, *Cancer Causes Control*, 2012, **23**, 399–404.
21. C. Luaces-Cubells, S. Mintegi, J. J. Garcia-Garcia, E. Astobiza, R. Garrido-Romero, J. Velasco-Rodriguez and J. Benito, *Pediatr. Infect. Dis. J.*, 2012, **31**, 645–647.
22. M. Verma, *Ann. N. Y. Acad. Sci.*, 2003, **983**, 170–180.
23. H. Kuper, H. O. Adami and D. Trichopoulos, *J. Intern. Med.*, 2000, **248**, 171–183.
24. D. M. Parkin, *Int. J. Cancer*, 2006, **118**, 3030–3044.
25. H. zur Hausen, *Eur. J. Cancer*, 1999, **35**, 1174–1181.
26. C. P. Wild and P. C. Turner, *Mutagenesis*, 2002, **17**, 471–481.
27. H. Sun, A. Zhang, G. Yan, C. Piao, W. Li, C. Sun, X. Wu, X. Li, Y. Chen and X. Wang, *Mol. Cell. Proteomics*, 2013, **12**, 710–719.
28. F. L. van de Veerdonk, P. C. Wever, M. H. Hermans, R. Fijnheer, L. A. Joosten, J. W. van der Meer, M. G. Netea and P. M. Schneeberger, *J. Infect. Dis.*, 2012, **206**, 197–201.
29. S. F. Graham, O. P. Chevallier, D. Roberts, C. Holscher, C. T. Elliott and B. D. Green, *Anal. Chem.*, 2013, **85**, 1803–1811.
30. P. Vanek, O. Bradac, P. DeLacy, K. Saur, T. Belsan and V. Benes, *Spine*, 2012, **37**, 1645–1651.
31. M. Verma, *Curr. Opin. Clin. Nutr. Metab. Care*, 2013, **16**, 376–384.
32. B. K. Dunn, M. Verma and A. Umar, *Ann. N. Y. Acad. Sci.*, 2003, **983**, 1–4.
33. M. Verma, *Curr. Genomics*, 2012, **13**, 308–313.
34. M. Faryna, C. Konermann, S. Aulmann, J. L. Bermejo, M. Brugger, S. Diederichs, J. Rom, D. Weichenhan, R. Claus, M. Rehli, P. Schirmacher, H. P. Sinn, C. Plass and C. Gerhauser, *FASEB J.*, 2012, **26**, 4937–4950.

35. B. Poudel, A. Mittal, R. Shrestha, A. K. Nepal and P. S. Shukla, *Asian Pac. J. Cancer Prev.*, 2012, **13**, 2149–2152.
36. I. Casanova-Salas, J. Rubio-Briones, A. Fernandez-Serra and J. A. Lopez-Guerrero, *Clin. Transl. Oncol.*, 2012, **14**, 803–811.
37. L. C. Soliman, Y. Hui, A. K. Hewavitharana and D. D. Chen, *J. Chromatogr. A*, 2012, **1267**, 162–169.
38. L. Murphy and R. W. Watson, *Nat. Rev. Urol.*, 2012, **9**, 464–472.
39. R. Kuner, J. C. Brase, H. Sultmann and D. Wuttig, *Methods*, 2013, **59**, 132–137.
40. S. K. Martin, T. B. Vaughan, T. Atkinson, H. Zhu and N. Kyprianou, *Oncol. Rep.*, 2012, **28**, 409–417.
41. L. Ng, N. Karunasinghe, C. S. Benjamin and L. R. Ferguson, *N. Z. Med. J.*, 2012, **125**, 59–86.
42. J. R. Prensner, M. A. Rubin, J. T. Wei and A. M. Chinnaiyan, *Sci. Transl. Med.*, 2012, **4**, 127rv123.
43. S. Souchelnytskyi, M. Lomnytska, A. Dubrovska, U. Hellman and N. Volodko, *Proteomics*, 2006, **6**(suppl. 2), 65–68.
44. P. D. Hayes, D. A. Payne, N. J. Evans, M. M. Thompson, N. J. London, P. R. Bell and A. R. Naylor, *Eur. J. Vasc. Endovasc.*, 2003, **26**, 665–669.
45. M. P. Endsley and M. Zhang, *Methods Enzymol.*, 2011, **499**, 149–165.
46. E. S. Lianidou, A. Markou and A. Strati, *Cancer Metastasis Rev.*, 2012, **31**, 663–671.
47. A. Alvarez Secord, K. M. Darcy, A. Hutson, Z. Huang, P. S. Lee, E. L. Jewell, L. J. Havrilesky, M. Markman, F. Muggia and S. K. Murphy, *Gynecol. Oncol.*, 2011, **123**, 314–319.
48. Z. J. Xie, G. Chen, X. C. Zhang, D. F. Li, J. Huang and Z. J. Li, *Asian Pac. J. Cancer Prev.*, 2012, **13**, 6145–6149.
49. M. E. Arellano-Garcia, S. Hu, J. Wang, B. Henson, H. Zhou, D. Chia and D. T. Wong, *Oral diseases*, 2008, **14**, 705–712.
50. C. A. Righini, F. de Fraipont, J. F. Timsit, C. Faure, E. Brambilla, E. Reyt and M. C. Favrot, *Clin. Cancer Res.*, 2007, **13**, 1179–1185.
51. Z. Yalniz, S. Demokan, Y. Suoglu, M. Ulusan and N. Dalay, *Mol. Biol. Rep.*, 2010, **37**, 3541–3545.
52. L. L. Gleich, J. Wang, J. L. Gluckman and C. M. Fenoglio-Preiser, *ORL J. Otorhinolaryngol. Relat. Spec.*, 2003, **65**, 193–198.
53. G. Cuda, A. Gallelli, A. Nistico, P. Tassone, V. Barbieri, P. S. Tagliaferri, F. S. Costanzo, C. M. Tranfa and S. Venuta, *Lung Cancer*, 2000, **30**, 211–214.
54. Z. Q. Yuan, B. Legendre, D. Q. Cai, J. Cao, J. Zhu and T. K. Weber, *Pathology*, 2009, **41**, 393–394.
55. A. I. Shemirani, M. M. Haghighi, S. M. Zadeh, S. R. Fatemi, M. Y. Taleghani, N. Zali, Z. Akbari, S. M. Kashfi and M. R. Zali, *Asian Pac. J. Cancer Prev.*, 2011, **12**, 2101–2104.
56. H. Danaee, H. H. Nelson, M. R. Karagas, A. R. Schned, T. D. Ashok, T. Hirao, A. E. Perry and K. T. Kelsey, *Oncogene*, 2002, **21**, 4894–4899.
57. S. Mourah, O. Cussenot, V. Vimont, F. Desgrandchamps, P. Teillac, B. Cochant-Priollet, A. Le Duc, J. Fiet and H. Soliman, *Int. J. Cancer*, 1998, **79**, 629–633.
58. U. Wahn, R. L. Bergmann and R. Nickel, *Clin. Exp. Allergy*, 1998, **28**(suppl 1), 20–21; discussion 32–26.
59. I. Okur, L. Tumer, F. S. Ezgu, E. Yesilkaya, A. Aral, S. O. Oktar, A. Bideci and A. Hasanoglu, *J. Pediatr. Endocrinol. Metab.*, 2013, 1–6.
60. M. Kucukazman, N. Ata, B. Yavuz, K. Dal, O. Sen, O. S. Deveci, K. Agladioglu, A. O. Yeniova, Y. Nazligul and D. T. Ertugrul, *Eur. J. Gastroenterol. Hepatol.*, 2013, **25**, 147–151.
61. *Duke Med Health News*, 2012, **18**, 4–5.
62. E. F. Kern, P. Erhard, W. Sun, S. Genuth and M. F. Weiss, *Am. J. Kidney Dis.*, 2010, **55**, 824–834.
63. A. E. Mehta, *Cleve. Clin. J. Med.*, 1995, **62**, 210–211.

64. D. P. Dosanjh, M. Bakir, K. A. Millington, A. Soysal, Y. Aslan, S. Efee, J. J. Deeks and A. Lalvani, *PLoS One*, 2011, **6**, e28754.
65. M. Sakamoto, *J. Gastroenterol.*, 2009, **44**(suppl 19), 108–111.
66. M. Di Luca, E. Grossi, B. Borroni, M. Zimmermann, E. Marcello, F. Colciaghi, F. Gardoni, M. Intraligi, A. Padovani and M. Buscema, *J. Transl. Med.*, 2005, **3**, 30.
67. M. Verma, G. L. Wright Jr, S. M. Hanash, R. Gopal-Srivastava and S. Srivastava, *Ann. N. Y. Acad. Sci.*, 2001, **945**, 103–115.
68. I. Simon, Y. Liu, K. L. Krall, N. Urban, R. L. Wolfert, N. W. Kim and M. W. McIntosh, *Gynecol. Oncol.*, 2007, **106**, 112–118.
69. V. Kashuba, A. A. Dmitriev, G. S. Krasnov, T. Pavlova, I. Ignatjev, V. V. Gordiyuk, A. V. Gerashchenko, E. A. Braga, S. P. Yenamandra, M. Lerman, V. N. Senchenko and E. Zabarovsky, *Int. J. Mol. Sci.*, 2012, **13**, 13352–13377.
70. J. Ren, H. Cai, Y. Li, X. Zhang, Z. Liu, J. S. Wang, Y. L. Hwa, Y. Zhang, Y. Yang and S. W. Jiang, *Expert Rev. Mol. Diagn.*, 2010, **10**, 787–798.
71. S. Lindstrom, F. Schumacher, A. Siddiq, R. C. Travis, D. Campa, S. I. Berndt, W. R. Diver, G. Severi, N. Allen, G. Andriole, B. Bueno-de-Mesquita, S. J. Chanock, D. Crawford, J. M. Gaziano, G. G. Giles, E. Giovannucci, C. Guo, C. A. Haiman, R. B. Hayes, J. Halkjaer, D. J. Hunter, M. Johansson, R. Kaaks, L. N. Kolonel, C. Navarro, E. Riboli, C. Sacerdote, M. Stampfer, D. O. Stram, M. J. Thun, D. Trichopoulos, J. Virtamo, S. J. Weinstein, M. Yeager, B. Henderson, J. Ma, L. Le Marchand, D. Albanes and P. Kraft, *PLoS One*, 2011, **6**, e17142.
72. K. K. Tsilidis, R. C. Travis, P. N. Appleby, N. E. Allen, S. Lindstrom, F. R. Schumacher, D. Cox, A. W. Hsing, J. Ma, G. Severi, D. Albanes, J. Virtamo, H. Boeing, H. B. Bueno-de-Mesquita, M. Johansson, J. R. Quiros, E. Riboli, A. Siddiq, A. Tjonneland, D. Trichopoulos, R. Tumino, J. M. Gaziano, E. Giovannucci, D. J. Hunter, P. Kraft, M. J. Stampfer, G. G. Giles, G. L. Andriole, S. I. Berndt, S. J. Chanock, R. B. Hayes and T. J. Key, *Am. J. Epidemiol.*, 2012, **175**, 926–935.
73. F. Li, S. Li, H. Chang, Y. Nie, L. Zeng, X. Zhang and Y. Wang, *Genet. Test. Mol. Biomarkers*, 2013, **17**, 656–661.
74. D. R. Taylor, *Thorax*, 2009, **64**, 261–264.
75. S. A. Peters, F. L. Visseren and D. E. Grobbee, *Nat. Rev. Cardiol.*, 2013, **10**, 12–14.
76. G. B. Lim, *Nat. Rev. Cardiol.*, 2012, **9**, 672.
77. J. Madden, C. M. Williams, P. C. Calder, G. Lietz, E. A. Miles, H. Cordell, J. C. Mathers and A. M. Minihane, *Annu. Rev. Nutr.*, 2011, **31**, 203–234.
78. N. Sattar, O. Scherbakova, I. Ford, D. S. O'Reilly, A. Stanley, E. Forrest, P. W. Macfarlane, C. J. Packard, S. M. Cobbe and J. Shepherd, *Diabetes*, 2004, **53**, 2855–2860.
79. A. Betriu-Bars and E. Fernandez-Giraldez, *Nefrologia*, 2012, **32**, 7–11.
80. Y. Shiferaw, A. Alemu, A. Girma, A. Getahun, A. Kassa, A. Gashaw, T. Teklu and B. Gelaw, *BMC Res. Notes*, 2011, **4**, 505.
81. G. B. Gerbi, T. Habtemariam, B. Tameru, D. Nganwa and V. Robnett, *AIDS Care*, 2012, **24**, 331–339.
82. T. Oni, H. P. Gideon, N. Bangani, R. Tsekela, R. Seldon, K. Wood, K. A. Wilkinson, R. T. Goliath, T. H. Ottenhoff and R. J. Wilkinson, *Clin. Vaccine Immunol.*, 2012, **19**, 1243–1247.
83. M. F. Humblet, M. Gilbert, M. Govaerts, M. Fauville-Dufaux, K. Walravens and C. Saegerman, *J. Clin. Microbiol.*, 2010, **48**, 2802–2808.
84. X. F. Wang, Y. W. Cao, Z. Z. Feng, D. Fu, Y. S. Ma, F. Zhang, X. X. Jiang and Y. C. Shao, *Mol. Biol. Rep.*, 2013, **40**, 779–785.
85. M. Silvestrini, G. Viticchi, C. Altamura, S. Luzzi, C. Balucani and F. Vernieri, *J. Alzheimer's Dis.*, 2012, **32**, 689–698.
86. H. M. Schipper, *Alzheimer's Dementia*, 2011, **7**, e118–123.
87. F. E. Ahmed, P. Vos, S. iJames, D. T. Lysle, R. R. Allison, G. Flake, D. R. Sinar, W. Naziri, S. P. Marcuard and R. Pennington, *Cancer Genomics Proteomics*, 2007, **4**, 1–20.

88. W. Chung, B. Kwabi-Addo, M. Ittmann, J. Jelinek, L. Shen, Y. Yu and J. P. Issa, *PLoS One*, 2008, **3**, e2079.
89. T. Shimbo, A. Tanemura, T. Yamazaki, K. Tamai, I. Katayama and Y. Kaneda, *PLoS One*, 2010, **5**, e10566.
90. D. J. Sargent, M. B. Resnick, M. O. Meyers, A. Goldar-Najafi, T. Clancy, S. Gill, G. O. Siemons, Q. Shi, B. M. Bot, T. T. Wu, G. Beaudry, J. F. Haince and Y. Fradet, *Ann. Surg. Oncol.*, 2011, **18**, 3261–3270.
91. W. Han, H. Kim, S. Y. Ku, S. H. Kim, Y. M. Choi, J. G. Kim and S. Y. Moon, *Gynecol. Endocrinol.*, 2013, **29**, 16–19.
92. M. G. Tektonidou and M. M. Ward, *Nat. Rev. Rheumatol.*, 2011, **7**, 708–717.
93. D. Plant, I. Ibrahim, M. Lunt, S. Eyre, E. Flynn, K. L. Hyrich, A. W. Morgan, A. G. Wilson, J. D. Isaacs and A. Barton, *Arthritis Res. Ther.*, 2012, **14**, R214.
94. I. Ibrahim, S. A. Owen and A. Barton, *Expert Rev. Clin. Immunol.*, 2012, **8**, 509–511.
95. R. Mogelvang, S. H. Pedersen, A. Flyvbjerg, M. Bjerre, A. Z. Iversen, S. Galatius, J. Frystyk and J. S. Jensen, *Am. J. Cardiol.*, 2012, **109**, 515–520.
96. C. C. Chen, J. L. Huang, C. J. Chang and M. S. Kong, *J. Pediatr. Gastroenterol. Nutr.*, 2012, **55**, 541–547.
97. K. B. Walhovd, A. M. Fjell, J. Brewer, L. K. McEvoy, C. Fennema-Notestine, D. J. Hagler Jr, R. G. Jennings, D. Karow and A. M. Dale, *AJNR Am. J. Neuroradiol.*, 2010, **31**, 347–354.
98. K. A. Kolquist, R. A. Schultz, A. Furrow, T. C. Brown, J. Y. Han, L. J. Campbell, M. Wall, M. L. Slovak, L. G. Shaffer and B. C. Ballif, *Cancer Genet.*, 2011, **204**, 603–628.
99. A. S. Patel, A. L. Hawkins and C. A. Griffin, *Curr. Opin. Oncol.*, 2000, **12**, 62–67.
100. J. M. Cowan, *Otolaryngol. Clin. North Am.*, 1992, **25**, 1073–1087.
101. Z. Gibas and L. Gibas, *Cancer Genet. Cytogenet.*, 1997, **95**, 108–115.
102. D. Geleick, H. Muller, A. Matter, J. Torhorst and U. Regenass, *Cancer Genet. Cytogenet.*, 1990, **46**, 217–229.
103. R. P. Pearlstein, M. S. Benninger, T. E. Carey, R. J. Zarbo, F. X. Torres, B. A. Rybicki and D. L. Dyke, *Genes, Chromosomes Cancer*, 1998, **21**, 333–339.
104. D. W. Chang, J. Gu and X. Wu, *Urol. Oncol.*, 2012, **30**, 524–532.
105. B. Ludewig, J. V. Stein, J. Sharpe, L. Cervantes-Barragan, V. Thiel and G. Bocharov, *Eur. J. Immunol.*, 2012, **42**, 3116–3125.
106. V. Narang, J. Decraene, S. Y. Wong, B. S. Aiswarya, A. R. Wasem, S. R. Leong and A. Gouaillard, *Immunol. Res.*, 2012, **53**, 251–265.
107. J. D. Bregman, *J. Am. Acad. Child Adolesc. Psychiatry*, 2012, **51**, 1113–1115.
108. L. M. Oberman, *Expert Opin. Invest. Drugs*, 2012, **21**, 1819–1825.
109. P. Karolina, R. Chlup, Z. Jana, K. D. Kohnert, P. Kudlova, J. Bartek, M. Nakladalova, B. Doubravova and P. Seckar, *J. Diabetes Sci. Technol.*, 2010, **4**, 983–992.
110. S. Jaijakul, J. A. Vazquez, R. N. Swanson and L. Ostrosky-Zeichner, *Clin. Infect. Dis.*, 2012, **55**, 521–526.
111. T. R. Powell, L. C. Schalkwyk, A. L. Heffernan, G. Breen, T. Lawrence, T. Price, A. E. Farmer, K. J. Aitchison, I. W. Craig, A. Danese, C. Lewis, P. McGuffin, R. Uher, K. E. Tansey and U. M. D'Souza, *Eur. Neuropsychopharmacol.*, 2013, **23**, 1105–1114.
112. R. E. McCullumsmith, J. Hammond, A. Funk and J. H. Meador-Woodruff, *Curr. Pharm. Biotechnol.*, 2012, **13**, 1535–1542.
113. H. M. Ibrahim and C. A. Tamminga, *Curr. Pharm. Biotechnol.*, 2012, **13**, 1587–1594.
114. J. Mullenders, W. von der Saal, M. M. van Dongen, U. Reiff, R. van Willigen, R. L. Beijersbergen, G. Tiefenthaler, C. Klein and R. Bernards, *Clin. Cancer Res.*, 2009, **15**, 5811–5819.
115. S. Maier, C. Dahlstroem, C. Haefliger, A. Plum and C. Piepenbrock, *Am. J. PharmacoGenomics*, 2005, **5**, 223–232.
116. D. B. Longley, W. L. Allen and P. G. Johnston, *Biochim. Biophys. Acta*, 2006, **1766**, 184–196.

117. A. M. Tsimberidou, N. G. Iskander, D. S. Hong, J. J. Wheler, G. S. Falchook, S. Fu, S. Piha-Paul, A. Naing, F. Janku, R. Luthra, Y. Ye, S. Wen, D. Berry and R. Kurzrock, *Clin. Cancer Res.*, 2012, **18**, 6373–6383.
118. H. Wong and T. Yau, *Oncologist*, 2012, **17**, 346–358.
119. C. B. Weldon, J. R. Trosman, W. J. Gradishar, A. B. Benson 3rd and J. C. Schink, *J. Oncol. Pract.*, 2012, **8**, e24–31.
120. L. G. van Baarsen, C. A. Wijbrandts, D. M. Gerlag, F. Rustenburg, T. C. van der Pouw Kraan, B. A. Dijkmans, P. P. Tak and C. L. Verweij, *Genes Immun.*, 2010, **11**, 622–629.
121. S. Marsal and A. Julia, *Pharmacogenomics*, 2010, **11**, 617–619.
122. M. P. Grimaldi, S. Vasto, C. R. Balistreri, D. di Carlo, M. Caruso, E. Incalcaterra, D. Lio, C. Caruso and G. Candore, *Ann. N. Y. Acad. Sci.*, 2007, **1100**, 123–131.
123. N. Siitonen, L. Pulkkinen, J. Lindstrom, M. Kolehmainen, U. Schwab, J. G. Eriksson, P. Ilanne-Parikka, S. Keinanen-Kiukaanniemi, J. Tuomilehto and M. Uusitupa, *Cardiovasc. Diabetol.*, 2011, **10**, 83.
124. G. Ragia and V. G. Manolopoulos, *Pharmacogenomics*, 2012, **13**, 261–264.
125. P. D. Roy, M. Majumder and B. Roy, *Pharmacogenomics*, 2008, **9**, 311–321.
126. F. Martinelli-Boneschi, G. Giacalone, G. Magnani, G. Biella, E. Coppi, R. Santangelo, P. Brambilla, F. Esposito, S. Lupoli, F. Clerici, L. Benussi, R. Ghidoni, D. Galimberti, R. Squitti, A. Confaloni, G. Bruno, S. Pichler, M. Mayhaus, M. Riemenschneider, C. Mariani, G. Comi, E. Scarpini, G. Binetti, G. Forloni, M. Franceschi and D. Albani, *Neurobiol. Aging*, 2013, **34**, 1711.e7–1711.e13.
127. F. Listi, C. Caruso, D. Lio, G. Colonna-Romano, M. Chiappelli, F. Licastro and G. Candore, *J. Alzheimer's Dis.*, 2010, **19**, 551–557.
128. E. Dreussi, P. Biason, G. Toffoli and E. Cecchin, *Pharmacogenomics*, 2012, **13**, 1635–1650.
129. D. T. Merrick, *Chest*, 2012, **141**, 1377–1378.
130. C. Justenhoven, O. Obazee and H. Brauch, *Pharmacogenomics*, 2012, **13**, 659–675.
131. M. Nishiyama and H. Eguchi, *Adv. Drug Delivery Rev.*, 2009, **61**, 402–407.
132. T. Kohriyama, *Rinsho Shinkeigaku*, 2011, **51**, 179–187.
133. S. Malucchi, F. Gilli, M. Caldano, F. Marnetto, P. Valentino, L. Granieri, A. Sala, M. Capobianco and A. Bertolotto, *Neurology*, 2008, **70**, 1119–1127.
134. C. A. Braun Hashemi, Y. C. Zang, J. A. Arbona, J. A. Bauerle, M. L. Frazer, H. Lee, L. Flury, E. S. Moore, M. C. Kolar, R. Y. Washington and O. J. Kolar, *Mult. Scler.*, 2006, **12**, 652–658.
135. A. Miller, L. Glass-Marmor, M. Abraham, I. Grossman, S. Shapiro and Y. Galboiz, *Clin. Neurol. Neurosurg.*, 2004, **106**, 249–254.
136. J. Zhang, C. Wang, L. Li, Q. Man, L. Meng, P. Song, L. Froyland and Z. Y. Du, *Br. J. Nutr.*, 2012, **108**, 1455–1465.
137. S. Sour, M. Belarbi, D. Khaldi, N. Benmansour, N. Sari, A. Nani, F. Chemat and F. Visioli, *Br. J. Nutr.*, 2012, **107**, 1800–1805.
138. L. Ozick, P. Chander, A. Agarwal and A. Soni, *Am. J. Gastroenterol.*, 1989, **84**, 195–197.
139. A. Maisel, S. X. Neath, J. Landsberg, C. Mueller, R. M. Nowak, W. F. Peacock, P. Ponikowski, M. Mockel, C. Hogan, A. H. Wu, M. Richards, P. Clopton, G. S. Filippatos, S. Di Somma, I. Anand, L. L. Ng, L. B. Daniels, R. H. Christenson, M. Potocki, J. McCord, G. Terracciano, O. Hartmann, A. Bergmann, N. G. Morgenthaler and S. D. Anker, *Eur. J. Heart Failure*, 2012, **14**, 278–286.
140. F. P. Manfredsson, M. S. Okun and R. J. Mandel, *Curr. Gene Ther.*, 2009, **9**, 375–388.
141. D. C. Gilbert, A. J. Chalmers and S. F. El-Khamisy, *Br. J. Cancer*, 2012, **106**, 18–24.
142. E. D. Hsi, S. H. Jung, R. Lai, J. L. Johnson, J. R. Cook, D. Jones, S. Devos, B. D. Cheson, L. E. Damon and J. Said, *Leuk. Lymphoma*, 2008, **49**, 2081–2090.
143. P. C. Tsai, F. J. Hernandez-Ilizaliturri, N. Bangia, S. H. Olejniczak and M. S. Czuczman, *Clin. Cancer Res.*, 2012, **18**, 1039–1050.

144. M. Giachelia, M. T. Voso, M. C. Tisi, M. Martini, V. Bozzoli, G. Massini, F. D'Alo, L. M. Larocca, G. Leone and S. Hohaus, *Leuk. Lymphoma*, 2012, **53**, 411–416.
145. M. P. Chao, A. A. Alizadeh, C. Tang, J. H. Myklebust, B. Varghese, S. Gill, M. Jan, A. C. Cha, C. K. Chan, B. T. Tan, C. Y. Park, F. Zhao, H. E. Kohrt, R. Malumbres, J. Briones, R. D. Gascoyne, I. S. Lossos, R. Levy, I. L. Weissman and R. Majeti, *Cell*, 2010, **142**, 699–713.
146. J. N. Sampson, S. M. Boca, X. O. Shu, R. Z. Stolzenberg-Solomon, C. E. Matthews, A. W. Hsing, Y. T. Tan, B. T. Ji, W. H. Chow, Q. Cai, D. K. Liu, G. Yang, Y. B. Xiang, W. Zheng, R. Sinha, A. J. Cross and S. C. Moore, *Cancer Epidemiol., Biomarkers Prev.*, 2013, **22**, 631–640.
147. X. Zhao, J. Fritsche, J. Wang, J. Chen, K. Rittig, P. Schmitt-Kopplin, A. Fritsche, H. U. Haring, E. D. Schleicher, G. Xu and R. Lehmann, *Metabolomics*, 2010, **6**, 362–374.
148. E. P. Rhee and R. E. Gerszten, *Clin. Chem.*, 2012, **58**, 139–147.
149. J. M. Yuan, Y. T. Gao, S. E. Murphy, S. G. Carmella, R. Wang, Y. Zhong, K. A. Moy, A. B. Davis, L. Tao, M. Chen, S. Han, H. H. Nelson, M. C. Yu and S. S. Hecht, *Cancer Res.*, 2011, **71**, 6749–6757.
150. C. Abate-Shen and M. M. Shen, *Nature*, 2009, **457**, 799–800.
151. T. Zhang, X. Wu, C. Ke, M. Yin, Z. Li, L. Fan, W. Zhang, H. Zhang, F. Zhao, X. Zhou, G. Lou and K. Li, *J. Proteome Res.*, 2013, **12**, 505–512.
152. M. Chadeau-Hyam, T. M. Ebbels, I. J. Brown, Q. Chan, J. Stamler, C. C. Huang, M. L. Daviglus, H. Ueshima, L. Zhao, E. Holmes, J. K. Nicholson, P. Elliott and M. De Iorio, *J. Proteome Res.*, 2010, **9**, 4620–4627.
153. C. Gieger, L. Geistlinger, E. Altmaier, M. Hrabe de Angelis, F. Kronenberg, T. Meitinger, H. W. Mewes, H. E. Wichmann, K. M. Weinberger, J. Adamski, T. Illig and K. Suhre, *PLoS Genet.*, 2008, **4**, e1000282.
154. R. W. Blamey, I. O. Ellis, S. E. Pinder, A. H. Lee, R. D. Macmillan, D. A. Morgan, J. F. Robertson, M. J. Mitchell, G. R. Ball, J. L. Haybittle and C. W. Elston, *Eur. J. Cancer*, 2007, **43**, 1548–1555.
155. S. C. Lee, X. Xu, W. J. Chng, M. Watson, Y. W. Lim, C. I. Wong, P. Iau, N. Sukri, S. E. Lim, H. L. Yap, S. A. Buhari, P. Tan, J. Guo, B. Chuah, H. L. McLeod and B. C. Goh, *Pharmacogenet. Genomics*, 2009, **19**, 833–842.
156. C. Sotiriou and L. Pusztai, *N. Engl. J. Med.*, 2009, **360**, 790–800.
157. G. Ouyang, Y. Chen, L. Setkova and J. Pawliszyn, *J. Chromatogr. A*, 2005, **1097**, 9–16.
158. C. L. Silva, M. Passos and J. S. Camara, *Talanta*, 2012, **89**, 360–368.
159. D. Y. Wang, S. J. Done, D. R. McCready, S. Boerner, S. Kulkarni and W. L. Leong, *Breast Cancer Res.*, 2011, **13**, R92.
160. S. S. Tang and G. P. Gui, *Biomarkers Med.*, 2012, **6**, 567–585.
161. S. Srivastava, *Gastrointest. Cancer Res.*, 2007, **1**, S60–S63.

17 Genomics for Clinical Practice in Oncology

Vandana Sharma[a], Mohit Chowdhry[a], and Debmalaya Barh*[b,c]

[a]Department of Transfusion Medicine, Transplant Immunology & Molecular Biology, Indraprastha Apollo Hospitals, New Delhi-110076, India; [b]Centre for Genomics and Applied Gene Technology, Institute of Integrative Omics and Applied Biotechnology (IIOAB), Nonakuri, Purba Medinipur, WB, India; [c]Division of Bioinformatics and Computational Genomics, NITTE University Center for Science Education and Research (NUCSER), NITTE (Deemed to Be University), Deralakatte, Mangaluru, Karnataka, India
*E-mail: dr.barh@gmail.com

17.1. Introduction

Deciphering the information locked in a patient's genome and applying this information to identify clinical phenotypes, for both common and rare genetic disease, provides accurate diagnosis and disease outcome.[1,2] Moreover, evidence-based diagnosis and treatment in clinical settings (*e.g.* in cancer) have evolved rapidly and have become an integral part of clinical practice. With the development of cytogenetic techniques, such as karyotyping and fluorescence *in situ* hybridization analysis (FISH), and microarray and next-generation sequencing many diseases, including cancer, are no longer just diagnosed on the basis of histological morphology, identification of genetic aberrations and alterations has become an integral part of the diagnostic armamentarium.

Genomics and Clinical Diagnostics
Edited by David Whitehouse and Ralph Rapley

Published by the Royal Society of Chemistry, www.rsc.org

Cancer is a complex disease of heterogeneous origin and occurs due to interactions between multiple genetic and epigenetic factors. Cancer cells are genetically unstable and are different from normal cells because of altered DNA, leading to the development of oncogenes, resulting in accelerated growth and dysregulated gene function. The tumor-specific alterations at the genetic level include insertions, point mutations, deletions, amplifications and translocations, leading to genetic instability and heterogeneity predisposing to carcinogenesis. The cancer genes are broadly classified as oncogenes, tumor suppressor genes, mutator genes, cell death genes, modifier genes and others.[3]

Certain environmental factors trigger mutation "signatures" in certain cells, which become tumor cells, as a result of altered machinery. For example, the TP53 gene is mutated (a G > T transversion) in cancer and highly differs in smokers and non-smokers in lung cancer patients. The International Agency for Research on Cancer (IARC) P53 database is documenting all the information on the pattern of P53 mutations within different cancers since 1989.

Cancer risk is divided into single-gene disorders, with an absolute risk of developing a certain type cancer in an inherited Mendelian fashion, and, on the other hand, disorders arising due to the interaction of multiple genes with environmental factors. In this case, the predisposing mutation(s) are passed from one generation to another, and are therefore, germline mutations. Inherited mutations confer a different degree of penetrance and occur either in a dominant or a recessive way and may cause early- or late-onset disease, leading to marked variations in disease presentation within the cancer population.[4] A particular condition that does not fall into these two categories is genetic mosaicism, where a mutation occurs in the early stages of embryonic development, leading to the presence of the mutation in only a subset of cells.

17.2. Molecular Diagnostics

The completion of the Human Genome in 2003, paved the way for identifying the rare single-gene Mendelian disorders, prenatal screening and established the novel therapeutic targets in neoplasms (Figure 17.1). Recent advances in molecular diagnostics have restructured the conceptions of understanding, diagnosis and management of disease. The clinical decision making is superseded by molecular or genomic analysis and algorithms based on these investigations. According to

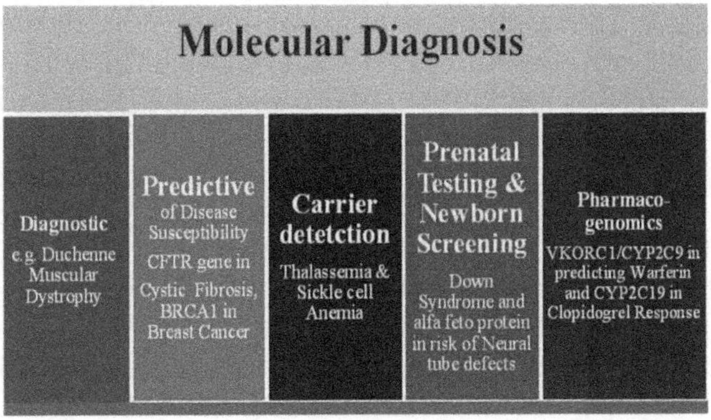

Figure 17.1 Molecular diagnosis using different types of biomarkers.

the US National Institute of Health working group and Biomarkers Consortium, Molecular diagnosis consists of laboratory investigations of a single or a set of precise and selective biomarkers (panel) predictive of disease susceptibility and therapy, with the intended use in a specific group of patients (Figure 17.1). Prognostic biomarkers evaluate the patient's overall outcome, such as probability of cancer outcome and progression. Predictive biomarkers evaluate the effects of pharmacotherapy. With the availability of techniques such as qualitative polymerase chain reaction-amplification refractory mutation system (PCR-ARMS) and restriction fragment length polymorphism (RFLP), real time PCR-TaqMan assays, nested PCR, FISH, capillary electrophoresis, sequencing/pyrosequencing, Sequenom, targeted gene panel sequencing and microarrays cancer detection and diagnosis has been improved.[5]

17.3. Cytogenetic Techniques

Numerous cytogenetic biomarkers are in use in the diagnosis and treatment of breast, head and neck, lung, liver and ovarian cancers. Cytogenetic techniques, such as are used in the diagnosis of lymphoid malignancies, have resulted in the expansion of these capabilities and become an invaluable adjunct by identifying the specific genetic changes in non-dividing interphase cells. The established chromosomal rearrangements in acute myeloid leukemia (AML) identified by FISH analysis, include translocation between chromosome 8 and 21 [t(8;21)(q22;q22), t(15;17)(q22;q12)] and inversion of chromosome 16 [inv(16)(p13.1q22)] or translocation [t(16;16) (p13.1q22)]. The other findings for a comprehensive risk assessment include translocations and gene fusions [t(8;21)(q22;q22)/RUNX1–RUNX1T1; t(15;17)(q22;q12)/PML–RARa; inv(16)(p13.1q22) or t(16;16)(p13.1;q22)/CBFB–MYH1] and mutation in both the CEBPA and NPM1 genes are associated with good clinical outcome, whereas t(1;22)(p13;q13)/RBM15–MKL1;inv(3)(q21q26.2) or t(3;3)(q21;q26.2) RPN1–EVI1; t(6;9)(p23;q34)/DEK–NUP214, the MLL gene rearrangement and mutations in both KIT and FLT3 are indicative of a less favorable prognosis.[6]

Chronic lymphocytic leukemia (CLL) is another disorder due to genetic mutations. The most common recurrent chromosomal abnormalities are trisomy 12, del(13q), del(11q), del(17p) and del(6q) (19–21). In chronic myeloid leukemia (CML), the presence of breakpoint cluster region–Abelson murine leukemia homolog 1 (BCR–ABL1) rearrangement is the diagnostic hallmark of diagnosis[7,8] where the cytogenetic analysis by FISH tests provides critical insights for therapeutic decisions. The chemoimmunotherapeutic regimen including fludarabine, cyclophosphamide and rituximab (FCR) has been observed to result in a better response among patients with trisomy 12 or del(11q), whereas in patients with 17p deletions no such benefit from FCR treatment was observed.

In addition to the treatment selection benefit, FISH testing also facilitates differential diagnosis between CLL and other types of small B-cell lymphoma or leukemia. For example, mantle cell lymphoma is morphologically similar to CLL but the characteristic cyclin D1 translocation, leads to a worse prognosis. The FISH test is the gold standard method of identifying various gene rearrangements and translocations as compared with other techniques.

17.4. From Single Gene Analysis to Multiple Gene Panel Testing

The concept of screening a patient harboring a hereditary susceptibility to develop a particular type of cancer for a mutation in a single gene or a finite number of genes associated with a particular syndrome, *e.g.*, a syndrome-specific gene panel *BRCA1* and *BRCA2* for hereditary breast and ovarian cancer and testing of mismatch repair genes for Lynch syndrome (MLH1, MSH2, MSH6, PMS2 and EPCAM), is based on the patient's clinical history and family history or pedigree analysis (https://www.asco.org). Now the scenario has changed and next-generation sequencing (NGS) offers testing for multiple genes associated with a susceptibility to a disease or cancer risk and its management. The gene panels are prepared including high-, moderate- and low-penetrance genes. The high-penetrance gene panel for risk of breast cancer includes screening for mutation detection in BRCA1, BRCA2, TP53, PTEN, CDH1 and STK11. The comprehensive cancer risk gene panel consists of more than 20 highly and moderately penetrant genes associated with risk for different cancers.

Genomic analysis with the help of whole-exome sequencing (WES) and whole-genome sequencing (WGS) has changed the molecular landscape by allowing a patient's entire genome to be investigated at improved resolution, thereby allowing the detection of medium to large genomic lesions.[9] Sequencing identifies the germline as well as the somatic mutations. Several clinical genomics platforms are available, *e.g.* Foundation Medicine, Ingenuity and Personalis, initiated for the implementation of genomics into clinical practice.[10,11] In 2015, the NIH's genetic testing registry catalogued 28 542 tests, including 4726 genes for the purpose of diagnosing any of 9927 conditions, including classical Mendelian diseases, such as Huntington's chorea, and also complex diseases, such as Alzheimer's. The drug response, such as sensitivity to the anticoagulant warfarin, better known as pharmacogenomics, was also mentioned. There are many companies which are direct-to-consumer tools like 23andMe, https://www.ancestry.com and many others which make the genomic information accessible to interested individuals. The disease-centric mutation databases have morphed into human disease variant databases, which provide important information for documenting clinical variation, such as the Human Gene Mutation Database (HGMD) and the hand-curated databases ClinVar and MutaDatabase.[10,12,13] Molecular profiling is a standard technique for classifying different types of tumors, with established guidelines from the College of American Pathologists[7] and the National Comprehensive Cancer Network, a few have been listed in Table 17.1.[8] Molecular profiling is critical for identifying and characterizing the unique somatic mutations that accrue in cancer cells.

In cancer diagnosis and treatment, The Cancer Genome Atlas and the International Cancer Genome Consortium have provided immeasurable insights into underlying genetic mechanisms, molecular types and subtypes of cancers.[14] Tools like the cBio Portal allow anyone to find out the mutation load of any given gene in all assayed cancer types. Therefore, cancer genomics have already been translated into clinical practice. An example is recurrent

Table 17.1 List of a few FDA approved diagnostic devices.[23]

Name of diagnostic kit/device	Disease/Indications	Mutations in gene	Drug
Cobas EGFR mutation test V2	NSCLC	EGFR	Pembrolizumab
Cobas KRAS mutation test, KRAS RGQ PCR KIT	CRC	KRAS	Erbitux/cetuximab/ vectibix, Gilotrif (afatinib)
BRCA analysis CDx™	Ovarian cancer	Germline BRCA variants	Lymarza™
INFORM HER-2/NEU, PATHVYSION HER-2 DNA Probe kit, HER-2CISH PharmDx Kit	Breast cancer – various indications including risk stratification in patients with lymph-node-negative disease	Her2/Neu	Herceptin (trastuzumab)
Vysis ALK breaks part FISH probe kit	NSCLC	ALK	Xalkori (Crizotinib)
COBAS 4800 BRAF V600 mutation test	Melanoma	BRAF	Zelboraf (vemurafenib)
The LeukoStrat® CDx FLT3 mutation Kit	AML	FLT3	Midostaurin (RYDAPT)
Abbott RealTime IDH2	AML	IDH2 (9 mutations)	Idhifa® (enasidenib),
MolecularMD MRDx® BCR-ABL test	AML	BCR-ABL1	Tasigna (Nilotinib)

mantle cell lymphoma, for which an integrative approach of genomics and transcriptomics coupled with extensive functional studies has helped in tailoring the treatment by establishing Bruton tyrosine kinas C481S as the cause of relapse after ibrutinib treatment.[15]

Another success story for the use of genomics in cancer diagnosis and treatment is in metastatic melanoma. Genomic screening of metastatic melanoma patients with BRAF V600 mutations predicts the increased sensitivity of cancer cells to BRAF inhibitors.[16]

Genomics has enabled determination of the compositions of various types of cancers and inference of the molecular pathology in distinct subpopulations. A female with positive family history of BRCA1 and BRCA2 (genes predisposing for breast and ovarian cancer) gene mutations may opt for genetic testing and predict her risk of developing the disease in her life time.

Recently, Cabanillas *et al.*,[17] developed a novel NGS diagnostic platform to detect (1) somatic genomic alterations associated with the response to approved targeted cancer therapies and (2) germline mutations predisposing to hereditary malignancies.

The researchers used next-generation sequencing libraries enriched in the exons of 215 cancer genes (97 for therapy selection and 148 for predisposition, with 30 informative for both applications), and selected introns from 17 genes involved in drug-related rearrangements, prepared from 39 tumors (paraffin-embedded tissues or cytologies), 36 germline samples (blood) and 10 cell lines using hybrid capture. This platform found single-nucleotide variants (SNVs) and insertions/deletions (indels) with sensitivity and specificity >99.5% (allelic frequency ≥0.1), as well as copy-number variants (CNVs) and rearrangements. Somatic testing identified tailored approved targeted drugs in 35 out of 39 tumors (89.74%), showing a diagnostic yield similar to that of leading commercial platforms. A somatic *EGFR*p.E746_S752delinsA mutation in a mediastinal metastasis from a breast cancer prompted its anatomopathologic reassessment, its definite reclassification as a lung cancer and its treatment with gefitinib (partial response sustained for 15 months). Testing of 36 germline samples identified two pathogenic mutations (in *CDKN2A* and *BRCA2*).

17.5. Tailored Therapy

Customized or tailored therapy on the basis of genomic profiling helps to identify variants in drug-metabolizing genes, drug transport genes and genes responsible for initiating the response at the required target site, altering the drug response and dose. Pharmacogenomics has helped in reducing drug-related adverse effects significantly, *e.g.* for anticancer drugs where drug resistance can be predicted, epileptic drugs, antiplatelet drugs and many other drugs. This has improved health care. Therefore, the genes conferring risk for susceptibility to a disease in an individual may predict the response to certain drugs. The FDA has already approved tests for pharmacogenomics for many drugs, such as the anticoagulant warfarin, antiplatelet clopidogrel and many anticancer drugs. A few other examples of targeted therapy are trastuzumab in breast cancer, gefitinib in non-small cell lung cancer and imatinib in chronic myeloid leukemia.

17.6. Key Enablers for the Integration of Genomics into the Health Care System

To harness the enormous potential of genomics in the health care system, certain factors must be taken care of: genomic literacy, confidentiality and privacy.

17.6.1. Genomic Literacy

Genomic information is personal and its integration into diagnostics is a very crucial aspect. Genomic literacy at the customer level is important because the google-centric customers are also not able to interpret the results of genomic analysis. Genetic literacy is defined as 'sufficient knowledge and appreciation of the principles of genetics to allow informed decision-making for personal well-being'.[18,19] For example, unfamiliarity with gene nomenclature, symbols denoting alleles (for instance, *1), and use of terminology (such as carrier, heterozygosity) will probably increase patient's confusion.[19] The use of scientific terminology may also create challenges for patients. This may lead to non-adherence and non-compliance with recommended interventions and can increase anxiety in patients. Therefore, the information should be presented in a format understood by the patient so that it should not affect medical care. Several studies have found that the content of genetic testing reports, especially for clinical genome sequencing, are generally focused as per the needs of the clinicians, although patient preferences are also an important element in the reporting of incidental findings.[19,20]

One of the cornerstones of the ethical conduct of diagnostics and research in genomics is informed consent. Informed consent ensures that subjects are aware of the risks and benefits of the investigation and make a voluntary decision about participating in it (National Commission for the Protection of Human Subjects of Biomedical and Behavioral Research, 1979).[21] The patient or person should be informed about DNA or biological specimens to be stored and used for the required purpose clearly.

To promote participation of the public in genomic medicine the US Department of Health and Human Services has launched the Personalized Health Care Initiative Policy to facilitate patient empowerment through genomics as a central priority of future health-care programs (Personalized health care. 2008).[22]

17.6.2. Privacy, Confidentiality and Security

The genomic data consists of an individual's sensitive information and should be secured appropriately. Sharing of genomic information requires security measures and exchange of information should be strictly taken care of, especially when the information is stored in electronic health records. The personal health information is subjected to legislation, such as the Personal Health Information Protection Act, Personal Information Protection and Electronic Documents Act and the Health Insurance Portability and Accountability Act (HIPAA) and Genetic Information Nondiscrimination Act.

17.7. Innovation

Genomics is a cutting edge, innovative science and provides the opportunity for developing novel methods to benefit customers. It is a new dimension of health care and requires coordinated working of public- and private-sector stakeholders.

17.8. Policy Leadership

Policy making is very crucial step for clinical genomics and diagnostics, related to security, privacy, reliability and applicability of genomic data to stratified specific subpopulations from different geographical areas and need to be addressed.

17.9. Informed Consent

To provide all the required information for the diagnostic tests to be carried out, informed consent is essential prior to genetic testing. The individual to be tested needs to understand the technical, medical, social and financial complexities of genomics testing and should sign the informed consent form. Genetic counseling prior to or post testing should be provided to the patient.

17.10. Conclusion

While the full absorption of applications of genome-guided diagnostics and precision medicine has yet to be realized, the major three applications are the diagnosis of single-gene disorders and multifactorial disorders and pharmacogenomics, which currently serve as the basis for the concept of genome-informed precision medicine. Genomic biomarkers have transformed the diagnosis of cancer diagnosis and classification, prognosis and, most importantly, molecularly guided treatment. Clinically, the value of genomic information is directly proportional to its immediate implementation for a clinical condition. The social, ethical and economic aspects must be taken into account.

References

1. E. A. Worthey, *et al.*, *Genet. Med.*, 2010, **13**, 255.
2. M. N. Bainbridge, *et al.*, *Sci. Transl. Med.*, 2011, **3**, 87re3.
3. I. R. Kirsch, Oncogenesis, in *Surgery of Infants and Children: Principles and Practice*, ed. K. T. Oldham, R. B. Foglia and P. M. Colombani, Lippincott-Raven, Philadelphia, 1996, pp. 549–567.
4. S. A. Frank, *Nat. Rev. Genet.*, 2004, **5**, 764.
5. S. Sethi, S. Ali, P. A. Philip and F. H. Sarkar, *Int. J. Mol. Sci.*, 2013, **14**, 14771.
6. C. Bachas, G. J. Schuurhuis, D. Reinhardt, U. Creutzig, Z. J. Kwidama, C. M. Zwaan, M. M. van den Heuvel-Eibrink, E. S. J. M. De Bont, S. Elitzur, C. Rizzari, V. de Haas, M. Zimmermann, J. Cloos and G. J. L. Kaspers, *Br. J. Haematol.*, 2014, **166**, 902.
7. S. A. Smoley, S. R. Brockman, S. F. Paternoster, R. G. Meyer and G. W. Dewald, *Cancer Genet. Cytogenet.*, 2004, **148**, 1.
8. Y. Zhang and J. D. Rowley, *Clin. Lab. Med.*, 2011, **31**, 687.
9. S. H. Katsanis and N. Katsanis, *Nat. Rev. Genet.*, 2013, **14**, 415.
10. P. Vijay, B. R. Alexa, A. B. R. McIntyre, C. E. Mason, J. P. Greenfield and S. Li, *Crit Rev Eukaryot Gene Expr.*, 2016, **26**(2), 97–113.
11. C. J. Patel, A. Sivadas, R. Tabassum, T. Preeprem, J. Zhao, D. Arafat, R. Chen, A. A. Morgan, G. S. Martin, K. L. Brigham, A. J. Butte and G. Gibson, *Genome Med.*, 2013, **5**, 58.
12. P. D. Stenson, *et al.*, The Human Gene Mutation Database: 2008 update, *Genome Med.*, 2009, **1**, 13.
13. S. Bale, *et al.*, MutaDATABASE: a centralized and standardized DNA variation database, *Nat. Biotechnol.*, 2011, **29**, 117.
14. G. Ciriello, M. L. Miller, B. A. Aksoy, Y. Senbabaoglu, N. Schultz and C. Sander, Emerging landscape of oncogenic signatures across human cancers, *Nat. Genet.*, 2013 Oct, **45**(10), 1127–1133.
15. D. Chiron, M. Di Liberto, P. Martin, X. Huang, J. Sharman, P. Blecua, S. Mathew, P. Vijay, K. Eng, S. Ali, A. Johnson, B. Chang, S. Ely, O. Elemento, C. E. Mason, J. P. Leonard and S. Chen-Kiang, *Cancer Discovery*, 2014 Sep, **4**(9), 1022–1035.
16. P. B. Chapman, A. Hauschild, C. Robert, J. B. Haanen, P. Ascierto, J. Larkin, R. Dummer, C. Garbe, A. Testori, M. Maio, D. Hogg, P. Lorigan, C. Lebbe, T. Jouary, D. Schadendorf, A. Ribas, S. J. O'Day, J. A. Sosman, J. M. Kirkwood, A. M. M. Eggermont, B. Dreno, K. Nolop, J. Li, B. Nelson, J. Hou, R. J. Lee, K. T. Flaherty and G. A. McArthur, *N. Engl. J. Med.*, 2011, **364**, 2507.
17. R. Cabanillas, M. Diñeiro, D. Castillo, P. C. Pruneda, C. Penas, G. A. Cifuentes, Á. de Vicente, N. S. Durán, R. Álvarez, G. R. Ordóñez and J. Cadiñanos, *Mol. Genet. Genomic Med.*, 2017, **23**, 336.
18. B. V. Bowling, E. E. Acra, L. Wang, M. F. Myers, G. E. Dean, G. C. Markle, C. L. Moskalik and C. A. Huether, *Genetics*, 2008, **178**, 15.
19. S. B. Haga, R. Mills, K. I. Pollak, C. Rehder, A. H. Buchanan, I. M. Lipkus, J. H. Crow and M. Datto, *Genome Med.*, 2014, **6**, 58.
20. C. A. Brownstein, A. H. Beggs, N. Homer, B. Merriman, T. W. Yu, K. C. Flannery and P. Dworzynski, *Genome Biol.*, 2014, **15**, [R53].

21. National Commission for the Protection of Human Subjects of Biomedical and Behavioral Research, *The Belmont Report: Ethical Principles and Guidelines for the Protection of Human Subjects of Research*, Dep. Health Educ. Welfare, Washington, DC, 1979.
22. Personalized health care. 2008 https://www.hhs.gov/myhealthcare.
23. https://www.fda.gov/MedicalDevices/ProductsandMedicalProcedures/InVitroDiagnostics/.

Subject Index

Locators in **bold** refer to tables; those in *italic* to figures